Radar and EW Modeling in MATLAB® and Simulink®

For a complete listing of titles in the
Artech House Radar Library,
please turn to the back of this book.

Radar and EW Modeling in MATLAB® and Simulink®

Carlos A. Dávila
Glenn D. Hopkins
Gregory A. Showman

ARTECH HOUSE
BOSTON | LONDON
artechhouse.com

Library of Congress Cataloging-in-Publication Data
A catalog record for this book is available from the U.S. Library of Congress.

Cover design by Joi Garron

ISBN-13: 978-1-63081-906-4

Accompanying software to this book can be found at: https://us.artechhouse.com/Assets/downloads/Davila_906.zip

685 Canton Street
Norwood, MA 02062

10 9 8 7 6 5 4 3 2 1

To my dear wife Gloria. Thank you for your unconditional love and for always believing in me even when I didn't.

C.A. Dávila

This book is dedicated to my parents, my wife Rose, and my children Mary and Willis. Thank you all for your patience and support.

G.D. Hopkins

To my wife, Beth, for whom "radar fun" is no fun at all, I owe you a dance. Maybe two.

G.A. Showman

Contents

Foreword

The effective development of radar and electronic warfare systems—engineering topics of significance across the globe—requires an understanding of underlying physics and a disciplined approach to modeling, simulation, and analysis. This book details the necessary framework, progressing from essential discussion and mathematical descriptions to MATLAB® and Simulink® implementation. Another great feature of the book is that it provides MATLAB and Simulink code.

Radar and EW Modeling in MATLAB® and Simulink® is the work of three noted experts in the field. I have had the privilege of working with Carlos Dávila, Glenn Hopkins, and Gregory Showman for more than 20 years, and I have learned a good deal from each. I am sure this will also be the case for readers of this book!

The book itself is logically organized into key topics building one upon the other. The beginning of the book is devoted to radar, starting with the radar range equation, an innocent-looking equation with many nuances and room for misunderstanding in the absence of clarifying discussion. The radar section of the book also describes modeling of hardware that allows the radar to actively interrogate its environment as it collects necessary information on targets of interest. Radar waveforms represent the connection between the hardware and receive processing necessary to generate radar products. In this vein, the book expertly describes waveform modeling, followed by important signal processing steps whose quality affects radar performance, such as range-Doppler processing, constant false alarm rate (CFAR) detection threshold setting, and angle estimation using monopulse. These topics are then followed by detailed descriptions of tracker modeling, leading into unique chapters on engagement modeling and simulation. A key feature of this book is that it presents voltage-level models, providing the detail necessary to analyze real radar design with requisite engineering rigor.

The latter chapters of the book offer readers an essential understanding of canonical electronic warfare topics: electronic attack, electronic protection, and electronic support. This is a good approach, as radar and electronic warfare go hand-in-glove. This is a book that I will definitely value in my technical library!

William L. Melvin, PhD
Georgia Institute of Technology
October 2023

Preface

This book is based on "Modeling and Simulation of Radar Systems" (MSRS), a defense technology training course offered by Georgia Tech Professional Education (GTPE). We created this course and have been teaching it for the last 20 years as part of GTPE's defense technology portfolio. Both the course and the book offer a cohesive treatise of the modeling and simulation (M&S) of key radar components and related topics (antennas, waveforms, radar cross section, etc.) as stand-alone modules. Moreover, we also demonstrate an approach for stitching together all these modules to form an end-to-end engagement simulation, suitable for analysis and performance prediction. MATLAB and Simulink are the modeling environments adopted for this purpose: the former due to its powerful capabilities and ubiquity in the engineering and scientific community, and the latter for its user-friendly graphical interface that enables students to quickly put together simple models and gain insight into relevant concepts.

The target audience consists of practicing engineers and scientists with some familiarity of basic radar concepts. Most topics start by presenting a mathematical representation of the concept, and then implement it in either a MATLAB and/or Simulink example; however, the modeling principles may be applied in the reader's programming language of choice.

All the MATLAB and Simulink example software developed for this book is available online at https://us.artechhouse.com/Assets/downloads/Davila_906.zip, along with two online-only appendixes that serve as refreshers on selected MATLAB and Simulink features used throughout the book.

Chapter 1 introduces basic modeling terminology and concepts, including the U.S. Department of Defense (DoD) M&S pyramid. In that context, most of this book is firmly entrenched in the engineering level of the pyramid (i.e., "waveform-level" modeling), eventually venturing into the engagement level with "one versus one" or "one versus few" scenarios. Chapter 2 focuses on the radar range equation (RRE) as the foundation for radar M&S and analysis. Chapters 3 to 14 focus on relevant topics ranging from radar subsystems, environmental phenomenology, signal processing, and related functionality. Each of these chapters roughly represent a single course lecture; the exception being Chapter 3 on antennas, in which Glenn Hopkins masterfully merged three long lectures; it is by far the longest chapter in the book. It all comes together in Chapter 15, where a full end-to-end engagement simulation is unveiled. This simulation is composed of subsystem and component models from all previous chapters.

Chapters 16 to 18 delve into electromagnetic (formerly electronic) warfare (EW). The contents are based on "Basic EW Modeling," yet another GTPE course that shares some core material with MSRS. The three major "pillars" of EW—electromagnetic attack (EA), electromagnetic protection (EP), and electromagnetic support (ES)—are covered in Chapters 16, 17, and 18, respectively.

I am grateful to Martie Goulding who reviewed the original manuscript. His insightful comments and suggestions resulted in a much better final product.

As I look back on my career, I would be remiss if I did not acknowledge a number of esteemed colleagues, whose support I deeply appreciate and will always treasure. My first job right out of college was at Hughes Aircraft Company-Missile Systems Group in Canoga Park, CA, where Darrell Hibbs took the time to teach me how to be a good radar analyst. He was my boss but also my mentor—and to this day, my friend. At the Georgia Tech Research Institute (GTRI), I was blessed with several additional mentors who taught me not only about radar and EW, but also about performing high-level applied research: Sam Piper, Bill Melvin (who graciously agreed to write this book's foreword), and Byron Keel, to name a few. I must also acknowledge my two coauthors, Glenn Hopkins and Greg Showman, both titans in their fields, and on whose shoulders I stand. I am honored to call them colleagues, but even more so to call them friends.

Carlos A. Dávila

CHAPTER 1

Introduction

1.1 Basic Concepts and Terminology

The DoD M&S Glossary [1] defines a model as a "physical, mathematical, or otherwise logical representation of a system, entity, phenomenon, or process." Similarly, it defines a simulation as a "method for implementing a model over time." The two terms are closely intertwined, and are often used interchangeably. Furthermore, the M&S Glossary categorizes simulations into live, virtual, or constructive (LVC):

- *Live simulation:* Live simulation involves real people operating real systems. Military training events using real equipment are good examples of live simulations.
- *Virtual simulation:* A virtual simulation involves real people operating simulated systems. Virtual simulations inject human-in-the-loop in a central role by exercising motor control, decisions, or communication skills. Flight simulators used to train pilots are good examples of virtual simulations.
- *Constructive simulation:* A constructive simulation includes simulated people operating simulated systems. Real people stimulate such simulations but are not involved in determining the outcomes. All computer simulations are examples of constructive simulations.

Many simulation environments are hybrids of all three categories. One example is a hardware-in-the-loop (HWIL) setup where computer-generated signals stimulate real systems. Live components such as operators monitoring the system response may be added to the constructive elements of the simulation. This book focuses on constructive modeling and simulation (M&S).

1.2 The M&S Pyramid

The DoD M&S community also popularized the concept of the M&S pyramid, which categorized models and simulations based on their degrees of fidelity and scope, which are inversely related. The DoD M&S pyramid was first postulated in the mid 1990s, and by the early 2000s, it had been widely established as the standard hierarchy of simulations across DoD [2, 3]. Although several variations and excursions have been postulated, arguably the most common version is presented in Figure 1.1. Each level of the pyramid may be characterized by the level

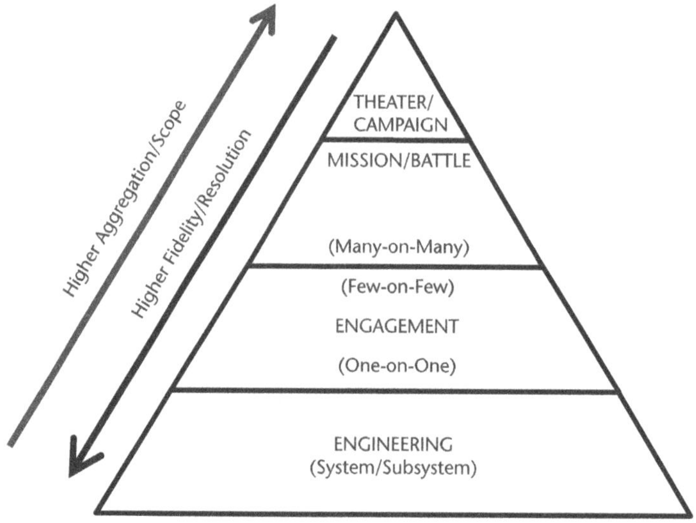

Figure 1.1 The DoD M&S pyramid.

of fidelity, the level of scope (or aggregation), and the length of time being simulated, among other factors.

Engineering-level M&S is typically physics-based and of the highest fidelity relative to other levels. Computer simulation of radar and EW systems at the engineering level usually involves generating waveform samples that are processed as digital signals. Due to the prevalence of in-phase and quadrature (I/Q) demodulation in radar receivers, these are often referred to as I/Q-level models. The scope of the model may be just a single radar system, or even a single subsystem, such as the signal processor, but with very detailed representation. The simulated time duration of an engineering-level radar or EW model tends to be in the order of a second or less, consistent with a radar timeline. The majority of the models presented in this book may be considered engineering-level models.

The interaction between two or more systems is captured at the next higher level of the pyramid, the engagement level. The entities being modeled may be as few as two (for example, one radar versus one target or one radar versus one jammer). The engagement may be more complex, however, and involve "few versus few." The size of numbers that constitute "few" varies, but it is typically in the order of tens. The simulated duration of the engagement may be approximately a few minutes or less. A good example of an engagement-level simulation is Brawler. Originally called the tactical air combat (TAC) Brawler, Brawler is a U.S. government–owned, comprehensive simulation tool that provides a detailed representation of air-to-air combat involving multiple aircraft in the visual- and beyond-visual-range arenas. The Defense Systems Information Analysis Center (DSIAC) maintains and distributes Brawler [2].

Mission-level M&S involves "many versus many" entities, potentially hundreds. The duration of the mission is in the order of minutes to hours, and by the sheer number of interactions involved, the fidelity on each model may be considerably lower to allow reasonable runtimes. Mission-level models are usually effects-based as opposed to physics-based. As an illustration, in the context of radar models, signals are not simulated at the sample (or I/Q) level. The effect of a radiated signal reflected from a target may be represented by a relatively simple calculation to determine whether the reflection is visible to the radar (i.e., whether the target is detected).

The premier mission-level simulation environment currently used in DoD is arguably the Advanced Framework for Simulation, Integration, and Modeling (AFSIM). It was originally developed by Boeing and later transferred to the Air Force Research Laboratory (AFRL) [3]. It has been widely adopted by many services in addition to the Air Force, although several other competing/complementary mission-level environments are in use, such as the Next Generation Threat System (NGTS) [4], the Extended Air Defense Simulation (EADSIM) [4, 5] and the older Suppressor [6].

Campaign-level M&S (also called theater-level) sits at the top of the pyramid. It features the highest level of aggregation (potentially thousands of entities) and the longest simulated duration, in the order of days or weeks. At the same time, it has the lowest level of fidelity. Lookup tables often define the behavior and capabilities of the simulated entities, with data generated by lower-level simulations. (This is also true at the mission level.) Wargaming and tactics development often utilize campaign-level M&S. The best example of a campaign-level environment that comes to mind is the Synthetic Theater Operations Research Model (STORM), developed and used by the Air Force Studies, Analyses and Assessments Office (AF/A9) [7]. Table 1.1 summarizes the preceding discussion.

There are no hard boundaries between levels, and one often finds simulation hybrids where different portions of the environment are categorized at different levels. The focus of this book is on engineering-level models, aggregating up to a one-versus-one (or one-versus-few) engagement model.

Table 1.1 Key Features Between M&S Levels

Level	*Fidelity*	*Scope*	*Timeline*	*Example*
Engineering	Highest fidelity, physics-based	Single system/ subsystem	< 1 second	Most examples in this book
Engagement	Variable, combination physics and effects-based	One-on-one to few-on-few systems	Seconds to minutes	Brawler
Mission	Low-fidelity, mostly effects-based	Many-on-many	Minutes to hours	AFSIM
Campaign	Lowest fidelity, effects-based	Force-on-force	Hours to days	STORM

1.3 Radar M&S

Engineering-level radar M&S involves suitable representation of the signal environment, which may be quite complex. Figure 1.2 depicts the typical radar environment in a military context.

A radar system radiates electromagnetic (EM) energy into the environment, which reflects from objects encountered along its path and returns back to the radar very much like echoes from acoustic waves. Some of these objects are of interest to the radar and are referred to as targets. The radar processes these returns to extract useful information from these targets (e.g., position and velocity estimates) and must do so in the presence of receiver noise, an internal interference signal inherent in all electronic systems. Moreover, there are often other objects in the environment that are not of interest to the radar but that will nevertheless cause return signals back to the radar. These unwanted signals, referred to as clutter, are, in general, a nuisance and another source of interference with the radar's ability to handle target returns. There are additional sources of man-made interference generated external to the radar, whether intentionally or not. Other emitters in the environment such as TV or radio stations—and even other radars—are sources of unintentional EM interference (EMI). There may also be *intentional* sources of interference, especially in a combat environment, that purposely attempt to impede the radar from finding and properly handling targets. This is known as electromagnetic attack (EA), or jamming.

To model such a complex environment properly, a structured and methodical approach is highly desirable. A recommended first step is to define a high-level framework that covers all relevant areas; the simulation developer may then revisit each area as needed to add the necessary fidelity. Figure 1.3 illustrates an example framework. The block diagram of simulation components follows multiple signal paths from their origin to their destination.

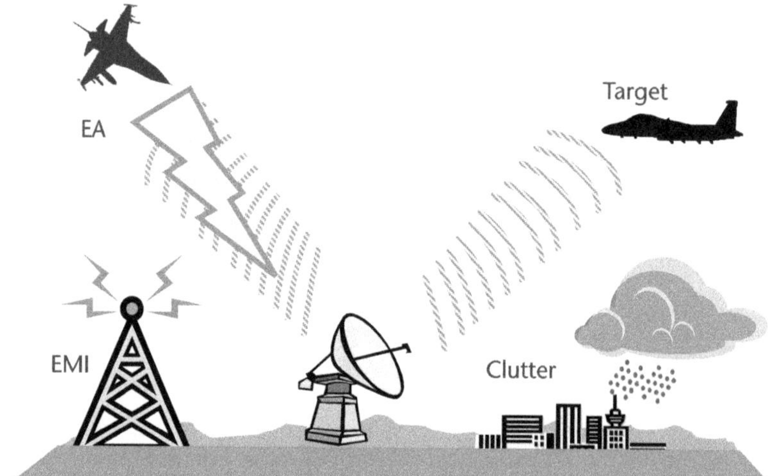

Figure 1.2 The radar environment at a glance.

Figure 1.3 is a suitable roadmap for developing a radar simulation from scratch. The developer may start with one signal path—for example, the target's—and evaluate the modeling requirements at each block (e.g., decide what type of antenna model is needed or what propagation effects need to be modeled). Simple, low-fidelity antenna and propagation models such as fixed gains and losses may initially be used as a placeholder just to get the simulation operational. They may subsequently be replaced by better models, in a manner consistent with the simulation objectives but also within development schedule and cost constraints. This crawl-walk-run approach enables a functional, even if basic, simulation to be up and running early in the development cycle. It may then be improved incrementally over time until all requirements are met (or until the project runs out of time and/or funding).

Figure 1.4 shows an alternate way of visualizing a radar/EW simulation in the context of a one-on-one engagement between a radar and a jammer-carrying target. Each of the radar, target/jammer, and propagation models may be at the engineering level, and the engagement consists of simulating the signal interactions between them over an arbitrary time span. Figure 1.4 also includes a nonexhaustive listing of key features and functions associated with the blocks that form each model, as a form of checklist for the developer.

This book focuses on the development and implementation of an engagement simulation in the spirit of Figure 1.4, following the incremental approach described earlier.

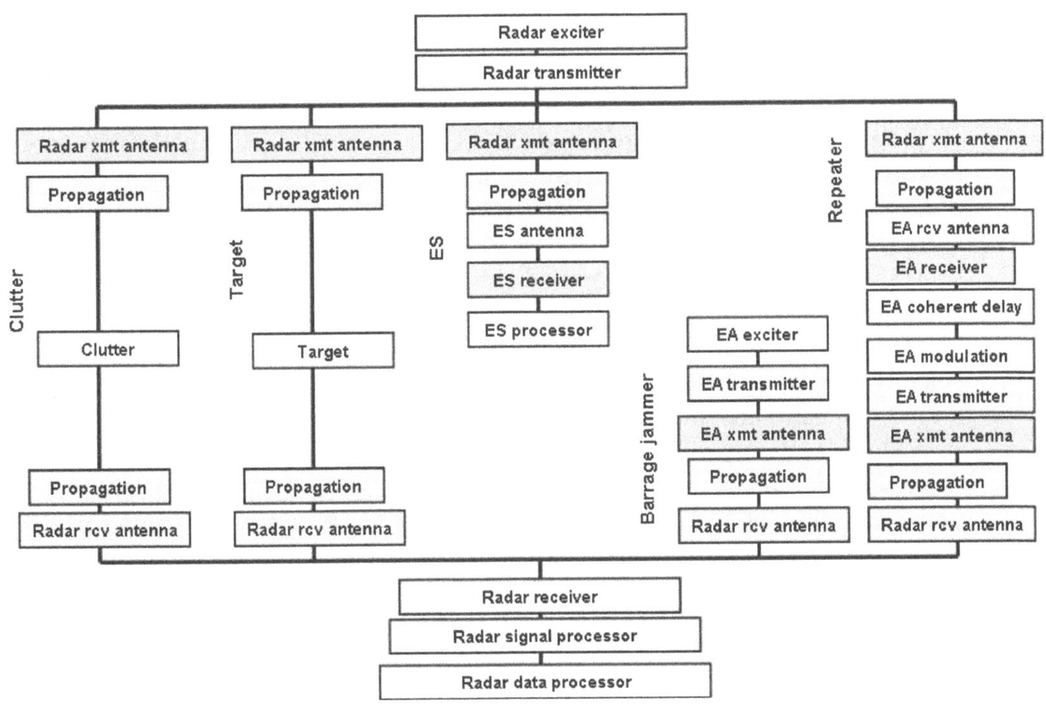

Figure 1.3 Typical M&S signal paths.

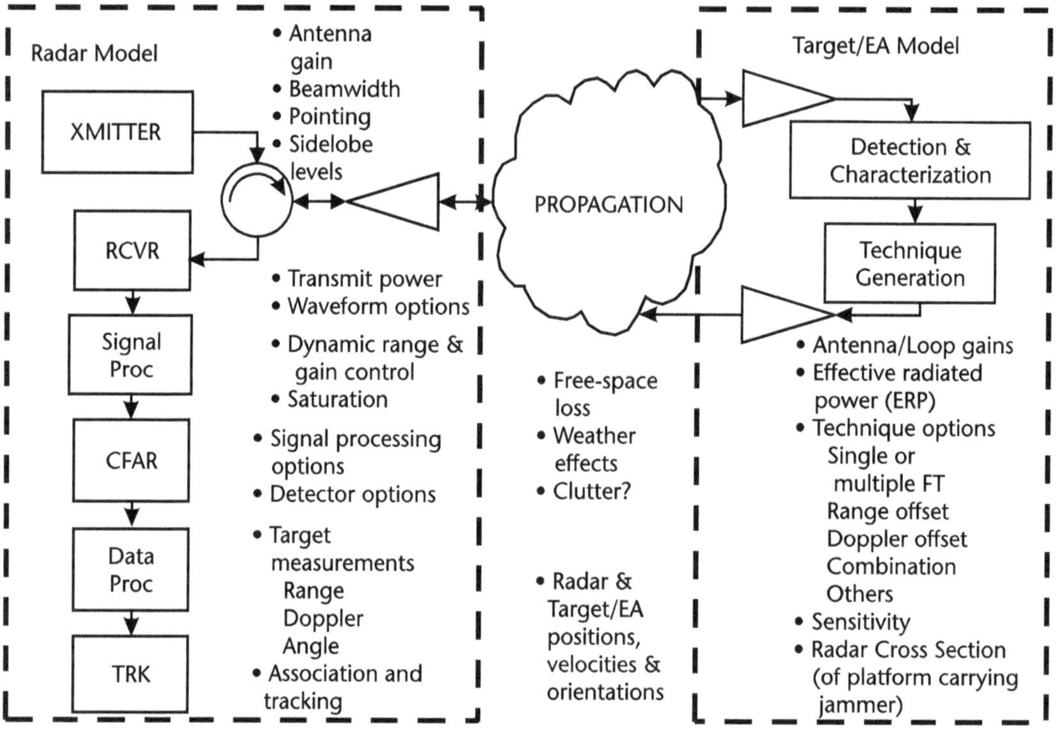

Figure 1.4 Top-level overview of the engagement model developed and presented in this book.

1.4 Concluding Remarks

Figure 1.5 shows a simplified block diagram for a typical pulse-Doppler radar system; Table 1.2 defines the acronyms in Figure 1.5.

The analog portion of the receiver takes the received signal and downconverts to a low (baseband) frequency in multiple stages. This is known as a superheterodyne receiver. The last downconversion stage splits the signal into I and Q channels, in a process called I/Q demodulation. The baseband signal is then digitized and processed in the range and Doppler dimensions, forming a range-Doppler map (RDM) that is sent to a detector to identify the target location(s) within the RDM. The data processor makes measurements on the detected target (typically range, velocity, and angle) and passes the information to a tracking filter to maintain target state estimates over time.

All the system components, processing, and terminology used here and in Table 1.2 are covered more thoroughly in Chapters 2 to 14 as we present the different models. The devil is in the details, however, and a key to understanding why models are defined in a certain way is to understand the assumptions that were made when defining them. The British statistician George Box famously stated that "all models are wrong, but some are useful" [8]. Two models of the same process may exhibit significant discrepancies, yet this does not mean either model is invalid—it may just be that their assumptions are different. Appendix A provides a selection of simple but very common examples of differing assumptions or implementations that may lead to confusion or discrepancies.

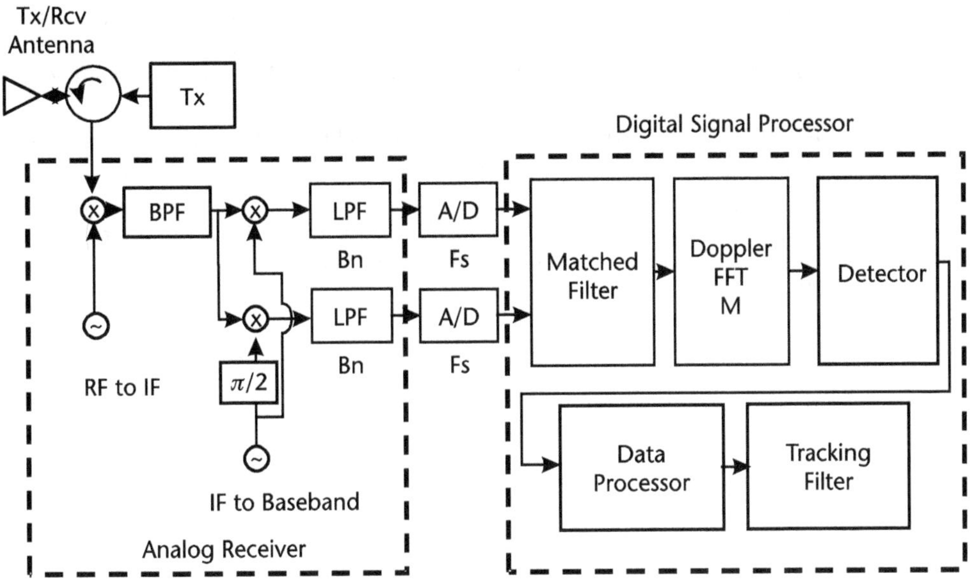

Figure 1.5 Generic block diagram of a pulse-Doppler radar system.

Table 1.2 Acronyms in Pulse-Doppler Radar Block Diagram (Figure 1.5)

Acronym	*Definition*
Tx	Transmit or transmitter
Rcv	Receive or receiver
RF	Radio frequency
IF	Intermediate frequency
BPF	Bandpass filter
LPF	Low-pass filter
Bn	Bandwidth
A/D	Analog-to-digital converter
Fs	Sampling frequency
FFT	Fast Fourier transform

References

[1] Modeling and Simulation (M&S) Glossary, Department of Defense; Modeling and Simulation Coordination Office (www.msco.mil), Alexandria, VA, October 1, 2011.

[2] https://dsiac.org/models/.

[3] West, T. D., and Birkmire, B., "AFSIM: The Air Force Research Laboratory's Approach to Making M&S Ubiquitous in the Weapon System Concept Development Process," *CSIAC Journal*, Vol. 7, No. 3, December 2019, pp. 50–55.

[4] Bestard, J. J., "Air Force Research Laboratory Innovation: Pushing the Envelope in Analytical Wargaming," *CSIAC Journal*, Vol. 4, No. 3, November 2016, pp. 1217.

[5] https://www.tbe.com/missionsystems/eadsim.

[6] Whitehurst, R., J. Phipps, and V. Kowalenko, "A Programmer's Reference to the Suppressor Simulation System," DSTO-GD-0130, Defence Science and Technology Organisation (DSTO) Aeronautical and Maritime Research Laboratory, Melbourne, Australia, February 1997.

[7] Bickel, W. G., "Improving the Analysis Capabilities of the Synthetic Theater Operations Research Model (STORM)," master's thesis, Naval Postgraduate School, Monterey, CA, September 2014.

[8] Box, G. E. P., "Robustness in the Strategy of Scientific Model Building," in *Robustness in Statistics* (R. L. Launer and G. N. Wilkinson, eds.), New York: Academic Press, 1979, pp. 201–236.

CHAPTER 2

The Radar Equation

2.1 Introduction

The radar equation, also known as the radar range equation (RRE), is the most fundamental tool used to assess radar performance. The 2008 IEEE Standard Radar Definitions manual [1] describes it as a "mathematical expression that relates the range of a radar at which specific performance is obtained to the parameters characterizing the radar, target, and environment." In its purest form, the RRE calculates the power of a target signal return received at the radar as a function of the aforementioned parameters; however, this quantity in and of itself conveys little information on performance. A more useful form is to normalize this received power by the noise power in the radar receiver, thereby resulting in a target signal-to-noise power ratio, or signal-to-noise ratio (SNR) for short. The SNR is a fundamental quantity in assessing system performance, as it is key to target detection and calculation of related measures of performance (MOP) in radar systems such as detection range, as well as measurement accuracies, which in turn drive target tracking. Section 2.2 walks the reader through the development of the radar equation and maps its components to the propagation of EM energy from the radar transmitter to the target, and back to the radar receiver. This is an insightful exercise of particular benefit to readers new to the field, and it will fully define the canonical form of the equation. Section 2.2 also introduces variations of this basic form, including the search and track forms of the radar equation; eventually, however, it returns to the basic form, to set the stage for subsequent modeling discussions and implementations in Sections 2.3 (MATLAB) and 2.4 (Simulink).

2.2 Derivation of the Radar Equation

2.2.1 Received Target Power

A step-by-step walkthrough over the terms of the RRE maps quite nicely over the sequence of events that form the radar timeline. Consider a radar transmitter that radiates EM energy equally in all directions, as depicted in Figure 2.1.

This notional model corresponds to an isotropic antenna. As the EM wave radiates away from the transmitter, the wavefront forms an ever-growing sphere centered on it. The transmitter only has a finite amount of power that it can radiate,

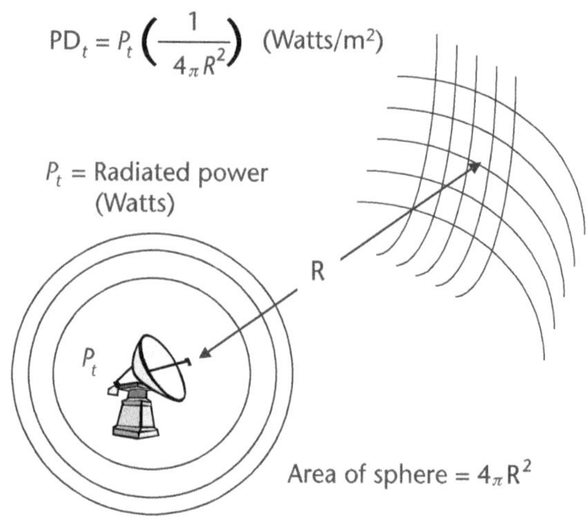

Figure 2.1 Power density of an isotropic emitter at a distance R.

however. This transmitted peak power, denoted P_t, is equally distributed over the area of the sphere. We can therefore define a transmitted power density (power per unit area) given by

$$PD_t = \frac{P_t}{4\pi R^2} \tag{2.1}$$

where R is the radius of the sphere. The denominator in (2.1) shows that the radiated power decreases inversely proportional to the squared of the distance from the transmitter as it propagates through the atmosphere. This term is therefore referred to as a spherical propagation or spherical spreading loss.

We now add a few additional terms to (2.1). First, we recognize that transmitters are not 100% efficient, so there are some losses as the total power available for transmission is delivered to the antenna load. Some of the power is dissipated as heat, for example. The total losses are usually in the order of a few decibels; however, it is important to represent them so calculations will not be overly optimistic. A transmit loss term L_t is added to (2.1) to account for these nonideal effects. Moreover, radar antennas are not isotropic; rather, they concentrate the transmit energy in a preferred direction, characterized by a peak gain G_t, which is defined relative to the isotropic case. The energy distribution of the radar antenna across space (the antenna gain pattern) is detailed in Chapter 3; for now, it is sufficient to state that the radiated power density in the antenna look direction—that is, the direction of maximum gain—is scaled by G_t, which in general is much greater than 1. Last, as the EM wave propagates through the atmosphere, it suffers additional losses beyond the spherical spreading, due to interactions with the propagation medium. The main effect is attenuation due to absorption by water vapor and other gases in the atmosphere; these are covered in Chapter 4. We denote the one-way atmospheric attenuation loss as L_{atmos}. Adding all these additional terms to (2.1) results in

$$ERPD = \frac{P_t G_t L_t L_{\text{atmos}}}{4\pi R^2} \tag{2.2}$$

The term $P_tG_tL_t$ in the numerator of (2.2) is often referred to as the effective radiated power (ERP). We thus refer to (2.2) as the effective radiated power density (power per unit area) (ERPD).

Figure 2.2 shows our notional scenario with the addition of all these terms. At some distance R from the transmitter, and in the antenna look direction, the ERPD is given by (2.2).

Now consider the presence of a target at that location, such as the aircraft shown in Figure 2.2. In this case, the sphere radius R represents the range from the radar to the target; it is a *slant* range (straight-line distance) as opposed to a ground range.

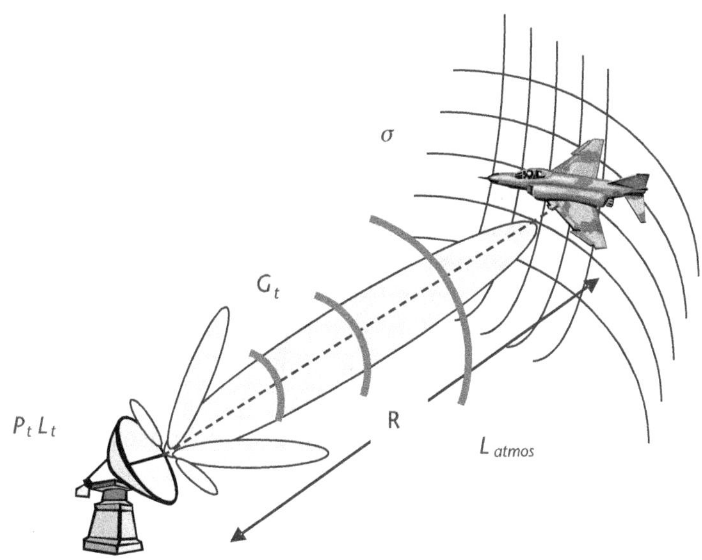

Power reflected by target (W)

ERPD = Power density at target (W/m²)

ERP (W)

$$P_{reflected} = P_t G_t L_t \left(\frac{L_{atmos}}{4\pi R^2} \right) \sigma$$

Figure 2.2 Reflected power from the target in the direction of the radar.

This object intercepts the transmitted wave and scatters its energy in all directions. A small fraction is reflected back to the radar, however. The total reflected power will be a product of the ERPD incident on the target, and a target-dependent quantity known as the radar cross-section (RCS), denoted by the Greek symbol σ. RCS is covered in Chapter 5, but for the purposes of this discussion, it is precisely the scaling factor that describes the strength of the reflected wave. It has units of area, such that the product (ERPD $*$ σ) has units of power.

The EM field reflected from the target propagates in a manner analogous to an antenna, in the sense that it too experiences a spherical propagation loss identical to that described by (2.1). It will also experience atmospheric losses on its way back to the radar. We can therefore define the power density at the input of the radar receiver (i.e., input to the receive antenna) as

$$PD_{\text{rcv}} = (ERPD\sigma)\left(\frac{L_{\text{atmos}}}{4\pi R^2}\right) = \frac{P_t G_t L_t L_{\text{atmos}}^2 \sigma}{(4\pi)^2 R^4} \tag{2.3}$$

This is a power density (power per unit area), because the ratio (σ/R^4) has units of inverse area (1/m^2, for example). The total power into the receiver is the product of this power density and the effective capture area of the receive antenna, A_{eff}. This term is defined as

$$A_{\text{eff}} = \frac{G_r \lambda^2}{4\pi} \tag{2.4}$$

where G_r is the gain of the receive antenna in the direction of the target (the peak gain in the current example), and λ is the wavelength of the radar transmit signal. The available power going into the receiver is the product of (2.3) and (2.4). These relationships are illustrated in Figure 2.3.

Once inside the receiver, however, there are additional losses due to various sources, just as there are on the transmit side. Receive losses are system-specific, and they may include contributions from both analog as well as digital components in the receiver chain and processor. For the purposes of this discussion, all loss sources are grouped into a single loss term L_r. Finally, depending on where this power is defined relative to the signal processing chain, there may be additional scaling to account for signal-processing gains. Example contributors are matched filter/pulse compression gains and integration gains. We represent these terms as G_{sp}. If the receive power is defined at the input to the signal processor, we simply set G_{sp} to 1. The end result is

$$P_r = \left(PD_{\text{rcv}}\right)\left(A_{\text{eff}}\right) L_r G_{sp} = \frac{P_t G_t G_r L_t L_r L_{\text{atmos}}^2 \lambda^2 \sigma G_{sp}}{(4\pi)^3 R^4} \tag{2.5}$$

Variations on this basic form include consolidating all transmit and receive losses into a single system losses term, $L_{\text{sys}} = L_t L_r$. If the transmit and receive antenna gains have the same value G, one may write $G_t G_r = G^2$ instead.

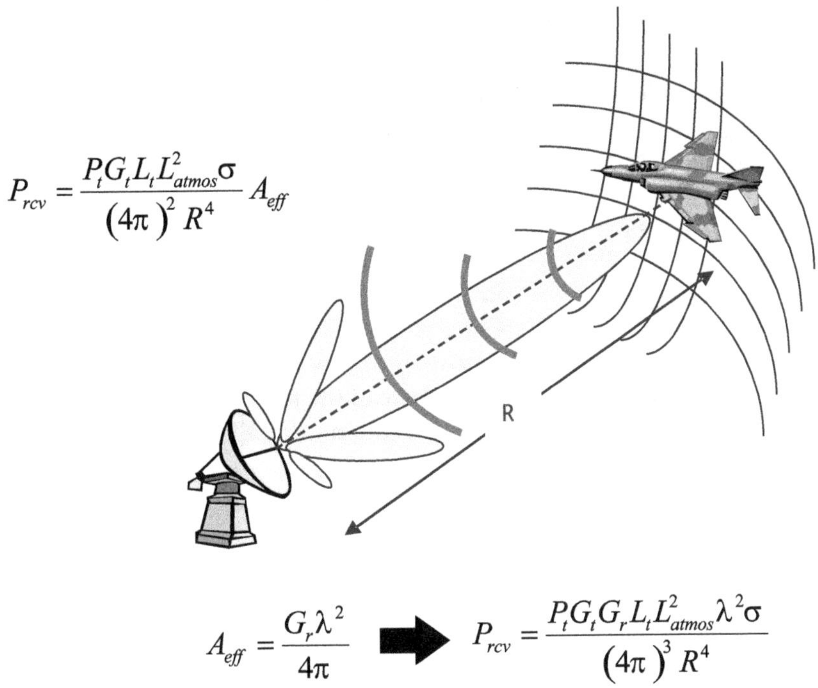

Figure 2.3 Received target power at the radar receiver.

As previously noted, (2.5) by itself provides limited information on the radar system performance. Target detection, for example, depends not on the absolute receive target power level, but on the target power relative to the receiver noise power. We therefore normalize (2.5) by the predicted noise power.

2.2.2 Noise Power Definition

Receiver noise power is defined [2–4] as

$$P_n = \left(kT_s\right)B_n = N_0 B_n \tag{2.6}$$

where $N_0 = kT_s$ is the noise power spectral density (PSD), or power per unit frequency, and B_n is the receiver noise bandwidth. The constant k is Boltzmann's constant (1.38×10^{-23} J/K), and T_s is the system temperature.

We digress momentarily from the discussion on RRE to present modeling options for system temperature. The system temperature includes the temperature of the antenna as well as the temperature of the receiver itself. The simplest model for system temperature is [5]

$$T_s = T_0 F \tag{2.7}$$

where T_0 is the standard temperature of 290K (16.9 °C or 62.3 °F), close to room temperature. F is the receiver noise factor, which is a measure of the additional noise a real receiver generates, relative to an ideal, noiseless receiver. The noise factor is defined relative to the standard temperature T_0, and is always $F \geq 1$ based on that definition. When expressed in units of decibels, it is referred to as the noise figure, although the terms noise figure and noise factor are often used interchangeably.

Equation (2.7) is the model we assume throughout this book. For completeness, two alternate models are presented. Nathanson [4] offers a model that accounts for the antenna and receiver temperatures separately, given by

$$T_s = T_a + T_e, \quad T_e = T_0(F - 1) \tag{2.8}$$

where T_e is the effective temperature of the receiver. If we assume that $T_a = T_0$, then (2.8) is identical to (2.7); however, this model allows the antenna temperature to be other than room temperature.

Yet another, more elaborate model is provided by Blake [2], defined as

$$T_s = T_a + T_{tr}\left(L_r - 1\right) + L_r T_e \tag{2.9}$$

which accounts for the thermal temperature T_{tr} and loss L_r of the transmission line between the antenna and the receiver. The effective receiver temperature T_e is defined the same as in (2.8). Assuming a lossless transmission line ($L_r = 1$), (2.9) reduces back to (2.8).

2.2.3 The SNR Equation

Returning our attention to the target received power, normalization by noise power results in the SNR given by

$$SNR = \frac{P_r}{P_n} = \frac{P_t G_t G_r L_{\text{sys}} L_{\text{atmos}}^2 \lambda^2 \sigma G_{sp}}{(4\pi)^3 R^4 k T_0 F B_n} \tag{2.10}$$

The signal-processing gain in (2.10) must now be redefined as an SNR gain instead of simply a signal power gain, given that the noise does experience some gain through the signal processor as well. To summarize, all the components in (2.10) are defined in Table 2.1.

SNR values can have a very large dynamic range, meaning that the values can vary from very large to very small. It is therefore often convenient to specify SNR in decibels, defined as

$$SNR\ (\text{dB}) = 10\log_{10}\left(\frac{P_r}{P_n}\right) \tag{2.11}$$

where the argument inside the parentheses in (2.11) is the unitless quantity defined by (2.10).

Table 2.1 Summary of RRE Components

Symbol	*Description*	*Comments*
SNR	Signal-to-noise ratio	Power ratio (unitless)
P_r	Receive signal power	Units of watts (W)
P_n	Receiver noise power	Given by kT_0B_nF
P_t	Transmit peak power	Units of watts (W)
G_t	Transmit antenna gain	Assumed peak gain (unitless)
G_r	Receiver antenna gain	Assumed peak gain (unitless)
L_{sys}	System losses	Combination of transmit and receive losses = L_tL_r
L_{atmos}	One-way atmospheric losses	Range and frequency-dependent
λ	Wavelength	Units of meters (m)
σ	RCS	Units of squared meters (m^2)
G_{sp}	Signal-processing gain	System-dependent SNR gain
R	Radar-to-target range	Slant range, units of meters (m)
k	Boltzmann constant	1.38e-23 J/K
T_0	Nominal system noise temperature	290K
F	Noise factor	Called noise figure when in decibels
B_n	Noise bandwidth	Typically assumed matched to waveform bandwidth

It was briefly mentioned in Section 2.1 that an alternate name for (2.10) is the radar range equation, and the reason should now be obvious. Given all radar parameters from Table 2.1, this equation provides the SNR achieved at a specific range; conversely, it may also be used to calculate the range at which some SNR value is achieved for a given target RCS. Solving (2.10) for R results in

$$R = \left(\frac{P_t G_t G_r L_t L_r L_{\text{atmos}}^2 \lambda^2 \sigma G_{sp}}{SNR(4\pi)^3 kT_0 FB_n} \right)^{1/4} \tag{2.12}$$

As an example, assume that a 13-dB SNR is required for target detection, and all other values in (2.12) are known. The detection range is then calculated using (2.12), with the value for detection SNR applied in linear units (which is 20 in this example, since $10 * \log_{10}(20) = 13$ dB).

Equation (2.12) provides a convenient short-form notation for SNR calculation. Setting both SNR and σ in (2.12) to 1, denote the resulting range as R_0. By definition, this will be the range at which the radar achieves a 0-dB SNR for a target RCS of 1 m^2. We may now use R_0 to calculate the SNR for any arbitrary range and/or any arbitrary RCS using the simplified expression

$$SNR\text{ (dB)} = 10\log_{10}\left(\sigma \left(\frac{R_0}{R} \right)^4 \right) \tag{2.13}$$

There are multiple variations to the RRE form shown in (2.10), as the radar parameters may be combined in several ways. As an example, one common assumption is that the noise bandwidth B_n is matched to the waveform bandwidth, which for an unmodulated square pulse of width τ is approximated as $B_n = 1/\tau$. We may thus rewrite the ratio P_t/B_n as $P_t\tau$, which is the energy of a single pulse. As covered in Chapter 7, we can define an average power over the entire pulse repetition interval (PRI) as

$$P_{\text{avg}} = P_t d_t = P_t\left(\frac{\tau}{PRI}\right) \tag{2.14}$$

where d_t is defined as the duty factor or duty ratio and represents the fraction of the time the radar is transmitting. Furthermore, consider the case where the signal processing consists of the coherent integration of N pulses. In the absence of windowing or other losses, the signal-processing gain G_{sp} is simply equal to N. With these relationships in mind, we may rewrite (2.10) as

$$SNR = \frac{P_{\text{avg}} G_t G_r L_{\text{sys}} L_{\text{atmos}}^2 \lambda^2 \sigma N PRI}{(4\pi)^3 R^4 k T_0 F} = \frac{P_{\text{avg}} G_t G_r L_{\text{sys}} L_{\text{atmos}}^2 \lambda^2 \sigma T_{\text{dwell}}}{(4\pi)^3 R^4 k T_0 F} \tag{2.15}$$

Notice that on the right-hand side (RHS) of (2.15) we define the product $N*PRI$ as the dwell or observation time T_{dwell} of the radar. Equation (2.15) is the average power form of the radar equation, and it explicitly shows that the SNR increases the longer the radar observes the target (i.e., the more return pulses it integrates).

We will use the RRE as the fundamental framework in developing an end-to-end radar simulation in Simulink. Section 2.3 presents a MATLAB representation of the RRE, implemented via a graphical user interface (GUI). This simple tool will prove insightful in showing the dependencies of the RRE parameter values in the calculated SNR.

Before moving on to the modeling discussion, we conclude with Section 2.2.4 discussing a couple additional variations of the RRE that prove useful in radar analysis.

2.2.4 Search and Track Forms of the Radar Equation

Search and track are fundamental radar operations, and often the system has dedicated operational modes for the implementation of these functions. During target search, the radar must cover a region in space defined by a 2-D angular extent (azimuth and elevation), quantified by a solid angle Ω in units of steradians. Steradians are proportional to squared degrees, so the solid angle may be thought of as an angular area. The radar must point its antenna to a location within this search zone, collect and process data, then move the antenna to the next position and repeat the process. If we define the antenna beamwidths in units of radians, then the number of beam positions M required to cover the search region is approximately

$$M = \frac{\Omega}{\theta_{AZ}\theta_{EL}} \tag{2.16}$$

The total search time required to cover the search region is then

$$T_s = MT_{dwell} \tag{2.17}$$

where T_{dwell} is the dwell time for a single beam position, as defined in (2.15). Also, the product of the beamwidths in the denominator of (2.16) can be approximated as

$$\theta_{AZ}\theta_{EL} \approx \left(\frac{\lambda}{L_{AZ}}\right)\left(\frac{\lambda}{L_{EL}}\right) \approx \frac{\lambda^2}{A_{eff}} \tag{2.18}$$

where L_{AZ} and L_{EL} are the antenna physical dimensions in the azimuth and elevation dimensions.

Using (2.4) and (2.16) to (2.18), we rewrite (2.15) as

$$SNR = \left(P_{avg}A_{eff}\right)\left(\frac{L_{sys}L_{atmos}^2}{(4\pi)kT_0F}\right)\left(\frac{\sigma}{R^4}\right)\left(\frac{T_s}{\Omega}\right) \tag{2.19}$$

Given a minimum required SNR for reliable detection at a single dwell, we may rearrange (2.19) as

$$\frac{P_{avg}A_{eff}L_{sys}}{\left(4\pi kT_0\right)F} \geq \left(\frac{SNR_{min}}{L_{atmos}^2}\right)\left(\frac{R^4}{\sigma}\right)\left(\frac{\Omega}{T_s}\right) \tag{2.20}$$

This is known as the search form of the RRE. The terms on the RHS of the inequality are user requirements, which in general are outside of the control of the radar designer. The terms on the left-hand side (LHS) (i.e., the power-aperture product, system losses, and noise figure) are terms that can be designed to meet the performance requirements by satisfying the inequality. The denominator terms in parenthesis are constants that could be on either side of (2.20). This equation offers a good first-order summary of the performance trades involved in a radar search mode. It shows that SNR depends on the power-aperture (PA) product and is independent of wavelength for a given search region and search time requirements.

The track form of the RRE follows a similar development. Assume the radar needs to track N_t targets. Assuming a fixed dwell time, the total time to update the track information on all targets is ($N_t * T_{dwell}$). The track update rate is therefore

$$r = \frac{1}{N_tT_{dwell}} \tag{2.21}$$

Using (2.4) and (2.21), we again rewrite (2.15) as

$$SNR = \left(\frac{P_{\text{avg}} A_{\text{eff}} G_t L_{\text{sys}}}{(4\pi)^2 kT_0 F} \right) \left(\frac{L_{\text{atmos}}^2 \sigma}{R^4 (rN_t)} \right) \tag{2.22}$$

which, for a minimum SNR required to maintain tracking performance, is rearranged as

$$\frac{P_{\text{avg}} A_{\text{eff}} G_t L_{\text{sys}}}{\left((4\pi)^2 kT_0 \right) F} \geq \left(\frac{SNR_{\text{min}}}{L_{\text{atmos}}^2} \right) \left(\frac{R^4}{\sigma} \right) (rN_t) \tag{2.23}$$

Equation (2.23) is one variant of the track form of the RRE, showing that performance depends on the power-aperture-gain (PAG) product, independent of wavelength. An alternate form is

$$\frac{P_{\text{avg}} A_{\text{eff}}^2 L_{\text{sys}}}{\left(4\pi kT_0 \right) F\lambda^2} \geq \left(\frac{SNR_{\text{min}}}{L_{\text{atmos}}^2} \right) \left(\frac{R^4}{\sigma} \right) (rN_t) \tag{2.24}$$

This version is known as the power-aperture-squared (PA^2) form, and is no longer independent of wavelength.

2.3 MATLAB Model

The MATLAB implementation is based on (2.10), the peak power form of the RRE. It is effectively a simple SNR calculator where the user enters the input parameter values and presses a button, and the resulting SNR is displayed in decibels. The calculator is started by typing "SNRCalc" at the MATLAB command prompt. The GUI is shown in Figure 2.4.

The edit boxes for the input parameters display default values, which the user may change manually. Each input box features some level of error-checking, in that the GUI will display an error message if the parameter entered is not a number. This check is currently very limited, however, as it will not fully validate whether the entry makes sense (e.g., flag a negative value when the amount should be positive). A more robust interface could be implemented with some additional effort.

Upon pressing the "calculate" button, the corresponding SNR value is displayed in the output box (22.78 dB for the default input values). If the range to target is specified as a vector, the box will still show the calculated SNR for the first element of the range vector; but in addition, the calculator will generate a plot of SNR versus range, as illustrated in Figure 2.4. Finally, the "reset" button returns the input boxes to their default values, and the output SNR box is zeroed out.

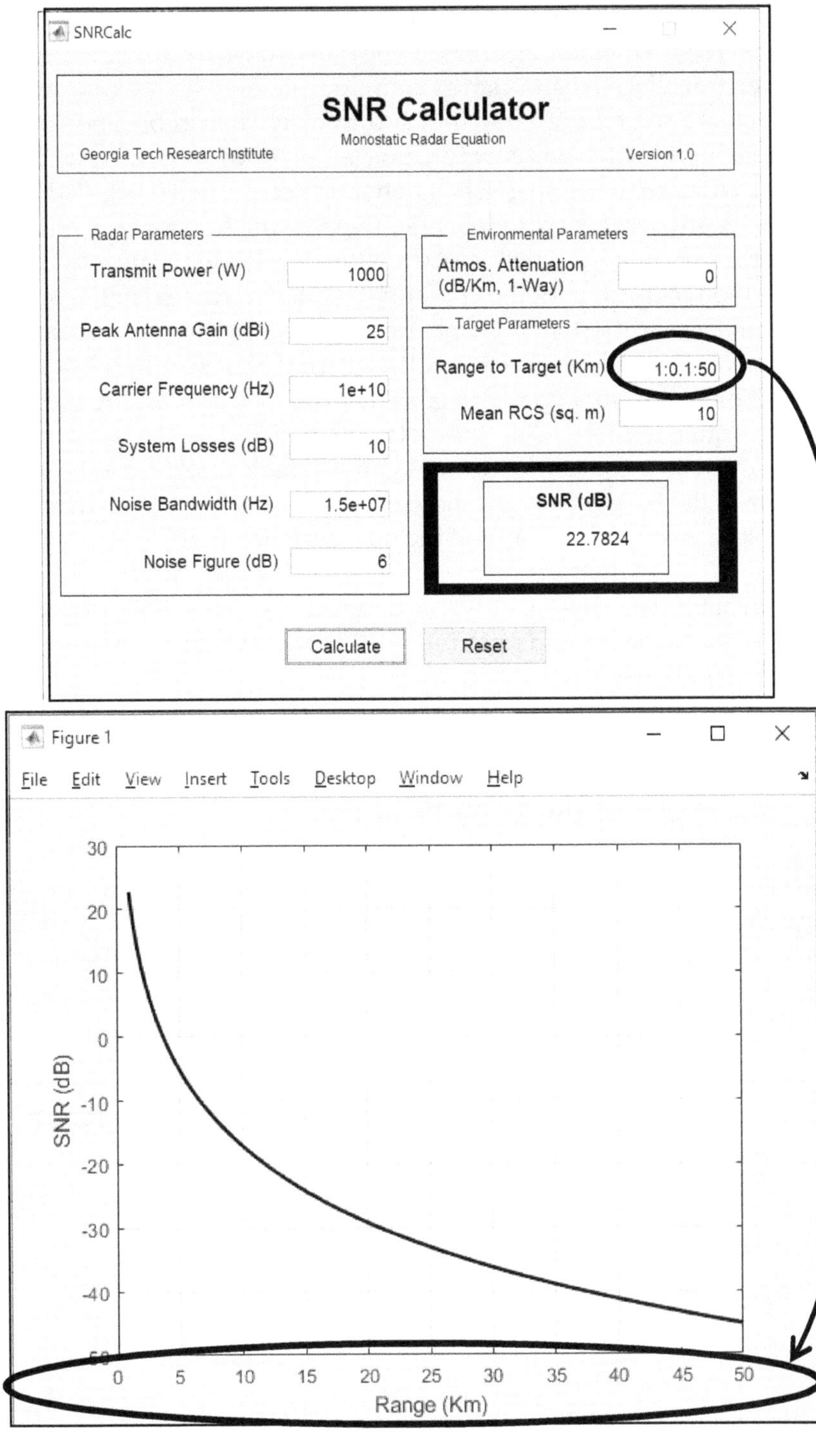

Figure 2.4 SNR Calculator GUI and resulting SNR versus range plot.

A number of assumptions have been made to simplify the calculations. The notional radar system is assumed to be monostatic, meaning the transmitter and receiver are collocated and share a common antenna. The peak antenna gain input is therefore used for both transmit and receive. This ignores potentially having a different illumination window function (amplitude taper) on receive than on transmit. In addition, the atmospheric attenuation coefficient (specified in decibels per kilometer one-way) is set independently from the frequency input. In reality, the former is very much a function of the latter. It is left to the user to enter the correct attenuation coefficient for the operating frequency. As a potential upgrade, a drop-down menu of different atmospheric conditions (clear air, fog, rain at a specific rain rate, etc.) could be added, and the internal logic would select the attenuation coefficient from a lookup table consistent with the input frequency.

As simple and limited as this tool is, it can provide a newcomer to the field with a quick way to explore what-ifs and gain valuable insight into how SNR changes relative to the baseline value when input parameters change. Example questions that might be quickly examined include the following:

- What if the transmit power is doubled? (Answer: SNR increases by 3 dB.)
- What if the antenna gain is × dB higher/lower? (Answer: SNR is 2× dB higher/lower.)
- What if the range is cut in half? (Answer: SNR increases by 12 dB.)

2.4 Simulink Model of the Radar Equation

We now turn our attention to a Simulink implementation of the same SNR calculation from (2.10). The final form of the Simulink model is shown in Figure 2.5, where the model has already been executed, and the calculated SNR is displayed as 22.78 dB, the same value calculated by the MATLAB model using the default values.

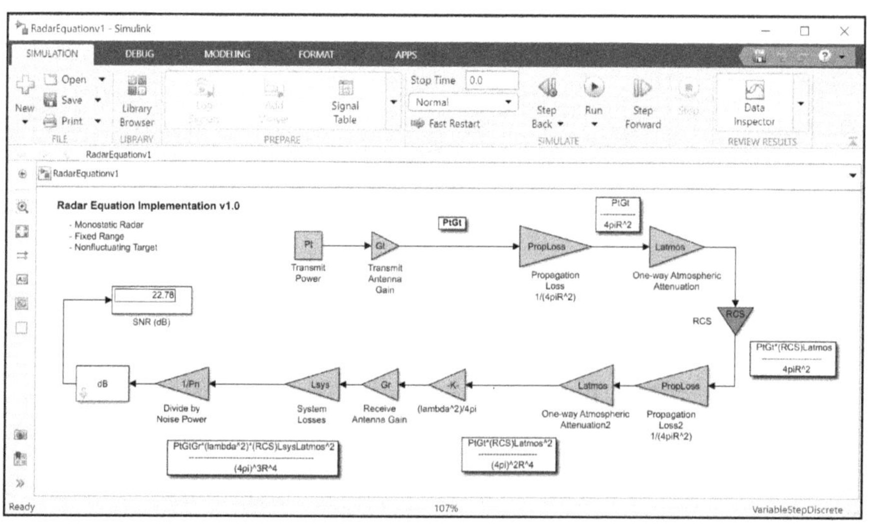

Figure 2.5 Simulink implementation of the RRE.

This model is called RadarEquationv1 and consists of a "Constant" block as the starting point, used to represent the transmit power. This is followed by multiple "Gain" blocks used for the remaining factors of the RRE. The model ends with a "dB Conversion" block and a "Display" block to show the final result.

The steps to create RadarEquationv1 from scratch are listed as follows:

1. Create an init function (a new script m-file in this case);
2. Create a new model file;
3. Add Simulink blocks;
4. Make connections;
5. Add annotations (optional);
6. Set block parameters;
7. Set stop time;
8. Add Init Callback function to the model callbacks;
9. Execute!

Figure 2.6 shows the m-file used for the init function (step 1) and a couple examples for setting block parameters (step 6). Steps 2–5 are covered in the online Simulink appendix.

The parameters entered in step 6 are the variables defined in the init function (e.g., "Pt" for transmit power). Most of them are visible as the labels on the blocks themselves. Notice that on the init function, the values used are the same as the

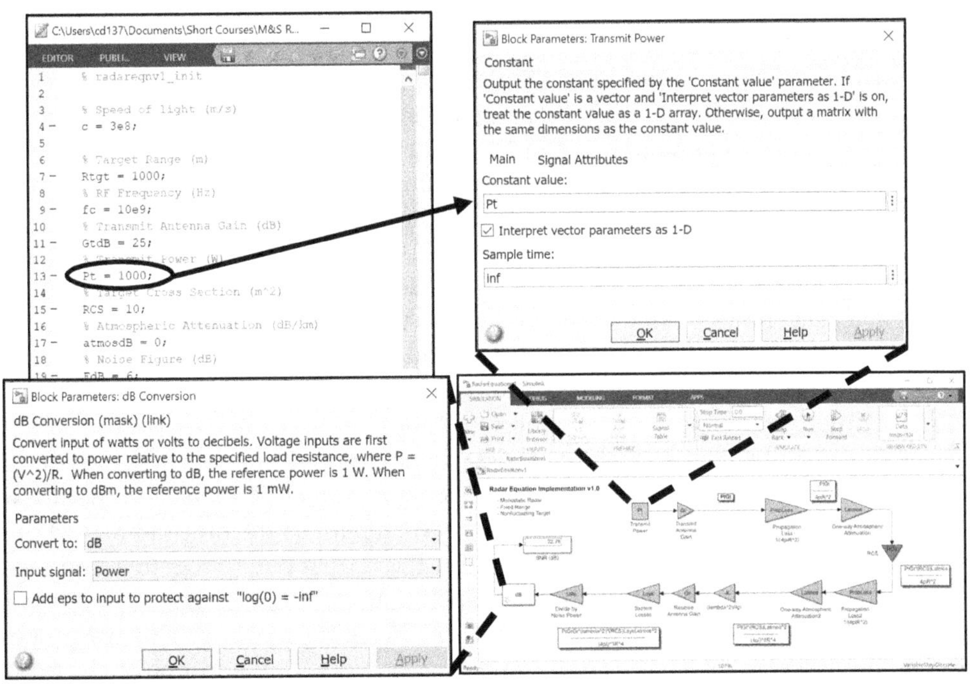

Figure 2.6 Initialization m-file (upper left) and selected Block Parameter windows of RadarEquationv1.

default values in the MATLAB model, which should come as no surprise given that they both result in the same SNR output.

Since the model only needs to run one time step to calculate the SNR, we set the simulation "Stop Time" parameter to 0 (step 7). In step 8, we need to tie the init m-file from Figure 2.6 to the simulation initialization callback function ("InitFcn") so that the variables will be loaded to the model. The steps are also in the online Simulink Appendix, but repeated here for convenience.

Right-click anywhere in the model background, then select "Model Properties" from the menu that appears. This opens the "Model Properties" window. Select the "Callbacks" tab, and then select (click) "InitFcn". Type the name of the init function ("radareqnv1_init") in the whitespace to the right and press "OK" to close the window. After saving your changes, the model is now ready to execute by pressing the green "Run" button in the "SIMULATION" Toolstrip tab.

The color-coding on the blocks is optional but is intended as a visual aid for logical block groupings: All radar-related blocks are green in the accompanying software; all propagation-related blocks are blue in the software; and all target-related blocks are red in the software. This model will be further developed into a more comprehensive engagement simulation between a radar and a target. As we add complexity to this basic block diagram, the color-coding will assist with model readability.

Section 2.3 described how to define a vector of target ranges in the MATLAB model and use it to create a plot of SNR versus range. Similarly, the next two excursions over the basic Simulink model accomplish the same functionality. Versions 2 and 3 of the radar equation model are shown in Figure 2.7.

In RadarEquationv2, the range to the target is a scalar that changes every simulation time step. The collection of blue blocks at the bottom of the model window show the start range "Rstart" in a "Constant" block. This initial value decreases every simulation time step by a range step "deltaR." The range-dependent, one-way spherical propagation and atmospheric attenuation losses are calculated based on the resulting range at a particular time and applied to the main signal path at the top of the model. At the end of the signal path, the scalar received power values are collected into a vector using a "Buffer" block, and sent to an "Interpreted MATLAB Fcn" block for SNR calculation and plotting. RadarEquationv3 is an alternate implementation in which the target range is defined as a vector from the beginning, and the vector is inserted into the signal path when the range-dependent losses are applied. This can be observed in Figure 2.7(b) where the "Signal Dimensions" and "Wide Nonscalar Lines" options are turned on in the model. The "Buffer" blocks at the end of RadarEquationv2 are not needed in this version.

On both of these models, the signal out of the RCS block is split into two. One of the branches maintains the same return path as the original RadarEquationv1 model, while the second branch skips the return path altogether and feeds right into the "Matrix Concatenation" block that collects the signals and hands them over to the "Interpreted MATLAB Fcn" block for plotting. This was done so that the power level at the target could also be plotted, in addition to the power back at the radar receiver. The output plots created by both of these models are shown in Figure 2.8.

(a)

(b)

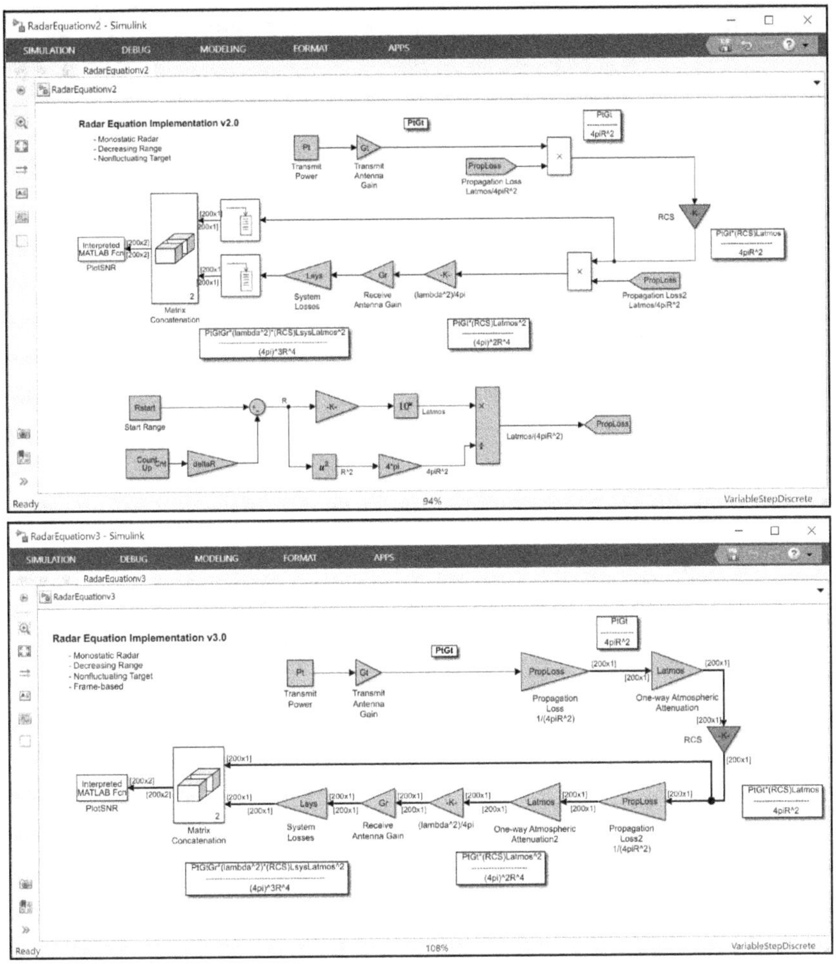

Figure 2.7 (a) RadarEquationv2: scalar version of multirange calculation and (b) RadarEquationv3: vector version.

The plot in Figure 2.8(a) shows the power versus range at the target and at the receiver; notice the large signal loss in the return path from the target back to the radar. In addition, it is worth noticing that the two curves have different range dependencies. The power at the receiver exhibits the familiar $1/R^4$ dependency, while the power at the target falls off as $1/R^2$ since it is a one-way power. The plot in Figure 2.8(b) is SNR versus range (same as the MATLAB model plot).

2.5 Concluding Remarks

The relevance of the RRE to the design and analysis of radar systems cannot be understated. It is ubiquitous throughout this book; we quantify the role of SNR in detection performance in Chapter 11, and in tracking performance in Chapter 13.

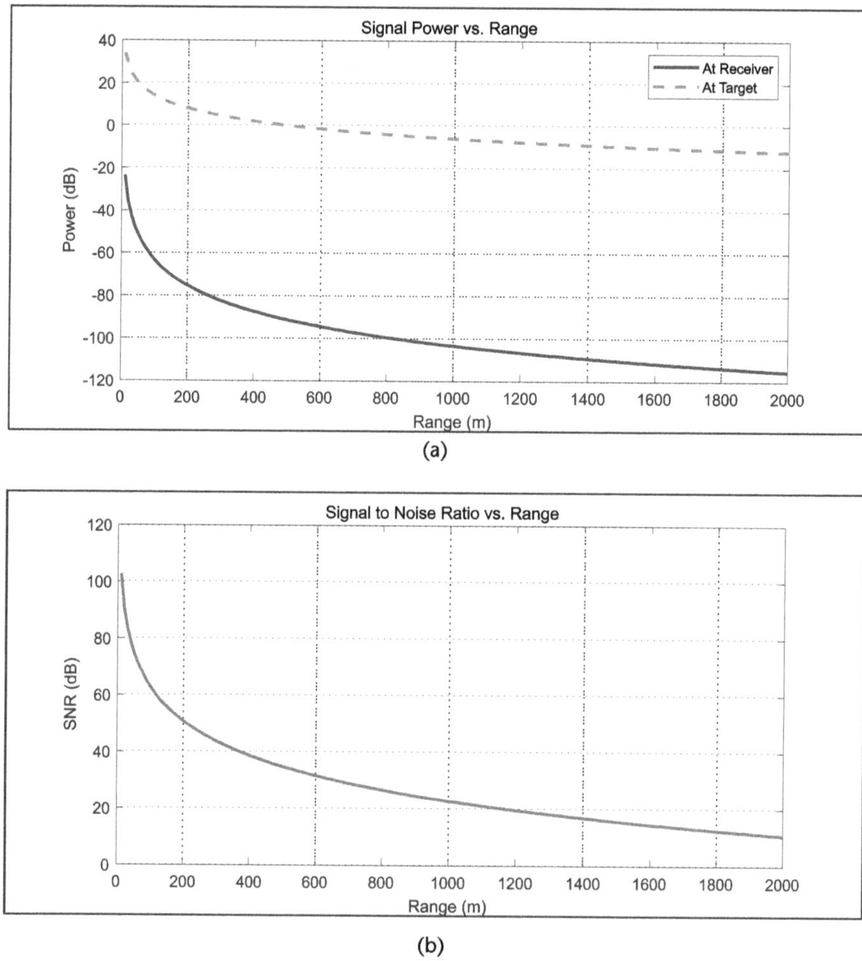

Figure 2.8 Signal power versus range (a) and SNR versus range (b).

Moreover, we use other forms to calculate clutter-to-noise ratio (CNR) in Chapter 6 and jammer-to-noise ratio (JNR) or jammer-to-signal ratio (JSR) in Chapter 16.

As the basic Simulink model presented here is expanded into a more comprehensive engagement simulation, it is worth revisiting the system losses L_{sys} and the sources that compose it. In the previous discussion we glossed over individual loss sources. Several references [2, 5, 6] provide detailed discussions and a breakdown of relevant loss sources. There is no universal standard for loss sources, so the reader must exercise care in ensuring that losses are accounted in the right RRE factor and avoid double-booking. For example, some references include atmospheric losses in L_{sys}, and some list it as a separate term, like we did in (2.10). Terms stemming from the processing of the signals may be included either in L_{sys} or in the signal processing gain G_{sp}. A classic example is the SNR loss associated with applying amplitude weights to slow-time samples prior to Doppler filtering. For N coherently integrated pulses in a radar dwell, the SNR gain is N if and only if uniform weights are applied (i.e., $w_n = 1$ for $n = 1$ to N). For any other nonuniform weighting function, the SNR gain is given by

$$G_{sp} = \frac{\left(\sum_{n=1}^{N} w_n\right)^2}{\sum_{n=1}^{N} w_n^2} = \left(\frac{\left(\sum_{n=1}^{N} w_n\right)^2}{N\sum_{n=1}^{N} w_n^2}\right) N = L_{sp} N \tag{2.25}$$

On the right side of (2.25), we have split the actual gain into the uniform weights gain (N) and an SNR loss term relative to the uniform case (L_{sp}). Again, there is no universal convention and some authors may include this loss in the total system loss L_{sys}.

Moreover, when developing a time-based simulation that performs these signal-processing functions, the SNR losses or gains will be inherently accounted for in that process, and the user should remove these contributors from the total system loss L_{sys} applied elsewhere in the model. Similarly, other operational losses such as beam-shape loss (reduced antenna gain due to the target not being in the center of the antenna main beam) and range/Doppler bin straddling (target return splitting its energy over contiguous cells in the range and/or Doppler dimension) may likewise be removed from L_{sys}, if they are properly accounted for in the simulation.

The next several chapters focus on specific subsystems and aspects of the RRE where higher-fidelity models will be presented as stand-alone simulations. Chapter 15 returns to the Simulink RadarEquationv1 model and presents an upgraded version that incorporates all the subsystem models into a full end-to-end engagement simulation.

References

[1] IEEE Standard Radar Definitions, IEEE Std 686-2008; IEEE Aerospace and Electronic Systems Society—Sponsored by the Radar Systems Panel, May 21, 2008.

[2] Blake, L. V., "Prediction of Radar Range," in *Radar Handbook* (M. I. Skolnik, ed.), Boston: McGraw-Hill, 1990, pp. 2.5–2.56.

[3] Skolnik, M. I., *Introduction to Radar Systems*, Second Edition, New York: McGraw-Hill, 1980, pp. 18–19.

[4] Nathanson, F. E., *Radar Design Principles*, Second Edition, Mendham, NJ: Scitech, 1999, pp. 54–57.

[5] Scheer, J. S., "The Radar Range Equation," in *Principles of Modern Radar, Vol I: Basic Principles* (M. A. Richards, J. A. Scheer, and W. A. Holm, eds.), Raleigh, NC: Scitech Publishing Inc., 2010, pp. 64–72.

[6] Belcher, M. L., "Multifunction Phased Array Radar Systems," in *Principles of Modern Radar, Vol III: Radar Applications* (W. L. Melvin and J. A. Scheer, eds.), Raleigh, NC: Scitech Publishing, Inc., 2014, pp. 257–259.

CHAPTER 3

Antennas

3.1 Introduction

The RRE has two variables, G_t and A_{eff}, which are determined solely by the system antennas, whether for radar or electromagnetic warfare (EW). In addition to these key terms, several additional antenna performance parameters such as the main beam beamwidth, ψ in azimuth and elevation, and the peak and average sidelobe levels (SLLs), SLL_{peak} and SLL_{avg}, also impact the system performance. Accurate results from any system simulation therefore drive requirements for accurate and appropriate simulations of the antenna radiation patterns. The word "appropriate" is deliberately chosen here, because there are numerous options for the numerical approaches to develop antenna simulations. Some of these approaches are very complex and rigorous and require significant computational resources and simulation runtimes. Other approaches use significant shortcuts and simplifications. While the shortcuts may offer much faster calculation speeds, their accuracy can be lacking.

This chapter presents what is needed from antenna simulations and the breadth of technical approaches to the simulation of antenna performance. Our goal is to summarize the simulation options and to offer a simple modeling path that can provide sufficient accuracy without the onus of more detailed approaches that would be much more costly in terms of development time, computational electromagnetic solvers, and simulation runtime.

Section 3.2 discusses antenna fundamentals and several key simplifications that are generally used, beginning with a basic introduction of antenna functionality and presentation of the nomenclature of directivity pattern features.

The antenna directivity pattern describes the angular dependence in two-dimensional angle space of the power levels transmitted or received. Calculation of the directivity pattern versus angle is the key product to be provided from the antenna model to the system-level radar or EW simulations.

Multiple different antenna architectures are also discussed within the following sections. Radar and EW systems have widely varying requirements including gains, beamwidths, SLLs, pattern adaptability, transmit power-handling, and frequency coverage. This range in requirements results in antennas that have significantly different sizes and architectures. Accordingly, the chapter covers techniques for simulations of the antenna directivity patterns for both the higher-gain, narrow-beamwidth radar antennas as well as the wider-beamwidth antennas used in electromagnetic support (ES) and EW systems.

3.2 Antenna Basics

An antenna has three roles in any system, as shown in Figure 3.1. The first is to transmit and/or receive electromagnetic signals. These signals, which are most often composed of sinusoidal oscillations at millions (megahertz) or billions (gigahertz) of cycles per second (hertz), are generically called radio frequency (RF) waves. The second role for the antenna is to perform this radiation or reception efficiently via a design that takes into account appropriate impedance matching between the antenna and its transmission line that connects to the transmitter/receiver. This matching refers to both the characteristic impedance and wave-propagating mode of the interconnecting transmission line. The third role for the antenna is to provide an appropriate directivity pattern, maximizing and minimizing radiation at different angles as desired in space.

The first and second goals are determined by the skill of the antenna developer and subsequent manufacturer. Many available texts provide the necessary theory and design information for the myriad variations of antenna architectures that have been successfully employed in systems. Balanis [1, 2], Kraus and Marhefka [3], and numerous other authors provide thorough resources to cover these topics. For the engineer tasked with modeling a radar or EW system, one can assume (if the antenna was or will be fielded in a system) that the first two goals have been sufficiently met by the antenna developer. While it is helpful for system modelers to be familiar with related topics including the expected efficiency of specific architectures, they do not have to perform simulations to the level required to accurately determine the actual antenna input impedance. The antenna-simulation task for the system modeler focuses on the third goal of any antenna: that of accurately and efficiently determining the amplitude and phase versus angle of the radiated fields—the directivity pattern. From this point forward, we assume that all of the antennas under discussion have been well-designed and impedance-matched. However, before we continue to the directivity pattern–simulation approaches, let's review a few additional background details.

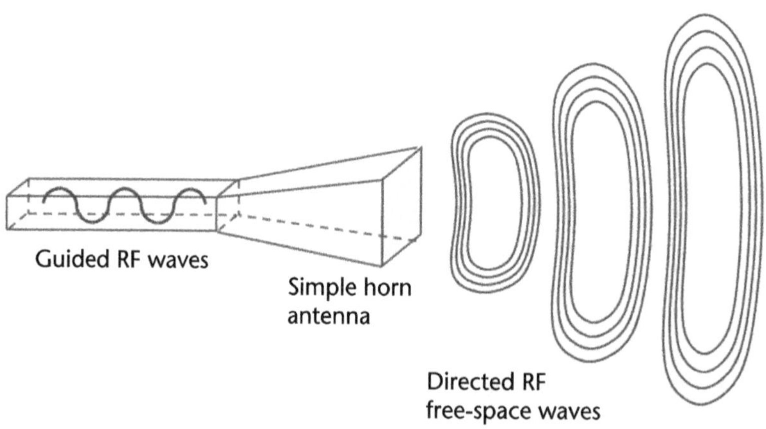

Figure 3.1 Fundamental functionality of an antenna.

Antennas perform their tasks in a reciprocal manner. If an antenna's directivity pattern is known for the transmit mode, you can safely assume it is the same for the receive mode. If antenna A in Figure 3.2 is identical to antenna B, their directivity patterns are the same, irrespective of whether they are transmitting or receiving. The impact to the modeler is that only one directivity pattern needs be calculated for a radar system that both transmits and receives. There are two caveats with this statement. The first is that there are some architectures of passive phased arrays that utilize ferrite phase shifters. Ferrites can exhibit some nonreciprocity, and hence their directivity patterns can exhibit small amounts of nonreciprocity. Usually this is not significant, and it is ignored in system-level modeling. The second caveat is that radar antennas with monopulse tracking beams actually have multiple simultaneous beams on receive, some of which are not used for transmit. These additional beams are important enough to warrant Chapter 9, which is dedicated to monopulse and its use in systems and simulation in models.

In our proceeding discussion of the procedure to calculate the radiated or received fields, we use the term radiated fields. Keep in mind that the directivity pattern is the same for transmit or receive.

3.3 Directivity Pattern Basics

The directivity pattern of an antenna is defined as the power distribution of the radiated fields versus angle in space. Therefore, it is necessary to discuss angle coordinate systems. Since there are many different global and local coordinate systems in use by designers and system engineers, it is crucial to understand the appropriate coordinate systems and to use the correct coordinate transformations. Many in the antenna community utilize theta-phi (θ, ϕ) spherical coordinates, while many radar and EW system engineers use azimuth-elevation (az, el) spherical coordinates. These are similar, with simply a 90° rotation of the downrange vector. The direction downrange from the antenna is most often referred to as the z-axis in a Cartesian coordinate system, with the antenna positioned in the x-y plane. Figure 3.3 shows the angle orientations with respect to the antenna plane for both

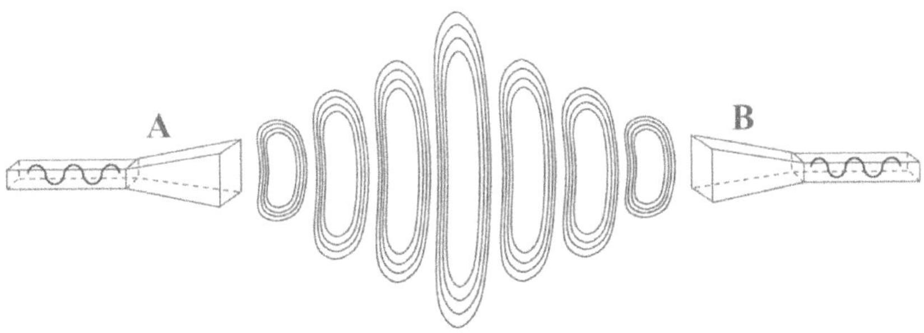

Figure 3.2 Antennas are reciprocal, and their directivity patterns are the same for transmit or receive.

the (θ, ϕ) and (az, el) coordinate systems. It is important to note that most antenna simulations use the convention that the z-axis is normal to the plane of the antenna. When incorporated into system simulations, it is critical that the antenna pattern coordinates are transformed, if needed, into the system simulation coordinates.

Antenna directivity patterns of amplitude versus angle space can be plotted in several formats. Slices or linecut plots through the three-dimensional (3-D) data are common, and they can be displayed in either polar or Cartesian formats, as shown in Figure 3.4. Less common, but often used by antenna engineers are 3-D contour or flattened contour projection plots, which are also presented in Figure 3.4.

Antenna system designers also frequently use the sine-space coordinate system, also referred to as direction cosines, which is essentially a graph of the full forward hemisphere of azimuth-elevation angles projected onto the plane beneath the hemisphere. Its maximum extents span from −1 to +1, representing the far horizons in both respective directions. Unfortunately, the antenna community uses multiple naming conventions for sine-space including (u, v)-, (Kx, Ky)-, and (Tx, Ty)-space. With regret, the authors will continue this trend of inconsistency in the pattern calculations and plots that follow. The equations for the sine-space coordinates are shown as (3.1) and (3.2). Although this conversion is simple, it is needed often enough that MATLAB provides the functions ***azel2uv*** and ***uv2azel*** [4].

$$
\begin{aligned}
u &= \sin(\theta)\cos(\varphi) \\
v &= \sin(\theta)\sin(\varphi) \\
w &= \cos(\theta)
\end{aligned}
\tag{3.1}
$$

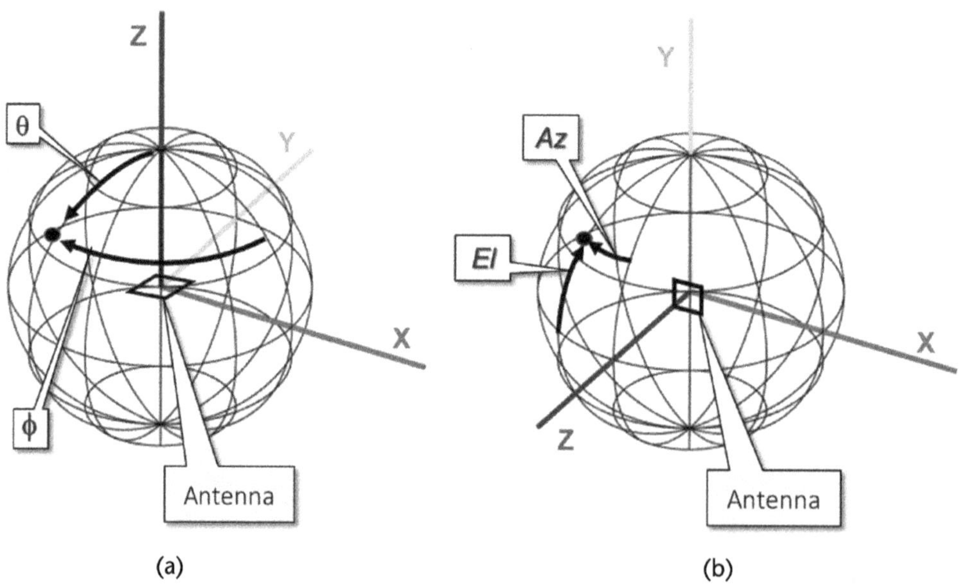

Figure 3.3 (a) Theta-phi and (b) azimuth-elevation spherical coordinate systems showing antenna orientation, the downrange direction Z, and the pattern coordinate conventions.

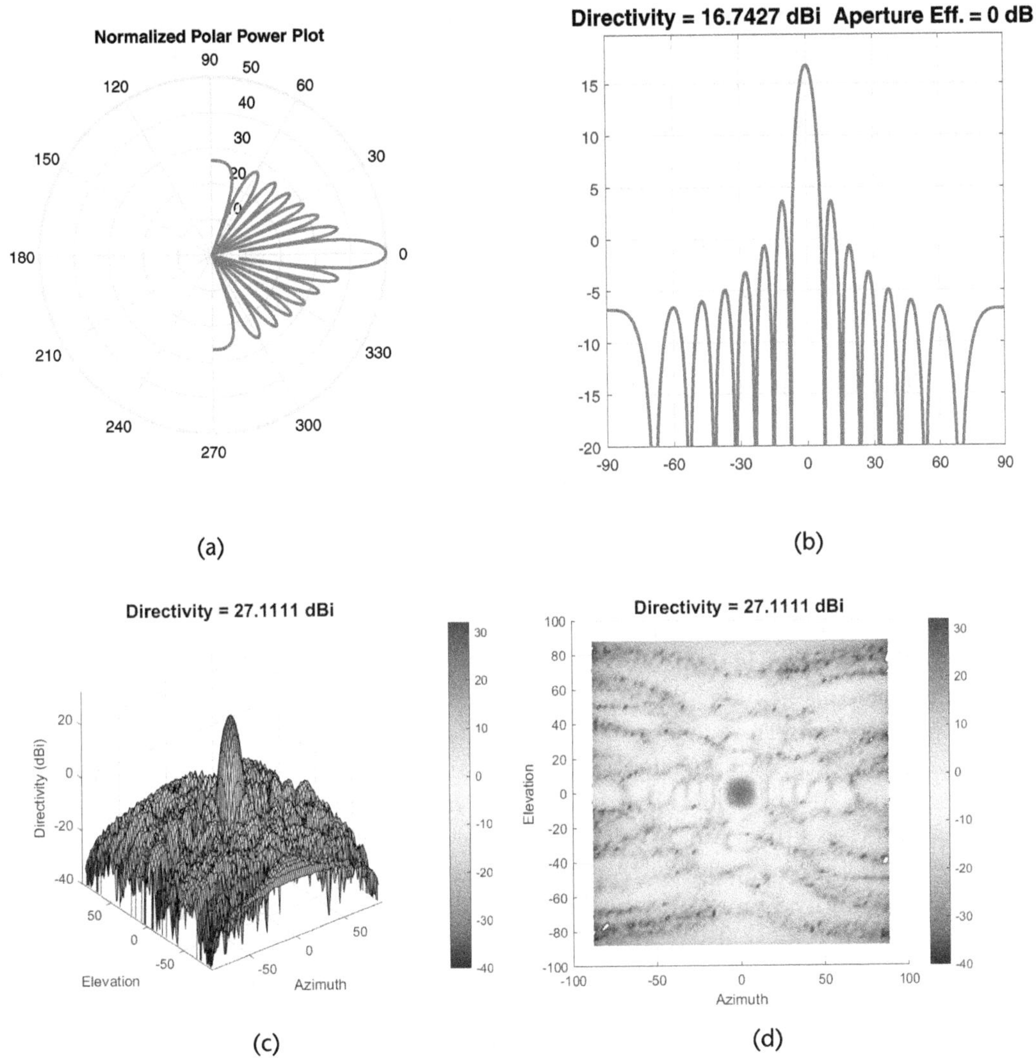

Figure 3.4 Common formats for antenna directivity pattern plots, including (a) polar linecut, (b) Cartesian linecut, (c) 3-D contour, and (d) planar contour plots.

$$
\begin{aligned}
u &= \cos(el)\sin(az) \\
v &= \sin(el) \\
w &= \cos(el)\cos(az)
\end{aligned}
\tag{3.2}
$$

It is important to note that the antenna patterns are usually collected or measured with respect to the boresight angle of the antenna itself and that antennas are often installed with tilt angles relative to their host vehicle and/or relative to the system or global coordinate systems. It is especially common to tilt antennas up in elevation from the horizon for ground- or ship-based systems that have a field of view centered somewhere between the horizon and zenith. Air-to-ground radars may be tilted down in elevation relative to the horizon. To include the tilt in the antenna field of view in sine-space coordinates, we use the sine-space conversion of (3.3), which includes the elevation tilt-back angle, θ_{tilt}, from Corey [5]. Note that the center of the sine-space coordinates, (0, 0) is the angle to the boresight (normal) to the antenna face.

$$
\begin{aligned}
u &= \cos(el)\sin(az) \\
v &= \sin(el)\cos(\theta_{tilt}) - \cos(el)\cos(az)\sin(\theta_{tilt}) \\
w &= \sin(el)\sin(\theta_{tilt}) + \cos(el)\cos(az)\cos(\theta_{tilt})
\end{aligned}
\tag{3.3}
$$

The plots of Figure 3.5 show a field-of-view angle coverage of ±45° in azimuth and 0° (horizon) to 60° in elevation. The center image depicts the antenna with a notional elevation tilt-back angle with the field-of-view plotted in sine-space coordinates including the tilt-back angle.

Readers can refer to Masters and Gregson [6] for a good summary of coordinate systems used in antenna calculations and measurements.

As detailed in coming sections, we break the antenna into many small parts and calculate the antenna directivity pattern by summing the radiated fields—specifically the electric fields, E, in units of volts/meter—from each small part of the antenna to each angle in space. While we always sum the E-fields, we describe the directivity pattern in terms of the magnitude of the power, which is proportional to the magnitude of the electric fields squared, either EE* or $|E^2|$. The power pattern in linear units often has a very wide range of power levels, from very large magnitudes in the main beam to very low levels in the pattern sidelobes. Therefore, we most often present this data in decibel form so that the wide range of values is compressed and more readable on graphs. Keep in mind as we progress through antenna models that we always calculate the sum of the E-fields versus angle, convert to power, and present in decibels, as shown in Figure 3.6.

Numerous pattern artifacts are important to the radar and EW system engineer. These include the main beam peak directivity or gain, the main beam angular width or beamwidth, the peak SLL, and the average or root-mean-square (RMS)

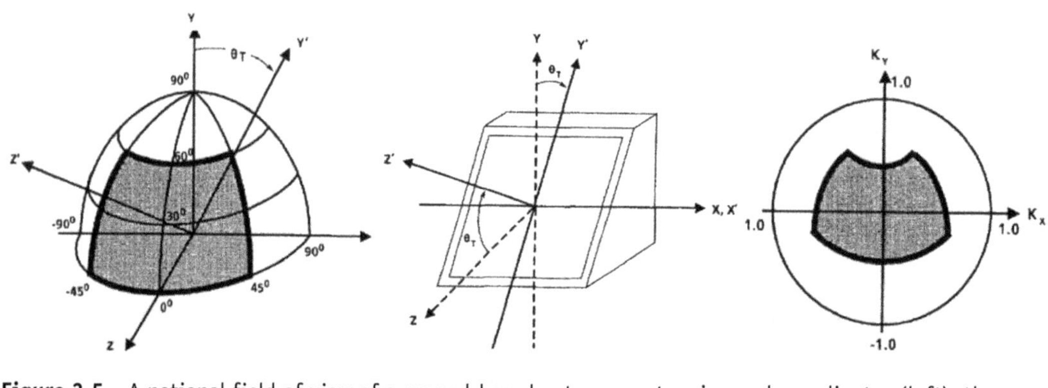

Figure 3.5 A notional field of view of a ground-based antenna system in az-el coordinates (left), the antenna aperture with a tilt-back angle relative to the horizon (center), and the field of view as plotted in sinespace (right). (*Source:* [5], ©1985 IEEE. Reprinted with permission.)

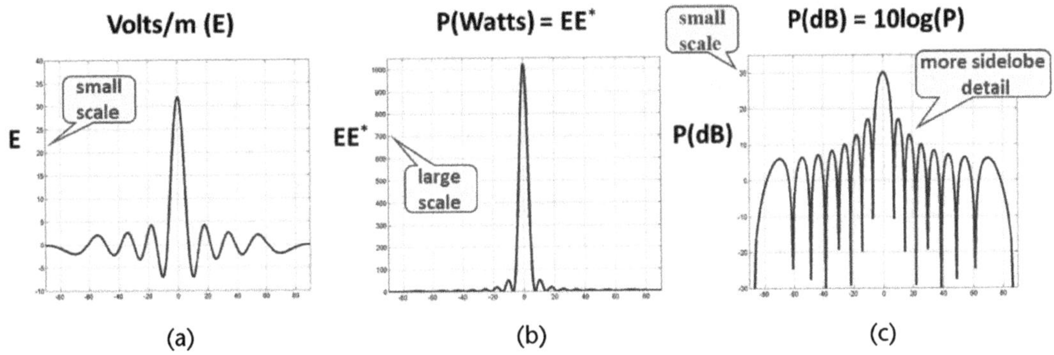

Figure 3.6 The antenna directivity pattern sequence of summing the E-fields versus (a) angle, (b) converting to power, and (c) presenting in the decibel scale.

SLLs. All of those values are primarily of importance for the single desired sense of polarization. For some systems, understanding of the same pattern features for the orthogonal polarization, commonly called the cross-polarization, may also be of interest in some EW engagements.

3.4 Fields and Frequencies

To simplify the directivity pattern models, we reduce the math required by using a few assumptions. First, we know that when we speak of a signal at a single frequency we are referring to a sinusoidally time-varying EM wave. In reality, there are both electric, E-, and magnetic, H-fields that simultaneously exist, both in a plane orthogonal to the direction of the wave travel, as shown in Figure 3.7. However, there are few antenna simulations that require the use or knowledge of the magnetic fields in the directivity pattern models. We know that the magnetic fields exist, and we can calculate them if and when needed—but the need rarely occurs. Therefore, most system simulations and all of the upcoming directivity pattern calculations focus solely on the summation of the radiated E-fields in the calculations of the directivity patterns.

The E-field is (3.4), wherein the sinusoidal variation in time and space is represented by the Euler formula, $e^{j\omega t}$, the phasor, where ω equals $2\pi f$, where f is the frequency in hertz.

$$E = \text{Re}\left\{E_0 e^{j\omega t}\right\} = E_0 \,\text{Re}\left\{\cos(\omega t) + j\sin(\omega t)\right\} \tag{3.4}$$

For most of our simulations, the waveform used is many wavelength cycles long—so long that constant waveform (CW) approximations hold. This means that the antennas are designed for, and essentially operate in, a CW mode and that transient time-based analyses are usually never completed. For this reason, we do not need to perform any time-domain nor transient analyses for our models—but this is a significant simplification. Since antenna directivity patterns are

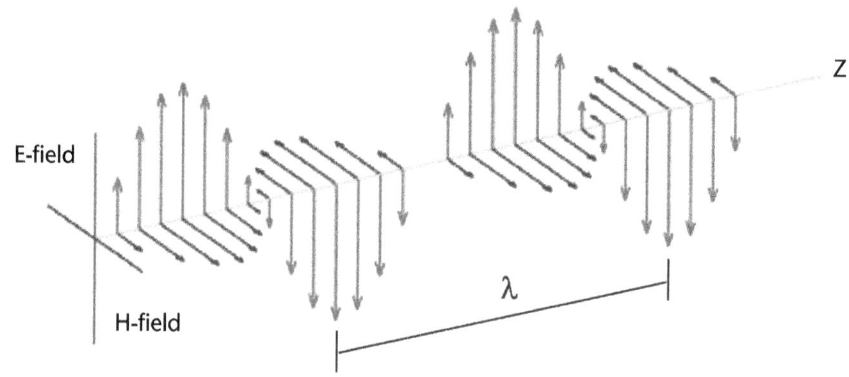

Figure 3.7 Plot of the sinusoidally varying electric and magnetic fields, detailing the definition of a wavelength, λ.

frequency-dependent, we perform frequency-domain calculations one frequency at a time. In all of the antenna models that follow, we solve the frequency-domain solutions where the sinusoidal variation in time is implied, and we simplify our input signals as $E = E_0$.

While some texts use the letter i to represent the imaginary number, most electrical engineering texts use the letter j to represent the imaginary number. Also note that some texts use $e^{-j\omega t}$ in place of $e^{j\omega t}$. Either nomenclature works. Our guidance is, foremost, to be consistent and to conform to the convention that a short electrical length or delay results in a small negative phase shift.

Any given waveform, be it for radar, EW, or communications, is not a single sinusoidal tone but rather a sum of simultaneous frequency or spectral components. However since antenna directivity patterns are frequency-dependent, the patterns must be solved at multiple discrete frequencies. It is up to the system user to determine whether the change in pattern versus the change in frequency warrants multiple frequency simulations. For all of the antenna directivity patterns calculated in this text, the model assumes operation at a single frequency.

3.5 Polarization

Antenna systems usually operate with a single sense of polarization, although some operate with multiple simultaneous polarizations. The polarization is defined as the orientation of the electric field. It can be linear and oriented along either the x- or y-axis, which we refer to as the horizontal and vertical, respectively. The wave shown in Figure 3.7 would be defined as having vertical linear polarization. The electric field can also be tilted at 45° relative to the x- and y-axes, which is referred to as slant or slant-45° polarization. Each of these linear polarizations has an orthogonal polarization (e.g., y- is orthogonal to x-, and slant-135° is orthogonal to slant-45°).

The polarization can also rotate in a circle as a wave propagates downrange. This is termed circular polarization, and it can rotate in a clockwise [right-hand circular polarization (RHCP)] or counter-clockwise [left-hand circular polarization (LHCP)] direction. Per the IEEE definition [7], the direction of rotation is as if viewed from behind the source antenna looking downrange. The antenna that generates the circularly polarized wave has to be able to support two orthogonal linear polarizations and provide a 90° phase delay between the two linear polarizations. The two senses of circular polarization are orthogonal to each other. The electric field components and their relative phasing are presented in (3.5), where $\vec{E_x}$ and $\vec{E_y}$ are unit vectors on the x- and y-axes [8]. Note that multiplying by j is analogous to rotating the phase of that vector by +90°.

$$E_{\text{RHCP}} = \frac{1}{\sqrt{2}}\left(\vec{E_x} - j\vec{E_y}\right); \quad E_{\text{LHCP}} = \frac{1}{\sqrt{2}}\left(\vec{E_x} + j\vec{E_y}\right) \tag{3.5}$$

Originally used in the optical field, a Jones vector is a vector describing the complex values of the E-field vectors along the x- and y-axes. This vector inherently describes the polarization of the fields for any linear or circular orientation as defined by the vector [9].

If the antenna is single linearly polarized (single is used to differentiate from dual linearly polarized systems), two terms are often used to denote the orientation of the directivity pattern linecut. The pattern E-plane is defined as the linecut parallel to the E-field vector polarization. The pattern H-plane is defined as the linecut parallel to the H-field vector polarization.

3.6 Isotropic Antenna Pattern

A very useful concept in modeling antenna directivity patterns is that of the isotropic radiator. The isotropic radiator cannot and does not exist. If it did exist, it would be an infinitesimally small source of radiation with no physical extent—a point source as shown in Figure 3.8. Its radiation pattern would be that of a perfectly spherical expanding wave, with equal magnitude in all directions. Its directivity pattern would be given by (3.6), where there is some initial electric field excitation, E_0, with the expanding phasor term that is spatially time-varying. Note that the time variation has been omitted for simplicity, so this pattern calculation can be thought of as a single snapshot in time.

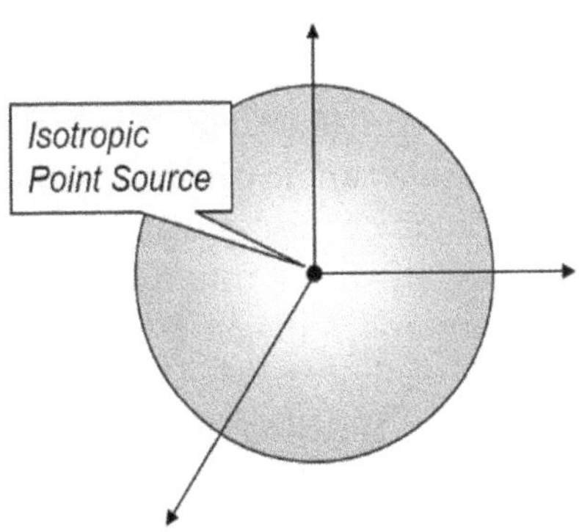

Figure 3.8 The isotropic point source radiator concept and its spherical directivity pattern.

$$E(\theta,\phi,r) = \frac{E_0}{r}e^{-j\frac{2\pi}{\lambda}r} \tag{3.6}$$

Note that as the pattern is that of a perfect sphere, there is no dependence upon θ or ϕ. Equation (3.6) does include the variation of the fields with downrange distance, r. The electric field decreases as $1/r$, and the phase varies as $(2\pi/\lambda)r$. Convention dictates that the range terms r are usually not included in the antenna directivity pattern. The range term is included in either the one-way Friis' transmission equation or the two-way RRE, presented in Chapter 2. Because of this convention, the range term r is not included in any of our antenna directivity pattern calculations henceforth. With the removal of all range-dependent terms, our equation for the directivity pattern for the isotropic radiator simply becomes $E = E_0$ (i.e., the fields are equal in all directions and proportional to the initial excitation).

3.7 Directivity and Gain

For any real antenna that has physical extent, we calculate the directivity pattern by integrating (or summing at discrete points) all of the radiated fields from all source regions in the antenna to a single point in angle space. The directivity pattern has an angle-dependent distribution, which is determined by how the fields either constructively or destructively add at each angle. Some antennas, referred to as low-gain, are designed to have low field strength spread over very broad angular regions. These patterns have two general distribution shapes, that of omnidirectional or hemispherical coverage, as shown in Figures 3.9(a, b), respectively. Although the name omnidirectional sounds like it would be appropriate for isotropic coverage, it defines a pattern that is equal in a plane, not equal at every angle in a sphere. The omnidirectional pattern has a null in the axis orthogonal to the plane in which it has equal power at all angles. Other antennas, referred to as high-gain, are designed to have the radiated fields focused in very narrow angular regions as shown in Figure 3.9(c). We call this high-gain region the main beam or mainlobe of the pattern. Each antenna is designed for its specific application, where the mainlobe beamwidth is sized to fit the angle coverage requirements for the system. Figure 3.10 presents a linecut of a high-gain pattern, detailing pattern artifacts including the peak directivity, the 3-dB main beam beamwidth, the peak SLL, and the average SLLs.

The numerical definition of directivity is given in (3.7), and it is defined as the radiated power intensity, $P_I(\theta,\varphi)$, which has the units of power per solid angle (watts/steradian) at any given angle divided by the power intensity averaged over all angles. With the division or normalization of the average power intensity, the amount of RF signal in watts input into the antenna is effectively removed from the directivity. What remains is the *angle dependency* of where spatially radiated power is concentrated. Directivity has values at all angles, but what most engineers mean when they talk about "the directivity" of an antenna is the peak directivity, D_0. The peak directivity of (3.8) is the maximum value of the power normalized by the average directivity radiated over all angles. Note that directivity is a dimensionless ratio, having no units.

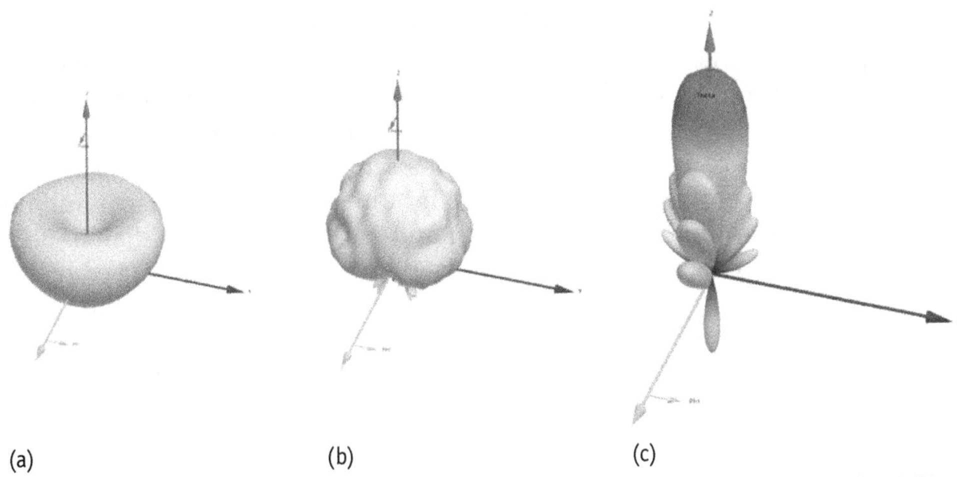

Figure 3.9 3-D antenna directivity patterns including (a) the low-gain omnidirectional and (b) hemispherical, as well as (c) high-gain, with a well-defined main beam and sidelobe structure.

$$D(\theta,\phi) = \frac{P_I(\theta,\phi)}{\frac{1}{4\pi}\int_{\phi=0}^{\phi=2\pi}\int_{\theta=0}^{\theta=\pi} P_I(\theta,\phi)\,\partial\theta\,\partial\phi} \tag{3.7}$$

$$D_0 = \frac{P_{I\ \max}}{\frac{1}{4\pi}\int_{\phi=0}^{\phi=2\pi}\int_{\theta=0}^{\theta=\pi} P_I(\theta,\phi)\,\partial\theta\,\partial\phi} = \frac{P_{I\ \max}}{P_{I\ \text{average}}} \tag{3.8}$$

An approximation for the peak directivity for larger-aperture antennas is given by (3.9). Approximations for the main beam 3-dB beamwidths, ψ_{azimuth} and $\psi_{\text{elevation}}$, at the boresight angle normal to the antenna aperture are given by (3.10) and (3.11) for a rectangular aperture; these equations introduce two interrelated

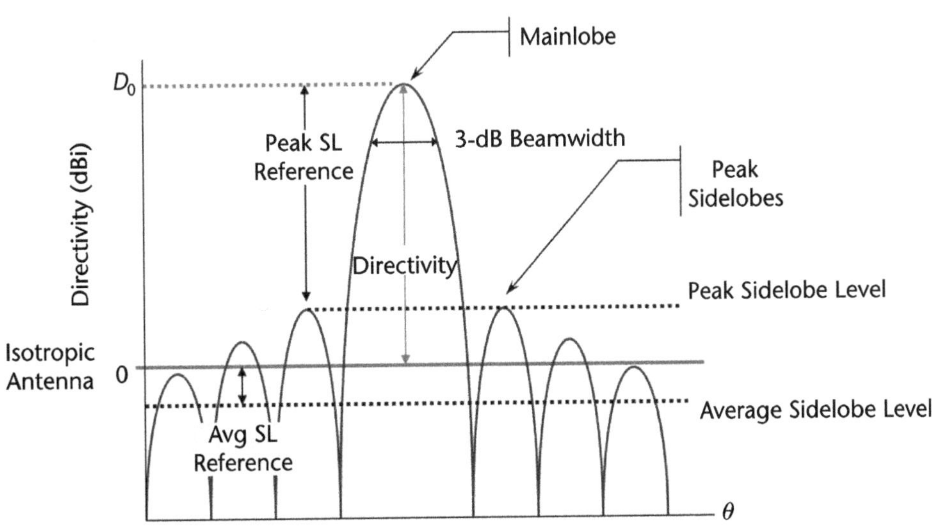

Figure 3.10 Directivity pattern features.

terms, that of the aperture efficiency, η, and the beam broadening factor, α. The aperture efficiency equals 1 for an antenna with uniform (meaning equal across the aperture) amplitude and phase and with no errors. With amplitude tapers and errors the aperture efficiency is less than one. Amplitude tapers are used to reduce SLLs, which is detailed in subsequent sections. The beam-broadening factor is approximately equal to 0.88 for uniform amplitude taper and no errors. With a nonuniform amplitude taper and errors, α is greater than 0.88. A in (3.9) is the cross-range area of the antenna aperture. Cross-range is used to denote the planar aperture area orthogonal to the downrange normal or boresight angle of the antenna. L (for length) in (3.10) and (3.11) represents the length of the aperture in one plane (i.e., either width or height). Note that some authors define $A_{\text{effective}}$ as the product of A and $\eta_{\text{aperture efficiency}}$.

$$D_0 = \frac{\eta_{\text{aperture efficiency}} 4\pi A}{\lambda^2} = \frac{4\pi A_{\text{effective}}}{\lambda^2} \tag{3.9}$$

$$\psi_{\text{azimuth}} = \alpha_x \frac{\lambda}{L_x} \text{ (units of radians)} = 57.3\alpha_x \frac{\lambda}{L_x} \text{ (degrees)} \tag{3.10}$$

$$\psi_{\text{elevation}} = \alpha_y \frac{\lambda}{L_y} \text{ (units of radians)} = 57.3\alpha_y \frac{\lambda}{L_y} \text{ (degrees)} \tag{3.11}$$

Figure 3.11 shows examples of amplitude distributions across an antenna aperture. One amplitude distribution is uniform, and the other is of a –30-dB Taylor-tapered amplitude distribution. (See Section 3.12.2 for more on Taylor tapering.) The amplitude distributions result in different directivity patterns with different $\eta_{\text{aperture efficiency}}$ and α values. Table 3.1 presents values for $\eta_{\text{aperture efficiency}}$ (in decibels) and α for different amplitude tapers. The tapers start at uniform (SLL of 13.2 dBmb) and progress through several Taylor tapers with lower first sidelobe levels. dBmb identifies that the signal is specified in dBs relative to the peak level of the main beam. Note that while antenna designers can design for lower sidelobe levels in some architectures, the prices to be paid include slightly lower directivity

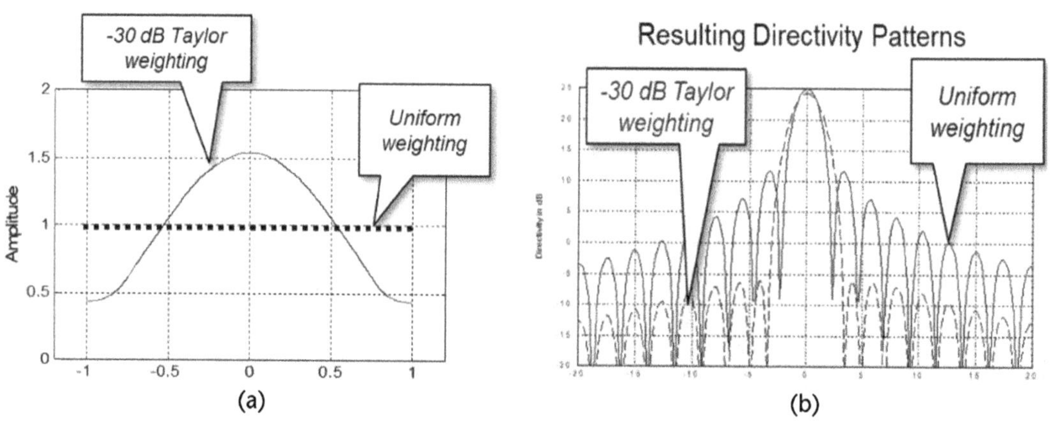

Figure 3.11 Uniform and tapered amplitude distributions (a) and resulting directivity patterns (b).

Table 3.1 Values of α and η for Different Amplitude Weightings for a Line-Array Antenna or a Single Dimension in a Rectangular Antenna

SLL Relative to the Main Beam (dBmb)	*Beamwidth-Broadening Factor* α *(unitless)*	*Directivity Aperture Efficiency Loss* $\eta_{aperture\ efficiency}$ *(dB)*
−13	0.88	0.0
−20	0.98	−0.22
−25	1.05	−0.46
−30	1.12	−0.70
−35	1.18	−0.95
−40	1.25	−1.18
−45	1.3	−1.39

(the $\eta_{\text{aperture efficiency}}$) and a broader main beam (given by the α) as compared to uniform weighting.

Note that the values for α and $\eta_{\text{aperture efficiency}}$ in Table 3.1 are for a line-source antenna only. Values for a 2-D rectangular or circular aperture are different, and the total aperture efficiency losses are greater.

Figure 3.12 illustrates an important conclusion from the beamwidth equations, (3.10) and (3.11): If the antenna is symmetric, the main beam will be circularly

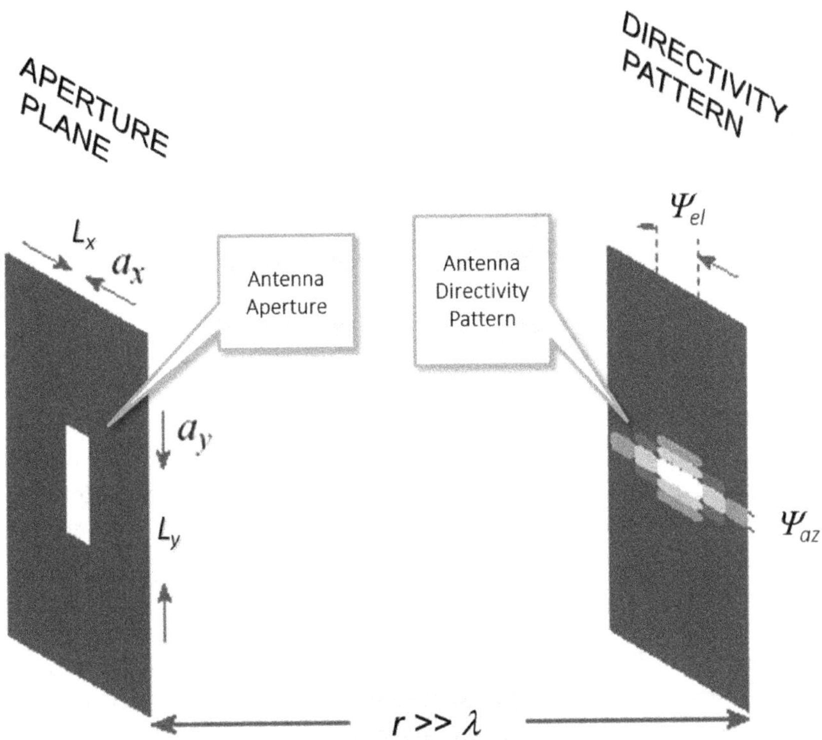

Figure 3.12 Illustration of the inverse relationship between aperture extent and beamwidth in the azimuth and elevation planes.

symmetric. If the beam is desired to be oval, that can be achieved via a nonsymmetric aperture shape. A common application for asymmetric beams in radars is in acquisition systems that rotate mechanically and sweep a full 360° in azimuth utilizing a vertical fan beam.

In the RRE, we use the terms for power input into the antenna and for the antenna gain, but not directivity. The transmit antenna gain is equal to the directivity after it is reduced by all losses after the power amplifier in the system. These total losses can include contributions from ohmic and reflection losses. In detailed radar signal budget analyses, separate line items for factors such as transmission line and radome losses may not be included specifically in reducing the antenna gain. The intent here is to ensure that all losses are included in the simulation. In (3.12), the loss term, L, is represented by a fraction less than 1, and as with directivity, gain has a value at every angle. When engineers refer to "the gain" of the antenna, they mean the peak gain or G_0. Note that some authors combine the loss as another multiplicative factor within $A_{\text{effective}}$.

$$G_0 = \frac{\eta_{\text{aperture efficiency}} 4\pi A L_{\text{total loss}}}{\lambda^2} = \frac{4\pi A_{\text{effective}} L_{\text{total loss}}}{\lambda^2} \tag{3.12}$$

One last consideration regarding directivity involves the relative comparisons between any antenna and the isotropic antenna. If we evaluate (3.8) for an isotropic antenna, the ratio of $P_{I_{\text{max}}}$ to $P_{I_{\text{average}}}$ equals 1; again this is because the directivity of an isotropic antenna is the same at all angles, and therefore its maximum is the same as its average. If converted to decibels, the directivity for an isotropic radiator is then 0 dB. This is commonly referred to as 0 dBi, or 0 dB relative to an isotropic antenna. Any antenna that is well impedance-matched and that focuses energy into a main beam has a peak gain greater than 0 dBi. Because of this, we use 0 dBi as a reference for which to compare the directivity of all antennas. Note that antennas can have negative gains, less than 0 dBi, but this is because their gain is less than their directivity because of losses. Subresonant or very electrically small antennas usually have negative gains.

3.8 Modeling Approaches

The primary goal of this antenna-modeling discussion is to present appropriate uses of simplifications for the model development and execution. There are many numerical approaches for calculating antenna directivity patterns. The most simplistic utilizes many approximations, in the extreme case providing the modeler a single-line equation. The most computationally complicated is through solution of the set of Clerk-Maxwell equations, which are covered in Section 3.8.2.

The authors have coined terms of different orders of complexity in simulation approaches—describing the relative level of simulation detail and the expected resulting accuracies, referred to as first-, second-, or third-order. With the simplifications come risks of inaccuracy, and we will attempt to convey those risks. First, Section 3.8.1 presents several first-order models. Then, Section 3.8.2 presents the third-order approach, which usually provides the highest fidelity but at the highest cost in development and simulation runtime. Last, Section 3.8.3 presents the second-order approach. Second-order models are recommended for many MATLAB-based simulations, because they provide a reasonable trade between accuracy, flexibility, and fast execution speed.

3.8.1 First-Order Antenna Models: Closed-Form Equations and Measured Data

The simplest path to calculating the directivity pattern is to use closed-form equations describing the directivity pattern of simple antenna architectures. Equations (3.9) to (3.11) are included in the category of first-order models. Although many details are omitted in these, they provide an easy path to an approximation of an antenna's main beam.

For the directivity pattern models of simple antennas, engineers have studied the distributions of fields and currents and have developed concise models to calculate the radiated patterns. These are useful approximations for the main beam directivity; however, again, many details are omitted. First, antenna sidelobes are often dominated by errors in the antenna construction. The term "errors" is used to describe nonideal deviations in amplitude and phase in the antenna aperture. Therefore, the first-order models should never be considered accurate for the sidelobe regions. Second, since no losses are included, the peak directivities and beamwidths are only approximate. These losses include the standard ohmic and reflection losses, but they also include losses due to errors, which are covered in Section 3.15. That said, the first-order calculations are still used because of their simplicity and fast execution speeds. Sections 3.8.1.1–3.8.1.4 briefly present four closed-form models, those of the dipole, loop, rectangular-aperture, and circular-aperture antennas.

3.8.1.1 Dipole Antenna

The dipole is a resonant, two-armed, differentially fed, half-wavelength or $\lambda/2$ wire antenna. Its simplicity and high efficiency have led it to be used in many systems. It only provides a narrow frequency band of operation, so it is rarely used in EW systems. The current distributions near the resonant frequency are well-understood [10], and its radiated fields can be approximated using a single-line closed-form expression for a single axis in the E-plane, presented in (3.13). The first-order dipole model is provided in the function ***dipole.m***. Inputs include wavelength, λ, and l which is the total arm length in λ. Figure 3.13 shows the orientation of the pattern angle θ with respect to the dipole arms and a photograph of a $\lambda/2$ dipole. Figure 3.14 presents calculated patterns for the dipole with lengths of $\lambda/2$ in (a) and (b) and a length of $3\lambda/2$ in (c) and (d). Note that the gain is considered close to correct for only the $\lambda/2$ dipole. While the pattern shape is correct, the gain is incorrect for the latter $3\lambda/2$.

$$E(\theta) = \frac{\cos\left(\frac{kl}{2}\cos(\theta)\right) - \cos\left(\frac{kl}{2}\right)}{\sin(\theta)} \tag{3.13}$$

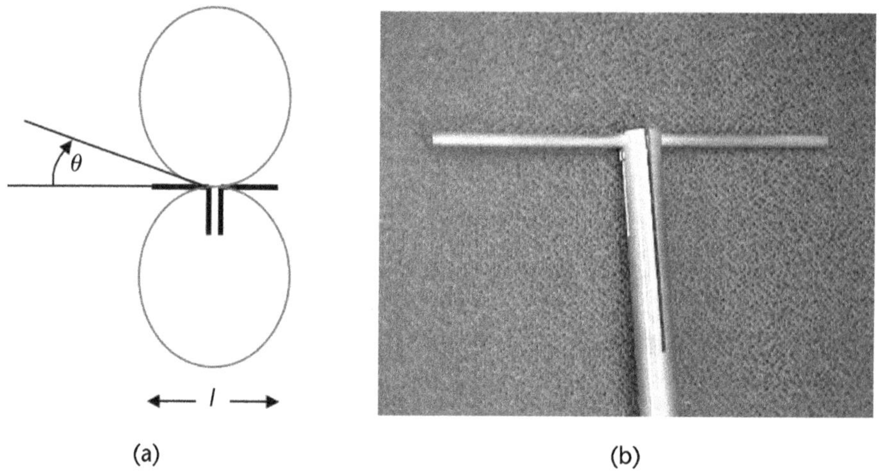

Figure 3.13 (a) Definition of pattern angle, θ, and (b) a photograph of a $\lambda/2$ dipole. (Photograph used with permission of GTRI.)

The negatives of using the first-order dipole model, ***dipole.m***, include that the polarization is assumed to be perfectly matched at all angles; that there is no information provided about the pattern for the cross-polarization; that the peak gain value is forced in the function to what a $\lambda/2$ dipole is known to achieve, 2.15 dBi; that reflection mismatch is not included, and hence the dipole is always assumed to be perfectly matched regardless of frequency; and last, that the antenna is magically suspended in midair with no other structure around it, including no transmission line feed.

3.8.1.2 Loop Antenna as a Surrogate for a Spiral or Sinuous Pattern

A loop antenna is resonant when its circumference is equal to one wavelength. Loop antennas are not commonly used in radar and EW systems. Low-gain, frequency-independent antennas including the common cavity-backed spiral and sinuous antennas are commonly used in EW systems. These antennas are referred to as frequency-independent, as they provide nearly constant gain and beamwidth across a very wide bandwidth. It is possible to purchase spiral antennas that provide 9:1 instantaneous frequency coverage (e.g., 2–18 GHz). The spiral antenna radiates a directivity pattern that is analogous to that of a simple loop with one exception. A loop radiates two primary lobes of energy, one in the forward and one in the backward direction. A spiral would do the same except that the backward radiation is deliberately suppressed. The most common form, which uses an absorber-filled cavity to absorb the backward-directed lobe, is referred to as a cavity-backed spiral. Figure 3.15 presents a photograph of both a cavity-backed and a microstrip-mode Archimedian spiral; it also shows the artwork from a logarithmic spiral with the resonant modes represented for high and low frequencies. The resonant modes automatically change diameter as the frequency is changed. This is often referred to as the active region of the frequency-independent antenna. Since the portion of the antenna actively involved in transmission or reception changes area as the

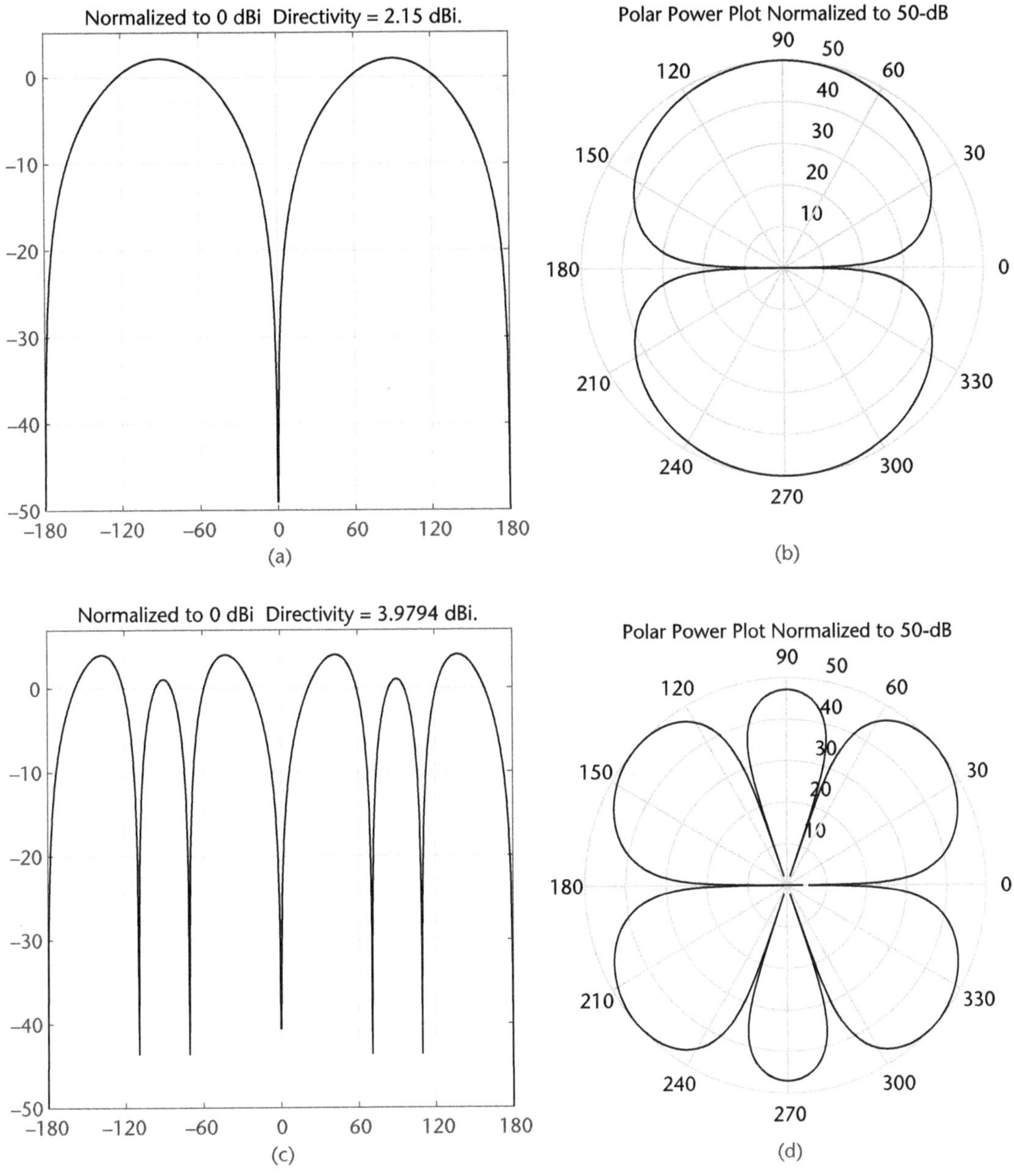

Figure 3.14 Directivity patterns calculated from dipole.m for a $\lambda/2$ (a, b) and for a $3\lambda/2$ dipole (c, d). Note that while the pattern is correct, the gain normalization is not correct for the $3\lambda/2$ dipole.

frequency is changed, the spiral antenna violates the directivity (and hence gain) predicted in (3.9), assuming the area used for the calculation was the total area of the spiral. The active region changing diameter with frequency is the mechanism whereby the frequency-independent antenna maintains constant gain, rather than having gain that increases with frequency as predicted by (3.9). The closed-form equation for a single loop is shown in (3.14), where J_1 represents the Bessel Function of the first kind of order one, $k = 2\pi/\lambda$, and a = radius of the loop in λ. A single-dimensional MATLAB model is provided in the function ***fourloops.m***,

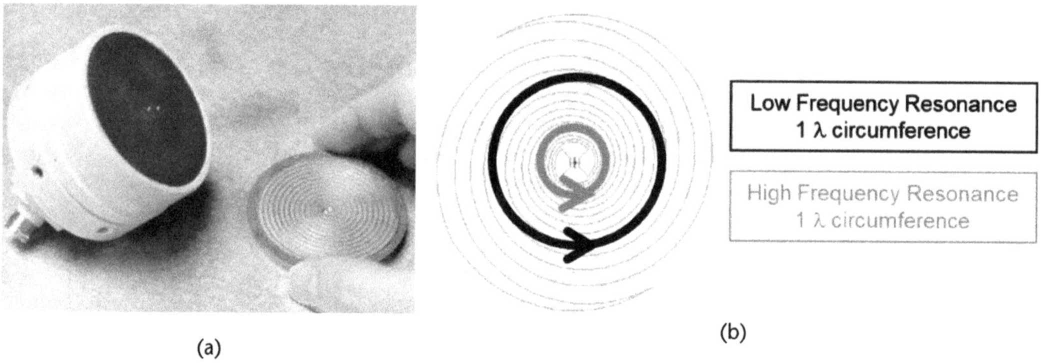

Figure 3.15 (a) A photograph of Archimedean spirals and (b) artwork illustrating the active region at smaller and longer wavelengths (higher and lower frequencies). (Used with permission of GTRI.)

and an example output of that model is presented in Figure 3.16. In Figure 3.16, four loops are modeled where each has the responsibility to point in a different quadrant around the full 360° azimuth circle.

$$E(\theta) = J_1\left(ka\sin\theta\right) \tag{3.14}$$

An application of this model would be to provide a first-order simulation of four spiral antennas placed around a host vehicle to provide detection capability as in the case of a radar warning receiver (RWR). The function ***rwr.m*** is an expanded version of the same simulation, but it offers 3-D angle calculations. Note that the output plot from rwr is 3-D and does permit slight tilts in the elevation pointing angle of the spirals.

3.8.1.3 Rectangular Aperture Antenna

The closed-form approximation for the radiation from a rectangular aperture antenna is probably the most widely used single-line equation for simulating directivity patterns. It comes from the fundamental relationship that the directivity pattern is the Fourier transform of the antenna aperture E-field distribution; see Skolnik [11]. This relationship was first published by Woodward and Lawson [12] and Booker and Clemmow [13]. The aperture field distribution could be mathematically described as a rectangle function, and the Fourier transform of the rectangle function is the sin(x)/x or sinc function, per Bracewell [14]. The relationship between the Fourier transform and the directivity pattern is detailed in Section 3.8.3.

$$E(\theta) = \frac{\sin\left(\frac{\pi L}{\lambda}\sin(\theta)\right)}{\frac{\pi L}{\lambda}\sin(\theta)} \tag{3.15}$$

The directivity pattern calculation is shown in single dimensional form in (3.15). The variable L represents the length of the antenna aperture and the pattern

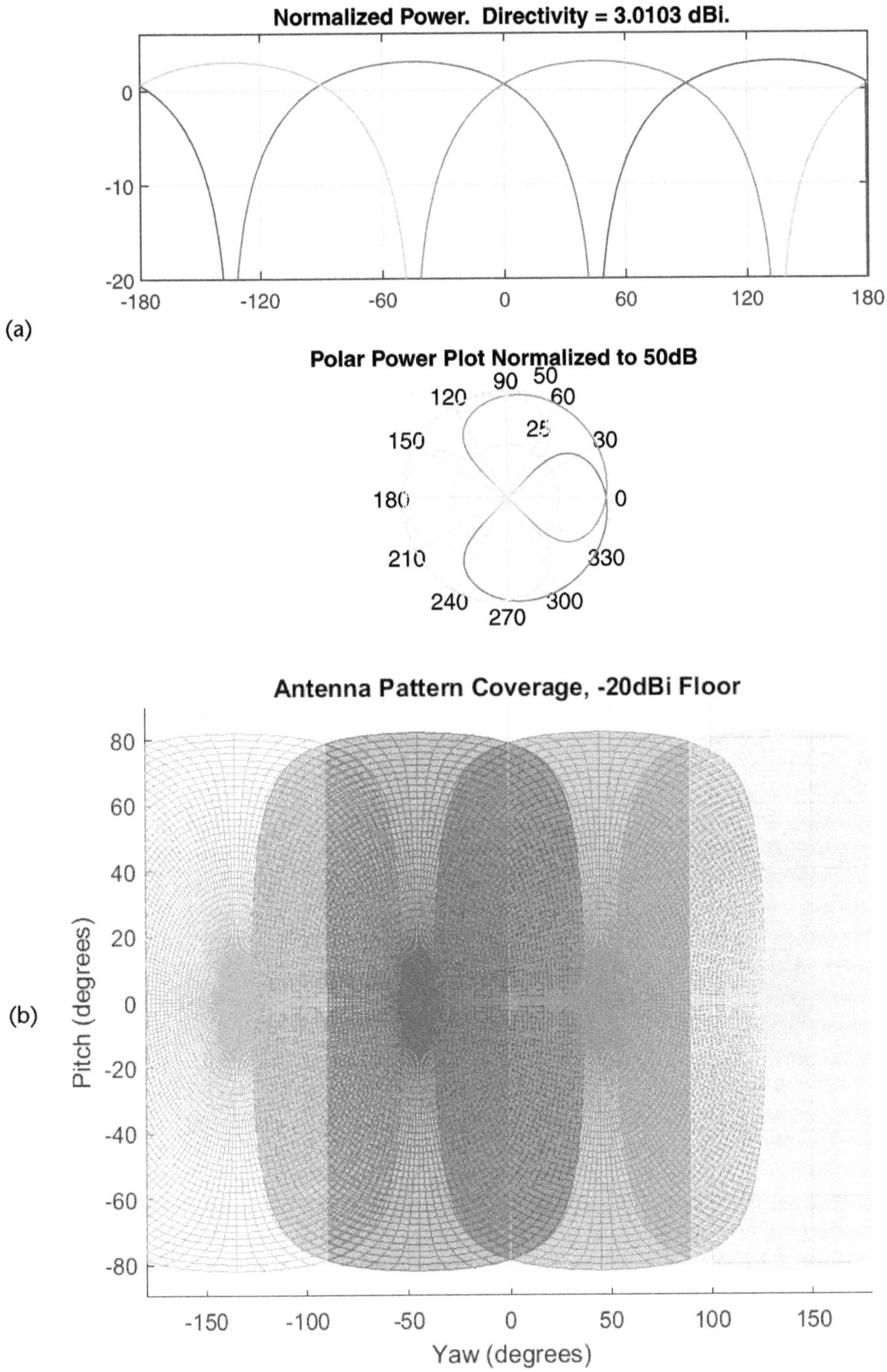

Figure 3.16 Directivity pattern plots output from the function ***fourloops.m*** (a) and ***rwr.m*** (b) for the case of four spiral antennas places at 90° intervals around a host platform.

equation is for a linecut in angle θ, which is in the same plane as the length, L. As an approximation, (3.15) is exactly correct as the antenna directivity pattern if and only if the following are true:

- The E-field amplitude and phase are uniform (nonvarying) across the entire extent of the aperture.
- No amplitude and phase errors are present in the amplitude and phase distributions.

Given these requirements and that every antenna has some amount of real-world errors, the sinc approximation is never exactly correct. The sinc pattern also has the highest sidelobes of all amplitude distributions, and because of this it is rarely used in radar antennas. However, sinc does provide a representative pattern, and its simplicity warrants frequent use in system models. The function ***rect_sinc.m*** provides an example of calculation of the first-order sinc antenna pattern. Output directivity pattern samples are presented in Figure 3.17. Advantages of this model, especially for antennas of multiple-wavelength extent include reasonable representation of the main beam directivity and beamwidths. Accuracy of the sidelobe levels is more suspect, as will be discussed in multiple subsequent sections. The peak sidelobe for the sinc function is −13.2 dB with respect to the main beam (dBmb). Note that the peak directivity calculations from ***rect_sinc.m***

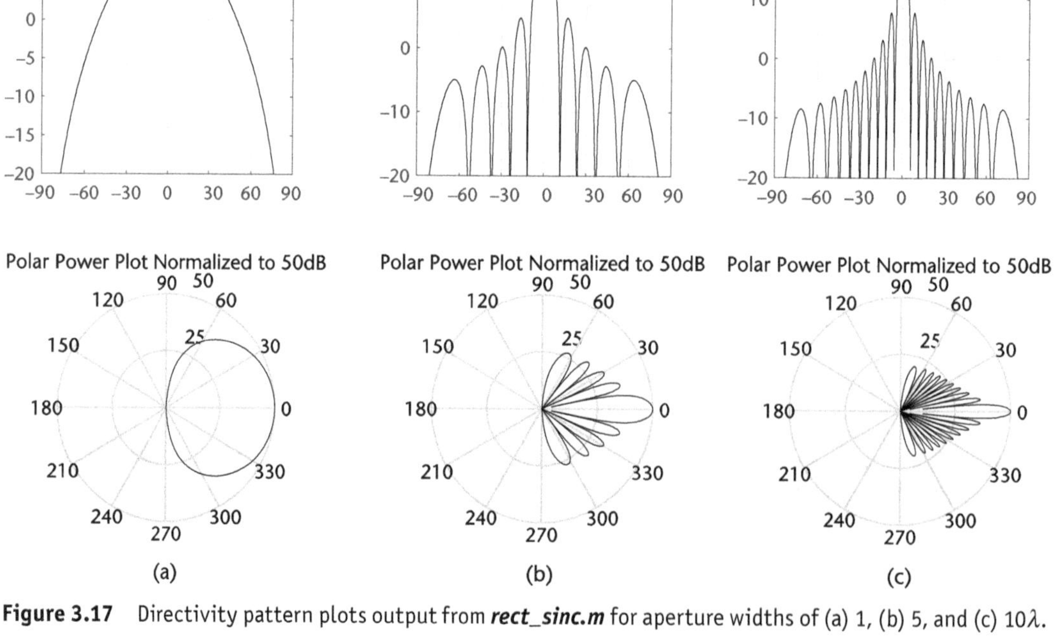

Figure 3.17 Directivity pattern plots output from ***rect_sinc.m*** for aperture widths of (a) 1, (b) 5, and (c) 10λ.

are normalized based on (3.8), using the aperture area and where $\eta_{\text{aperture efficiency}}$ is assumed to be equal to 1.

3.8.1.4 Circular Aperture Antenna

One of the negative results of employing a perfect rectangular antenna is the high-peak SLL of −13.2 dBmb. A simple way to reduce the peak SLL is to use a circular-shaped aperture in place of a rectangular one. With a circular aperture, the sinc directivity function is replaced by a Bessel function of the first kind, denoted by $J_1(x)$. The derivation of this relationship is presented in Lo and Lee [15]. The E-field directivity pattern for a uniform circular aperture is (3.16), where a is the radius of the circular aperture and $k = 2\pi/\lambda$.

$$E(\theta) = \frac{J_1\left(ka\sin(\theta)\right)}{ka\sin(\theta)} \tag{3.16}$$

The switch to the circular aperture reduces the peak sidelobe to −17.6 dBmb and results in a slightly broader beamwidth. The function ***circular.m*** simulates a single dimensional calculation of the directivity pattern from a circular aperture antenna. As with ***rect_sinc.m***, the peak directivity values of ***circular.m*** are based on normalizing to the peak directivity of (3.8). Example directivity patterns are shown in Figure 3.18.

3.8.1.5 Use of Measured Data

Use of measured antenna pattern data is another valid method for inclusion in a system simulation. The primary benefits to this approach include: (1) this data may already exist, (2) there is no concern with approximations as with many of the other models, and (3) changes in performance with the antenna installed in situ on a host platform are included in some cases. Among the negatives of using measured data are (1) that the user may have only one set of data from one antenna installed on one platform and that single sample may not be representative of the average system performance, and (2) measurement errors, if large, can introduce errors in the simulation that would not be representative of the average system performance. The risk of using data from a single sample is usually minimal. Figure 3.19 presents photographs of test-range data collection of antenna gain patterns.

3.8.2 Third-Order Antenna Models: Full-Wave Solutions to Clerk Maxwell's Equations

The single-line first-order directivity pattern approximations presented in the previous session have benefits of ease of utility, but they also have severe accuracy limitations. While the single-line sinc and jinc equation models might provide in-the-ballpark estimates for main beam gain and beamwidth, their estimates for sidelobe levels may be significantly inaccurate. Additionally, single-line models do not provide any cross-pol estimates. Low signal level pattern features, such

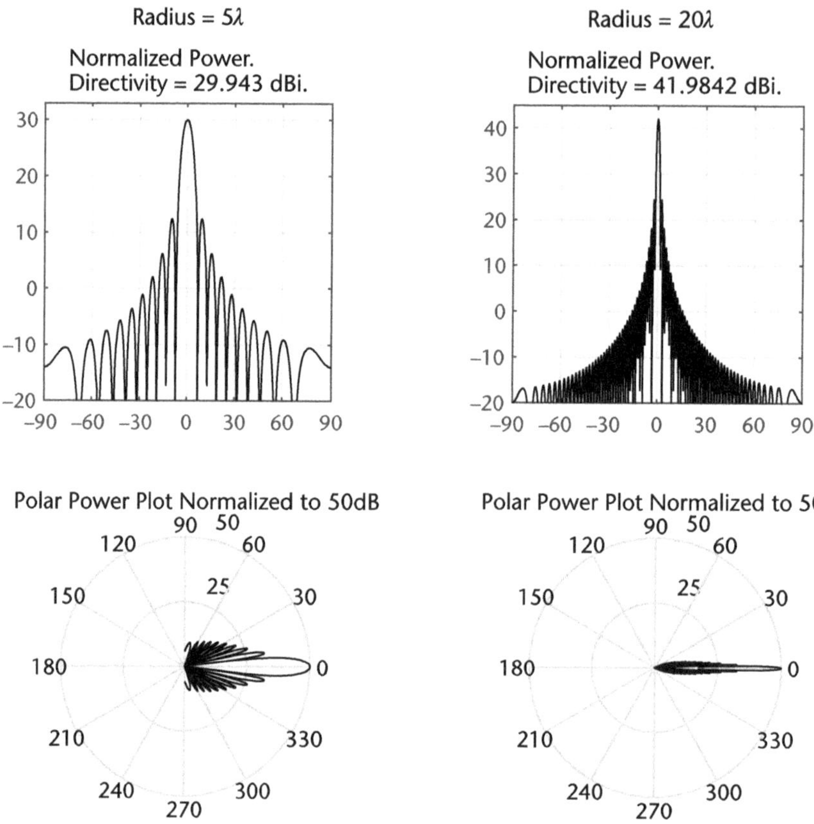

(a) (b)

Figure 3.18 Directivity pattern plots output from function ***circular.m*** of an ideal circular aperture of radii (a) 5λ and (b) 20λ.

(a) (b)

Figure 3.19 Antenna pattern data collection in test-range facilities including: a phased array in GTRI's planar near-field range (a), and in situ antenna testing on a host aircraft at the U.S. Army Electronic Proving Grounds on an outdoor compact antenna test range developed by GTRI (b). (Photographs courtesy of GTRI.)

as co-pol nulls and sidelobes as well as cross-pol, are often dominated by errors in the antenna hardware that are not accounted for in the simple single-line models. The direct solution of the set of Clerk Maxwell equations [16], provided in (3.17), provides more accuracy for pattern features, including full polarization fidelity and inclusion of real physical errors and their resulting impact on low signal levels in regions of pattern nulls and low sidelobes. Development of these models comes at a computational price in terms of complexity, runtime, and computer RAM (memory). The equations of (3.17) provide a simplified version of the Clerk Maxwell equations in their differential operator form, as presented by Oliver Heaviside [17, 18]. Interested readers should refer to [19] to learn about Clerk Maxwell and his grandparents' lives—and why these equations are named for him.

$$\begin{aligned} &\nabla \cdot D = \rho \\ &\nabla \cdot B = 0 \\ &\nabla \times E = -\frac{\partial B}{\partial t} \\ &\nabla \times H = J + \frac{\partial D}{\partial t} \\ &B = \mu H, \quad D = \varepsilon E \end{aligned} \tag{3.17}$$

The variables and units used in the Clerk Maxwell equations are defined as follows:

- E is the electric field intensity and is measured in (V/m);
- H is the magnetic field intensity and is measured in (A/m);
- D is the electric flux density and is measured in (q/m^2);
- B is the magnetic flux density and is measured in (V·s/m^2);
- ρ is the volume charge density in (q/m^3);
- ε is the permittivity of the media and is measured in (farads/m) or (A-s/V-m) or (q/V-m) ($\varepsilon_0 = 8.8542 \times 10^{-12}$ F/m);
- μ is the permeability of the media and is measured in (henrys/m) or (V-s/A-m) ($\mu_0 = 4\pi \times 10^{-7}$ H/m);
- σ is the conductivity of the media and is measured in units of (mhos/m) or (A/V·m).

The term media herein is used to mean any material that can include dielectric, magnetic, or conducting materials.

The direct solution of Clerk Maxwell's equations with the Helmholtz wave equation and with application of all boundary conditions for all materials in and around a given antenna is difficult. To facilitate this, several numerical-solution approaches or approximations have been developed. The common techniques for asymptotic and full-wave simulation approaches are described as follows:

- *Uniform theory of diffraction (UTD):* EM fields in space are solved based on geometrical analyses of propagation paths called rays, that reflect, diffract, and diverge, with no reference to currents.

- *Physical optics (PO):* EM fields in space are calculated from currents on surfaces. The currents are approximated as $2n \times H$, where H is the source magnetic field, and n is the vector normal to the surface.
- *Method of moments (MoM):* EM fields in space are calculated from currents on surfaces. The currents are found by solving Clerk Maxwell's equations in integro-differential form.
- *Finite-difference time-domain (FDTD):* EM fields in space are calculated by solving a differential form of Clerk Maxwell's equations. The EM fields can be used to compute the radiated fields.
- *Finite-element method (FEM):* EM fields in space are calculated by solving a variational form of Clerk Maxwell's equations (generally an inner product integral over a differential operator). Similar to FDTD, the EM fields can be used to compute the radiated fields.
- *Finite integration technique (FIT):* EM fields in space are calculated by solving an integral form of Clerk Maxwell's equations. These can be solved in either time domain (similar to FDTD) or frequency domain (similar to FEM).

The development and implementation of these simulation techniques is not covered here, as lengthy descriptions for the techniques can be found in many existing volumes. In particular, readers can find good summaries in Davidson [20] and in the GTPE course on the modeling and simulation of antennas [21].

These approaches are used in many antenna simulation packages, including commercially available ones, and most of the modern packages offer multiple solution techniques. Current commercially available packages along with their parent companies and the solution techniques for which they became best known are listed as follows:

- Electronic Desktop (previously known as the High Frequency Structure Simulator or HFSS) by Ansys Corporation, which uses the FEM and others;
- CST Studio Suite by Dassault Systèmes, which uses the FIT and others;
- FEKO by Altair, which uses the MoM and others;
- GRASP by Ticra, which uses the geometric theory of diffraction and the uniform theory of diffraction (UTD/GTD).

While we do not endorse any of the commercial packages listed, we are familiar with them and have used them with good results in our own design and development efforts. There are many additional commercially available sources of full-wave simulation packages that may likely be suitable as well.

Regarding the myriad of modeling approaches available, the question should be asked: When does development of a full-wave model make sense? Its cost is higher and the time required to both develop and simulate the model are more extensive than the much simpler closed-form equations presented in Section 3.8.1. The full-wave approach offers many simulation benefits originating from the fact that the entire geometry must be included in the full-wave model. While the first-order models are simpler, they include many assumptions that may not be valid. The full-wave models incorporate the specific hardware geometry and its interactions with the currents and fields. This can result in better, more faithful directivity patterns that include co- and cross-polarizations. The full-wave model should also provide significantly more accuracy for lower signal levels in the directivity pattern (i.e., the sidelobes, nulls, and cross-polarized levels). Another significant benefit of the full-wave approach is the ability to visualize the fields and currents in the antenna architecture. This insight is especially helpful for antenna designers as they debug and optimize designs. One last negative aspect of the full-wave models is their complexity and difficulty of implementation. Users must be knowledgeable about the limitations of any simulation approach and be diligent in the faithful representation of the antenna and its surrounding structures.

The following section details examples of where full-wave models would be required. Section 3.8.1.2 described the use of a single-line closed-form expression for the copolarized pattern for a spiral antenna. But what if we also need a model for its cross-polarized pattern? A closed-form expression for the spiral cross-pol pattern is not readily available, so the following example is that of a FEM model of a logarithmic spiral with an infinite balun feed. Figure 3.20 presents the FEM model architecture and calculated electric fields in the dielectric substrate between the balun feed and the spiral arms. Figure 3.21 presents measured co- and

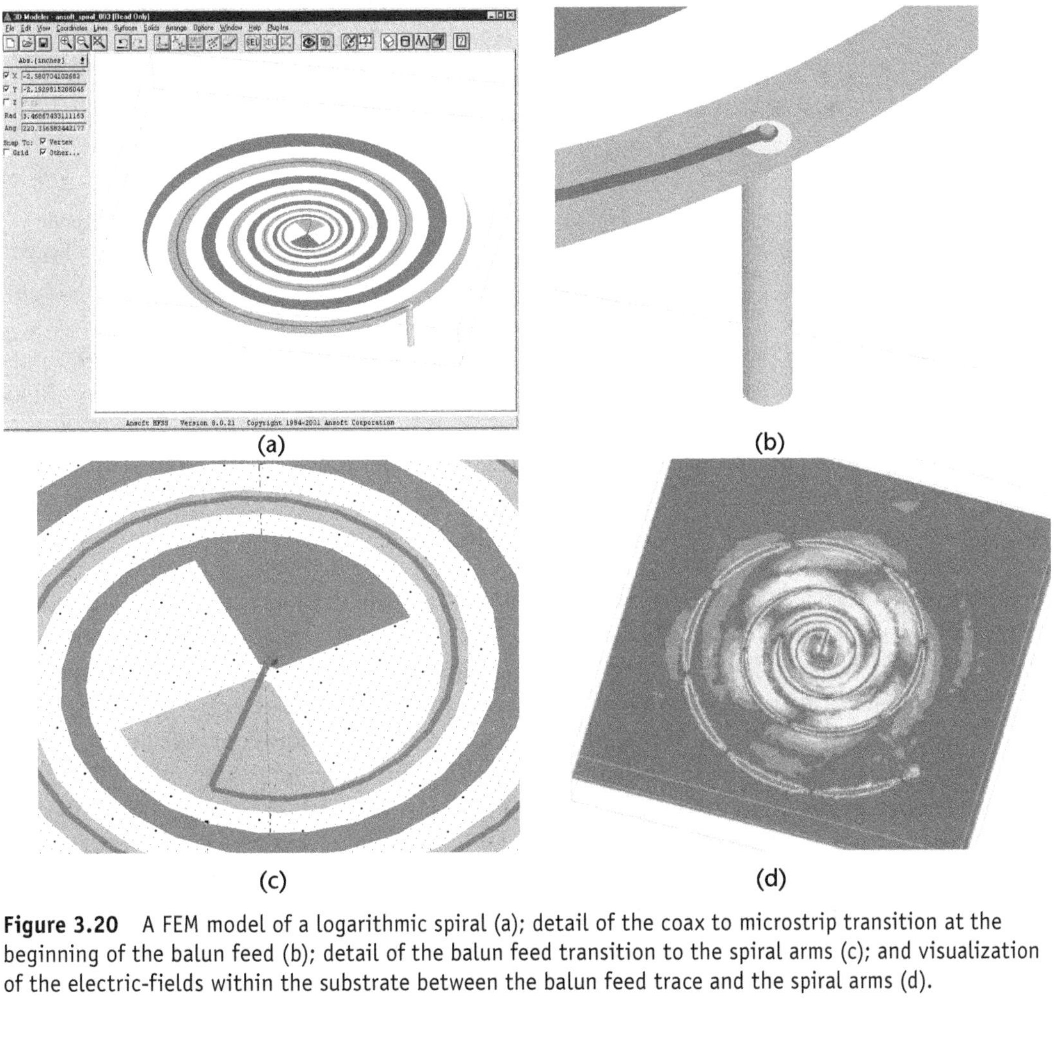

Figure 3.20 A FEM model of a logarithmic spiral (a); detail of the coax to microstrip transition at the beginning of the balun feed (b); detail of the balun feed transition to the spiral arms (c); and visualization of the electric-fields within the substrate between the balun feed trace and the spiral arms (d).

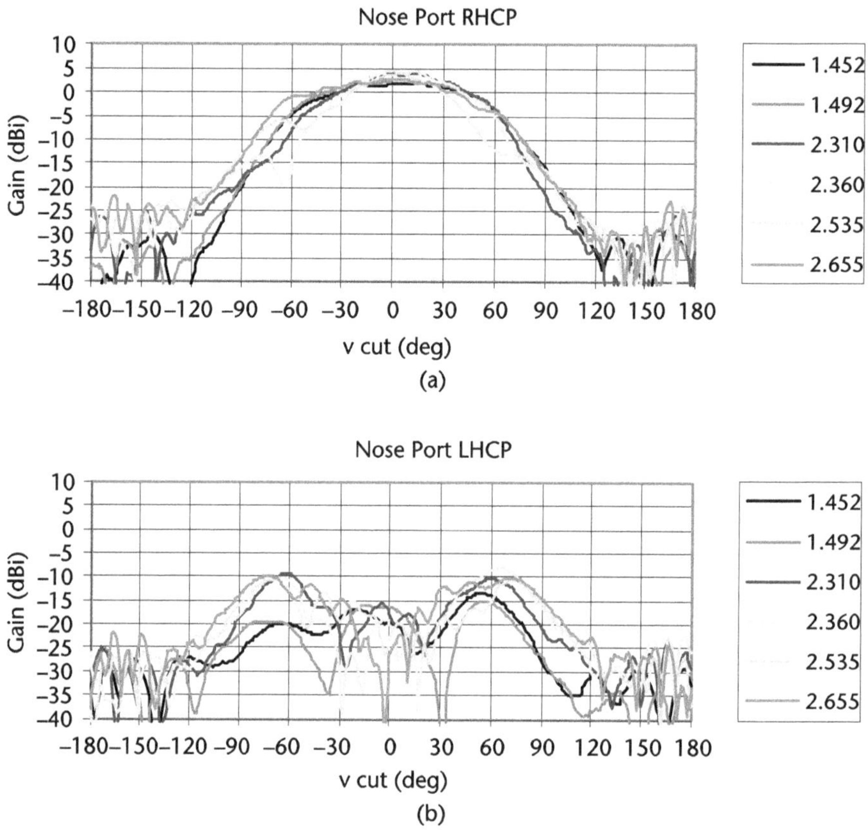

Figure 3.21 Measured copolarized (a) and cross-polarized (b) pattern data for the logarithmic spiral. Note the similarity in the co-pol pattern with that of the simple first-order model in Figure 3.16. No simple model is readily available for the cross-pol pattern.

cross-polarized patterns from the spiral hardware. Note that the copolarized pattern indeed has the shape and gain of the simple first-order patterns presented in Figure 3.16.

In Figure 3.17, patterns are presented for ideal rectangular aperture antennas. What if more accurate patterns are needed for an electrically short waveguide horn antenna? And what if its cross-polarized patterns are also desired? These are both examples where full-wave solutions are virtually required to accurately predict the desired horn antenna directivity patterns. Horn antennas exhibit two phenomena where their field distributions differ significantly from that of the ideal rectangular aperture. The first involves the amplitude distribution across the horn. Because of the interaction of the conducting walls with the linearly polarized E-field, the horn has uniform amplitude distribution in the E-plane (the plane parallel to the E-field vector), but it has an amplitude taper in the H-plane (the plane perpendicular to the E-field vector). These two different amplitude tapers result in significantly different first SLLs in the two principal planes of the directivity pattern. The second divergence from that of an ideal aperture distribution involves the phase front across the horn which results from the horn's phase center. The phase center is an ideal point source from which the fields spherically expand within the horn. The phase center is usually at the back or throat of the horn at the transition to the waveguide feed. The location of the phase center causes a spherical phase front that creates phase error at the aperture. The phase error fills in the pattern nulls, increases the SLLs, and if bad enough, causes partial destructive interference at the peak of the main beam resulting in reduced gain on boresight. All of these phenomena are accurately modeled using a full-wave method. Figure 3.22 shows a FEM model of a waveguide horn, a plot of its E-field magnitude illustrating the phase curvature across the horn aperture, its copolarized pattern that significantly deviates from the ideal sinc patterns presented in Figure 3.17, along with its cross-polarized pattern for which there is no good closed-form first-order model.

3.8.3 Second-Order Antenna Models: Fourier Transform Models

The simplicity of the first-order single-line equation models is often overshadowed by their inaccuracy. The accuracy of the third-order full-wave models is costly and time-consuming. Many modelers seek a middle ground in the trade between accuracy and expense. The middle ground solution is based on the Fourier transform which can be traced back to Christiaan Huygens' publication in 1690 [22]. Before Huygens, the prevailing theory was that light solely propagated as a ray. It originated by some unknown means, and when it met a boundary it

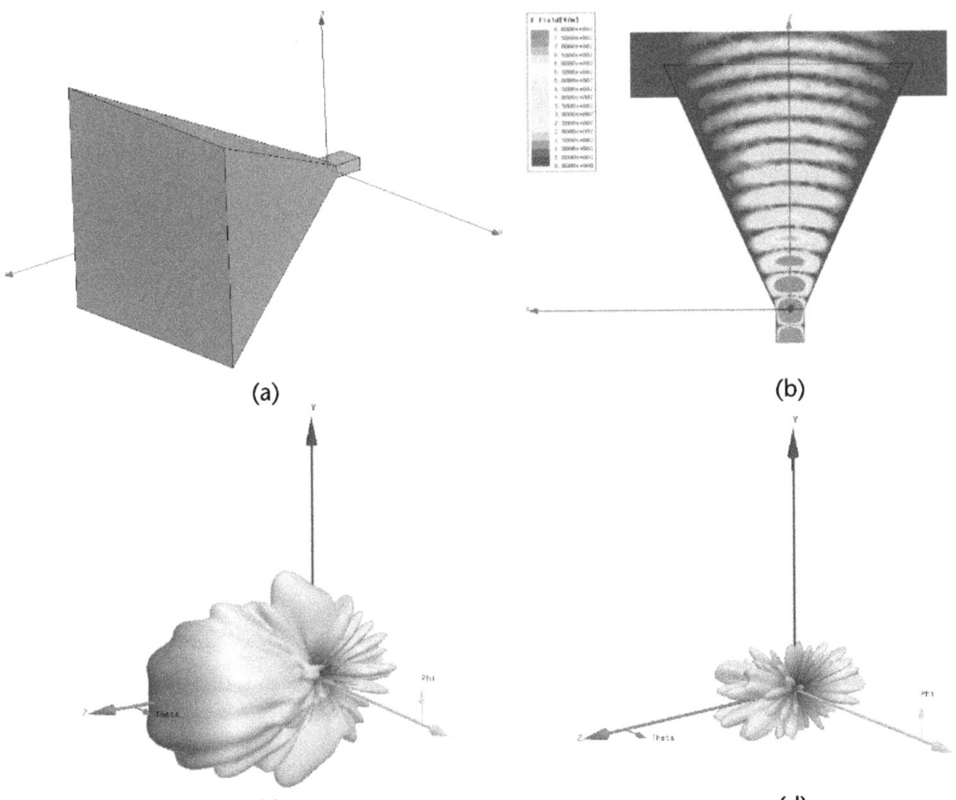

Figure 3.22 A FEM model of a waveguide horn antenna (a), a plot of the E-field magnitude within the horn (b), its copolarized directivity pattern (c), and its cross-polarized directivity pattern (d). Images courtesy of GTRI.

was either absorbed, reflected, or refracted. Light was essentially the only form of electromagnetics that our ancestors before Huygens could observe and study. The ray theory of electromagnetics had no explanation for diffraction, a phenomena that was observed in light at the time. Huygens postulated that light could be thought of as being composed of a very large number of individual point sources, each of which radiated a spherically expanding wave front. Figure 3.23 shows artwork describing the concepts of the multiple point sources from [22]. The light that resulted could be calculated by summing all of the contributions from all of the individual point sources at every direction in space. This approach works similarly for our calculation of patterns of electric fields rated by an antenna.

Huygens did not have the math language that we have today where we use Euler's Equation (3.4) and the resulting phasor, $e^{j\phi}$, to represent each of the point sources—the same point sources that we introduced as isotropic radiators in Section 3.6. As a simple example, Figure 3.24 presents an oversimplified model of a dipole antenna. The dipole has been replaced by two isotropic point sources that are separated on the horizontal axis by 0.5-λ spacing. Imagine that half the power is input into each and that their radiation is in phase with each other. Concentric circles have been drawn around the two point sources highlighting the 0° (solid) and 180° (dashed) equiphase rings. Note that because of the orientation of the two point sources and the fact that they are in phase with each other, the fields along the vertical axis always constructively add while the fields on the horizontal axis always destructively interfere. At angles in between, the fields exhibit partial constructive interference. Figure 3.24 shows the pattern that results if the fields are summed at every angle in a plane far away from the two sources. Note that it

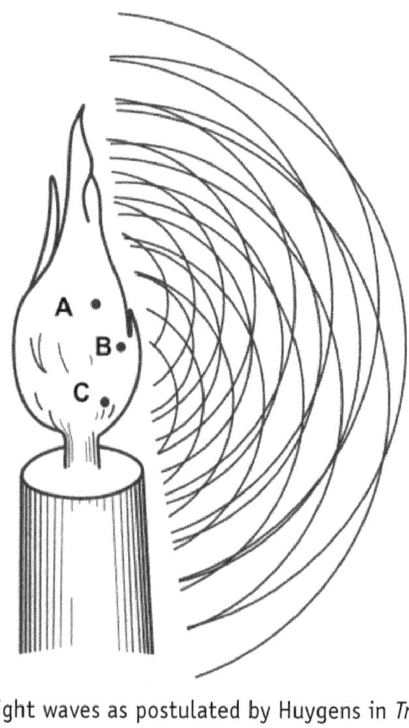

Figure 3.23 The origin of light waves as postulated by Huygens in *Treatise on Light*. (From: [22] (Public domain.)

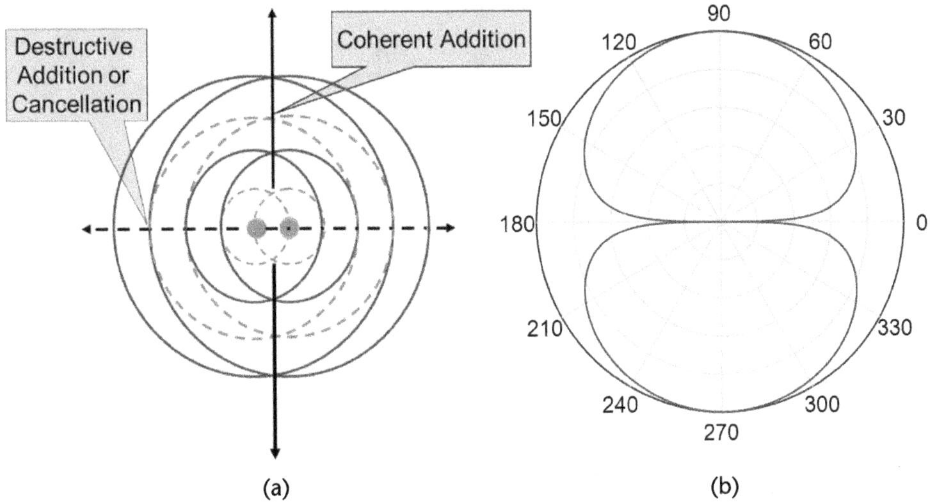

Figure 3.24 A model of a dipole antenna using two isotropic point sources spaced 0.5λ and fed in phase (a) and the resulting directivity pattern of the sum of the fields from the two sources versus angle (b).

is the same as the pattern presented earlier from the first-order equation results of Figure 3.14 for a dipole antenna. This is a simple example of how the Huygens point source concept can be used to simulate a simple directivity pattern.

The sum of a linear array of point sources, or elements, is given in (3.18). A drawing of the model is shown in Figure 3.25.

$$E(\theta) = \sum_{n=1}^{N} e^{jkn\Delta x \sin(\theta)} \tag{3.18}$$

In (3.18), $k = 2\pi/\lambda$, n is the point source index number, and Δx is the interelement spacing. (Note: The index of the summation can start with $n = 1$, $n = 0$, or $n = -N/2$. The patterns that result will be the same with the exception of where the phase center resides in the antenna coordinate system. For simplicity, we will use the lower limit of the summation as $n = 1$.)

Equation (3.18) mathematically states what Huygens proposed centuries ago—simply that the radiated fields can be represented by a summation of all of the electric fields from the spherically expanding point sources. Where they combine in angle-space constructively or destructively is simply based on the locations of the point sources, the relative phase differences to the summation point, and of course the wavelength of the frequency used. Equation (3.18) is important as it is a discrete form of the Fourier transform. Its use in modeling of antenna directivity patterns is very powerful. We can take a line source (or a 2-D aperture, which is covered in subsequent sections), discretize the antenna into multiple point sources, apply correct amplitudes and phases (covered in subsequent sections), and sum the fields to calculate the directivity pattern. Huygens did not have the math needed for the Fourier transform, yet he described its application approximately 130 years before Fourier. Use of Huygens' and Fourier's concepts were not reported for antenna pattern modeling until approximately 250 years after Huygens by Woodward and Lawson [12].

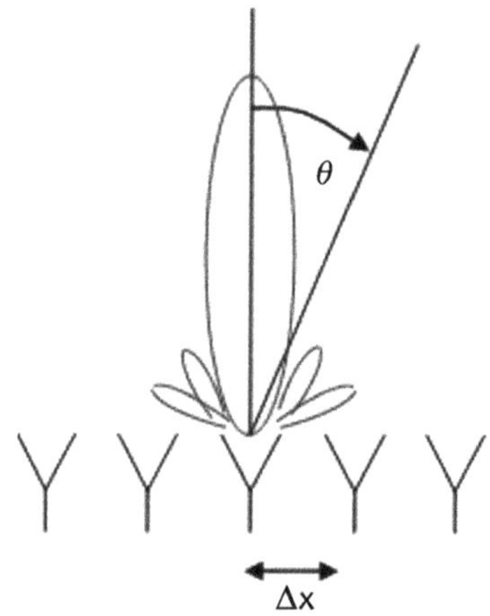

Figure 3.25 Configuration of a linear array of point source radiators, separated by an interelement spacing of Δx.

3.9 Fourier Transform Model Approaches

Our first Fourier transform model is ***array.m***. It simply provides a summation of a number of N point source elements arranged in a line, separated by the interelement spacing Δx, with the far-field directivity pattern calculated in a single linecut pattern in angle θ. If we are describing an actual array architecture, the individual antennas are often referred to as either elements or radiators. The line array architecture is as shown in Figure 3.25. This model is the discrete Fourier transform version that produces the same results as the previous single-line model, ***rect_sinc.m***, presented in (3.15) and its patterns in Figure 3.17. The equation used in ***array.m*** is that of (3.18). Sample patterns calculated using ***array.m*** are shown in Figure 3.26 where the interelement spacing is 0.5λ.

The ***rect_sinc.m*** and ***array.m*** functions are variants of the most commonly used simple antenna model—that of a sinc function. It is simple and incredibly expedient and provides an answer that is a reasonable approximation for the main beam region of the pattern. For rectangular apertures with uniform amplitude and phase and with no amplitude nor phase errors, sinc also provides an accurate representation for the sidelobe regions.

Note that the patterns of Figure 3.26 have an interelement spacing, Δx of 0.5λ. While this is the most common interelement spacing, it is not the only one that is ever used. In ***array.m***, if $\Delta x > 1\lambda$, multiple main beams will form. The additional main beams are called grating lobes. The topic of grating lobes is closely related to aliasing in sampling theory and the Nyquist-Shannon criteria [23, 24]. The array of point sources act as spatial samples in the aperture area. The extent of the sinespace coordinate system that extends from -1 to $+1$ (or $-90°$ to $90°$ in degrees) spans from horizon to horizon and is referred to as real space. Mathematically, spatial angles exist outside of or beyond real space, and we refer to those angles as being in imaginary space. Imaginary space is any angle in sinespace with a magnitude

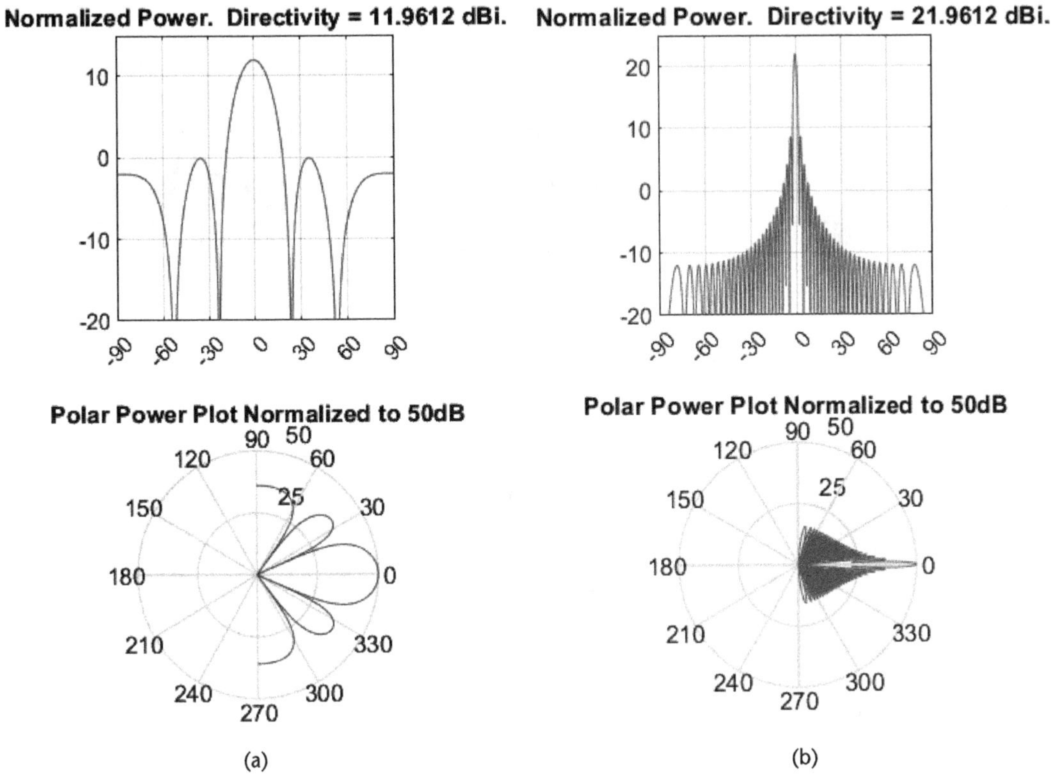

Figure 3.26 Patterns calculated using ***array.m*** for $N = 5$ elements (a) and $N = 50$ elements (b).

greater than unity. Aliasing can occur with our array antenna directivity patterns, forming simultaneous main beams both in real space and imaginary space. These infinitely numbered multiple beams form at angles of $\pm n\Delta x/\lambda$ (units in sinespace), where n can range from $-\infty$ to $+\infty$. Δx is the interelement sample spacing. For a single fixed-beam array antenna, the mainlobe forms in real space where $n = 0$ and the angle $= 0$. Grating lobes occur at angles of $n \neq 0$. A grating lobe can occur in real space if the $\Delta x \geq 1\lambda$. A grating lobe can exist in real space for a scanning array if the $\Delta x \geq 0.5\lambda$, depending on the maximum scan angle of the antenna. Figure 3.27 presents patterns from ***array.m*** for a linear array of 50 elements where the interelement spacing is varied from 0.5 to 1 to 2λ. Note that the grating lobes form at the horizon for the case with the spacing of 1λ. These grating lobe beams are not simply a mathematical concept. If an array was constructed with the interelement spacings as in Figure 3.27, multiple simultaneous beams would form. Grating lobe beams are undesirable for almost every radar system. They would be detrimental to system performance by confusing the angle of arrival and increasing problems with clutter, multipath, and jammers. Designers strive to avoid their formation.

As the Fourier transform relates the antenna aperture field distribution to the directivity pattern, knowledge of Fourier transform pairs is very useful. Figure 3.28 shows three examples of the application of Fourier transform pairs to antenna directivity patterns. A good reference for Fourier transform pairs is Bracewell [25].

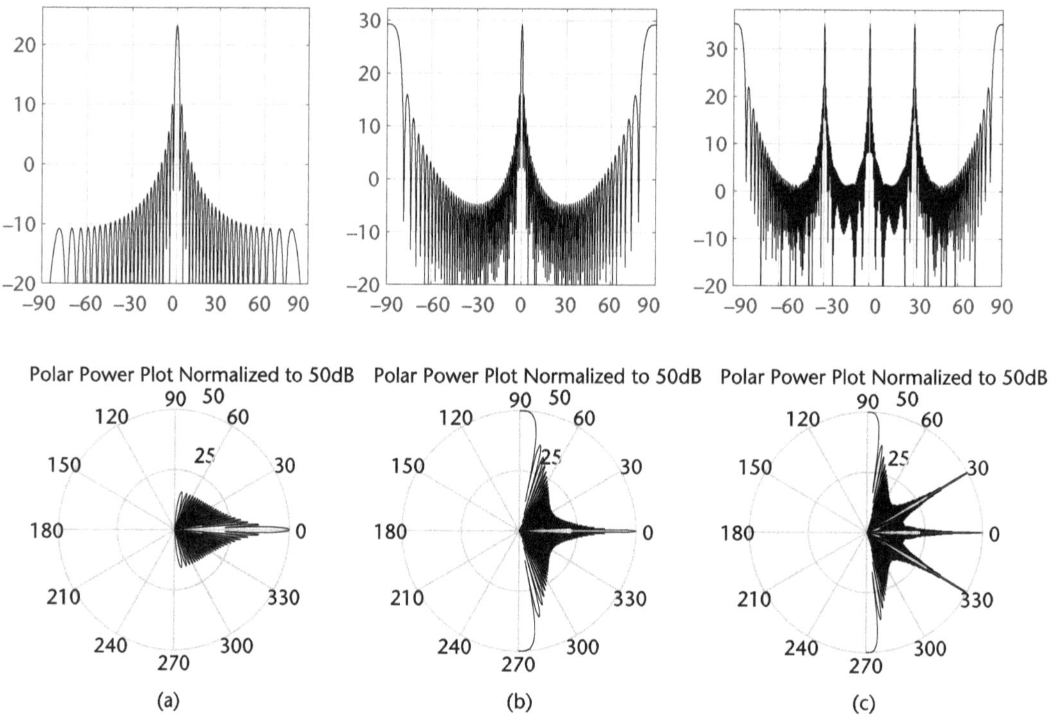

Figure 3.27 Directivity patterns from ***array.m*** where $N = 50$ and the interelement spacing is varied from 0.5 (a) to 1 (b) to 2λ (c).

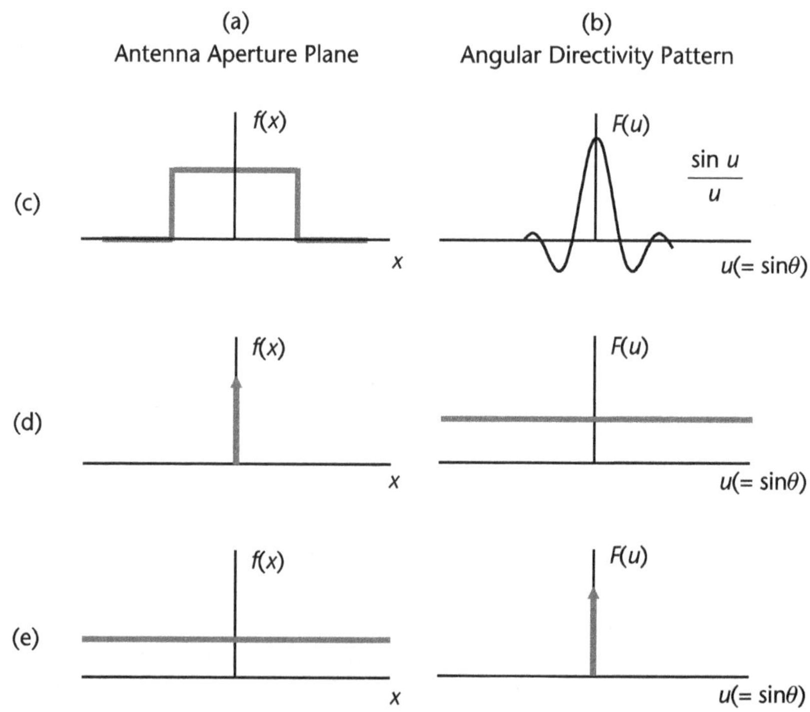

Figure 3.28 Fourier transform pairs relating (a) the antenna aperture amplitude distributions and (b) the resulting angular directivity pattern for (c) a rectangular line source, (d) an isotropic antenna, and (e) for an infinitely long antenna aperture.

Use of the Fourier transform in calculations of antenna directivity patterns is made even more powerful with a few simple extensions. These include incorporation of aperture amplitude and phase distributions, real element patterns, and phase-scanning for phased-array antennas. These extensions permit the user to define much more faithful pattern simulations than the sinc function with minimal computation expense. These extensions also support simulation of a very wide range of physical antenna architectures beyond that of a simple line array. The following sections will address these useful extensions and their incorporation into a 2-D antenna model.

3.10 Fourier Transform Peak-Directivity Normalization

Novice modelers have been known to calculate their first antenna directivity pattern using the Fourier transform, arriving at the erroneous conclusion that the peak electric field equals the number of elements, N, and that the peak directivity equals the number of elements squared, N^2, converted into decibels. This would be the correct array directivity for an architecture using uniformally excited isotropic radiators. The definition of directivity is (3.9), and the gain is (3.12). Use of those equations necessitates calculating the array pattern over an entire sphere with potentially fine angular resolution. Many modelers take a significant shortcut that is based on the approximation of peak directivity of the effective aperture, here in the case of a linear array, using (3.19).

$$\text{Peak Directivity} = \left(\frac{\text{Fourier Summation}}{\text{Number of Elements}}\right)^2 * \text{Aperture Directivity}$$

$$\text{Peak Directivity} = \left(\frac{\sum_{n=1}^{N} e^{jkn\Delta x \sin(\theta)}}{N}\right)^2 \frac{4\pi A_{\text{effective}}}{\lambda^2} \tag{3.19}$$

The directivity losses caused by any aperture efficiency are inherently included in the Fourier summation, and the summation normalized by the number of elements should be less than one if any aperture efficiency losses exist.

3.11 Fourier Transform Model for Antennas That Are Not Arrays

The Fourier transform technique is obviously very useful for modeling array antennas. The same technique is also very appropriate for simple models of other nonarray architectures—if the user is careful in applying appropriate amplitudes and phases across the antenna aperture. Some common antenna architectures, including horn and reflector antennas, do not provide uniform amplitude and phase across the antenna aperture. A priori knowledge of these amplitude and phase distributions can be added to a Fourier transform model if we use (3.20). Note that (3.20) is for a simple one-dimensional antenna and for a one-dimensional directivity pattern. In (3.20), two terms are added: α_n which is the electric-field magnitude, and φ_n which is the phase, both of which can have different values at each data point in the aperture. In this text, we will generalize the electric field amplitudes, referring to them simply as voltages.

$$E_{\text{total}}(\theta) = \sum_{n=1}^{N} a_n e^{j\varphi_n} e^{-jkn\Delta x \sin\theta} \tag{3.20}$$

Figure 3.29 presents different examples of reflector antennas with prime focus, offset fed, and Cassegrain feed orientations. In modeling of a reflector using the Fourier approach, several details must be incorporated. First, the usually circular aperture shape must be included in a 2-D model. Second, the amplitudes illuminating the reflector need to be calculated by modeling the antenna pattern of the feed horn. Figure 3.30 presents the geometry of a prime-focus parabolic reflector with definitions of the focal length/diameter (F/D) ratio. The F/D ratio sets the subtended angle of the feed to the reflector and hence the required beamwidth of the prime focus feed horn. Figure 3.30 presents an additional loss experienced by reflector antennas. Spillover loss is energy radiated by the feed that does not illuminate the reflector and hence is lost. Use of a wider beamwidth feed pattern results in greater spillover loss. Conversely, use of a narrower beamwidth feed pattern results in underillumination of the reflector and increased aperture efficiency loss. An amplitude taper is always expected on the reflector because the pattern of the feed horn has an angle-dependent amplitude. The feed amplitude taper is often described as a single term, the edge illumination or edge taper. The edge taper is the feed-pattern magnitude at the edge of the reflector, presented in decibels down from the peak of the feed pattern. Figure 3.31 presents the trade between spillover loss and aperture efficiency loss, where the horizontal axis is the reflector edge taper in decibels. Common edge tapers are usually designed to be in the 10–12 dB range, resulting in best-case combined efficiencies of just over 1 dB of loss.

Defocusing from the feed can occur, and these phases can be calculated using the geometry from the feed and by orienting the feed at the spatial focal point of

(a) (b) (c)

Figure 3.29 Photographs of reflector antennas: prime focus (a), offset focus (b), and Cassegrain (c). (Images courtesy of ViaSat.)

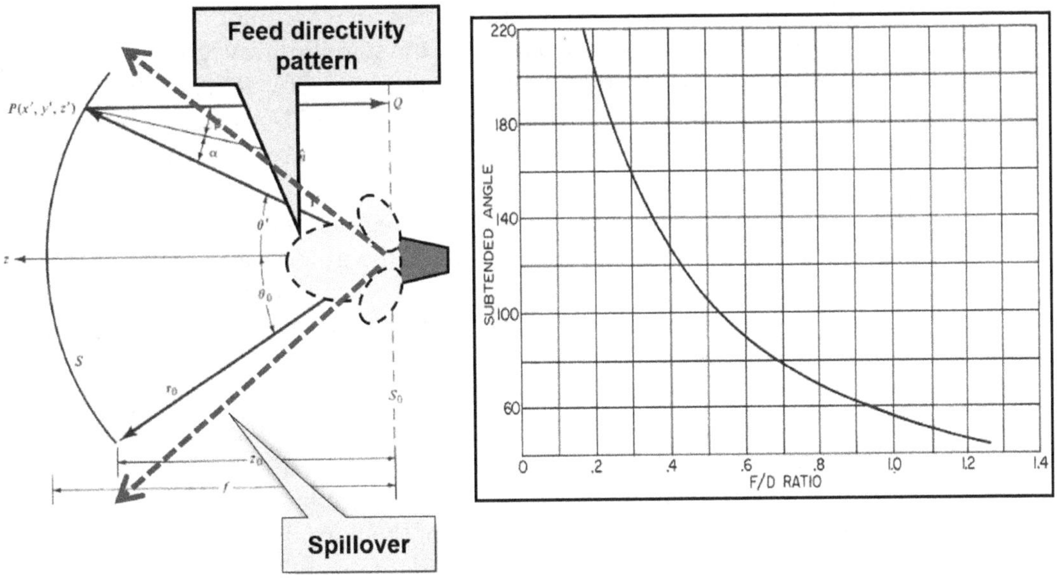

Figure 3.30 Geometry of a prime-focus feed reflector antenna, highlighting the focal length, diameter, feed pattern, and spillover loss.

the reflector. Blockage from the feed and its supporting struts can be modeled to a first order by zeroing points at the correct position in the aperture model of the reflector. Third-order details, such as diffraction from the edge of the aperture and from the feed and feed struts are not easily included in a Fourier transform model of a reflector antenna. While it is possible to utilize the Fourier transform model for a reflector antenna, better accuracy can be achieved via a more detailed model using either the UTD/GTD or MoM solutions.

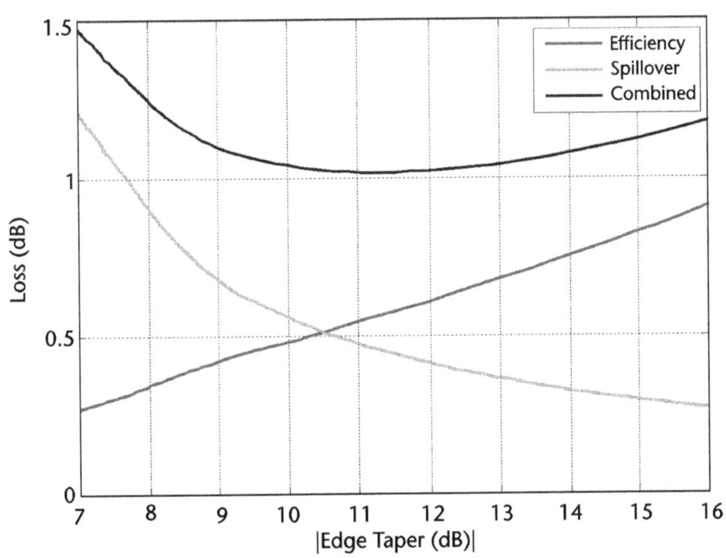

Figure 3.31 Plot of the aperture efficiency, spillover, and combined losses versus reflector amplitude edge taper.

Figure 3.32 presents photographs of variations of waveguide horn antennas. In these images, WR-XX is the nomenclature for the standard waveguide dimension. WR stands for waveguide rectangular, and in the case of WR-90, the 90 indicates that the waveguide inner broad wall dimension is 0.90″. As was introduced in Section 3.8.3, horn antennas provide rectangular apertures, but their amplitude and phase distributions are not uniform.

The electric field vector in a rectangular waveguide is linearly polarized with the polarization parallel to the narrow wall of the guide. Given the conducting boundary conditions of the metal walls, the electric field vector can exist normal to the broad walls of the waveguide. The electric field vector is, however, shorted out against the narrow walls of the waveguide. This causes two different amplitude distributions in the two principal plane directions of the waveguide. The direction parallel to the narrow wall has a uniform electric field distribution, and the direction parallel to the broad wall has a sinusoidal distribution with a peak in the middle and zeros at each end. That amplitude taper continues to exhibit throughout the physical taper or expansion of the horn. Figure 3.33 presents the electric field magnitudes in a WR-90 pyramidal horn with plots on the vertical and horizontal planes. The vertical plot shows that the electric field is uniform in the vertical direction. The horizontal plot shows that the electric field exhibits an amplitude taper where the field is forced to zero tangential to the side walls. Simulation of this amplitude can be achieved in a Fourier transform model by employing a sinusoid amplitude distribution in the horizontal direction while maintaining uniform magnitude in the vertical direction.

Another phenomenon exists within horn antennas that exhibits as a spherical phase error across the aperture. Every antenna has a point from which the wavefront appears to spherically expand. This is called the phase center of the antenna. With horn antennas, the phase center is usually near the transition from the waveguide into the horn. Figure 3.34 shows the spherical curvature of the phase fronts within a different horn antenna modeled using HFSS. The phase error at the face of the aperture can be observed to be greater than ½ wavelength! This

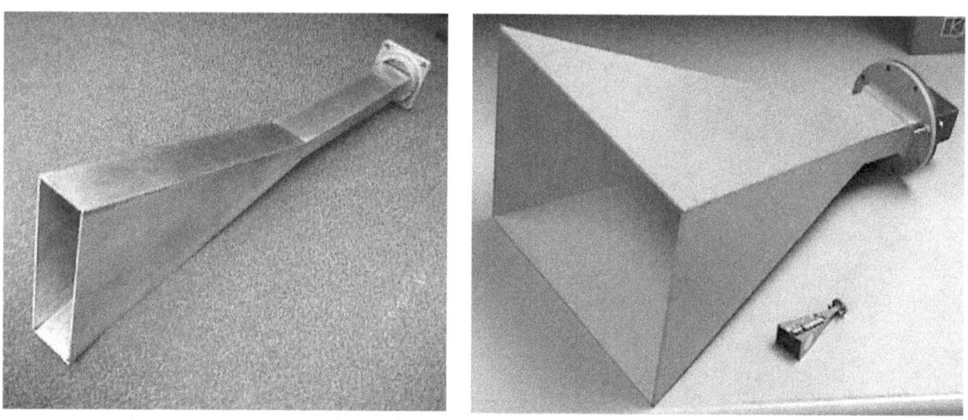

(a) (b)

Figure 3.32 Photographs of horn antennas: WR-90 E-plane sectoral horn and standard gain horns in (a) WR-12 and (b) WR-137.

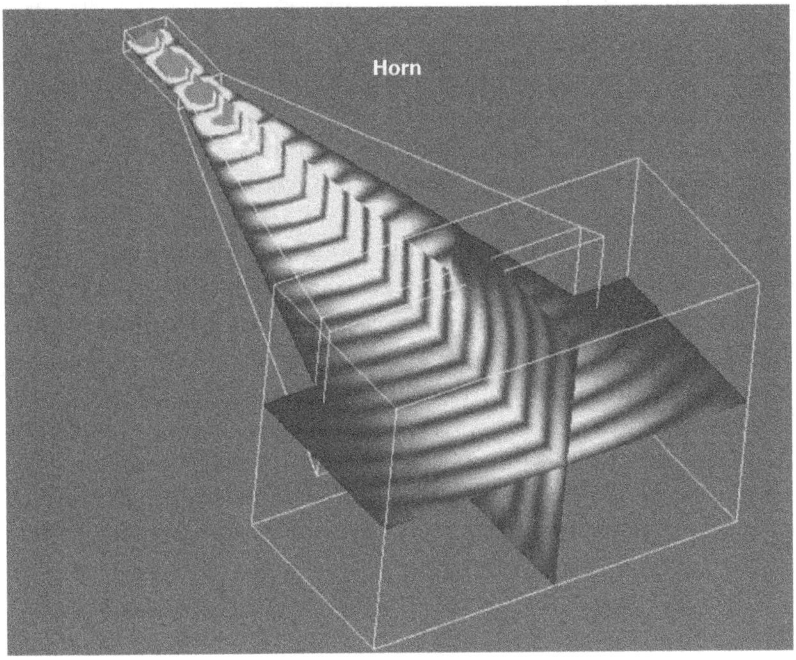

Figure 3.33 The distribution of the electric field magnitude in the horizontal and vertical planes within a WR-90 pyramidal horn antenna.

phase error causes destructive interference that results in reduced peak directivity, a misshaped main beam, and significantly higher sidelobes. Figure 3.34 also contains the pattern of the horn, illustrating the negative effects of the phase-defocusing on its directivity pattern. Note that the peak of the main beam is not at boresight, that the sidelobes are very high, and that the nulls have been filled in. Designers of horn antennas know that this phase-defocusing can be corrected by

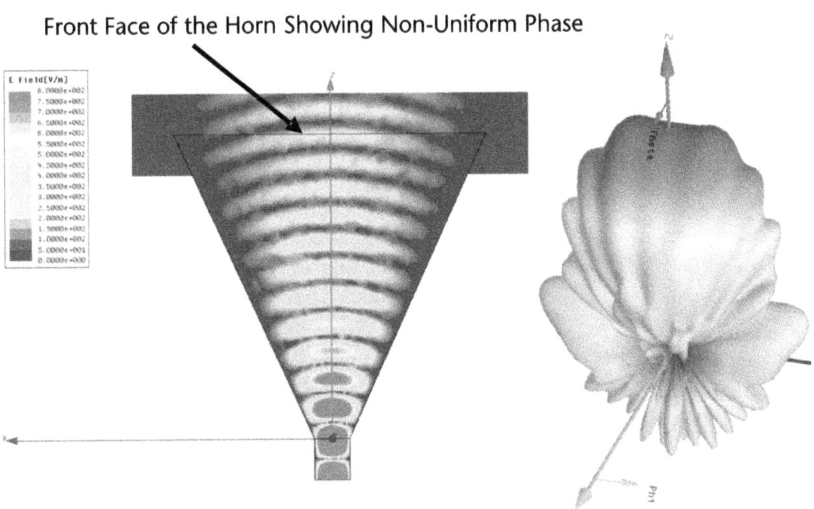

Figure 3.34 An FEM model of a waveguide horn antenna that demonstrates significant spherical phase defocusing at the horn aperture as demonstrated by the curvature of the phase front.

either making the horn much longer or by using a correcting lens inserted into the horn. These improvement techniques are not always needed as there are some applications where the patterns without phase correction are useful as is.

Given a priori knowledge of the horn amplitude and phase nonuniformity, these distributions can be simulated using a 2-D Fourier transform model by applying values of a_n and φ_n across the aperture.

A 2-D Fourier transform model was developed for the WR-90 horn shown in Figure 3.33. For it, the area of the aperture was sampled with spacing of less than ½ wavelength. Amplitude and phase tapers were calculated using simple sinusoid and spherical distributions, respectfully. Figure 3.35 presents aspects of the Fourier transform horn model including its sample points, its spherical phase error (upper) and its amplitude taper (lower) in only the horizontal plane. Figure 3.36 presents a comparison of the resulting directivity pattern calculated using HFSS (upper row) and the Fourier transform model (lower row). The images show a pattern side view of the E-plane (left) and the high resulting first sidelobe; the side view of the H-plane (center) with the very low first sidelobe, which is caused

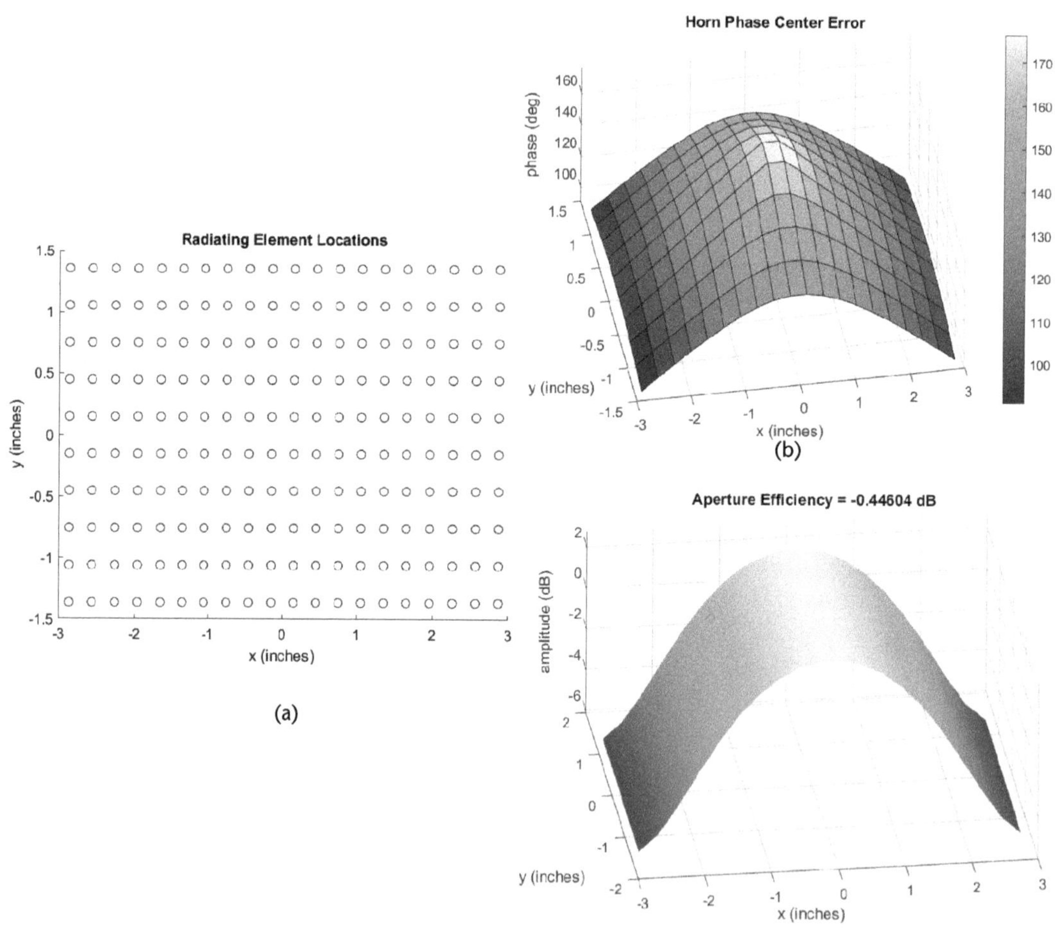

Figure 3.35 Details of the horn Fourier transform model: (a) the sample grid, (b) the spherical phase error, and (c) the sinusoidal amplitude taper in the horizontal direction only.

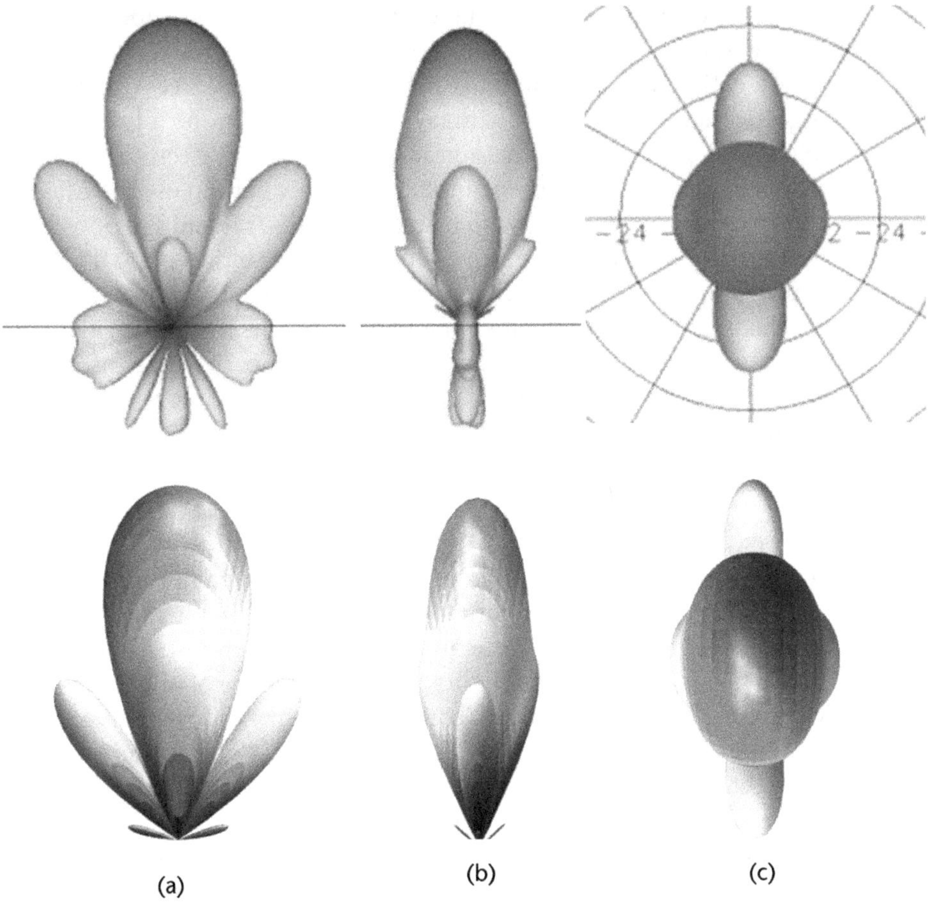

Figure 3.36 Directivity patterns of the WR-90 horn using different simulation approaches: (a) side view of the E-plane, (b) side view of the H-plane, and (c) nose-on or boresight view. The patterns in the upper row were calculated using HFSS FEM, and the patterns in the lower row were calculated from a simple Fourier transform model.

by the e-field amplitude taper in the H-plane; and the nose-on view (right). The Fourier transform model fails to accurately calculate any back lobes, but the calculations in the forward hemisphere are reasonably accurate for a quick and simple model. The extensions of appropriately applying amplitude and phase distributions with the Fourier transform elevate its utility from very rough approximations via first-order to significantly more accurate, yet still expedient, second-order modeling approaches.

3.12 Fourier Transform Modeling of Arrays

The Fourier transform modeling approach is ideally suited to model array antennas. Arrays use a multitude of individual low-gain radiators or elements to sample an aperature area. The amplitude taper across the array can be specified by the designer to reduce SLLs. Passive arrays add phase shifters at each element to facilitate electronic steering of the angle of the main beam. The very short time required to perform electronic steering makes beam-steering fast enough that arrays can multitask, for example, interleaving two tracks for a radar system or interleaving communication with two satellites, in both cases with the two functions separated by significant angle. Solid-state arrays or active electronically scanned antennas (AESAs) add small monolithic microwave integrated circuit (MMIC) amplifiers at every element. Getting these amplifiers closely behind every radiating element results in reduction of losses, which both increases transmitted power and reduces the system noise figure on receive. Both of these improve the SNR, thereby significantly improving system performance. And last, the emerging technology of digital beamforming at the per-element level results in even greater performance increases in simultaneity of multifunctionality. All of these performance improvements have been developed while the cost of array components and systems has been decreasing.

3.12.1 The Array Element Pattern and Its Effect

Unfortunately, Huygens or isotropic point sources do not exist other than as a mathematical concept. If we were to construct such a point source, it would have to be infinitesimally small and could not have any real transmission line feed. In real arrays, we use real radiators, usually of a size of approximately 0.5 × 0.5 wavelength. These real radiators, constructed out of dipoles, waveguides, printed circuit patches, or dielectric rods, have a peak gain at boresight and an element pattern that demonstrates an aperture projection loss at angles off broadside. The aperture projection versus angle results in an ideal element power pattern of $\cos(\theta)$. A $\cos(\theta)$ element pattern results in a gain reduction of a full −3 dB at 60° scan off broadside. This limits the maximum useful scan angle for most arrays to ±60° off broadside. Real radiators also experience reflections that reduce transmission efficiency. The radiator reflection coefficient usually increases with scan angle and hence reduces gain at far scan angles. This is commonly referred to as the problem of mutual coupling among array radiators. Because of the mutual coupling reflections versus scan, array simulations often use $\cos^p(\theta)$ for the power pattern, where p is an exponential power, on the order of 1.2–1.5 depending upon the actual element pattern degradation at 60°. Note that the element pattern loss of $\cos^p(\theta)$ is a power term. When we model the element pattern with the cosine function within the Fourier transform, remember that voltages are summed and not powers, and hence we use the square root of the cosine raised to a power function as the voltage multiplier.

Equation (3.21) presents the calculation for the directivity pattern for a linear array including the element pattern, assuming that every element pattern is different. If every element pattern is the same (or close enough to the same), then the element pattern, $E_{\text{element}}(\theta)$, can be brought outside of the summation and multiplied by (3.18), which is often referred to as the array factor. Equation (3.22) presents the directivity for the array where every element pattern is close enough to the same regardless of its position in the array.

$$E(\theta) = \sum_{n=1}^{N} E_{\text{element } n}(\theta) e^{jkn\Delta x \sin(\theta)} \tag{3.21}$$

$$E(\theta) = E_{\text{element}}(\theta) \sum_{n=1}^{N} e^{jkn\Delta x \sin(\theta)} \quad (3.22)$$

Figure 3.37 shows a block diagram of an array with multiple similar element patterns and plots of the element pattern, the array factor, and the array pattern, which is the product of the two. Note from the patterns of Figure 3.37 that while the array factor provides a consistent main beam independent of scan angle, the array pattern main beam experiences scan loss because of the element pattern. As the main beam loses directivity versus scan angle, its beamwidth broadens. Approximations for the directivity versus scan and the beamwidth versus scan are presented in (3.23). The reader can experiment with the element pattern as well as phase scanning of the main beam in Section 3.12.4 using ***array_phsteer.m***.

$$E_{\text{array}}(\theta) = E_{\text{peak}} \cos^{p}(\theta) \qquad \Psi_{\text{main beam}}(\theta) = \frac{\Psi_{\text{narrowest}}}{\cos^{p}(\theta)} \quad (3.23)$$

3.12.2 Amplitude Tapering for Sidelobe Reduction

As introduced in Section 3.2 and illustrated in Figure 3.11, amplitude tapering can be employed for the specific purpose of sidelobe reduction. Much has been written about the history and trade space among different numerical approaches to amplitude tapering (e.g., Mailloux [26] and Hansen [27]). Taylor presented a numerical

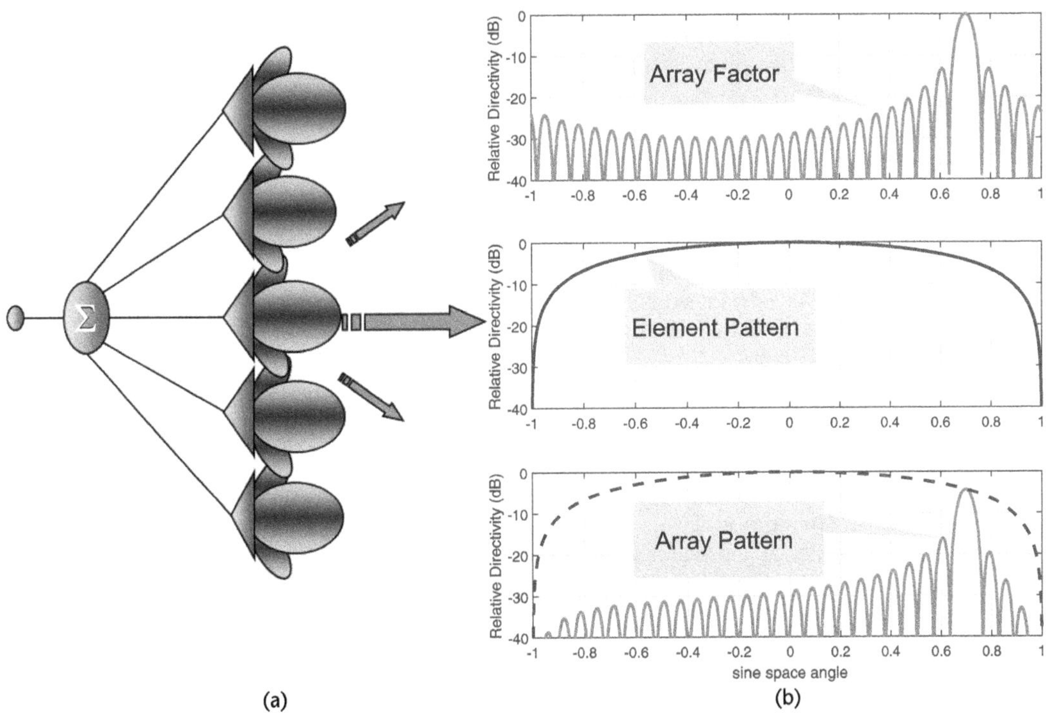

Figure 3.37 Block diagram of an array antenna showing (a) the impact of the element pattern directivity versus angle, and (b) plots of the element pattern, the array factor, and the array pattern.

approach to calculating E-field weights that maximize the aperture efficiency for a given peak first SLL for a line or rectangular array [28] and a circular array [29]. Taylor's approach is more efficient because his synthesis focuses on only the highest sidelobes. He observed that sidelobes for a sinc function fall off monotonically to low levels naturally and that the problem sidelobes were only the few high near-in ones. Two numeric inputs to the calculation for a Taylor function are the SLL, the value of the peak highest SLL desired with respect to the main beam peak, and $\bar{n}$ (often pronounced n-bar), the number of the high sidelobes on either side of the main beam for which the function will focus the SLL reduction. As SLLs are driven lower (i.e., from 20 to 40 dB), the values of n-bar need to increase from 2 or 3 to 5 or 6. Figure 3.38 shows a comparison between Dolph-Chebyshev and Taylor amplitude weights, their resulting patterns, and comments of the pattern differences between the two. Figure 3.39 shows the Taylor distributions for uniform, 25-dB SLL, and 40-dB SLL, with their resulting directivity patterns. Table 3.2 presents different sets of SLL and n-bar with their calculated beam broadening coefficient, α, and their aperture efficiency, η. Note that the values in Table 3.2 are calculated for a linear array only. Total efficiency losses are greater for a 2-D array.

To increase angle measurement precision, most modern radar arrays employ monopulse techniques, as published by Sherman [30]. In monopulse systems, multiple simultaneous beams are formed. The signal is transmitted on the sum (Σ) beam. Returned signals are received simultaneously on the Σ and Δ or difference beams. Three-dimensional systems have both Δ azimuth (ΔAz) and Δ elevation (ΔEl) beams. These beams are formed by using a combiner, often termed a monopulse comparator, as shown in Figure 3.40. The sum and difference patterns that

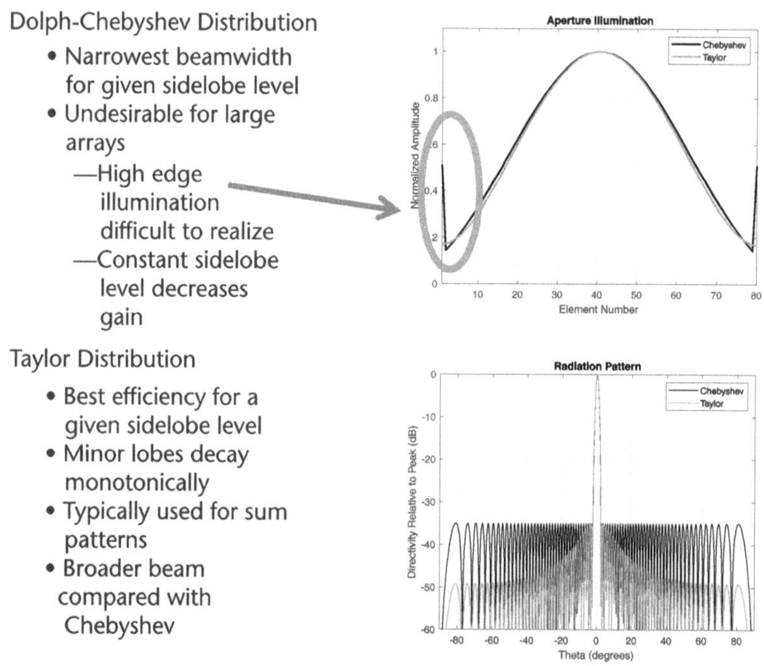

Figure 3.38 Comparison between Dolph-Chebyshev and Taylor weightings and their resulting patterns for an 80-element linear array with targeted −35-dB designed highest sidelobes.

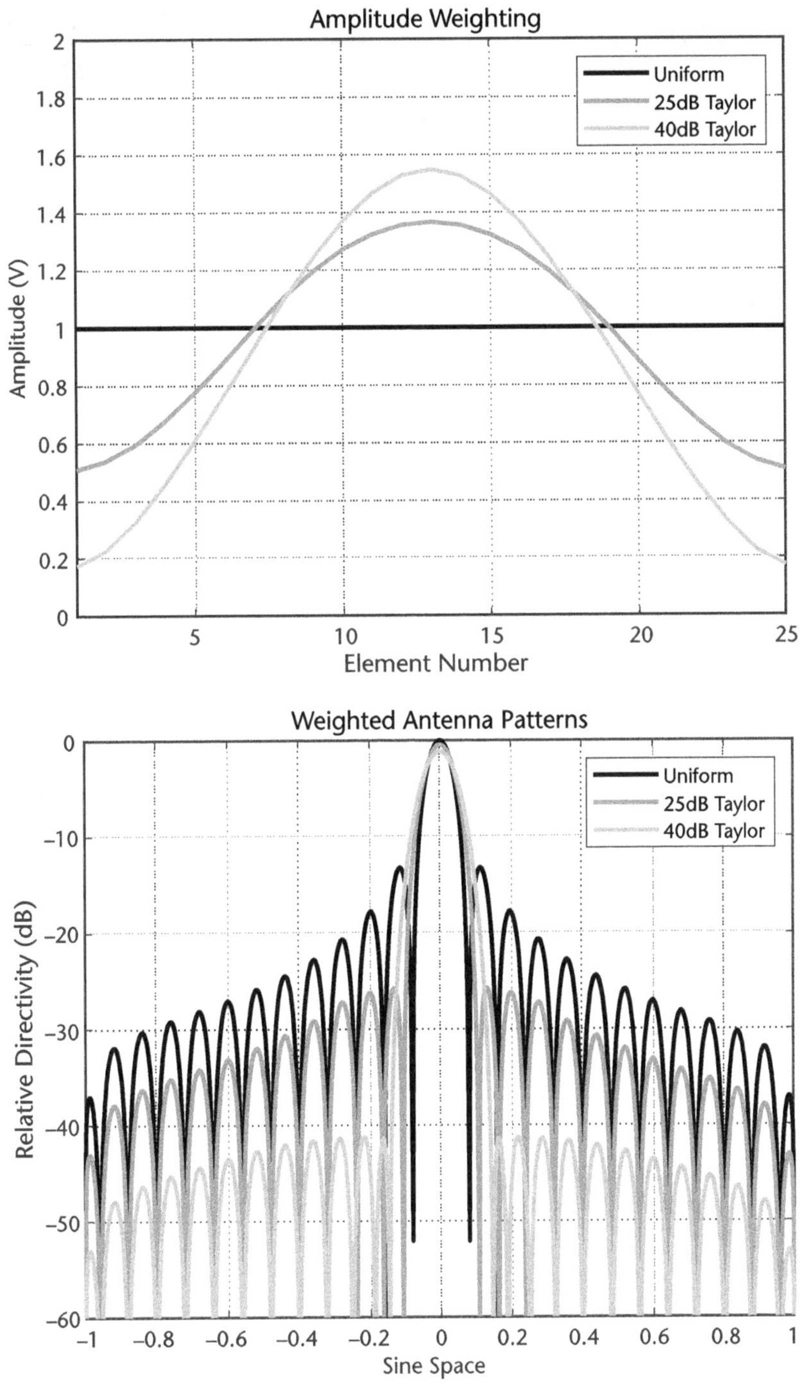

Figure 3.39 Taylor distributions and resulting directivity patterns for SLLs of 13 dB (uniform), 25 dB SLL and n-bar of 3, and 40 dB SLL and n-bar of 6.

Table 3.2 Taylor Amplitude Function Input Parameters and Resulting α and η

n-bar	*Sidelobe Level (dB)*	*Beamwidth* α *(unitless)*	*Aperture Efficiency* η *(dB)*
2	20	0.983	–0.218
3	25	1.049	–0.455
4	30	1.115	–0.704
5	35	1.179	–0.948
6	40	1.25	–1.178
8	45	1.3	–1.39

result are shown in Figure 3.41. Note that the difference pattern is split in the center (bifurcated). The differencing operation results in a very important 180–phase difference between the two mainlobes of the difference pattern. It is this phase difference, along with the very steep rate of amplitude change that provides more precise angle measurement. The monopulse comparator has been implemented in numerous ways including in waveguide, printed circuit, and digital form factors. More about the modeling and simulation approaches for monopulse can be found in Chapter 9.

The additional sharp discontinuity in the amplitude distribution of a uniformly illuminated difference pattern causes a significant increase in the difference pattern SLLs. Bayliss [31] took the approach used by Taylor for the sum beam and applied it to an optimum E-field amplitude distribution to achieve low SLLs for difference patterns.

array_amp.m includes the ability to model either uniform, Taylor, or Bayliss amplitude distributions and observe the resulting pattern SLLs. This calculation uses (3.20), where the a_ns are the voltage weights at each array element. Inputs include the number of elements; the interelement spacing; a flag to select whether uniform, Taylor, or Bayliss amplitudes are employed; and the SLL and n-bar for either the Taylor or Bayliss functions. ***taylor.m*** and ***bayliss.m*** are functions that calculate the requisite voltage weights, a_n, and are called by ***array_amp.m***. Figure 3.42 presents the weighting functions and directivity patterns simulated with a 15-element linear array with uniform weighting (top row), a 30-dB SLL n-bar of 5 Taylor taper (middle row), and a 30-dB SLL n-bar of 5 Bayliss taper (bottom row). Note that the peak value of the amplitude of the Taylor taper is 1, with the tapered amplitudes being less than 1. This is the way that the Taylor function commonly

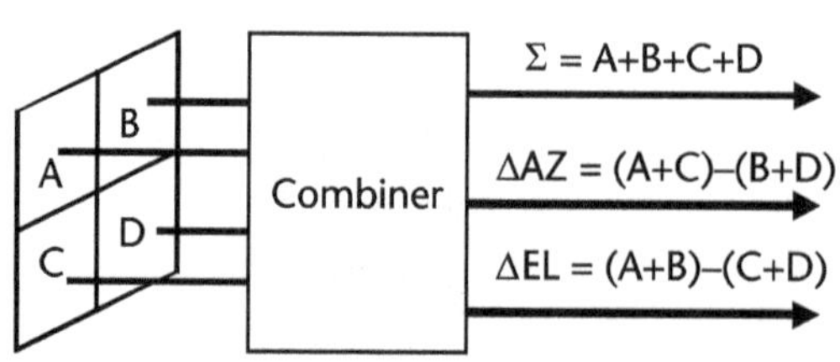

Figure 3.40 Block diagram for a monopulse combiner or comparator, which simultaneously generates the Σ, difference or delta azimuth (ΔAz), and the difference or delta elevation (ΔEl) beams.

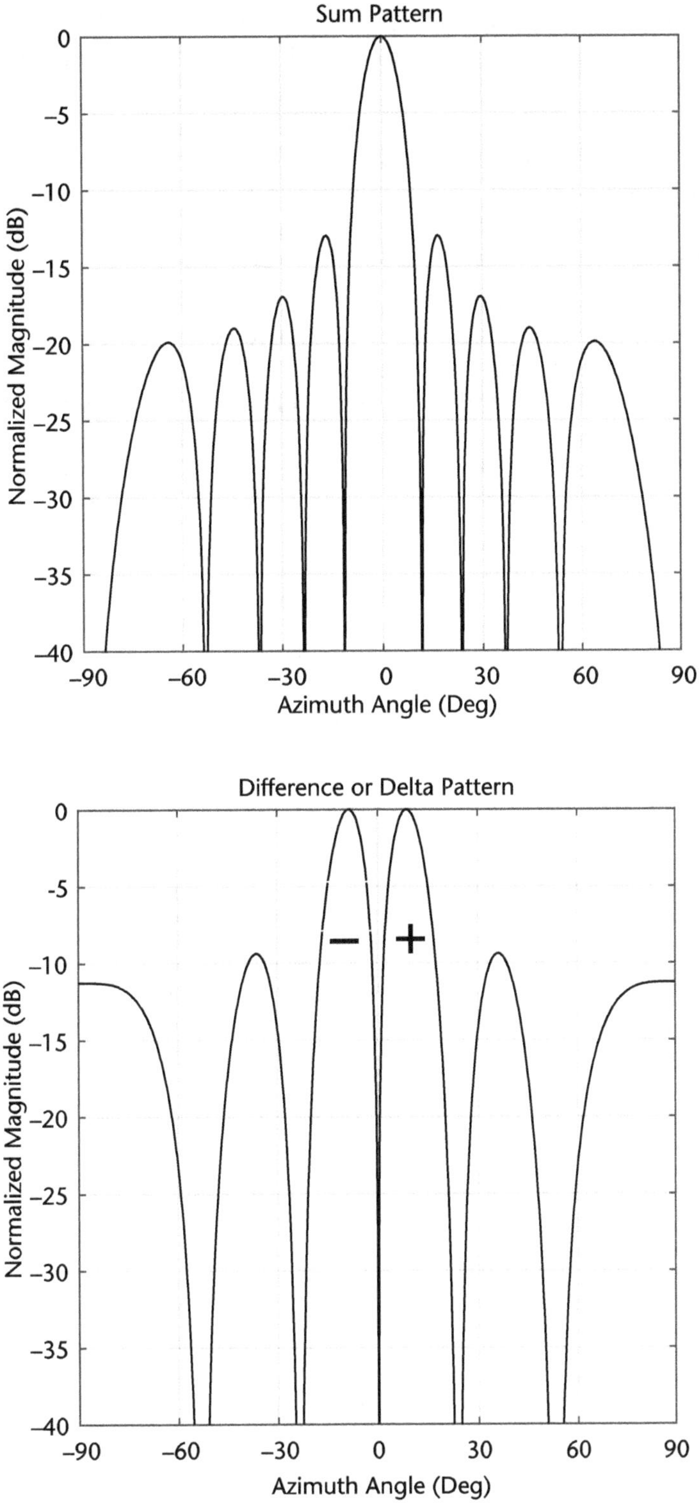

Figure 3.41 The sum and difference patterns generated in a monopulse comparator.

outputs the a_n amplitudes. More about this, the possible need for renormalization of the amplitudes, and the effect of this on the resulting aperture efficiency, η, follows in Section 3.12.2.

When modeling an array antenna, the simulation must faithfully model the architecture used in the hardware (or software in the case of a digitally beamformed array). The amplitude tapers can be created in four distinct array architectures, as shown in Figure 3.43. In Figure 3.43(a), the amplitude tapers can be created in the hardware beamformer by varying the voltage couplings between the different transmission lines. This is commonly implemented in waveguide or printed circuit hardware. In Figure 3.43(b), the feed horn pattern of a space-fed array generates the amplitude taper on the array lens. In Figure 3.43(c), the beamformer provides uniform amplitude while the attenuators apply the amplitude taper by selectively

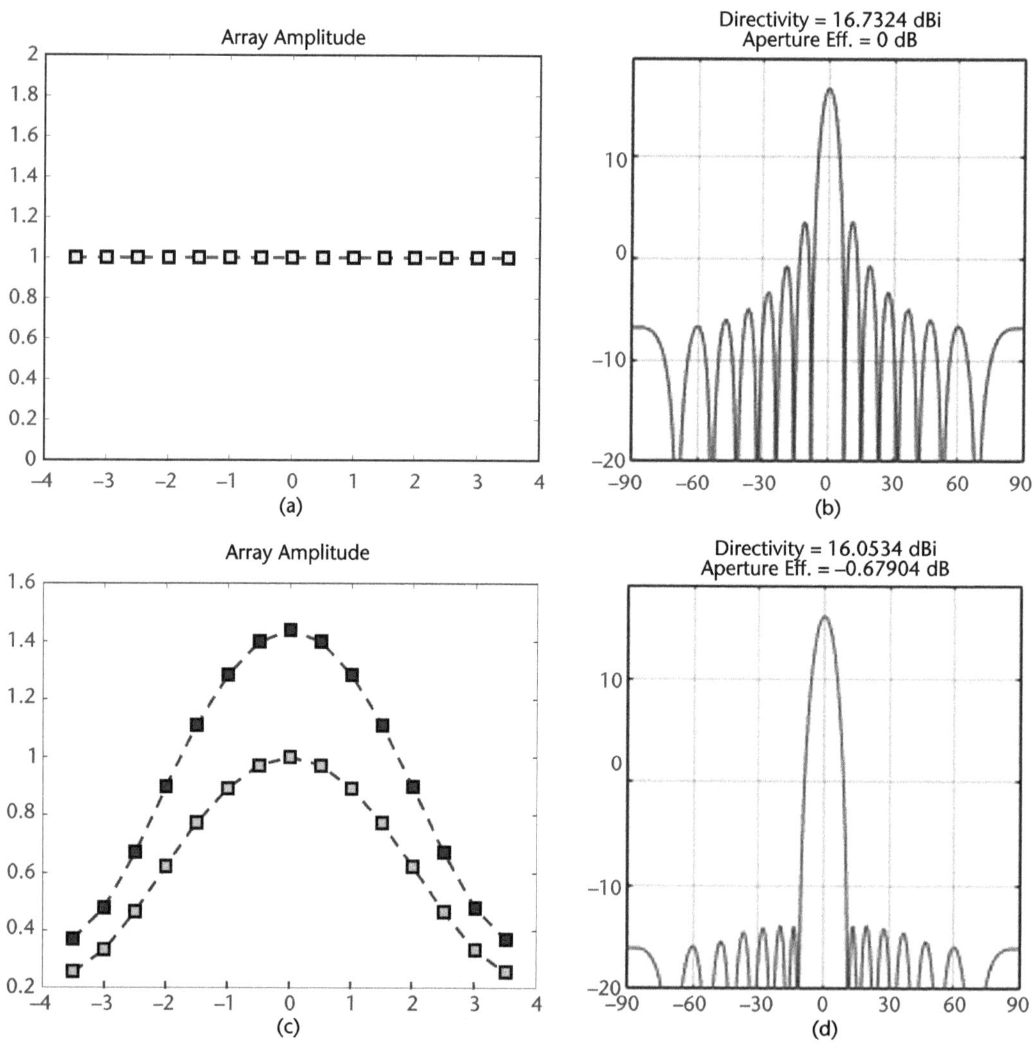

Figure 3.42 Plots of amplitude tapers and resulting directivity patterns calculated using **array_amp.m** for a 15-element linear array: (a) uniform amplitude, (b) the sinc directivity pattern that results from a uniform distribution, (c) a 30-dB SLL $\bar{n}$-bar of 5 Taylor taper, (d) the 30-dB SLL directivity pattern from the Taylor taper, (e) a 30-dB SLL n-bar of 5 Bayliss taper, and (f) the 30-dB SLL pattern from the Bayliss taper.

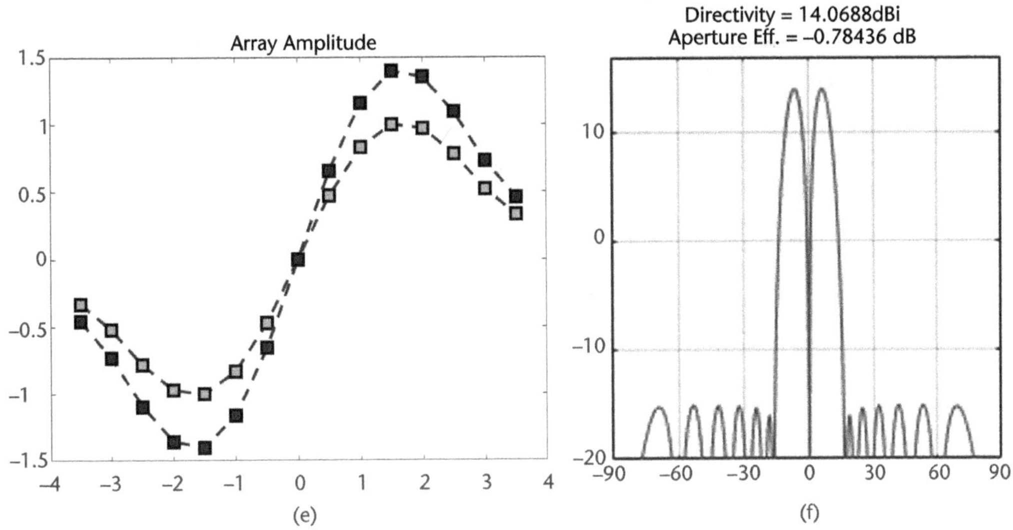

Figure 3.42 *(Continued)*

attenuating energy to some of the elements. Last, in Figure 3.43(d), a digital beamformer can apply arbitrary amplitudes as programmed. In the approaches of Figures 3.43(a, b), the amplitude tapers are permanently built into the hardware and are not easy to change. Conversely, in Figures 3.43(c, d) the amplitude taper is controlled by a computer and can be changed as needed. For Figures 3.43(a, c), if there is only a single hardware beamformer for the array, different simultaneous weightings for the sum and difference channels are not possible.

array_difftay.m demonstrates the case of a monopulse linear array that has only a single-hardware beamformer. If low sidelobes are a requirement, it is impossible using a single hardware beamformer to achieve low sidelobes for both the sum and difference patterns simultaneously. Taylor and Bayliss amplitude distributions are significantly different. What has been done in some array systems is to use the Taylor amplitude distribution for the sum beam and simply phase-difference the delta patterns array quadrants to generate delta patterns using the amplitudes of the Taylor function for all of the beams. This causes a sharp discontinuity in the amplitude taper for the difference patterns which results in high SLLs for the difference beams. The amplitude tapers (Σ is in light gray, and Δ is in dark gray) and the resulting sum and difference directivity patterns are shown in Figure 3.44. The sum pattern is well-behaved with the desired low SLLs. The difference pattern, while still very functional as a difference pattern, has much higher SLLs. Modeling of the amplitude tapers to match the actual hardware implementation is important in realizing accurate pattern SLL features.

3.12.3 Calculation of the η Aperture Efficiency

The aperture efficiency, η, is a measure of how much of the aperture is used as compared to the same aperture with a uniform taper. A uniform taper has an η of 1.0 (0 dB), while any amplitude taper other than uniform has an η of < 1.0 (some negative decibels). As mentioned in the previous section, the Taylor taper provides

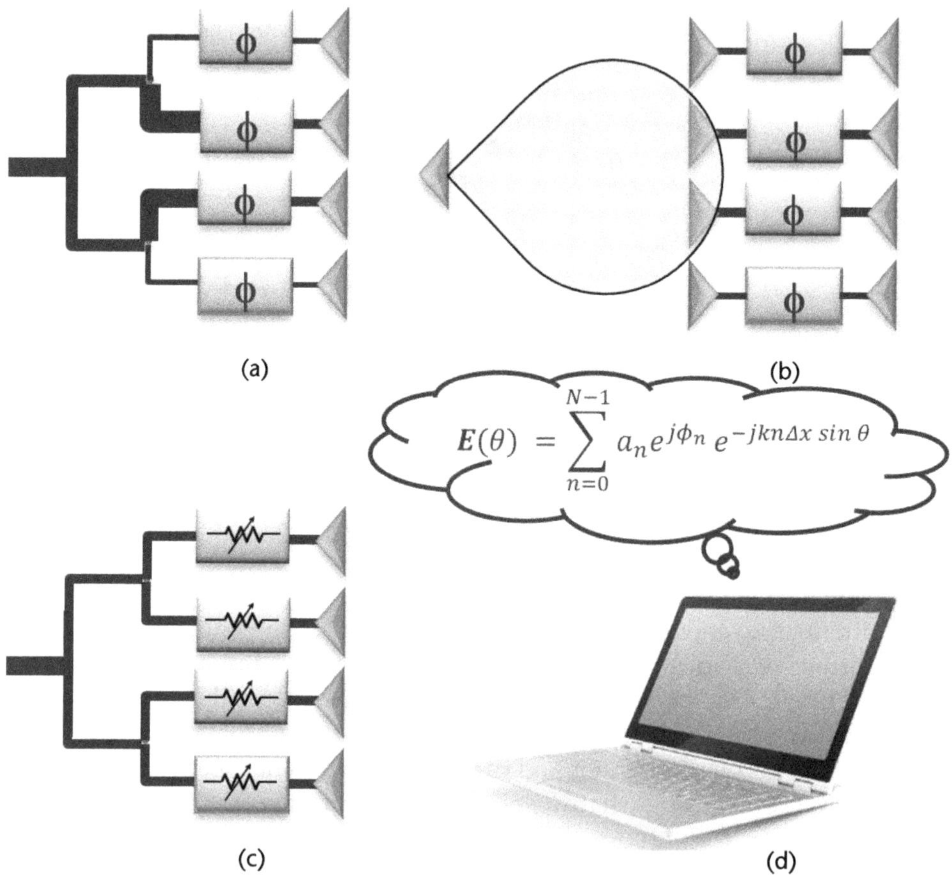

Figure 3.43 Different implementations of array amplitude tapers as used in different array architectures, with the tapers implemented in: (a) an analog beam former transmission line circuit, (b) the pattern of the space-fed feed horn, (c) digital attenuators in the elemental control modules, and (d) a digital beamforming algorithm.

E-field amplitudes that have a maximum value of 1.0, with all other values < 1.0. This is analogous to using attenuation to implement the Taylor taper, and it is only correct for simulating arrays that use attenuators to implement the taper. Figure 3.45 shows the output of the Taylor routine for a simple eight-element array with weights for uniform as well as Taylor tapers to achieve –25 to –65-dB first SLL. (And yes the authors realize that –65-dB first SLL is not realizable in a cost effective real array antenna.) Figure 3.45 also shows the resulting directivity patterns for that array with those weightings. The desired impact on the SLLs is achieved, but note the resulting loss in directivity caused by the significant aperture efficiency loss. Note also that if the power into each array were integrated, each taper would have a different total power radiated, with the uniform case being the highest.

In addition to being dependent upon the specific amplitude taper used, η is also dependent on how the amplitude taper was implemented in the hardware. Hopkins, et al., [32] described differences in implementation and proper calculation of the efficiency, η. Referring to Figure 3.43, for beamforming techniques the voltage removed from the outer elements is added back in to that of the central elements. In Figures 3.43(a, b, d), no power is wasted in resistive attenuation. To correctly use the outputs from either a Taylor or Bayliss routine for any of the three architectures in Figures 3.43(a, b, d), the user must renormalize the voltage weights so that the average voltage across the array equals 1.0. Equation (3.24) presents the method for renormalizing or redistributing the Taylor or Bayliss amplitudes, a_n, converting them to b_n. N in (3.24) is the number of elements in the array. Renormalization can be thought of as taking the power removed in the outer elements and reinserting it in the central elements of the array.

$$b_{n,\text{redistributed normalized}} = a_n \sqrt{\frac{N}{\sum_{n=1}^{N} \left(a_n\right)^2}} \tag{3.24}$$

The act of renormalization does not impact the sidelobe reduction, it only improves the overall efficiency as no power is being dissipated in attenuation. Figure 3.46 shows the same Taylor weights as in Figure 3.45, but renormalized. Note that the integral of the power in each amplitude distribution is now equal. The patterns shown in Figure 3.46 show similar sidelobe reduction but demonstrate significantly less main beam directivity loss that that of Figure 3.45.

Every array that uses an amplitude taper suffers from a reduction of gain because of the aperture efficiency loss. Those whose tapers are implemented via attenuation suffer an additional loss because of the attenuation. Figure 3.47

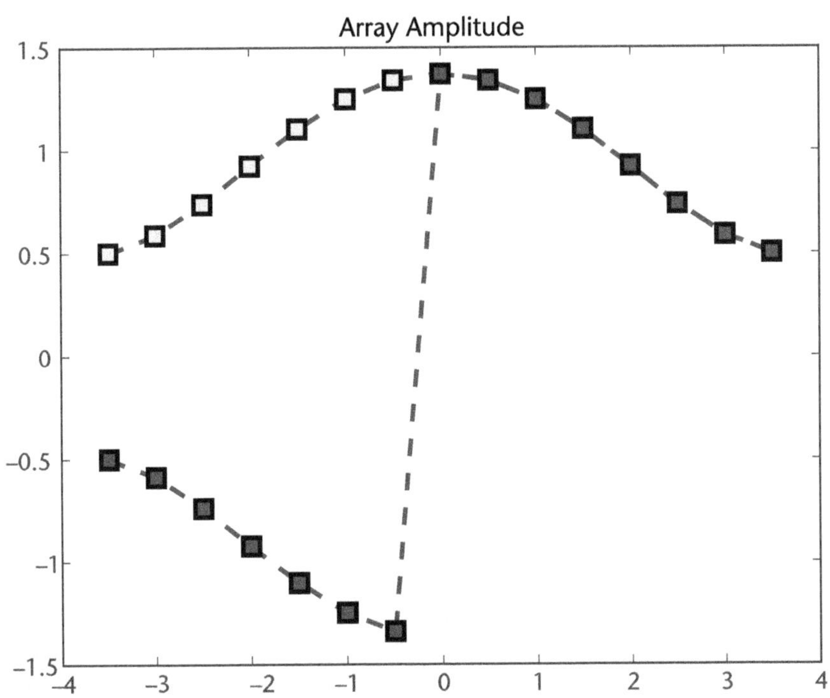

Figure 3.44 Antenna pattern simulation for sum and difference beams where a single analog beamformer is used with a Taylor amplitude taper.

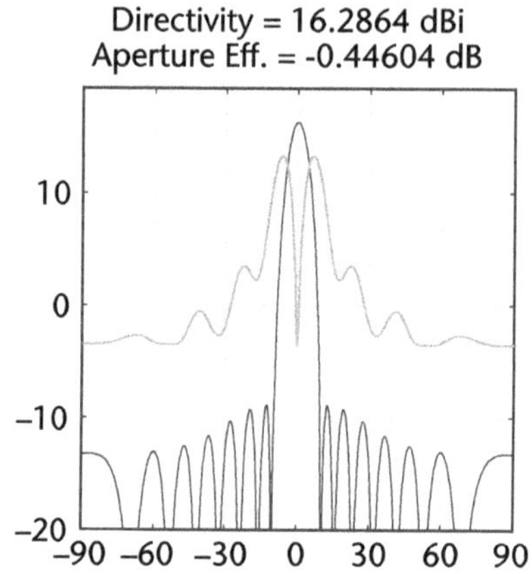

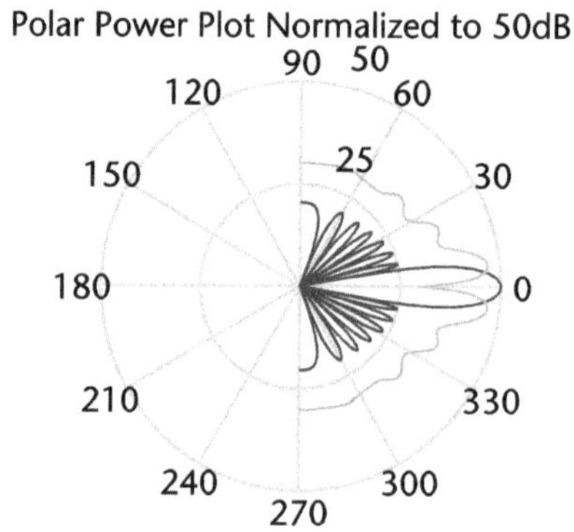

Figure 3.44 *(Continued)*

presents plots of the resulting efficiencies for both cases versus desired first SLL. The three plots in Figure 3.47 are the aperture efficiency (top trace); loss because of attenuation (middle); and the combination of the two (bottom), which is the case for the attenuated taper. Keep in mind that these losses are solely for a line array. The losses for a 2-D array are even greater, hence amplifying the need to correctly calculate the appropriate losses given the actual hardware implementation of the amplitude tapers.

One example of redistribution of power in a printed circuit beamformer was reported by Hopkins, et al. [33]. That design used unequal power dividers as described by Parad and Moynihan [34] to facilitate reduced SLLs with minimal losses. Figure 3.48 presents the relative Taylor amplitudes used in one half of the 16-element linear array combiner along with a plot of the E-field magnitude from an HFSS model of the beamforming circuit.

In Hopkins [32], the aperture efficiency, $\eta_{\text{aperture efficiency}}$, the attenuation, $\eta_{\text{attenuation}}$, (if the average voltage is not one), and the total efficiency η_{total} are derived. These are provided here in (3.25) to (3.27). In these, N is the total number of array elements. a_n are the voltage amplitude weights either (1) straight from the Taylor

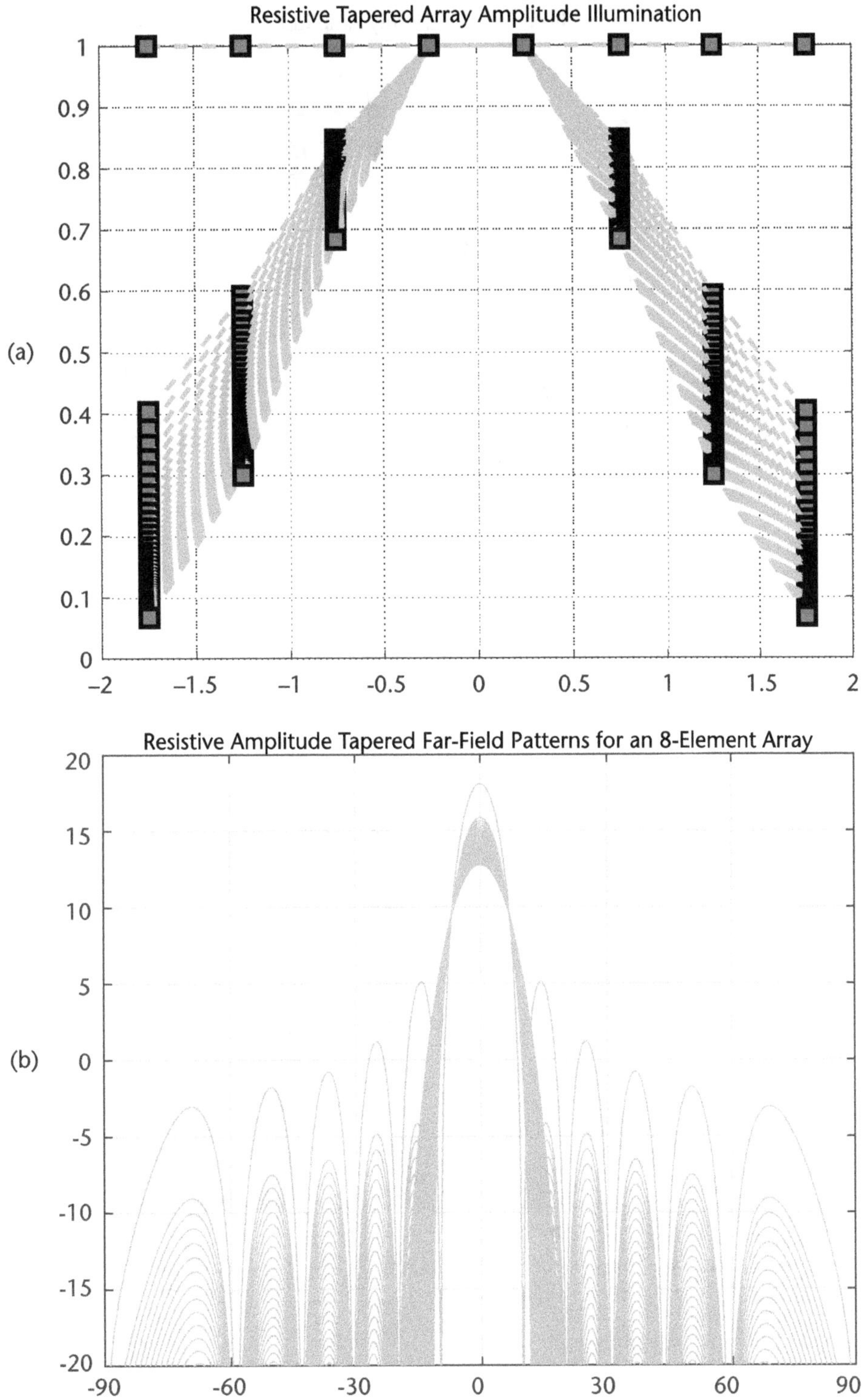

Figure 3.45 (a) Taylor amplitude distributions as output directly from the Taylor routine and (b) the resulting directivity patterns for an eight-element array (Source: [32], ©2007, IEEE. Reprinted with permission.)

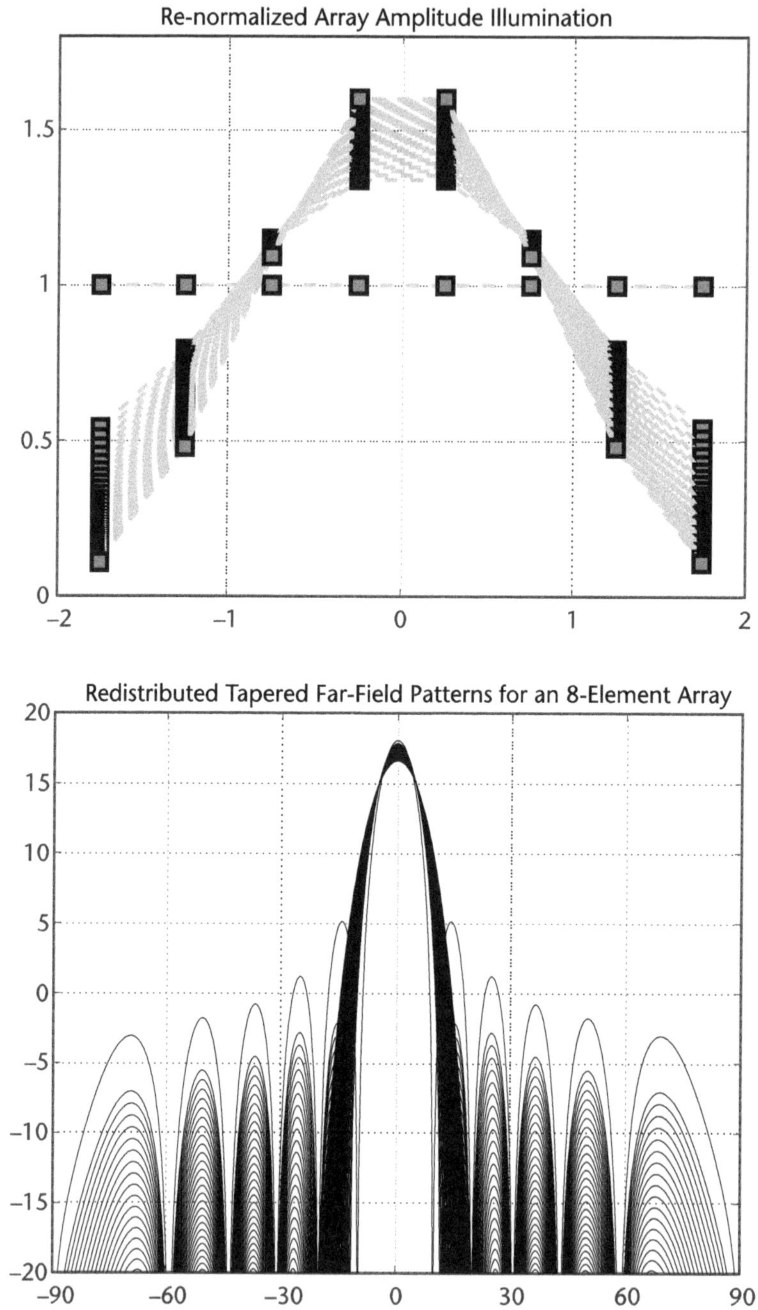

Figure 3.46 Renormalized or redistributed Taylor amplitude distributions and the resulting directivity patterns for an eight-element array. (Source: [32], ©2007 IEEE. Reprinted with permission.)

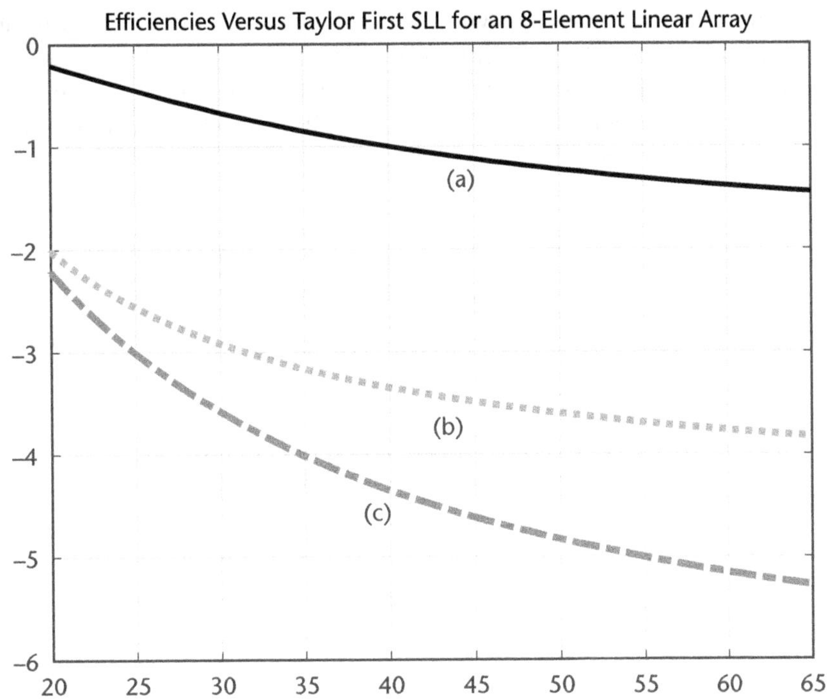

Figure 3.47 A comparison of (a) the η efficiency losses for the redistributed or renormalized implementation, (b) the loss because of attenuation alone, and (c) the combination of the two. (*Source:* [32], ©2007 IEEE. Reprinted with permission.)

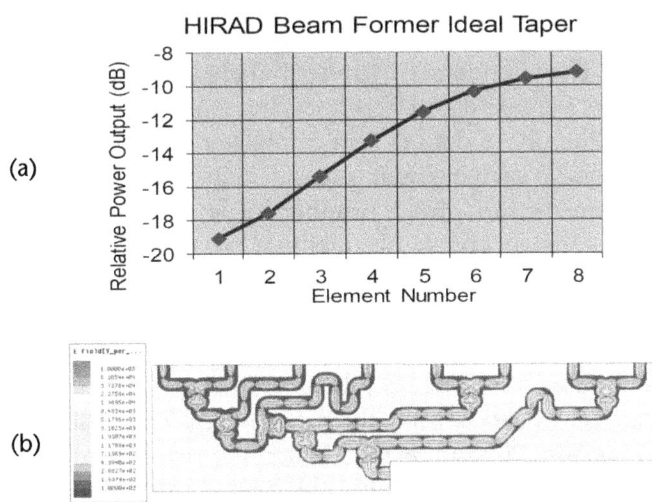

Figure 3.48 (a) Plot of the relative Taylor power amplitudes for one half of a 16-element linear array beam former, and (b) plot of the E-field magnitude from an HFSS model of the beam former circuit. (*Source:* [33], ©2015 IEEE. Reprinted with permission.)

or Bayliss routines if the weighting is implemented via attenuation or (2) the renormalized weights [b_n per (3.24)] if the weightings were implemented via redistribution of the voltages. The examples shown are using linear arrays only, but the efficiency calculations are equally valid for a 2-D array where N is the total number of elements.

$$\eta_{\text{aperture efficiency}} = 10\log_{10}\left[\frac{\left(\sum_{n=1}^{N} a_n\right)^2}{N\sum_{n=1}^{N}\left(a_n\right)^2}\right] \tag{3.25}$$

$$\eta_{\text{attenuation}} = 10\log_{10}\left[\frac{\sum_{n=1}^{N}\left(a_n\right)^2}{N}\right] \tag{3.26}$$

$$\eta_{\text{total}} = \frac{\left(\sum_{n=1}^{N} a_n\right)^2}{N^2} \tag{3.27}$$

3.12.4 Phase- or Time-Delay Scanned Arrays

The most beneficial attribute of a phased array is the ability to electronically steer the main beam. The array effectively tilts the flat phase front by delaying the phases at most of the elements. The amount of delay is based on the geometry shown in Figure 3.49. The interelement differential phasing, $\Delta\varphi$, for a 1-D array is (3.28). The phase delays are calculated by what is generally called a beam-steering computer (BSC), and the phase data is sent to phase-delaying control components, called phase shifters, at each element. The calculated phase delay in degrees can be any number from zero to a huge number for large phased-array antennas. It is not practical to construct phase shifters that can provide that very large phase range because of the transmission losses and size that would be required. Almost all phase shifters are designed to shift between 0° and 360°. The BSC calculates the ideal phase, using a single wavelength of the center frequency of the instantaneous bandwidth of interest. The BSC then subtracts off as many integer number of 360° wavelengths as needed until the remaining phase value is less than one wavelength or between 0° and 360°. This operation is referred to as taking modulo-2π of the phase (in radians). Having the phases calculated based on a single wavelength works perfectly at the single frequency that was used in the calculations. At all other frequencies, the beam actually squints or deviates from the desired steering angle, θ_{steer}. The effect of the beam squint can be significant if the beam is narrow and if enough instantaneous bandwidth is required. The solution to the beam squint problem is to use time delay in place of phase shift. Equation (3.29) presents the calculation for the interelement differential time delay, Δt, where c is the speed of light, and f is frequency in hertz. Emerging digital array systems may have the option of using either. As most systems in operation at present use

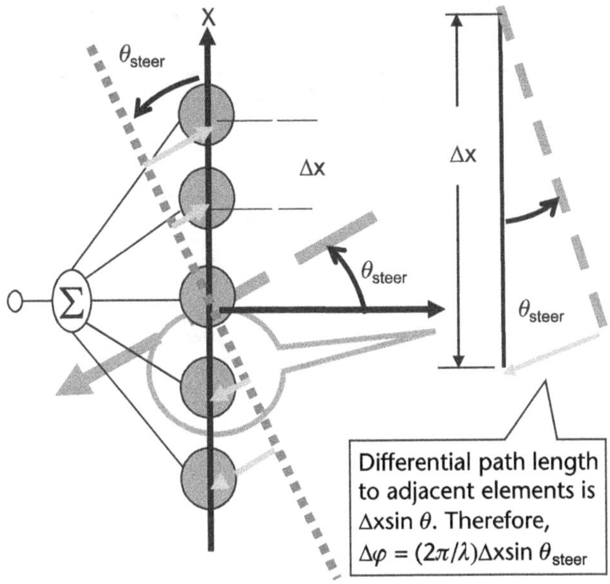

Figure 3.49 Linear array illustrating the phasing needed to steer a plane wave in the direction of θ_{steer}.

phase shift, all of the array models that follow in this chapter include simulation of the use of phase shifters.

$$\Delta\phi = \frac{2\pi}{\lambda}\Delta x \sin\left(\theta_{\text{steer}}\right)$$
$$\left(\Delta x \text{ in same units as } \lambda,\ \Delta\phi \text{ units in radians}\right) \tag{3.28}$$

$$\Delta t = \frac{1}{c}\Delta x \sin\left(\theta_{\text{steer}}\right)$$
$$\left(\Delta x \text{ units in same length units as } c,\ \Delta t \text{ units in seconds}\right) \tag{3.29}$$

Phase shifters have been implemented in several different ways including by switching transmission line lengths, changing the magnetic permeability μ_r in ferrite-filled waveguides, and switching different filters in MMIC phase shifters. Figure 3.50 presents drawings of the major classes of phase-shifter architectures. Hord [35] contributed to the development of ferrite phase shifters; Koul and Bhat [36] presented one of the best summaries of phase shifter technologies; and Harris and Sturdivant [37] presented transmit/receive (T/R) modules and their component technologies.

array_phsteer.m calculates the 1-D directivity pattern for a phased-array antenna, including the effect of the simulated element pattern discussed in Section 3.12.1. The key input parameters in this model are the interelement spacing; number of elements, N; and the desired angle of the main beam, thetasteer (θ_{steer}). The calculation for a single-dimensional steerable pattern for a linear array is presented in (3.30).

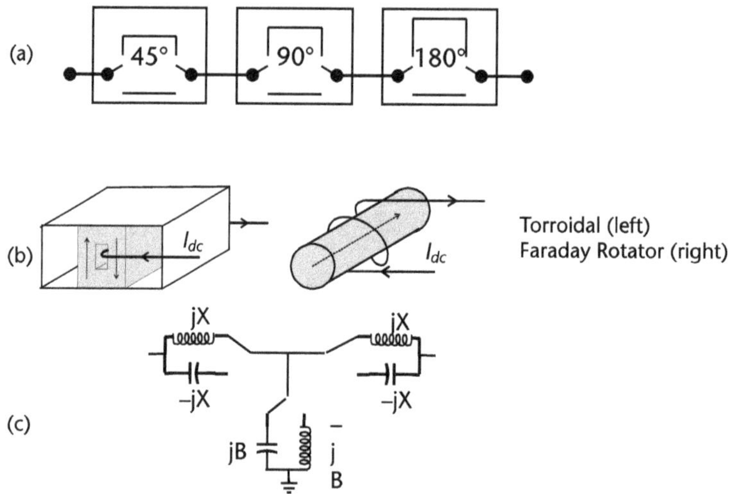

Figure 3.50 Major categories of array phase shifter architectures including (a) switched line length, (b) ferrite, and (c) switched filter.

$$E(\theta) = E_{\text{element}}(\theta)\sum_{n=1}^{N} e^{jkn\Delta x(\sin\theta - \sin\theta_{\text{steer}})} \tag{3.30}$$

As Re$\{ej^{(\text{argument})}\}$ is maximum when argument equals zero, then the pattern is maximum when $(\sin\theta - \sin\theta_{\text{steer}}) = 0$. Therefore the pattern field total is only maximum when $\theta = \theta_{\text{steer}}$. If phase shifters or time-delay components are controlled to replicate the plane wave normal to θ_{steer} per (3.28) or (3.29) then the pattern main beam angle is electronically steerable.

Figure 3.51 presents calculated patterns for a 25-element linear array with the main beam steered to 0° and 45° in azimuth. This model was run using a $\cos(\theta)$ element pattern. The scan loss and beam broadening caused by the element pattern effect is exhibited by the scanned pattern.

3.13 Multibeam Arrays

All RF systems benefit from increased gain (directivity) per either the Friis transmission or the RREs. With higher gain comes narrower main beams and the need to steer the beams. Some applications, such as electronic surveillance or RWR, however, have a requirement for constant surveillance. Steering a narrow beam in these applications is akin to tempting Murphy's law, given the risk that a threat may illuminate your platform during a time when the high-gain beam is steered away from the threat. Array systems that have multiple permanent main beams have been developed to meet the needs of both high-gain and constant surveillance.

To achieve multiple simultaneous beams, a multibeam beamformer is needed. This concept is shown in the block diagram in Figure 3.52. Numerous authors, including Rotman and Turner [38], Butler and Lowe [39], and Blass, developed approaches during the 1960s to provide for an efficient multibeam simultaneous

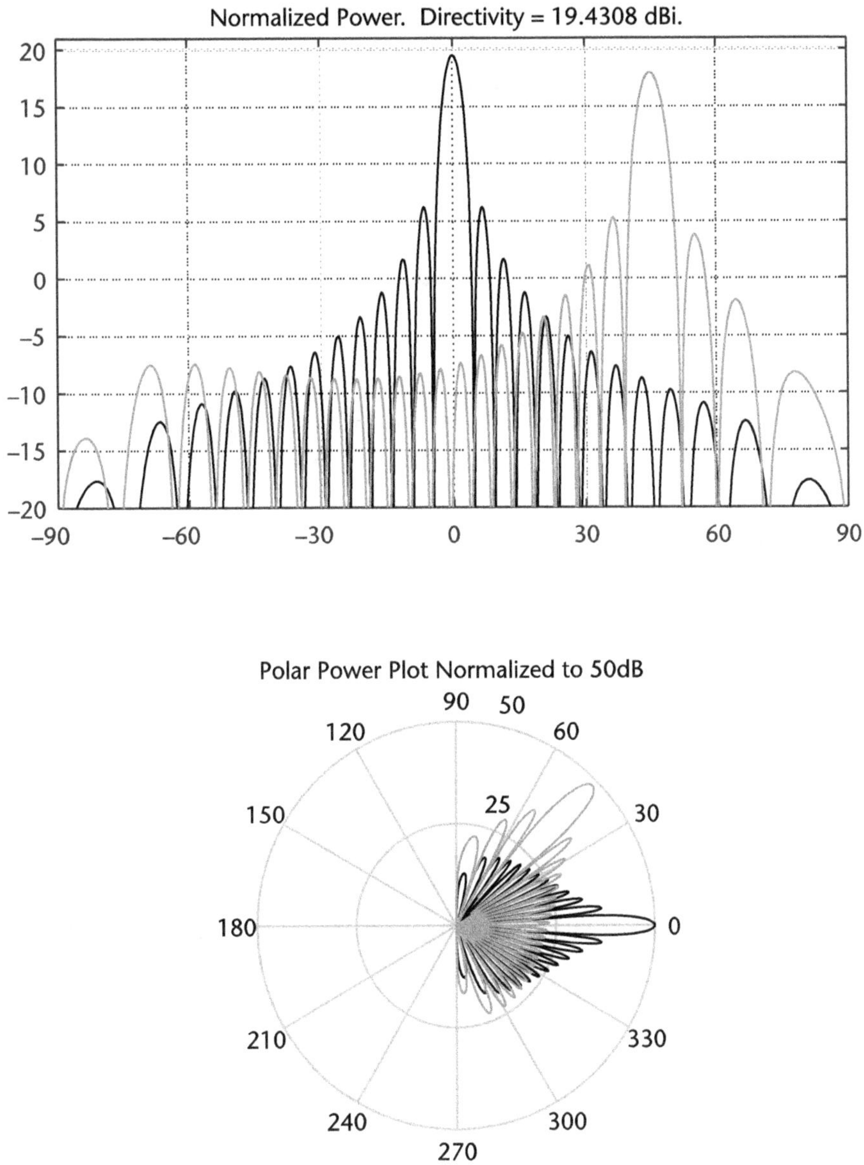

Figure 3.51 Plots of the directivity pattern calculated by ***array_phsteer.m*** for a linear array of 25 elements with the main beam scanned to 45°.

beamformer. Of these, the first two have seen significant use in real-world systems. Both, however, are much better suited to linear-array applications than to 2-D arrays. One such application of the Rotman-Turner lens was in the U.S. Navy's electronic warfare system, the AN/SLQ-32 V(1) – V(6). This system provides multiple simultaneous high-gain beams in the azimuth plane. Another difference between the Rotman-Turner and Butler-Lowe beamformers is that the Rotman-Turner can offer an odd or even number of beams, while the Butler-Lowe can only provide even numbers of beams that are a power of 2. Figure 3.53 shows an example of a Rotman-Turner lens fabricated in printed circuit form. This lens provides eight

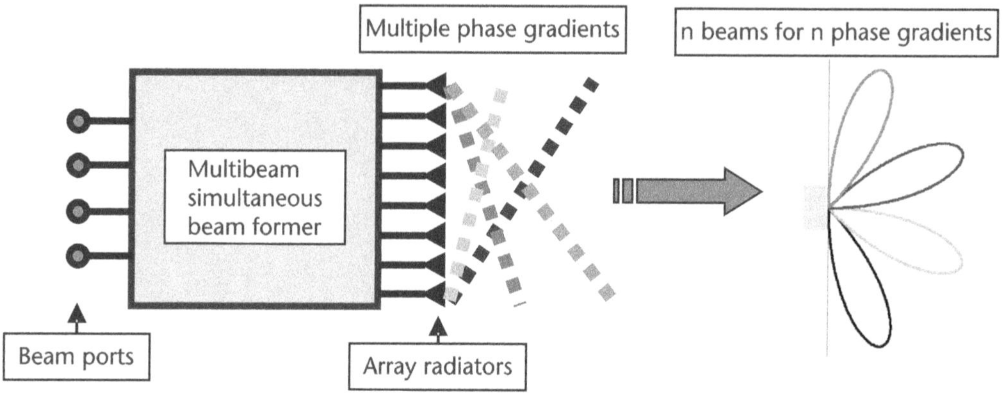

Figure 3.52 Block diagram of a multibeam beamformer and its multiple simultaneous steered beams.

beam ports for an eight-element array over a frequency band of 6–18 GHz. Plots of Figure 3.53 show the HFSS-predicted E-field magnitude distributions illustrating focusing of the lens on the beam-port side for multiple beams. Figure 3.54 presents the resulting multiple beam gain patterns with a comparison between measured data and modeled data from **multibeam_array.m. multi-beam_array.m** simply calculates eight different patterns, in this case steered to the same angles as the design of the Rotman-Turner lens of Figure 3.53.

3.14 Fourier Transform Models for 2-D Planar Phased Arrays

Almost all of the equations and array pattern discussions until this point in this chapter have been for line arrays, and the patterns were only calculated for one axis. The purpose of that was to simplify the math and to focus on the phenomenology being discussed. Now we transition to simulating planar arrays and pattern calculations in two dimensions, as these are very commonly used in radar and EW systems. Figure 3.55 presents a photograph of a state-of-the-art 2-D air defense AESA radar: the AN/SPY-6 developed by Raytheon Technologies for the U.S. Navy.

Figure 3.56 presents the geometry of a planar 2-D array. The array has interelement spacings of Δx and Δy as well as number of rows, n, and number of columns, m. The elements are all in the x-y plane and the z-direction is normal to the plane of the array. The E-field pattern calculation and the beam-steering angles are also now 2-D. The pattern calculation is shown in (3.31) in (θ, ϕ) coordinates and in (3.32) in (az, el) coordinates. The angle subscript s denotes the desired main beam-steering angle for each pattern.

$$E(\theta,\varphi) = E_{\text{element}} \sum_{M} \sum_{N} a_{mn} e^{j\frac{2\pi}{\lambda}\left\{m\Delta x\left(\sin\theta\cos\varphi - \sin\theta_s\cos\varphi_s\right) + n\Delta y\left(\sin\theta\sin\varphi - \sin\theta_s\sin\varphi_s\right)\right\}} \tag{3.31}$$

$$E(az,el) = E_{\text{element}} \sum_{M} \sum_{N} a_{mn} e^{j\frac{2\pi}{\lambda}\left\{m\Delta x\left(\sin az\cos el - \sin az_s\cos el_s\right) + n\Delta y\left(\sin el - \sin el_s\right)\right\}} \tag{3.32}$$

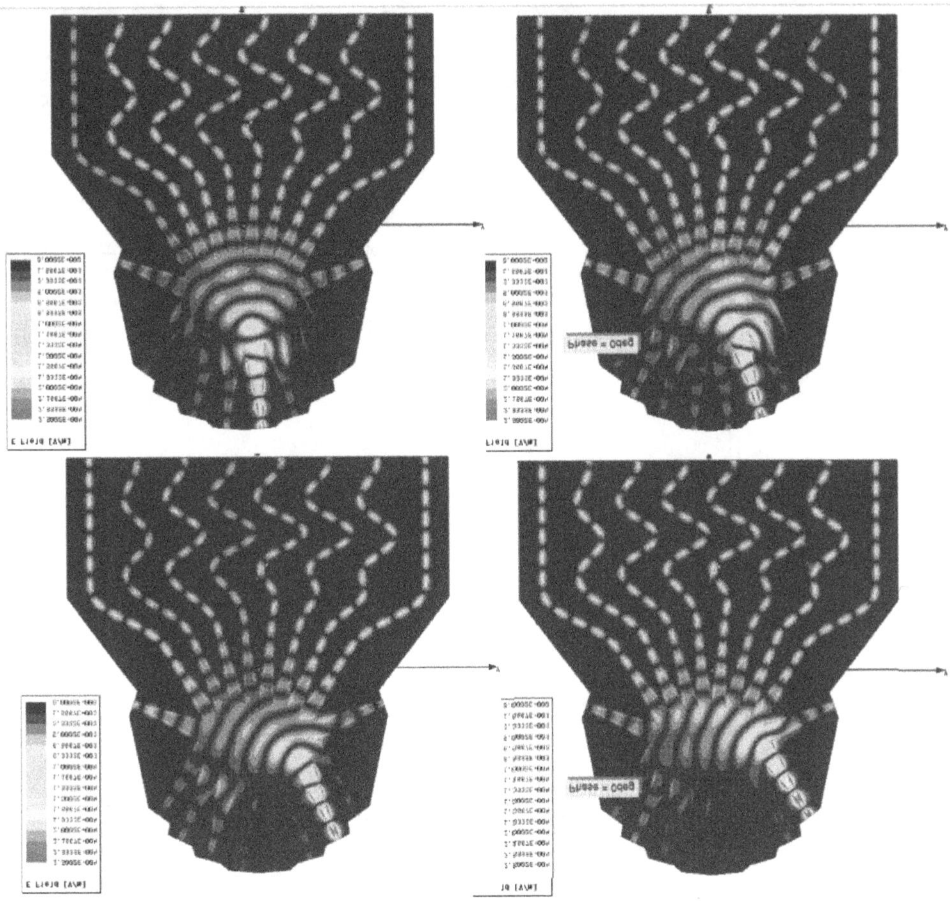

(a)

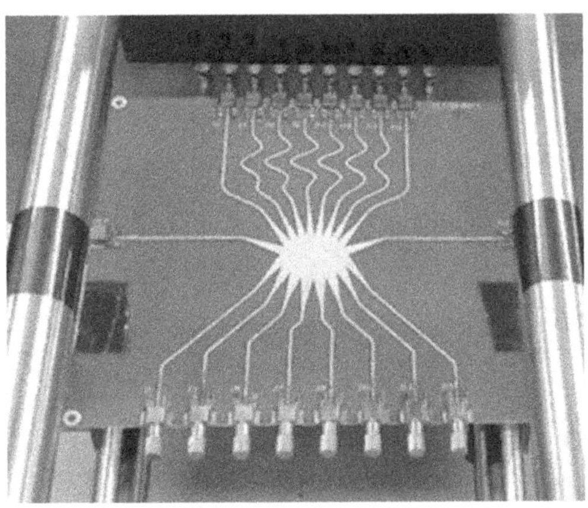

(b)

Figure 3.53 (a) Electric field plots for an 8 × 8 Rotman-Turner lens calculated using HFSS, and (b) a photograph of the printed circuit lens hardware. (Images courtesy of GTRI.)

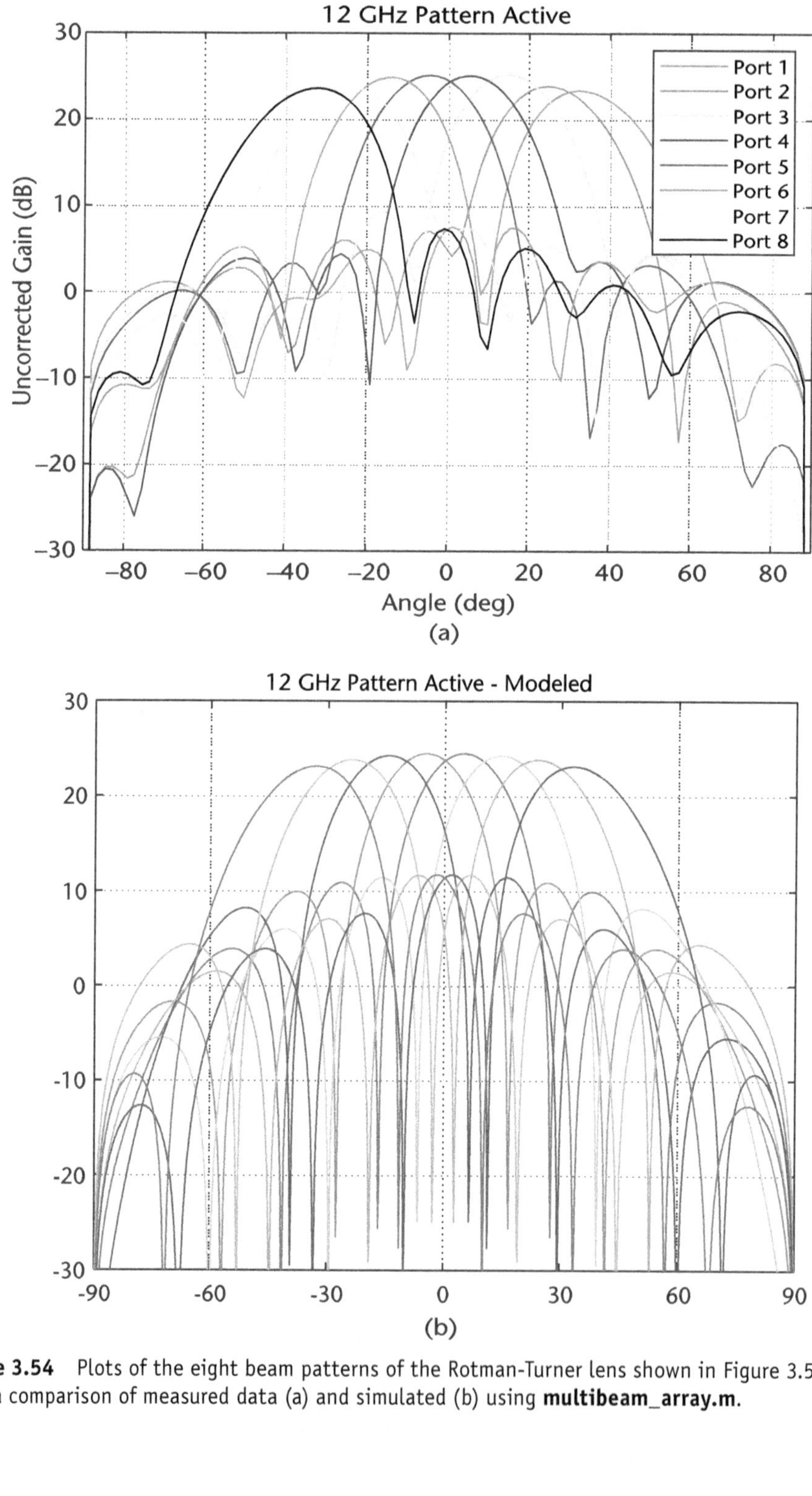

Figure 3.54 Plots of the eight beam patterns of the Rotman-Turner lens shown in Figure 3.53 with a comparison of measured data (a) and simulated (b) using **multibeam_array.m.**

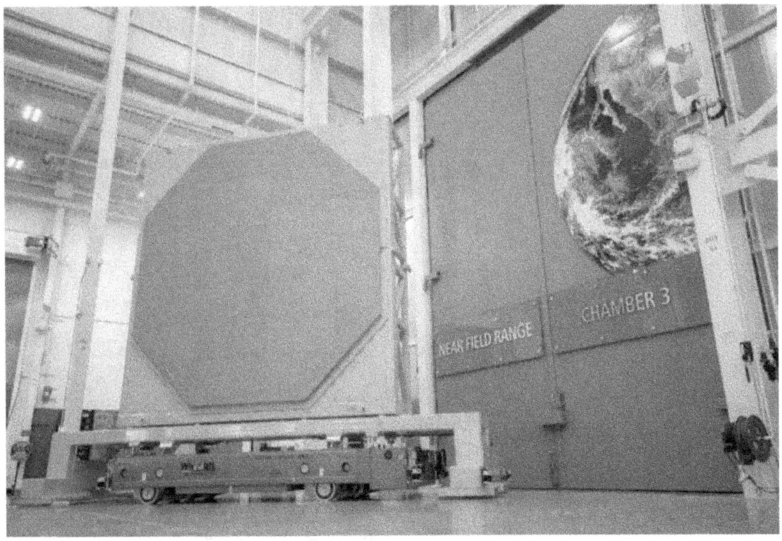

Figure 3.55 The U.S. Navy's SPY-6 integrated air defense missile defense radar being manufactured at the Raytheon Missiles & Defense radar development facility in Andover, Massachusetts. (Photograph courtesy of Raytheon Technologies.)

array_2D.m models 2-D patterns for 2-D planar arrays. It incorporates the previously presented details of the element pattern, amplitude weighting that is separable in both axes, and beam-steering in two dimensions. All of the pattern control variables are in the initial block of the function. Figure 3.57 presents pattern plots for a 10 × 10 element array with 0.5 × 0.5 wavelength interelement spacing with uniform amplitude weighting and the beam steered to boresight (0°, 0°) in

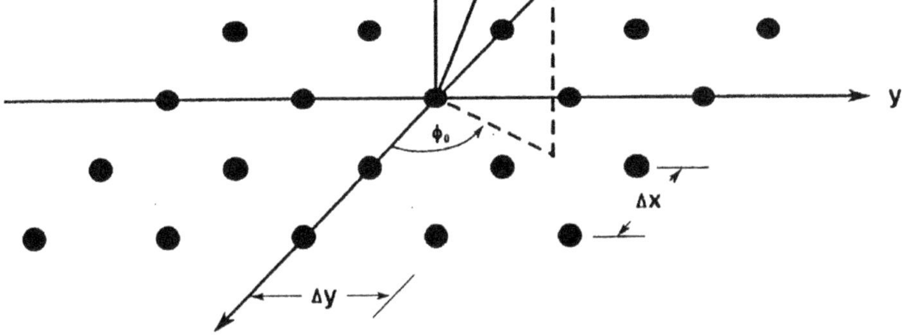

Figure 3.56 Geometry, spacing, and coordinate definition for a planar, 2-D array.

(a)

(b)

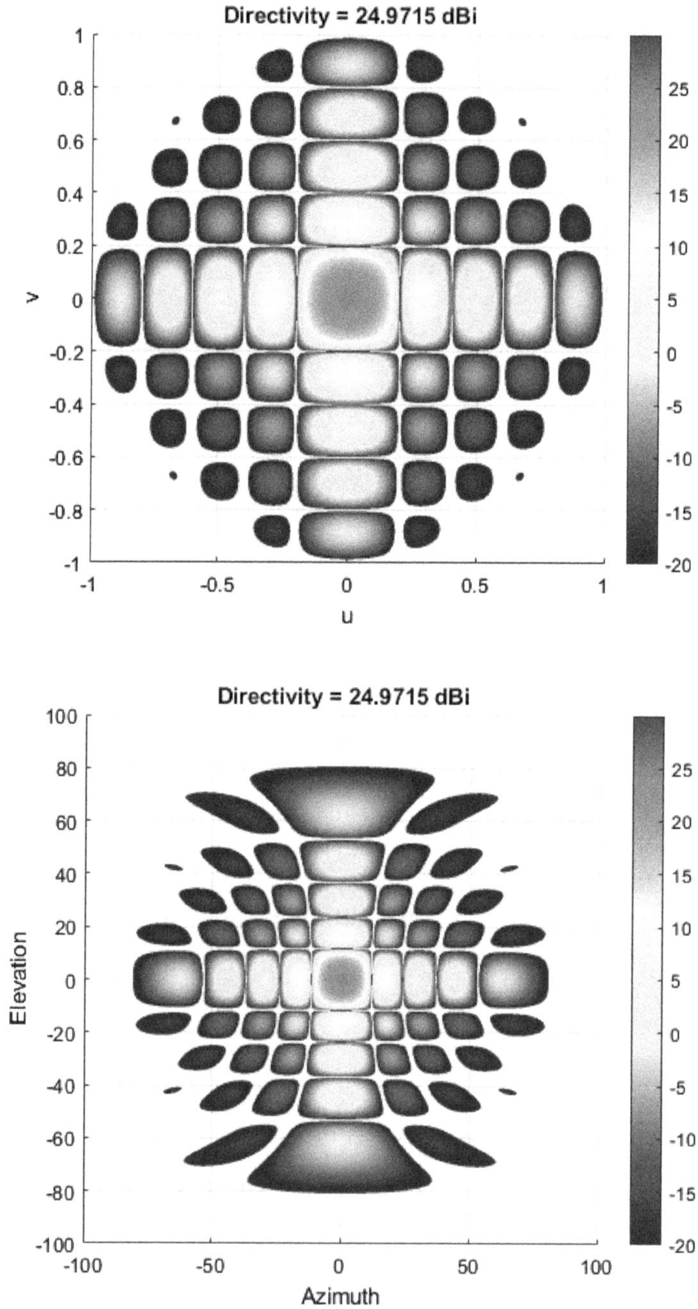

Figure 3.57 Plots of (a) the 2-D patterns from array_2D.m in sinespace, and (b) az-el coordinate systems.

both sinespace and az-el coordinates. The variable flag *uvorazel* determines which coordinate system is used. All of the 2-D plots can be rotated to show them in three dimensions. To do this, hover your computer mouse over the plot, and select the plot window control option that appears named Rotate 3-D. Figure 3.58 presents a rotated version of the planar plot of Figure 3.57 in sinespace coordinates.

Additional plots that open include a map of the element positions, as well as plots of the amplitude and phase across the aperture. These plots are presented in Figure 3.59. The uniform amplitude is plotted in units of power in decibels relative to uniform and the phase in units of degrees. The aperture efficiency, η in decibels, is calculated and included on the amplitude weighting plot. The amplitude and phase plots will become more useful as the extended function capabilities are exercised further. All three of the plots in Figure 3.59 are included as diagnostics for the user to be able to observe the array features being simulated.

To steer the main beam of the array, input the desired angles in either the pair of *usteer* and *vsteer* or in *azsteer* and *elsteer*, corresponding to the coordinate system selected for plotting. Figure 3.60 presents pattern plots for beam-steer angles where the beam is steered to (0.433, 0.866) in sinespace and (60°, 60°) in az-el coordinates. Figure 3.61 shows the phase across the array aperture used to steer the beams shown in Figure 3.60.

array_2D.m can also apply user-defined amplitude weightings, including uniform, Taylor, and Bayliss. Figure 3.62 presents pattern plots illustrating that the tapers can be implemented in either a single axis or both axes. This model uses two simultaneous linear tapers, one applied in each axis. Note the increased directivity loss with the amplitude tapered in both axes. The amplitude tapers can

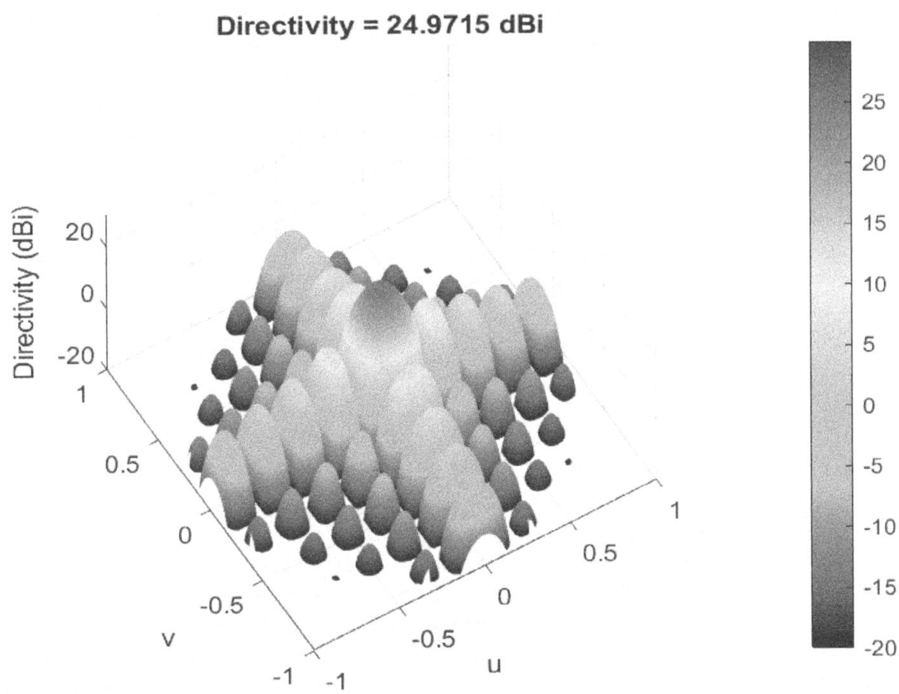

Figure 3.58 A rotated version of the sinespace plot of Figure 3.57.

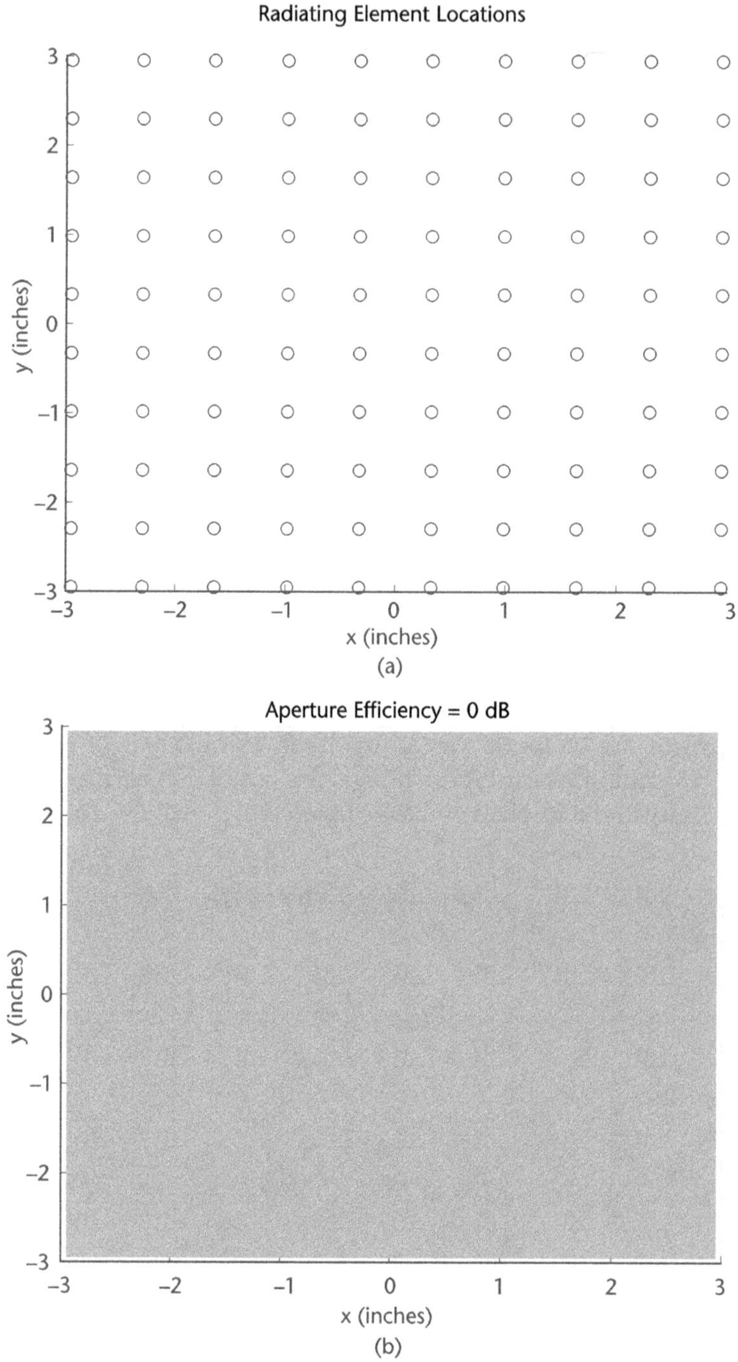

Figure 3.59 Plots of (a) the element lattice, (b) the uniform amplitude across the aperture, and (c) the flat phase across the aperture.

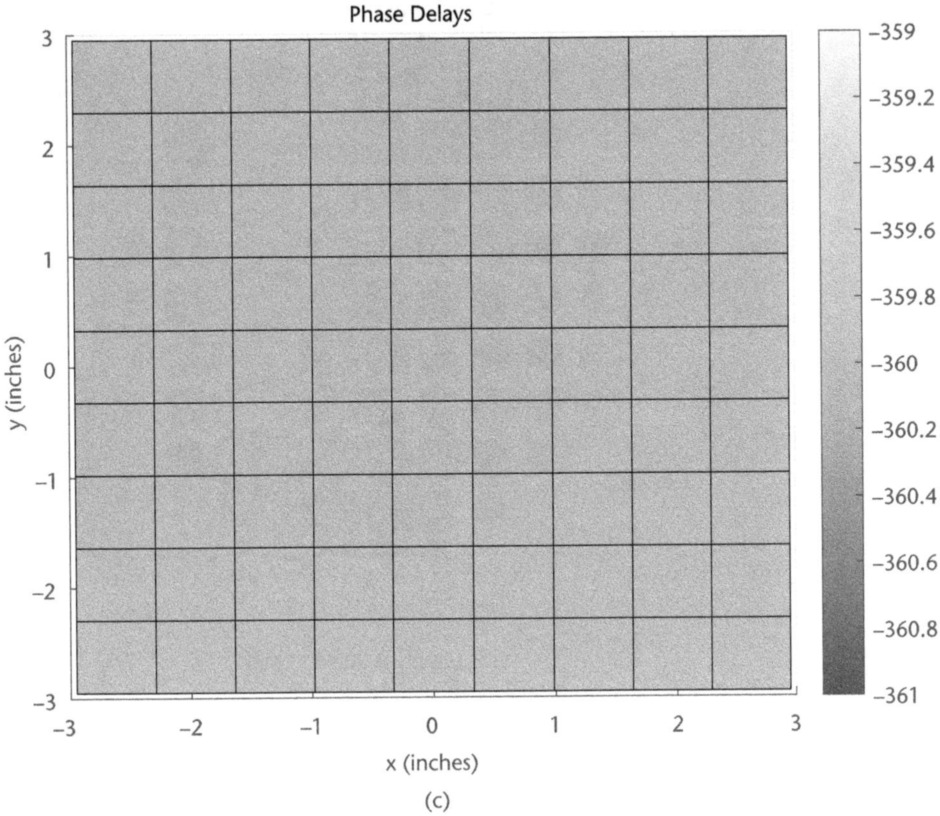

Figure 3.59 *(Continued)*

be visualized in both the array aperture amplitude plot and in the element lattice plot. The element lattice plot is 3-D as are all of the others. To see the amplitude taper in the element lattice plot, use the Rotate 3-D command in the plot window. Figure 3.63 shows the amplitude tapers from both plots for the Taylor SLL 30 dB and n-bar 4 in both axes. Figure 3.64 presents the amplitude distribution and patterns for a model that uses a Bayliss taper in the horizontal axis and a Taylor taper in the vertical axis. Note that the user input flag *renorm* assumes that the tapers are implemented using attenuation when set to zero and with renormalization when set to one.

array_2D.m also supports modeling of arrays with a circular shape. To simplify the model, the function always starts with a rectangular definition of the array elements. If the variable flag *rectorcircshape* is set to a value of 1, then the amplitudes of the elements that are outside of the ellipse defined by the maximum width and height of the array are set to zero. The elements in the corners are not removed, and although this is not the most efficient coding approach in terms of run time, it is useful to support flexibility of the function. The circular shape can be verified by looking at the amplitudes in the element lattice plot. Figure 3.65 presents a model of a 20 × 10–element square array converted to an elliptical array along with its directivity pattern.

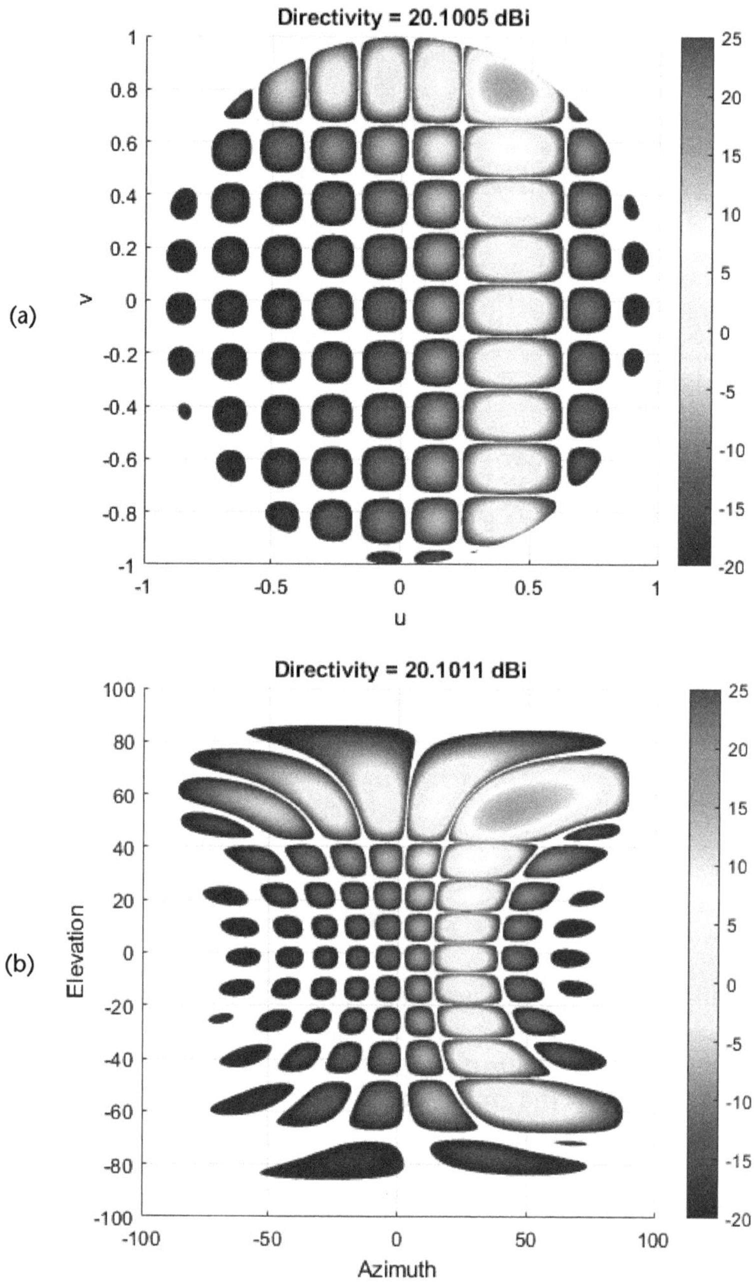

Figure 3.60 Plots of the same array with the main beam scanned to (a) (0.433, 0.866) in sinespace and (b) (60°, 60°) in az-el coordinates.

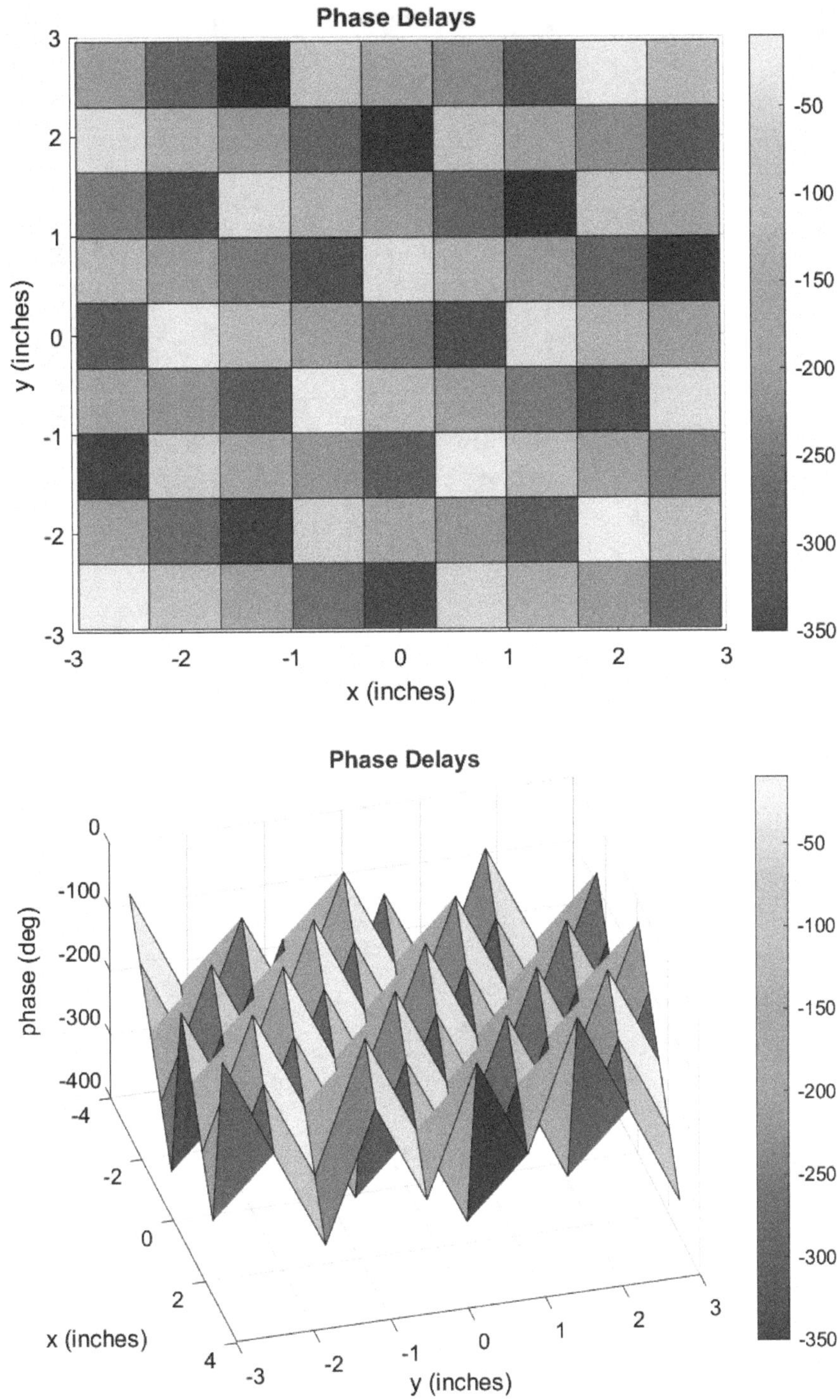

Figure 3.61 Plots of the ideal phases required to scan the beam to the positions shown in Figure 3.60. Note that these are modulo(2π).

(a)

(b)

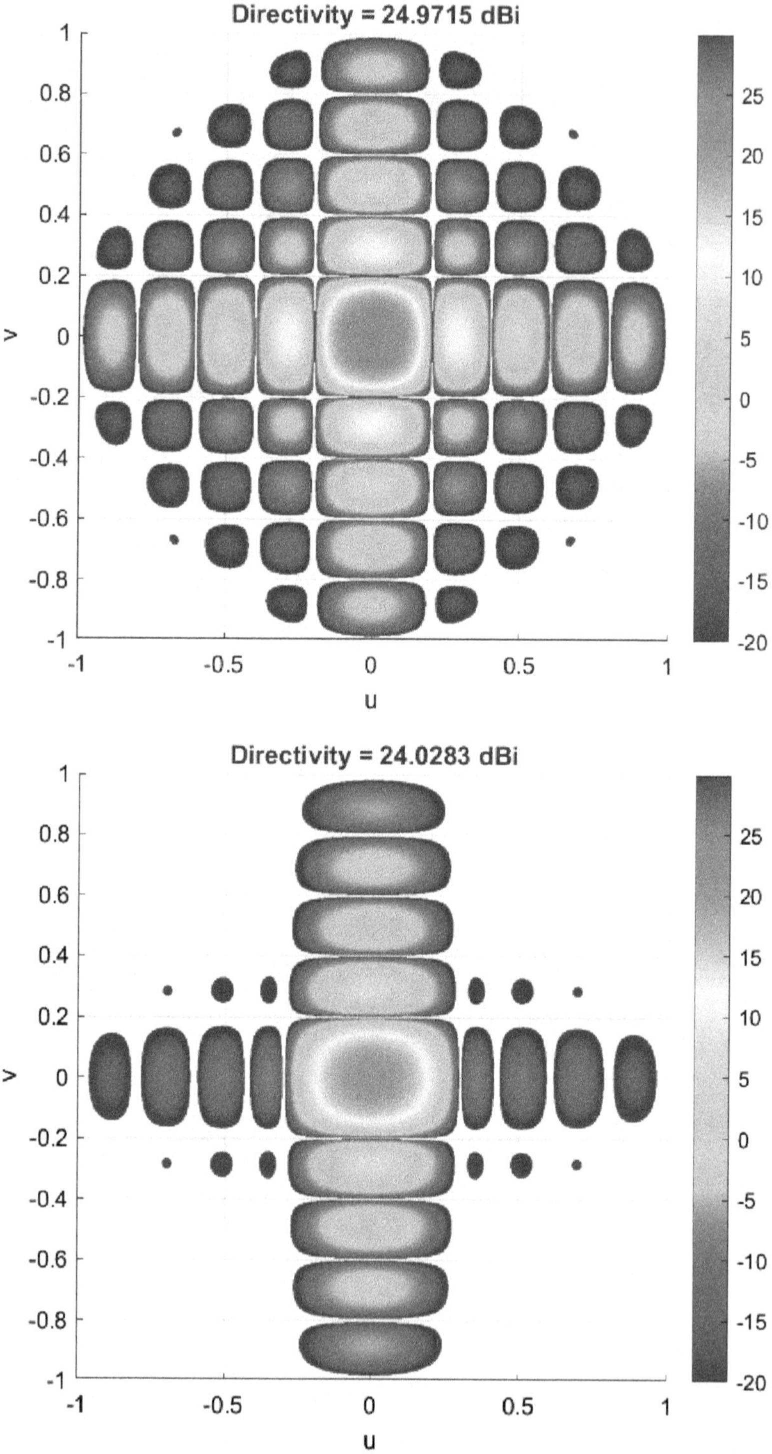

Figure 3.62 Plots illustrating the different tapers in each axis that the user can define in ***array_2D.m***. The patterns use amplitude tapers of (a) uniform in x- and y-, (b) Taylor in x- and uniform in y-, (c) uniform in x- and Taylor in y-, and (d) Taylor in both x- and y-directions.

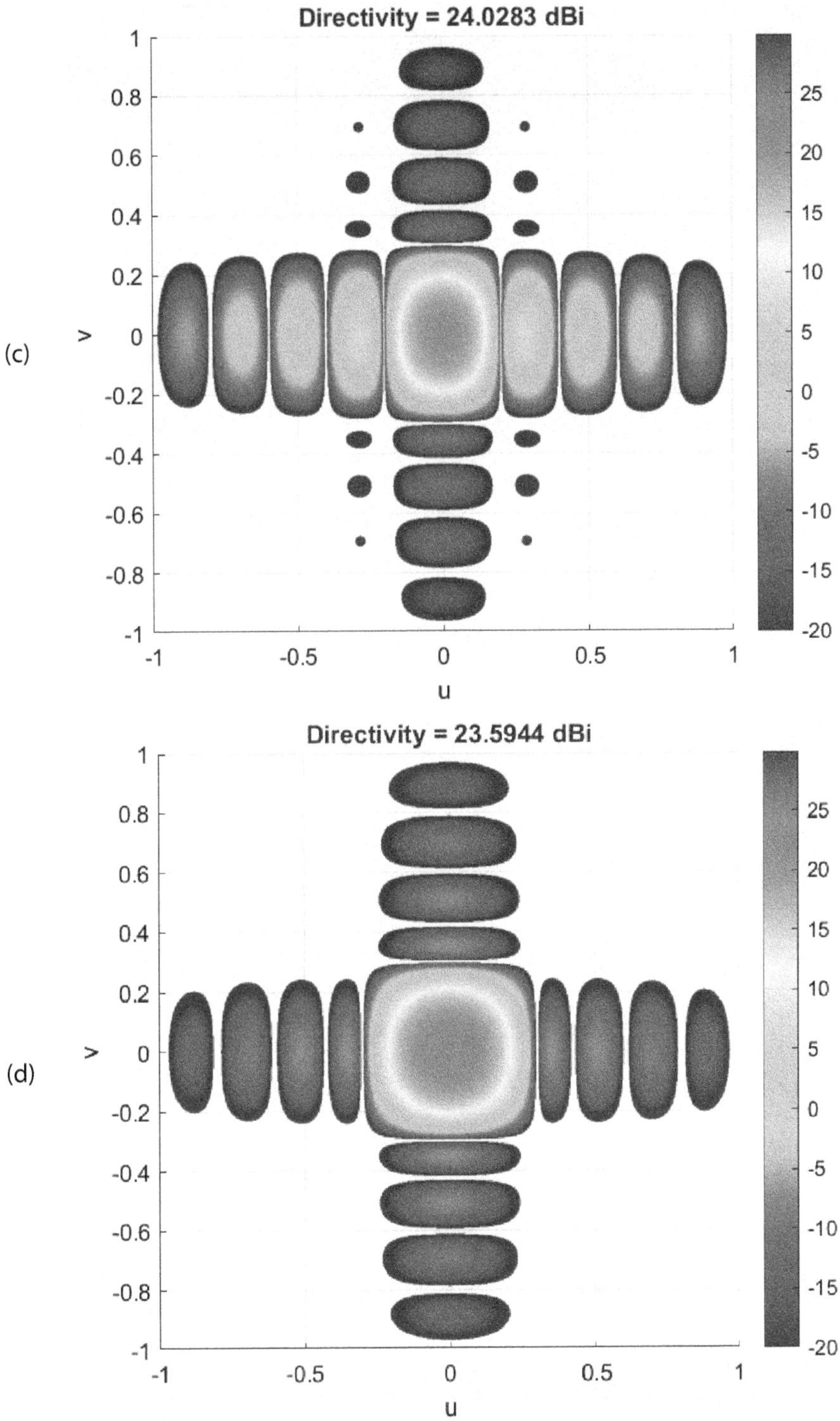

Figure 3.62 *(Continued)*

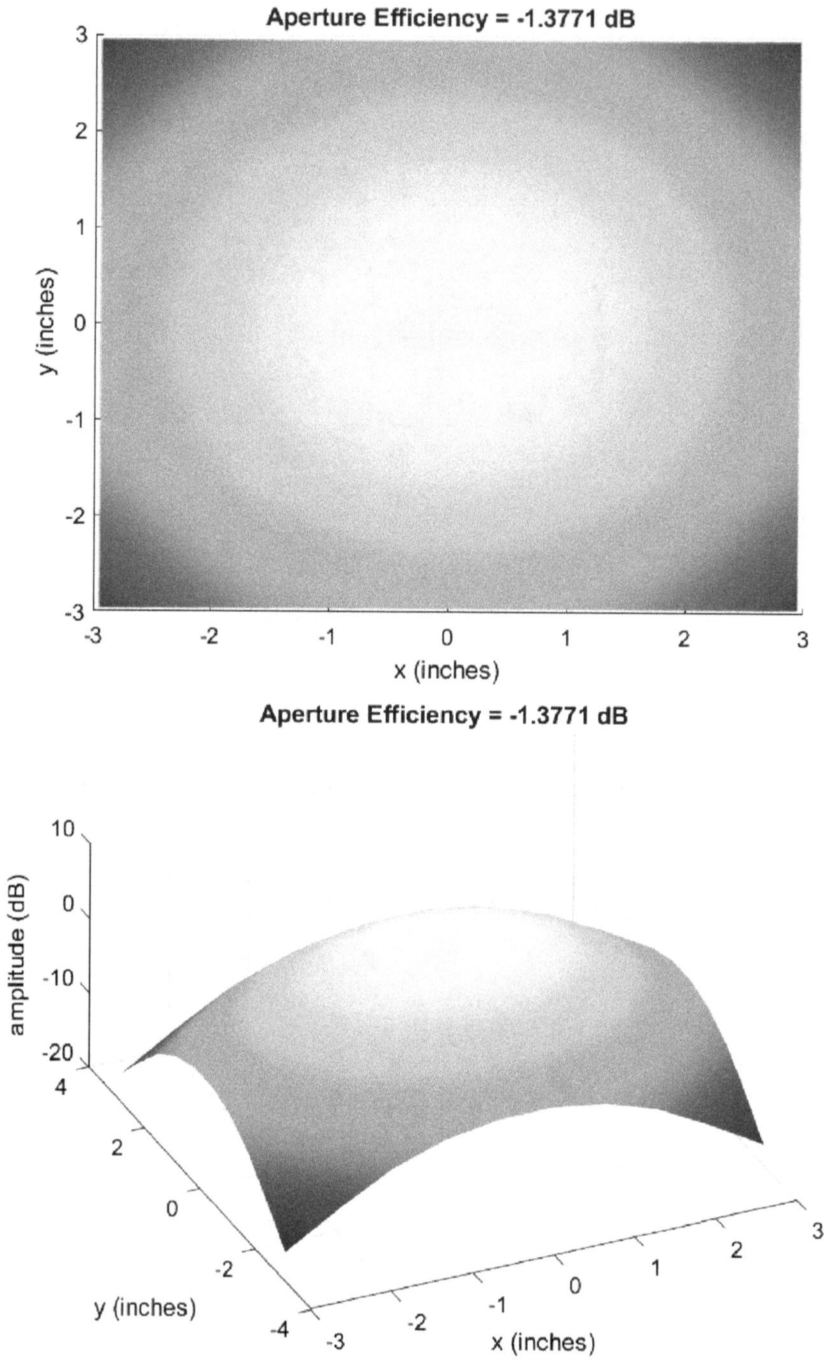

Figure 3.63 Plots showing the amplitude tapers in both element lattice and array amplitude plot windows from ***array_2D.m***. Note that the units of the amplitude as shown in the element lattice plot is in volts and that it is for the redistributed case, wherein the peak is > 1.

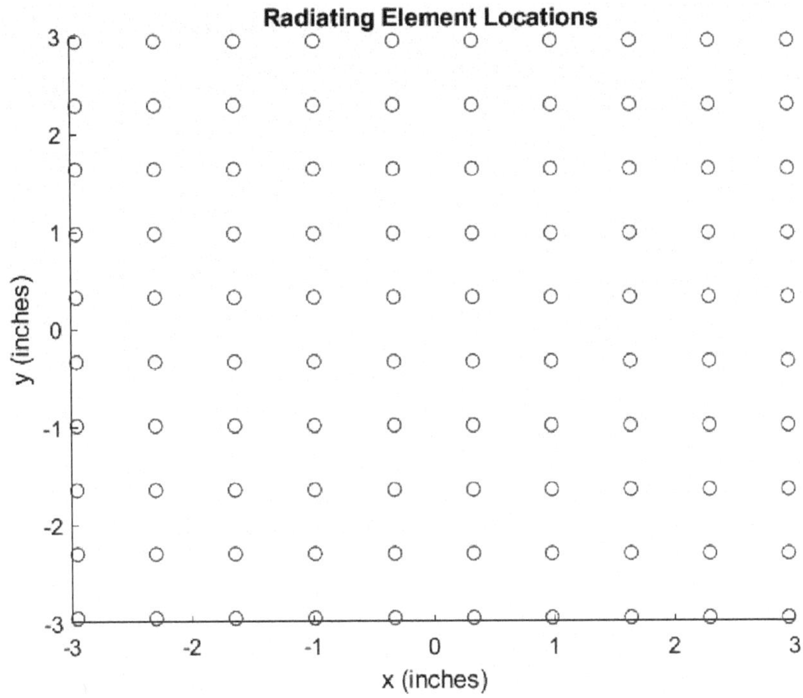

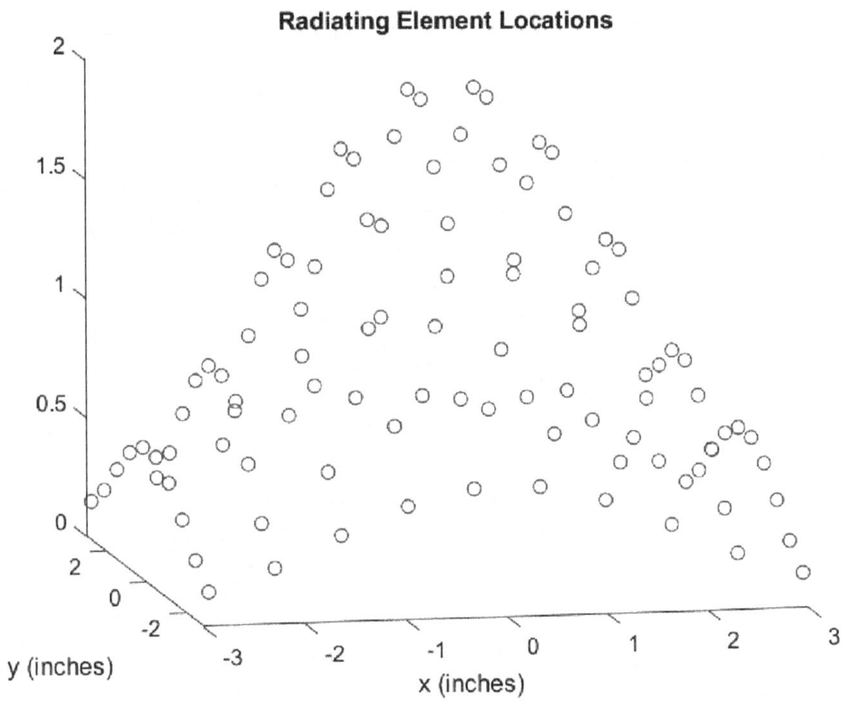

Figure 3.63 *(Continued)*

(a)

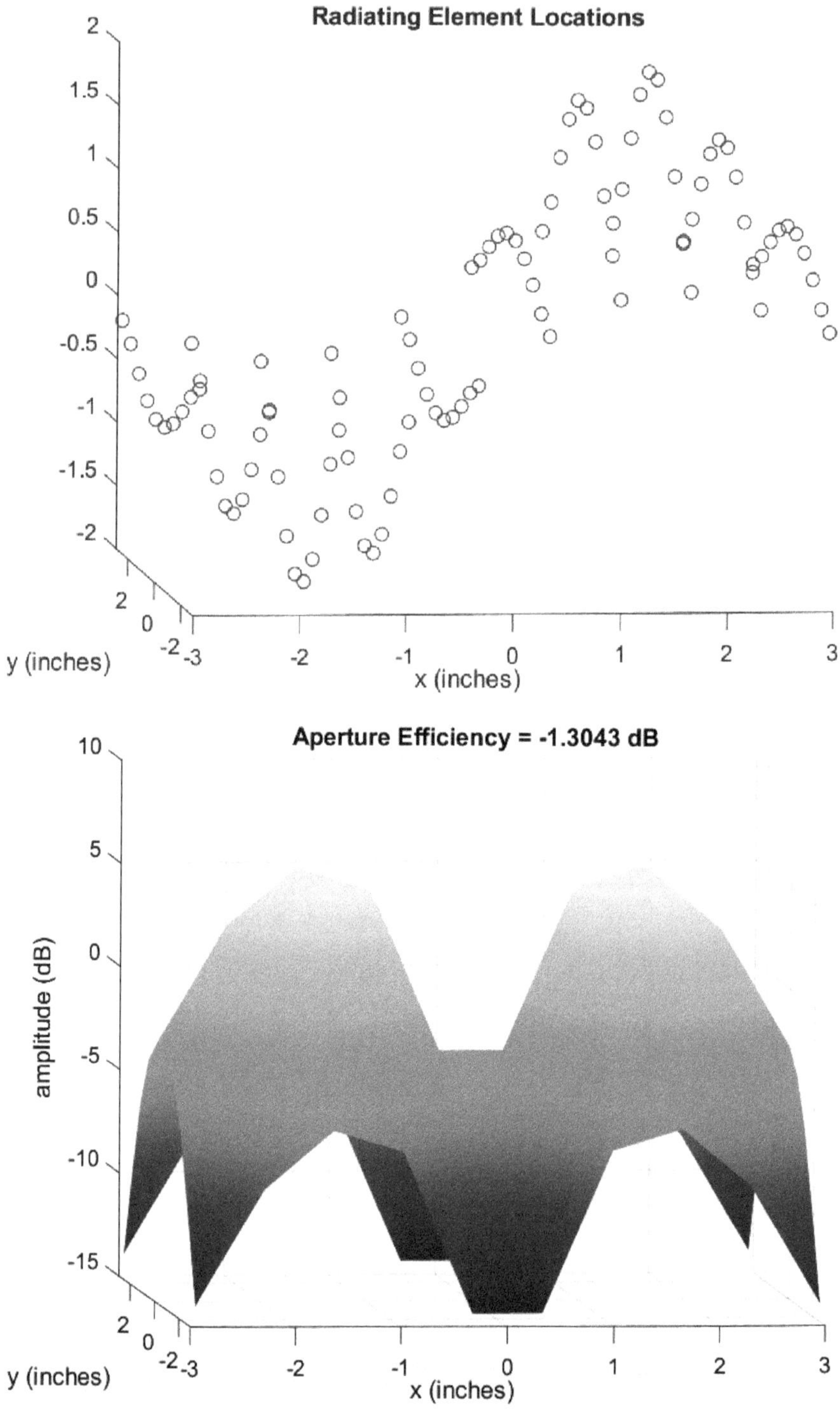

(b)

Figure 3.64 Plots of the amplitude tapers for a Taylor SLL 30 dB and n-bar 4 in the vertical axis and a Bayliss SLL 30 dB and n-bar 4 in the horizontal axis from ***array_2D.m***. The amplitude tapers as viewed from (a) the element lattice, and (b) from power magnitude in decibels. The directivity pattern is also shown in sinespace (c).

(c)

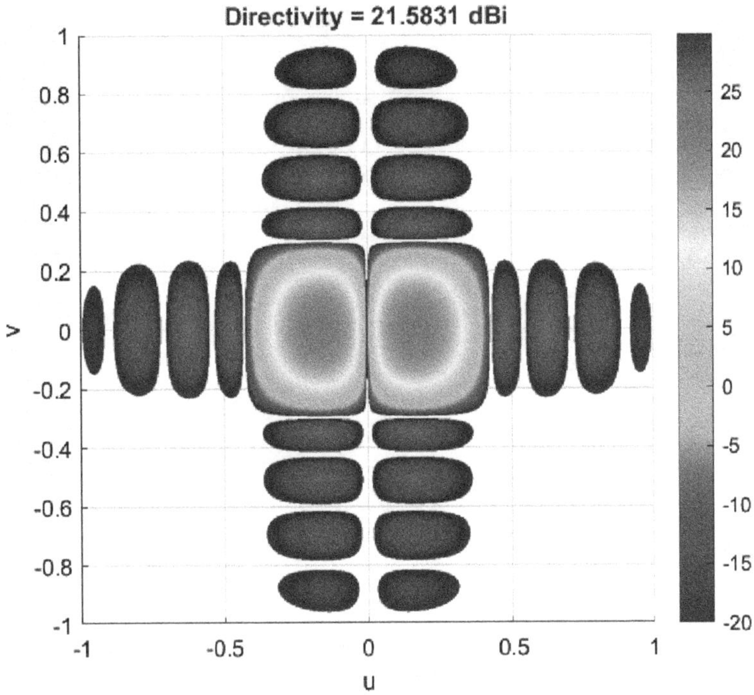

Figure 3.64 *(Continued)*

array_2D.m can also model element lattices arranged in hexagonal or triangular spacing. Note that this is achieved by starting with the elements in a rectangular arrangement and by displacing alternating columns. The variable flag *rectortrilattice* sets the function to simulate a triangular lattice when it is set to 1. Figure 3.66 presents models of a uniformly illuminated 1965-element circular aperture using the triangular lattice. This array started as a 50 × 50–element square array. Figure 3.66 shows the lattice from different perspectives as well as the pattern calculated in sinespace coordinates.

3.15 Modeling of Errors in Phased Arrays

Scanning-array antennas are constructed of multiple parallel RF circuit chains that can be quite complex with many constituent components. Every component in the array has the potential to introduce amplitude and phase errors. These errors can be either permanent or can vary with frequency, temperature, and time. If these errors are small, they do not significantly impact the main beam. If, however, the errors are large enough to impact the main beam, the negative effects result in peak gain loss and consequently beam-broadening, beam pointing angle error, and distortion of the main beam (i.e., misshapen with sidelobe shoulders). Errors, even if small, will modify the sidelobes and nulls of the directivity pattern, as the power levels in the sidelobes and nulls are significantly lower than that of the main beam. Negative effects from errors in the sidelobes include increased peak and

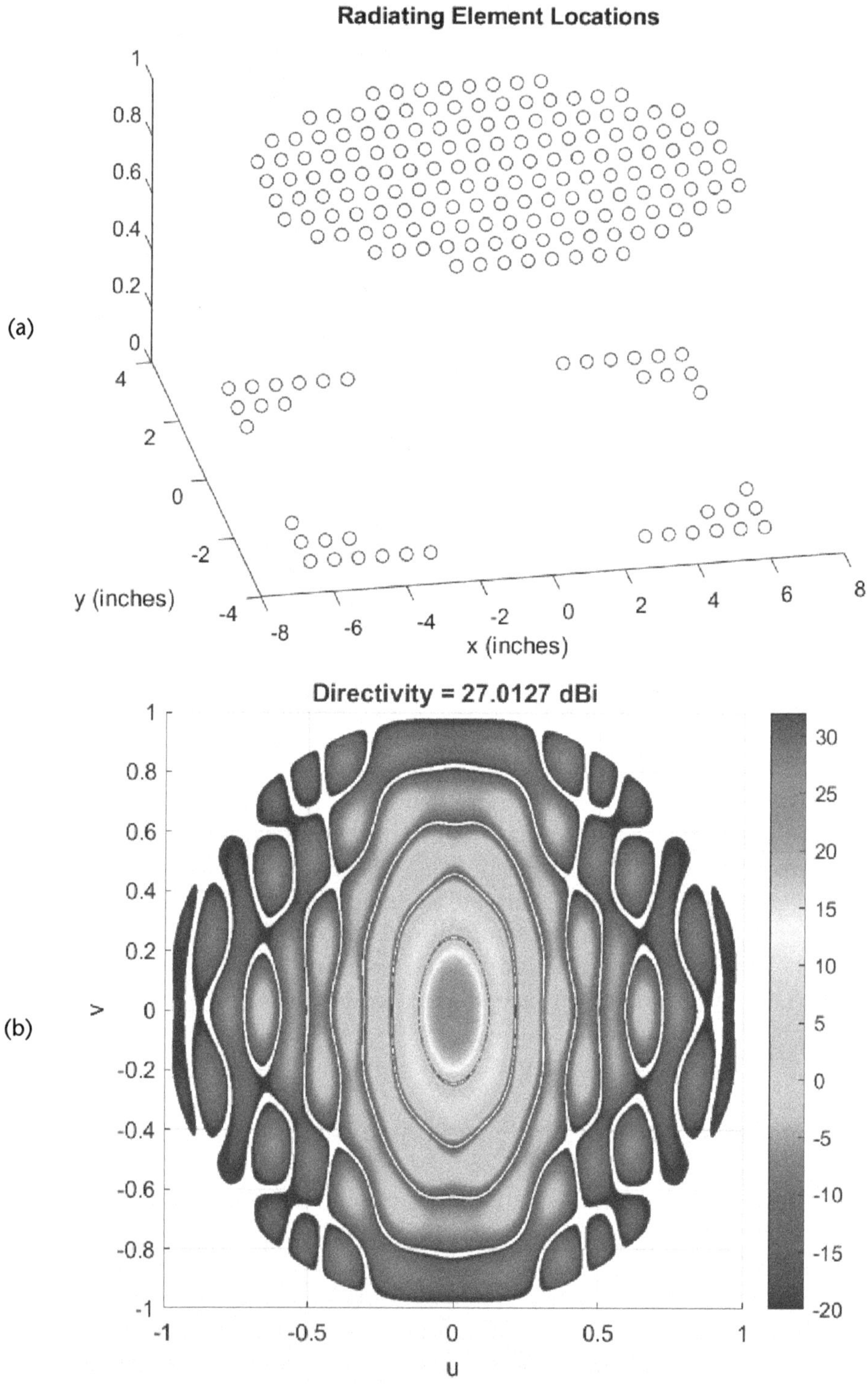

Figure 3.65 Plot of the element lattice and amplitudes of a uniformly illuminated elliptical array that started as a rectangular array of 20 × 10 elements (a) from ***array_2D.m*** and (b) the pattern calculated in sinespace coordinates.

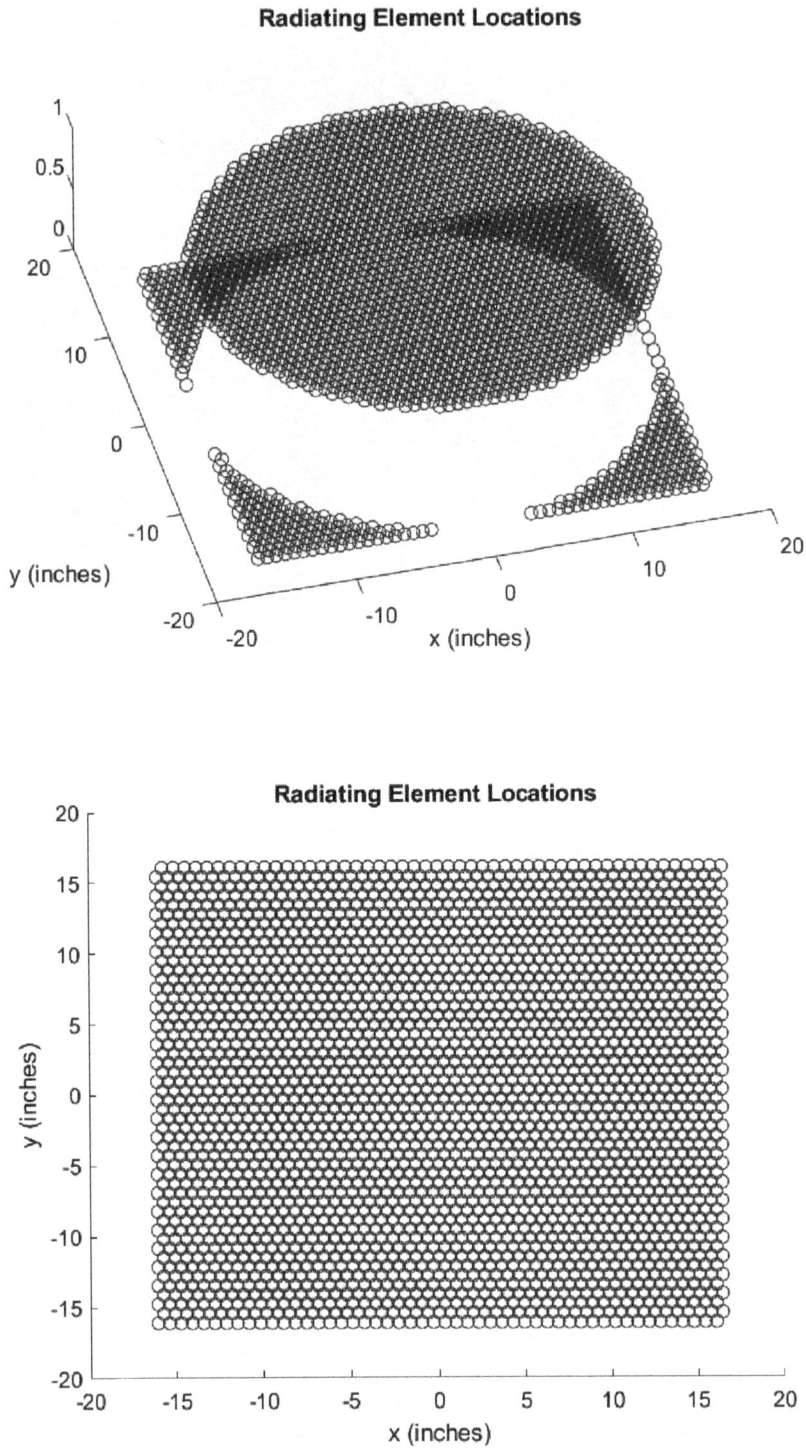

Figure 3.66 Plots of a uniformly illuminated circular array of 1965 elements using the triangular lattice from ***array_2D.m***.

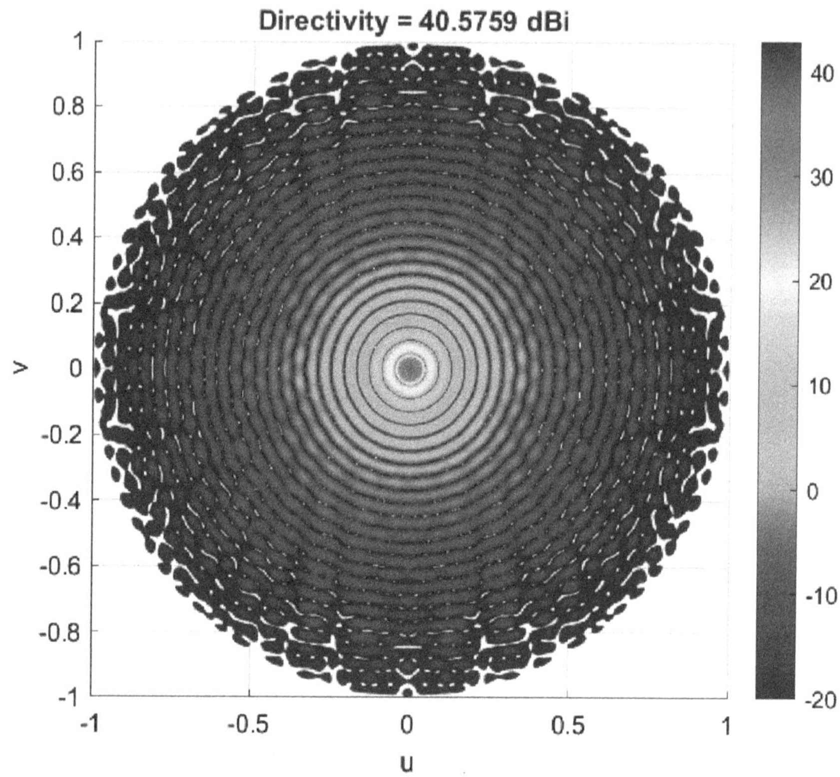

Figure 3.66 *(Continued)*

average SLLs and filling in of the pattern nulls. ***array_2D.m*** includes simulation capability of four aspects of errors commonly experienced in scanning arrays. These are presented in the following sections and include: quantization effects of the phase shifters and attenuators, random amplitude errors, random phase errors, and complete failures of active amplifiers.

3.15.1 Quantization of the Phase Shifter and Attenuator

In (3.28) and (3.29) the interelement phase values could be any value needed, as required to steer the beam to the desired angle in space. Those progressive values are then replicated across the array, modified by the modulo(2π) operation, resulting in phase maps similar to that shown in Figure 3.61. Almost all real-world phase shifters do not possess the capability of providing any arbitrary phase value between 0° and 360°. The reason is that control of the phase shifters is significantly simplified if fewer discrete phase values are used. Use of fewer discrete values is not of great detriment to the directivity pattern. Common control of phase shifters uses multiple binary bits to provide a quantized phase value. As shown in Figure 3.50(a), phase shifters usually offer a largest bit that provides a differential phase of 180°. Each subsequent bit provides half the phase resolution of the next larger bit. The drawing of the switched line-length phase shifter in Figure 3.50 would be referred to as a 3-bit phase shifter. Modern 6-bit phase shifters have a least-significant-bit resolution of 5.625°. The act of quantizing a linear phase ramp induces a stair-step phase error that is periodic across the array. The periodicity of this error causes coherent summing of these signals at different angles in the directivity pattern. These periodic error lobes are called quantization lobes.

Figure 3.67 presents the output from the model ***array_phsteer_quant.m***. In it, we revert back to a single-dimensional array and pattern. The model is of a uniformly illuminated 25-element linear array that utilizes ideal and 3-bit quantized phase shifters. The plot in Figure 3.67(a) shows the progression of the phase calculations. The long straight line is the ideal phase for steering the beam to 45°. The small circles represent the ideal phases after performing the modulo(2π) operation. The plus symbols are the phases selected from the eight possible phase states permitted by the 3-bit phase shifter. These are the quantized states. Given the vertical axis scale of thousands of degrees, none of the quantized values appear too different from the ideal values. The overlay of the two patterns of Figures 3.67(b, c) clearly shows the difference in the performance between ideal and quantized phase states. The main beam for both is virtually identical. The periodic higher sidelobes, the quantization lobes, are caused by the errors induced by quantization in this case by the 3-bit phase shifter.

The impact of the quantization lobes is dependent upon both the quantization number of bits as well as the number of elements, N, in the array. The larger N is, the lower the RMS SLLs are, including the quantization lobes. Miller [40] presented the approximations for the peak quantization lobe level with respect to the main beam (3.33), and the average SLL, (3.34). The variable *n*bits equals the integer number

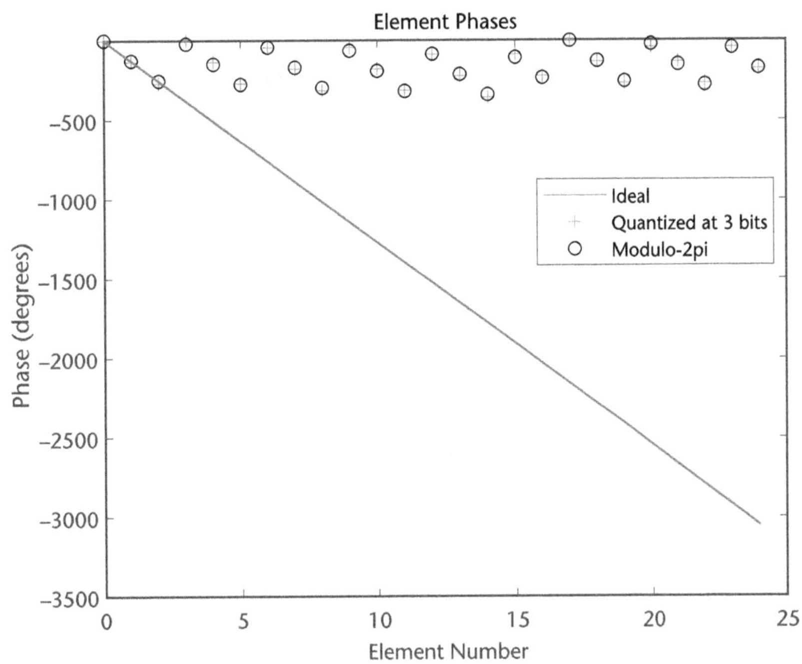

(a)

Figure 3.67 Plots from ***array_phsteer_quant.m*** for a linear array model of 25 elements. (a) shows the calculated phases including ideal, after modulo-2π, and after quantization; (b) and (c) present the directivity patterns for the ideal and quantized phase shifters in rectangular and polar formats.

(b)

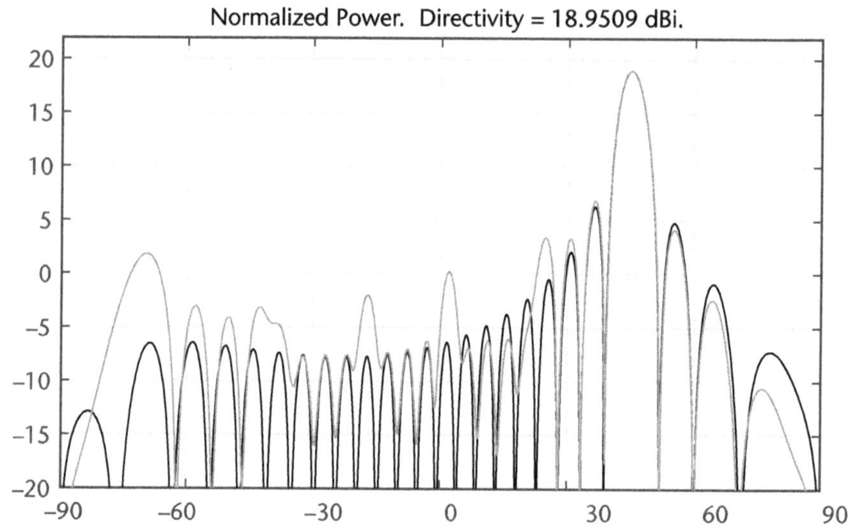

(c)

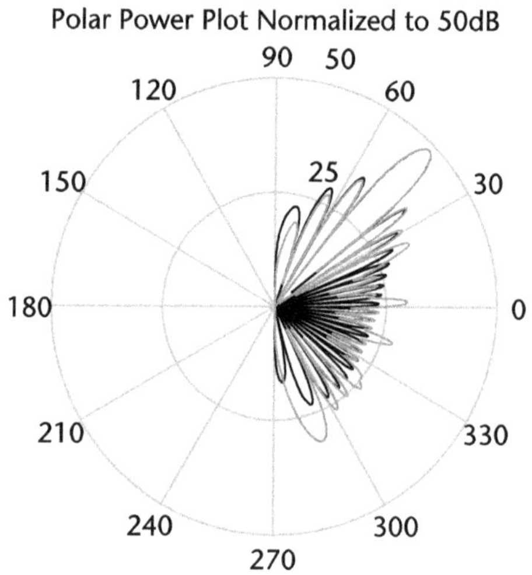

Figure 3.67 *(Continued)*

of bits for the phase shifter. Figure 3.68 presents a graph of the dependency of the RMS average SLLs (in decibels relative to the peak level of the main beam) versus the number of elements, N, in the array for different quantization number of bits, $n$bits. As can be seen, the larger the array is, the less the impact of any quantization lobes, even with relatively coarse phase shifter quantization.

$$\text{Peak Quantization Lobe (dBmb)} = -6n\text{bits} \tag{3.33}$$

$$\text{Average Sidelobe Level} = \frac{\pi^2}{3N2^{2n\text{bits}}} \tag{3.34}$$

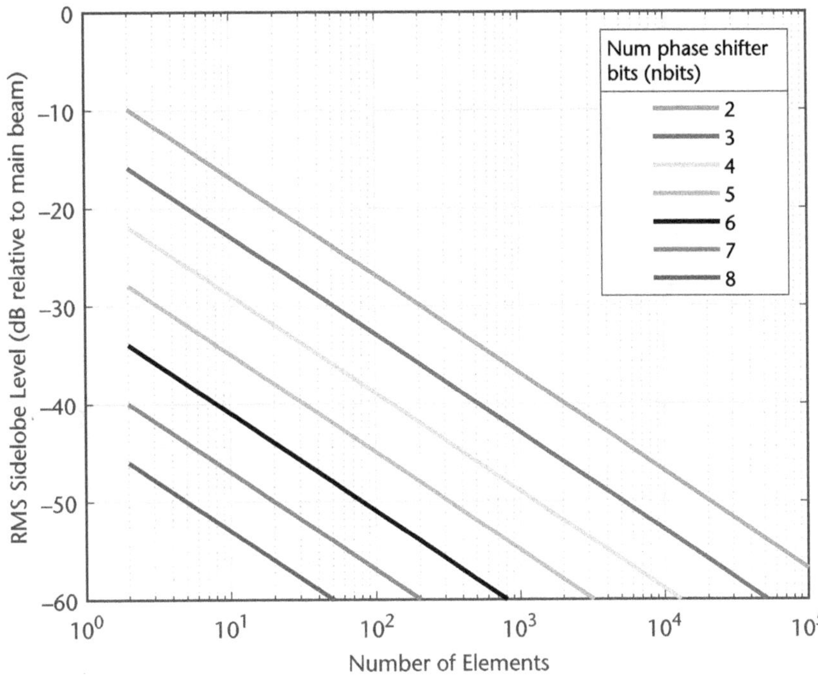

Figure 3.68 A plot of the dependency of the RMS average SLLs (in decibels relative to the peak level of the main beam) versus the number of elements, *N*, in the array for different quantization number of bits, *n*bits.

array_2D.m also provides quantization modeling for 2-D arrays. As shown in Figure 3.69, the quantization steps in the aperture phase can be more clearly seen when the beam is only slightly scanned. The phases and patterns of Figure 3.69 are for a 25 × 25–element uniformly illuminated array, scanned to 2° in azimuth using 3-bit phase shifters. In ***array_2D.m,*** the variable that controls the number of phase shifter bits is called *np*bits. If *np*bits is set to zero, the function will use the ideal phase values without quantization.

Quantization issues can also impact performance of switched attenuators in scanning arrays. As with phase shifters, it is impractical to build and control attenuators to provide any arbitrary attenuation value. Unlike the 180° bit in phase shifters, the most significant bit in attenuators must be specified by the component manufacturer. In ***array_2D.m,*** two variables are used as inputs to describe the attenuator, *na*bits, the number of attenuator bits, and *aLSB*, the value in negative decibels of the least significant bit in the attenuator. It is assumed that the increasing attenuator bits each have twice the attenuation of the previous bit. Note that the value for *aLSB* is specified in negative decibels, which is how the components are usually quoted by vendors. This value is converted to linear voltage units for use in the function.

Figure 3.70 presents an example simulation using ***array_2D.m*** where a 35 × 35–element array is scanned up to 25° in elevation. It used 3-bit phase shifters, which are what caused the larger than expected high sidelobes on the elevation axis only (because it was scanned in elevation only). It also used 4-bit attenuators to achieve an amplitude taper without the benefit of redistribution. Note the high

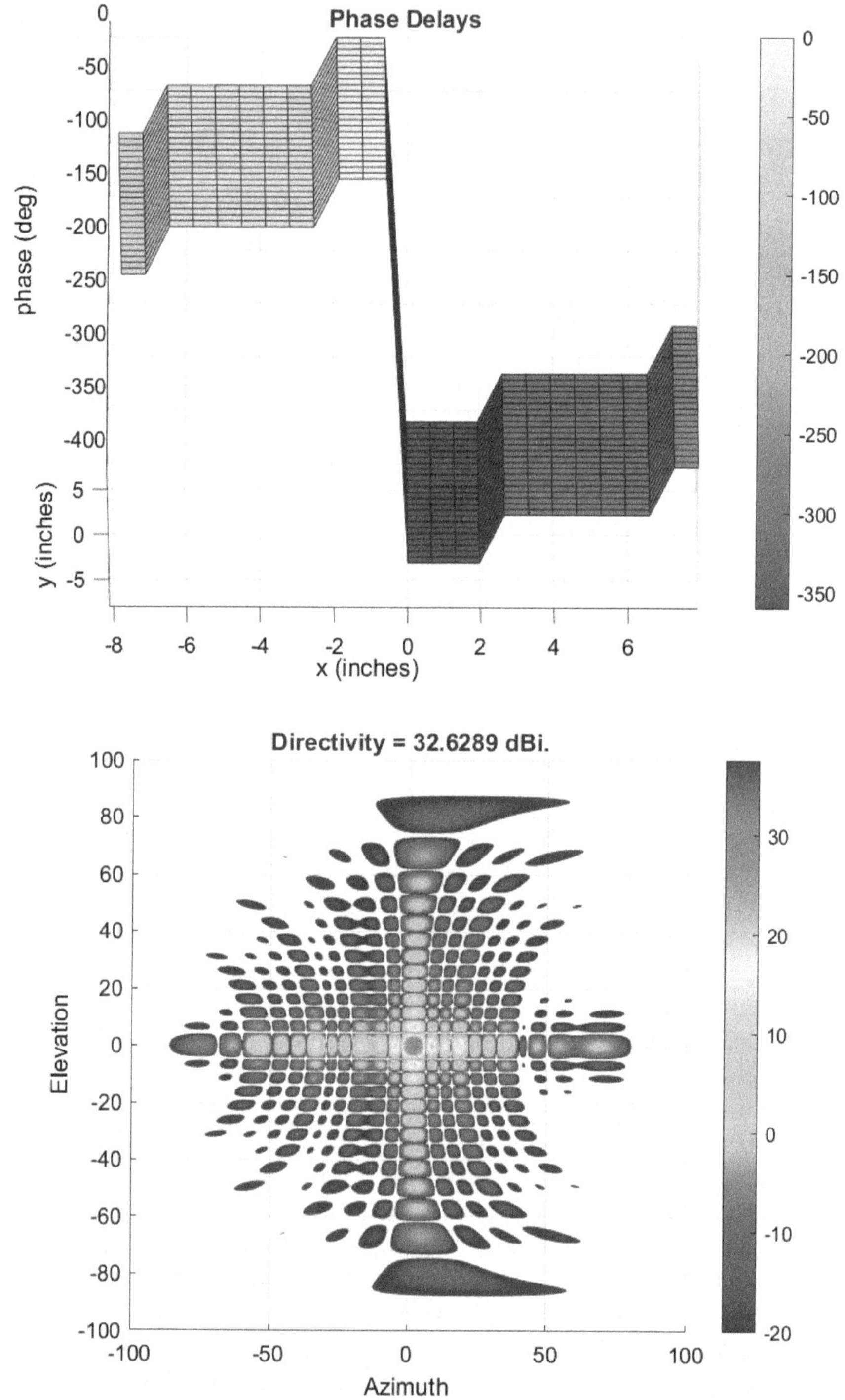

Figure 3.69 An example of phase quantization from ***array_2D.m*** for a 25 × 25–element array with phase shifter quantization at 3 bits and the main beam scanned to 2° in azimuth.

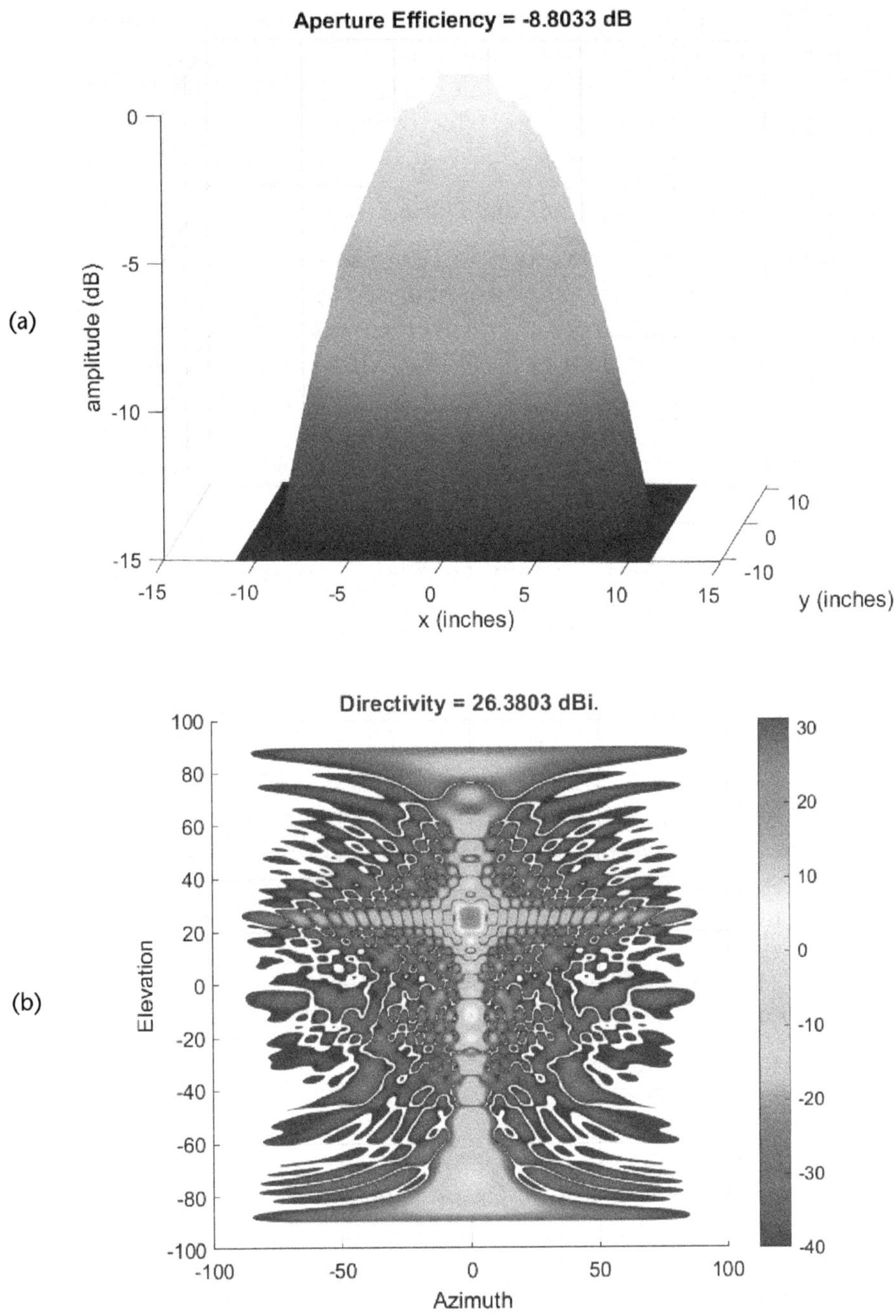

Figure 3.70 The Taylor SLL 40-dB and n-bar 4 taper (a) using 4-bit attenuators and the resulting pattern (b) from ***array_2D.m***.

aperture efficiency loss of over 8 dB. The relatively low resolution and high maximum-attenuation floor both degraded the low sidelobe performance. The image in Figure 3.70(a) shows the amplitude distribution, with its quantized, stepped appearance, and the amplitude floor, which can be observed at the outer array elements.

3.15.2 Random Amplitude and Phase Errors

Additional important errors to include, especially if rms SLL accuracy is desired, are random amplitude and phase errors. At the initial assembly of any array, amplitude and phase errors are highly likely to exist. These are often improved through an initial alignment (our term for first calibration). This alignment usually occurs in a planar near-field range. Figure 3.55 shows the SPY-6 AESA waiting outside multiple large near-field ranges at the Raytheon Technologies facility in Andover, Massachusetts. The initial alignment is usually required to improve the amplitude and phase flatness; however, the alignment is limited to the accuracy of the least significant bits of the phase shifter and attenuator. Therefore some residual amplitude and phase error always exists. The amplitude and phase errors usually exhibit as having normal or Gaussian distributions for amplitude and uniform distributions for phase. These can be modeled statistically, and ***array_2D.m*** supports modeling of these errors. The variable flag *errors* turns off errors with a value of zero and turns them on with a value of one. The variable *amprmserror* is used to input the RMS expected error in units of decibels. The function converts the value to volts for use in combining with the *Amp* amplitude matrix. The amplitude error simulation uses the gaussian random number generator *randn*. The phase variable is *phserrorrange*, and it uses the uniform random number generator *rand* in its selection of values. Figure 3.71 presents plots from ***array_2D.m*** detailing random errors and their impact on the array directivity pattern.

3.15.3 Amplifier Failure Errors in Active Arrays

AESAs use numerous small RF amplifiers on each element in the array. These can be in the form of high-power amplifiers, low-noise amplifiers (LNAs), or in drive amps. If any one of these amplifiers dies, the element is effectively turned off for either transmit, receive, or both functions. In large arrays with thousands of elements, it is very common for the AESA to be operated with some number of failed elements. In fact, it is common for brand new arrays to be delivered and pass acceptance testing with some number of failed elements. AESA system designers recognize the reality of failed elements and overdesign the systems to survive up to *X*% failures, where *X* is dependent upon the system designer's choices. The last error simulation included in ***array_2D.m*** is that of failed elements. The variable *ampfailures* is used to input the percentage of failed elements in any simulation. Uniform random numbers are drawn between 0 and 1 for each element in the array. If the draw is less than or equal to the *X*%, the amplitude of that element is set to zero. Thereby the failed elements are randomly distributed across the array.

Figure 3.72 presents a simulation from ***array_2D.m*** including AESA element failures for a circular array of 1,597 elements. The array used two Taylor SLL 25-dB

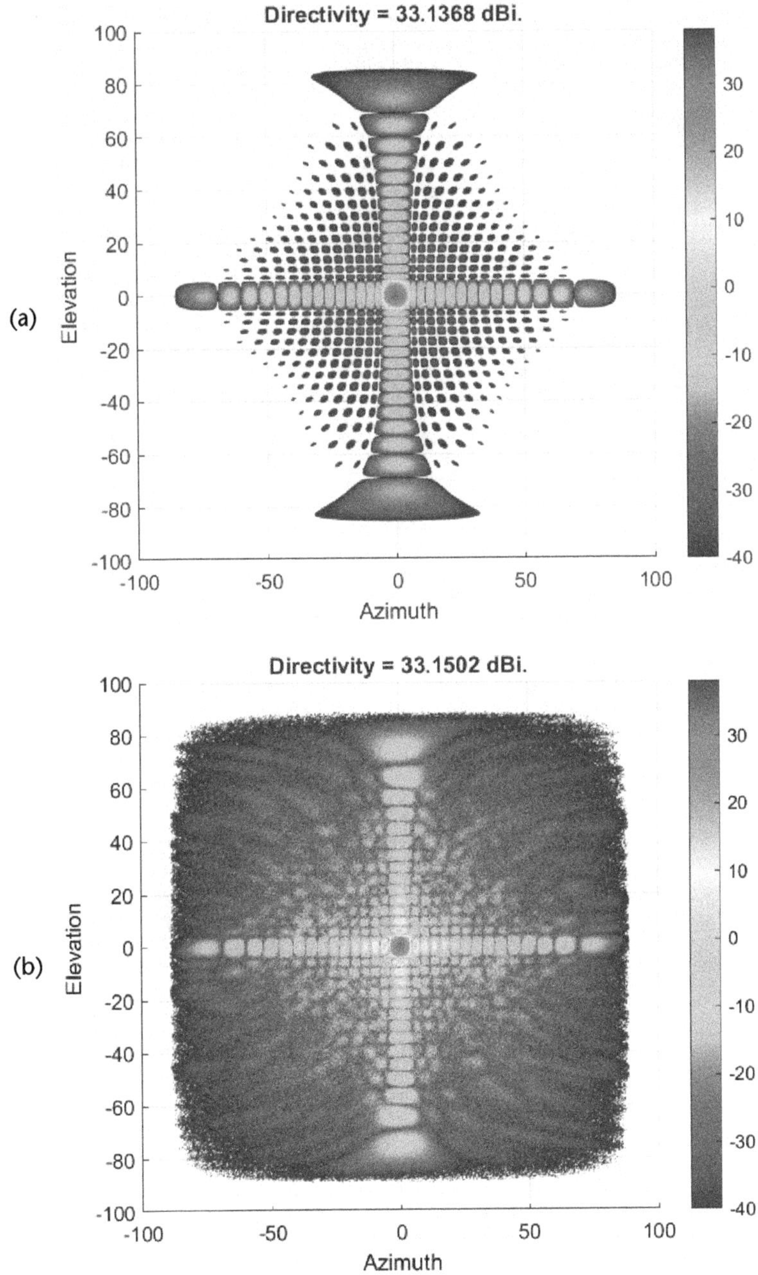

Figure 3.71 Plots from ***array_2D.m*** showing the impact of random errors on the directivity pattern for an array of 30 × 30 elements with an unsteered main beam, Taylor SLL 30-dB n-bar 4 in both axes, with no quantization errors. Patterns are plotted without errors (a) and with 1-dB RMS amplitude and 11.25° maximum phase error (b). The amplitudes and phases for the aperture with errors are shown in the plots (c–d).

(c)

(d)

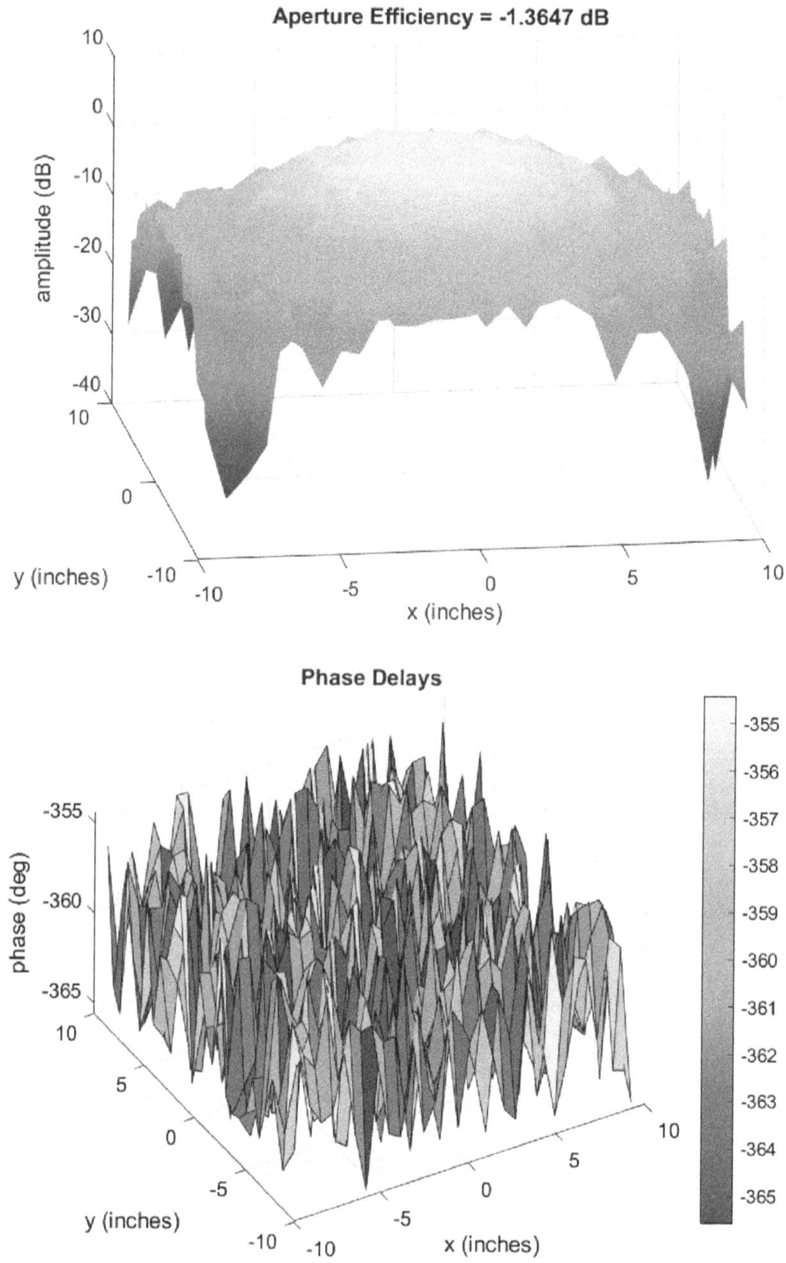

Figure 3.71 *(Continued)*

(a)

(b)

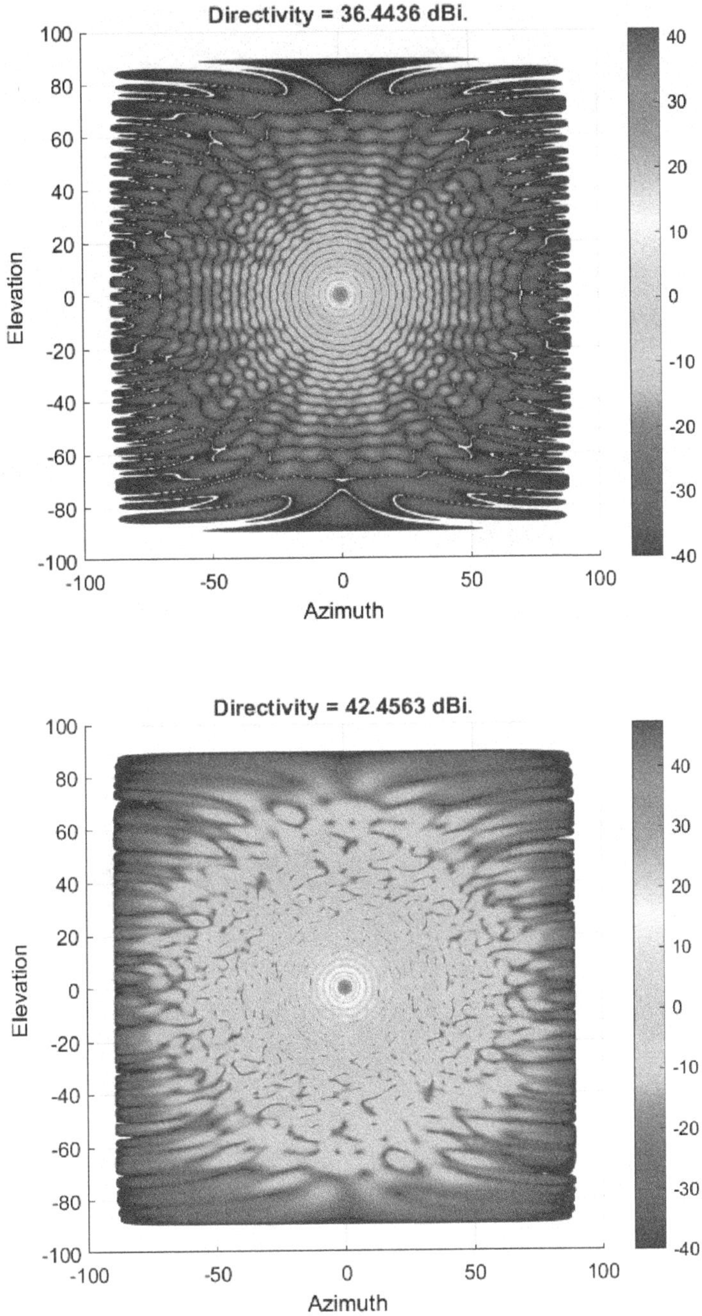

Figure 3.72 Plots from ***array_2D.m*** demonstrating the ability to simulate failed elements for a circular array of 1,597 elements with 5% failed. Patterns are presented without (a) and with (b) the failed element errors. The element map is shown in (c) and the amplitude taper in (d), both detailing the positions of the random draw of the 5% failed elements.

(c)

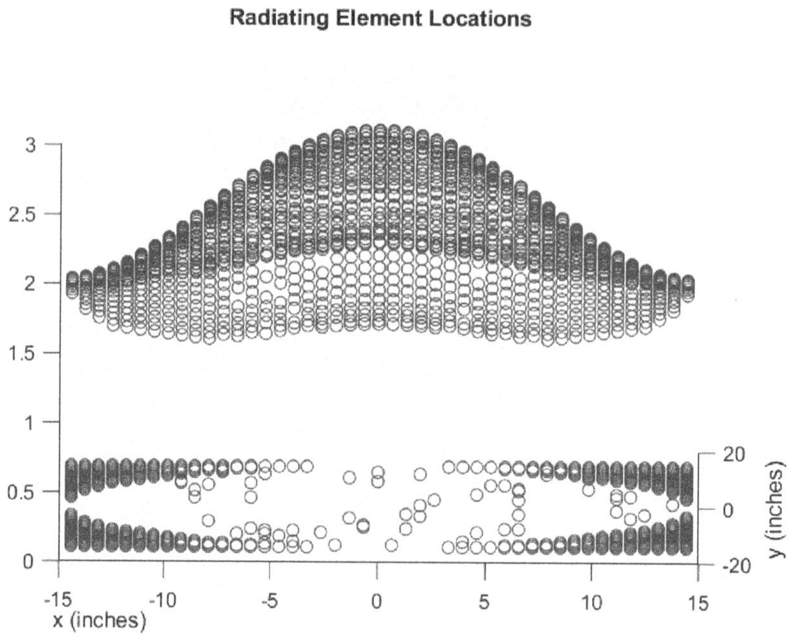

(d)

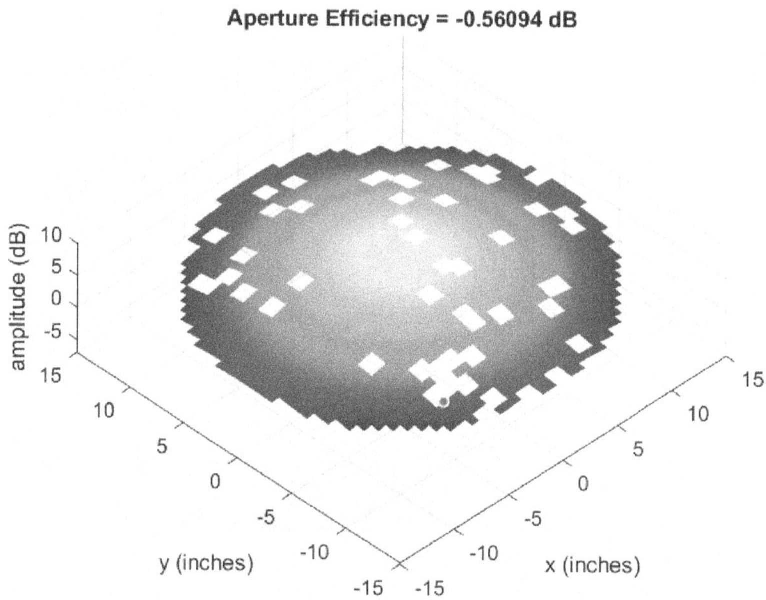

Figure 3.72 *(Continued)*

n-bar 4 tapers, with no other errors. The directivity patterns are plotted without and with 5% random-element failures.

3.16 Antenna Modeling Conclusions

Accurate modeling and simulation of antennas is important to achieving accurate simulation of radar and EW systems. There are many numerical paths that are possible to take in modeling antennas, and this chapter aims to describe the trade space among the major simulation options. The flexibility and expediency of the Fourier transform approaches make them common choices. This chapter provided an entry point for understanding approaches for developing antenna model components for radar and EW simulations.

References

[1] Balanis, C. A., *Antenna Theory: Analysis and Design*, Fourth Edition, Hoboken, NJ: John Wiley & Sons, 2016.

[2] Balanis, C. A., *Modern Antenna Handbook*, John Wiley & Sons, 2008.

[3] Kraus, J. D., and R. J. Marhefka, *Antennas for All Applications*, Third Edition, McGraw-Hill, 2003.

[4] https://www.mathworks.com/help/phased/ref/azel2uv.html.

[5] Corey, L. E., "A Graphical Technique for Determining Optimal Array Antenna Geometry," *IEEE Transactions on Antennas and Propagation*, Vol. AP-33, No. 7, July 1985, pp. 719–726.

[6] Masters, G. F., and S. F. Gregson, "Coordinate System Plotting for Antenna Measurements," *Antenna Measurement Techniques Symposium*, Vol. 32, 2007.

[7] *IEEE Standard, 211-2018, Definitions of Terms for Radio Wave Propagation.*

[8] Olin, I. D., "Polarization Characteristics of Coherent Waves," *Naval Research Lab Report NRL/FR/5317-12-10,210*, March 12, 2012.

[9] https://www.mathworks.com/help/phased/ug/polarized-fields.html.

[10] Balanis, C. A., *Antenna Theory and Design*, Harper & Row Publishers, 1982, p. 120.

[11] Skolnik, M. I., *Introduction to Radar Systems*, Second Edition, McGraw Hill Book Company, 1980, p. 231.

[12] Woodward, P. M., and J. D. Lawson, "The Theoretical Precision with Which an Arbitrary Radiation-Pattern May Be Obtained from a Source of a Finite Size," *Journal of the Institute of Electrical Engineers*, Vol. 95, Part III, No. 37, September 1948, pp. 363–370.

[13] Booker, H. G., and P. C. Clemmow, "The Concept of an Angular Spectrum of Plane Waves, and Its Relation to That of Polar Diagram and Aperture Distribution, *Proceedings of the IEE Part III: Radio and Communications*, Vol. 97, No. 45, January 1950, pp. 11–17.

[14] Bracewell, R. N., *The Fourier Transform and Its Applications*, Third Edition, McGraw–Hill, 1985.

[15] Lo, Y. T., and S. W. Lee, *Antenna Handbook Theory, Applications and Design*, Van Nostrand Reinhold Company, 1988, pp. 5–21.

[16] Clerk Maxwell, J., *A Treatise on Electricity and Magnetism*, 1873.

[17] Hunt, B. J., *The Maxwellians*, Cornell University Press, 1991.

[18] Heaviside, O., *Electromagnetic Theory*, The Electrician Printing and Publishing Co., London, Volume I (1893), Volume II (1899), and Volume III (1912).

[19] Garnett, W., J. Rautio, and L. Campbell, *The Life of James Clerk Maxwell (Illustrated)*, Independently Published, February 26, 2020, https://www.sonnetsoftware.com/resources/maxwell-bio.html.

[20] Davidson, D. B., *Computational Electromagnetics for RF and Microwave Engineering*, Cambridge, U.K.: Cambridge University Press, 2005.

[21] https://pe.gatech.edu/courses/modeling-and-simulation-antennas.

[22] Huygens, C., *Treatise on Light: In Which Are Explained the Causes of That Which Occurs in Reflection & Refraction* (French: Traité de la Lumière: Où Sont Expliquées les Causes de ce qui lui Arrive Dans la Reflexion & Dans la Refraction), 1690.

[23] Nyquist, H., "Certain Topics in Telegraph Transmission Theory," *Transactions of the American Institute of Electrical Engineers*, Vol. 47, No. 2, 1928, pp. 617–644.

[24] Shannon, C. E. "Communication in the Presence of Noise," *Proceedings of the Institute of Radio Engineers*, Vol. 37, January 1949.

[25] Bracewell, R. D., *The Fourier Transform and Its Applications*, Third Edition, McGraw-Hill Science/Engineering/Math, 1999.

[26] Mailloux, R. J., *Phased Array Antenna Handbook*, Norwood, MA: Artech House, 2005.

[27] Hansen, R. C., "Array Pattern Control and Synthesis," *Proceedings of the IEEE*, Vol. 80, No. 1, January 1992, pp. 141–151.

[28] Taylor, T. T., "Design of Line-Source Antennas for Narrow Beamwidth and Low Side Lobes," *IRE Transactions on Antennas and Propagation*, Vol. 3, No. 1, January 1955, pp. 16–27.

[29] Taylor, T. T., "Design of Circular Aperture for Narrow Beamwidth and Low Sidelobes," *IRE Transactions on Antennas and Propagation*, Vol. 8, No. 1, January 1960, pp. 17–22.

[30] Sherman, S. M., and D. K. Barton, *Monopulse Principles and Techniques*, Second Edition, Norwood, MA: Artech House, 2011.

[31] Bayliss, E. T., "Design of Monopulse Antenna Difference Patterns with Low Sidelobes," *The Bell System Technical Journal*, Vol 47, No. 5, May–June 1968, pp. 623–650.

[32] Hopkins, G. D., et al., "Aperture Efficiency of Amplitude Weighting Distributions for Sidelobe Level Control for Array Antennas," *IEEE Aerospace Conference*, 2007.

[33] Hopkins, G. D., et al., "Beam Former Development for the NASA Hurricane Imaging Radiometer," *IEEE Aerospace Conference*, 2015.

[34] Parad, L. I., and R. L. Moynihan, "Split Tee Power Combiner," *IEEE Transactions on Microwave Theory and Techniques*, Vol. 13, No. 1, January 1965, pp. 91–95.

[35] Hord, W. E., "Microwave and Millimeter Wave Ferrite Phase Shifters," http://www.magsmx.com/Micro_MMW_PS.pdf.

[36] Koul, S. K., and B. Bhat, *Microwave and Millimeter Wave Phase Shifters, Volumes I and II*, Norwood, MA: Artech House, 1991.

[37] Harris, H. M., and R. L. Sturdivant, *Transmit and Receive Modules for Radar and Communications Systems*, Norwood, MA: Artech House, 2015.

[38] Rotman, W., and R. Turner, "Wide-Angle Microwave Lens for Line Source Applications," *IEEE Transactions on Antennas and Propagation*, Vol. 11, No. 6, 1963, pp. 623–632.

[39] Butler, J., and R. Lowe, "Beamforming Matrix Simplifies Design of Electronically Scanned Antennas," *Electronic Design*, Vol. 9, No. 8, 1961, pp. 170–173.

[40] Miller, C. J., "Minimizing the Effects of Phase Quantization Errors in an Electronically Scanned Array," *Proceedings of the 1964 Symposium on Electronically Scanned Array Techniques and Applications, RADC-TDR-54-225*, Rome Air Development Center, Griffiss AFB, New York, 1964, pp. 17–38.

CHAPTER 4

Propagation

4.1 Introduction

A radar operates by injecting RF EM energy into space with its transmitting antenna and collecting echoes scattered off objects of interest, such as a target, with a receiving antenna. The process of traversing the distance from the transmitter to the target by the transmitted energy, and crossing the divide from the target to the receiver by the echo energy, is known as propagation.

Propagation subjects the outbound energy and return echoes to delays, amplitude scalings, and phase shifts. These effects may be due to intervening media, such as the Earth's atmosphere, or the presence of other objects, such as the Earth's surface. This chapter quantifies these phenomena, bounds their severity, derives closed-form expressions for predicting their levels, and discusses their impact on radar performance.

Here, monostatic radar operation (where transmit-and-receive antennas are colocated and frequently share the same physical aperture) is assumed. With monostatic radar, the outbound and return paths to and from the target are equivalent. Thus, only one path need be examined for propagation effects, with the understanding that the same effects apply to the remaining path. The monostatic concepts developed in this chapter are readily generalized to bistatic operation by treating the transmit and receive paths separately.

Two propagation effects are fundamental to all radar operation [1]. The first is the reliable mapping between target slant range and echo time delay. This relationship is codified in the appellation "radar," which originated as an acronym for "radio detection and *ranging*." The second are the geometric losses on the transmitter-to-target and target-to-receiver paths that manifest as the SNR dependence on range-to-the-fourth power, which figures prominently in the denominator of the RRE. These phenomena were described in detail in Chapter 2. This chapter introduces propagation effects that may or may not be important depending upon radar-target geometry and environmental conditions.

4.2 Radar Horizon

The Earth is mostly opaque to EM energy, so RF sensors are limited to operation above the local horizon. Should a target drop below the horizon, a radar cannot detect that target no matter its transmit power and aperture size. In truth, there is some diffraction at the horizon, so that power drops off faster than free-space calculations would predict for targets just above the horizon, and some nonzero echo energy will be scattered and make it back to the receiver for targets just below the horizon. However, these effects are limited to a small angular extent about the horizon and are significant only at very low frequencies. For the purposes of first-order sensor system design and prediction, the horizon acts as a brick wall, meaning all signal for targets above, and no signal for targets below.

A frequent question in radar design and performance prediction exercises is: How far out may one expect Earth surface returns? The motivation for this question may be a requirement to detect ground vehicles or ocean-going ships on the Earth's surface. Alternatively, the goal may be to quantify the range extent of ground clutter to drive waveform optimization. In both cases, the returns of interest have zero altitude. The geometry for this calculation appears in Figure 4.1.

The Earth's radius is denoted R_E and the radar altitude h_r. The Earth's surface is visible out to a range $R_{h,r}$, at the radar horizon. The radar horizon is established by a line-of-sight (LOS) vector from the radar that is tangent to the Earth, so that it just barely grazes the surface of the Earth. Mathematically, the LOS vector and the vector from the Earth's center at the radar horizon are normal to one another and form a right triangle, so Pythagorean's theorem gives [2]

$$R_{h,r} = \sqrt{\left(R_E + h_r\right)^2 - R_E^2} \tag{4.1}$$

Expanding the first term under the radical in (4.1) produces

$$R_{h,r} = \sqrt{2R_E h_r + h_r^2} \tag{4.2}$$

The Earth is an oblate spheroid whose exact radius is location-dependent, but it may be approximated as a sphere having a constant radius of 6,371 km.

Radar heights tend to be well under 20 km, even for high-altitude airborne systems. Therefore, $h_r << R_E$, and the first term in (4.2) dominates the second, so that a good approximation for the range to the radar horizon is

$$R_{h,r} \cong \sqrt{2R_E h_r} \tag{4.3}$$

While fine for ground-based and airborne radars, care must be taking in applying (4.3) to space-based radar. Satellites in low Earth orbit (LEO) may be as low as 400 km or as high as 1,200 km. The former is consistent with the conditions that led to (4.3), while the latter not so much.

Using the approximation in (4.3) and expressing radar altitude in feet (ft.) and range to the radar horizon in nautical miles (NM) gives

$$R_{h,r}(\text{NM}) \cong 1.064\sqrt{h_r(\text{ft.})} \tag{4.4}$$

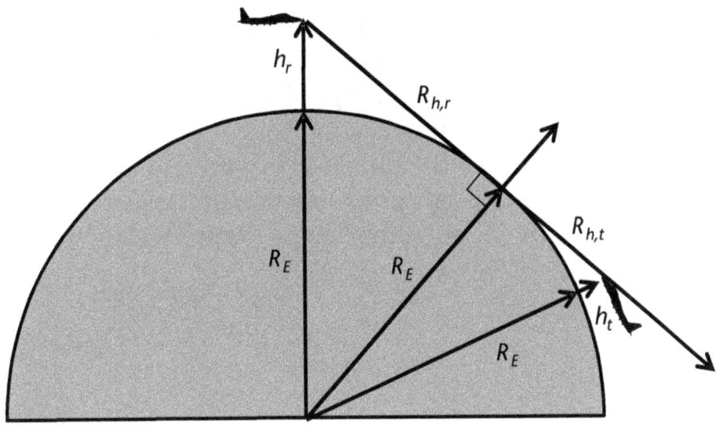

Figure 4.1 Range to radar horizon and maximum slant range to a target.

Equation (4.4) is a convenient rule-of-thumb for estimating slant range to the radar horizon given a radar height. For example, a ground-based radar on a knoll 100-feet-high has a radar horizon near 10 NM, and an airborne radar at 40,000 ft. has a radar horizon of approximately 200 NM.

Another common question for a proposed or existing radar system is: How far away can the radar see (have visibility to) a target at some altitude before the radar horizon intervenes? Figure 4.1 also shows the relevant geometry for a target height of h_t. At the maximum visible range, the LOS vector from the radar to the target just barely grazes the surface of the Earth; at any greater range, the target is beyond the radar horizon. Again, the vector from the Earth's center is normal to the LOS vector, and the maximum range is simply the sum of the ranges to the horizon for the radar $R_{h,r}$ and the target $R_{h,t}$

$$\begin{aligned} R_h(\text{NM}) &\cong R_{h,r} + R_{h,t} \\ &\simeq 1.064\sqrt{h_r(\text{ft.})} + 1.064\sqrt{h_t(\text{ft.})} \end{aligned} \tag{4.5}$$

Given a radar at 40,000 ft. and a target at 10,000 ft., the radar has visibility to that target out to a slant range of 200 NM + 100 NM = 300 NM.

The speed of light is a function of the propagating medium, being fastest in a vacuum and slower in air. Under standard atmospheric conditions, air density monotonically decreases as altitude increases, being densest at the Earth's surface and asymptotically approaching a vacuum at very high altitudes. Therefore, the speed of light is slowest at the Earth's surface and increases with altitude, approaching c (2.99792458 × 10^8 m/s) in the vacuum of space. As discussed in Section 4.6, this trend causes RF refraction, which imparts a mild downward curvature to propagating RF energy. At first glance, such a bowing might appear to decrease the range to the radar horizon: The LOS vector in a vacuum, Figure 4.1, would bend down and fall short of the radar horizon in air. However, a radar does not transmit and receive a single RF ray; it casts out and collects a continuum of rays over its transmit-and-receive elevation beamwidth. Figure 4.2(a) shows one such ray, at a slightly higher elevation angle than the ray in Figure 4.1, plunging beyond the geometric horizon. Refraction, then, increases the range to the radar horizon beyond that predicted by the simple free-space calculation in (4.4).

The influence of atmospheric refraction on radar horizon calculations is frequently captured by inflating the Earth's radius in (4.3) by four-thirds

$$R_h \cong \sqrt{2\left(\frac{4}{3}R_E\right)h_r} \tag{4.6}$$

The feet/NM rule-of-thumb in (4.4) becomes

$$R_h(\text{NM}) \cong 1.23\sqrt{h_r(\text{ft.})} \tag{4.7}$$

Atmospheric refraction increases the range to the radar horizon by about 20%.

Figure 4.2(b) is a diagram of the regular 3/3-Earth in dark gray and the 4/3-Earth in light gray. The Earth centers are displaced vertically so that the surfaces coincide at the nadir point below the radar. Moving away from nadir along the Earth's surface, the 4/3-Earth is seen to fall away from the radar more slowly than a proper-sized Earth, which provides some intuition for the impact of changes to the Earth's radius on range to the geometric radar horizon.

Plots of radar altitude versus radar horizon range for a 4/3-Earth appear in Figure 4.3, with lower altitudes and shorter ranges in Figure 4.3(a), and higher altitudes and longer ranges in 4.3(b). Figure 4.3(c) has the same information in a log-log plot, with callouts to everyday heights/altitudes.

Inflating the radius of the Earth serves as an exact correction to geometric calculations of the radar horizon when the increase in the speed of light with altitude is modeled as a linear function. Underlying the 4/3-Earth model, then, is a specific linear increase in the speed of light with altitude. In reality, the density of the atmosphere decays exponentially as altitude increases, so the speed of light asymptotically approaches its vacuum value with altitude. Thus, a linear

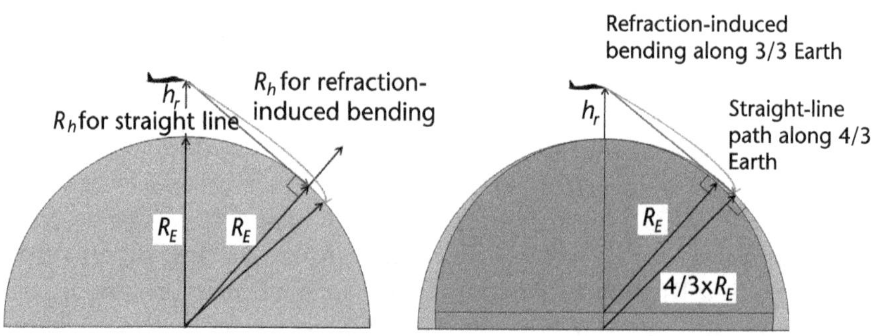

Figure 4.2 (a) A 4/3 Earth model and (b) model interpretations.

relationship between the speed of light and altitude is a good fit only at low altitudes in the Earth's atmosphere. Continuing such a linear function too high, an altitude will be reached above which the speed of light would exceed its vacuum value. In short, the 4/3-Earth model is appropriate only when the majority of the propagation path under consideration is through lower, denser Earth atmosphere. The 4/3-Earth is generally not appropriate for space-based radar, wherein most of the propagation path is through a near vacuum that induces negligible refraction. For high-altitude targets or radars, the 4/3-Earth should be applied with caution.

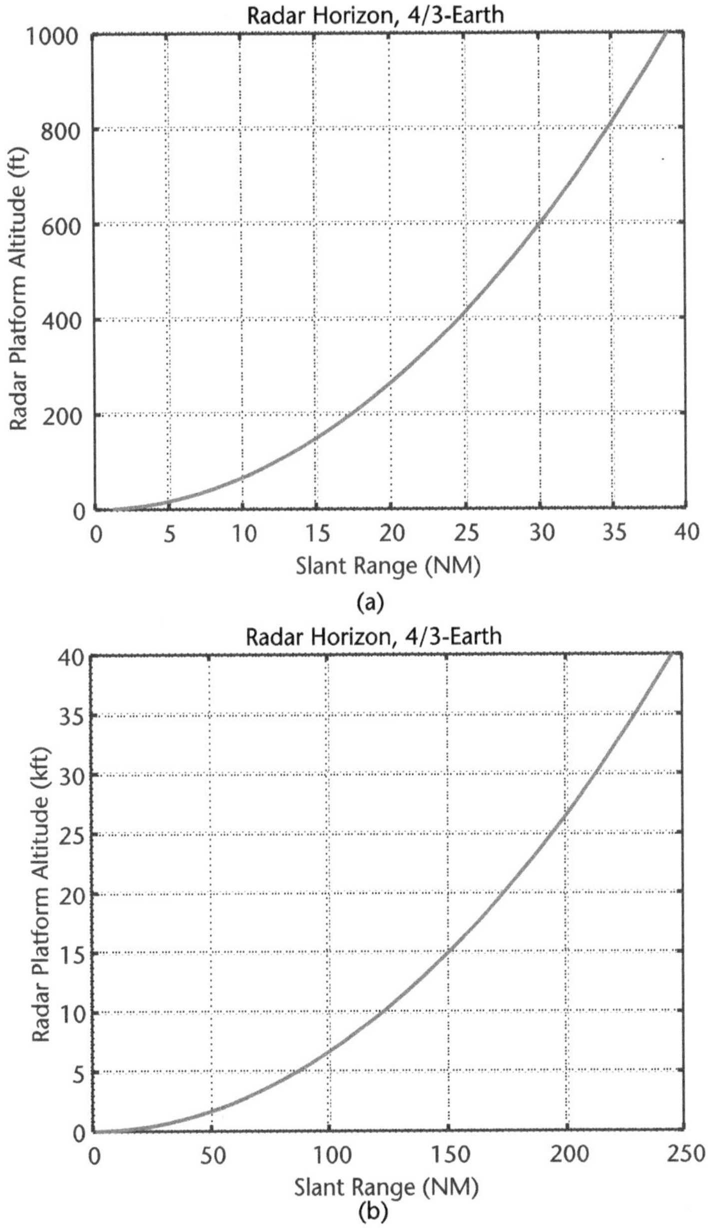

Figure 4.3 Radar horizon versus altitude: (a) lower altitudes and shorter ranges, (b) higher altitudes and longer ranges, and (c) same information in a log-log plot.

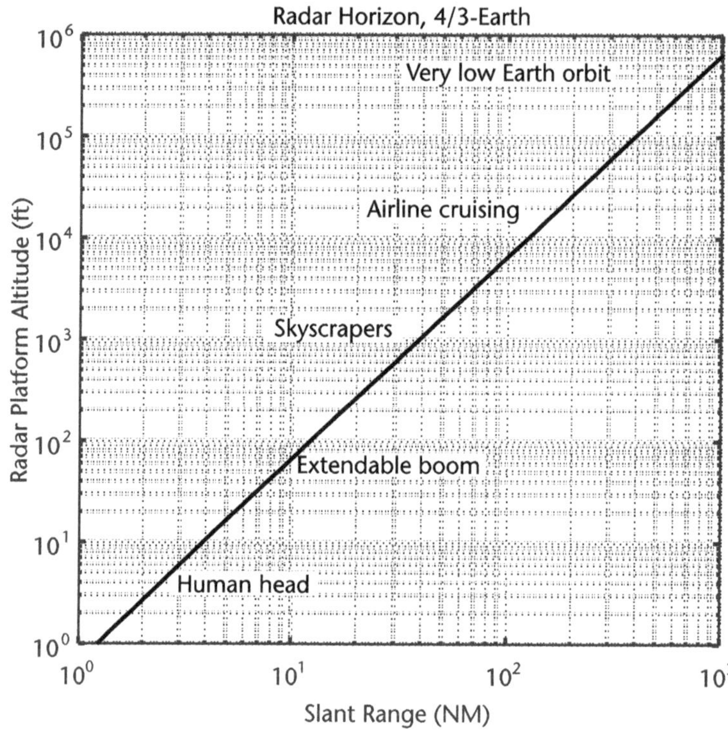

(c)

Figure 4.3 *(Continued)*

4.3 Atmospheric Attenuation

EM energy propagating through the Earth's atmosphere suffers an attenuation loss. Attenuation is the combination of two effects: (1) scattering of EM energy off the propagating path by particulates and aerosols and (2) extinction due to molecular and atomic interactions. Scattering can be significant at infrared and visible spectra, as wavelengths are comparable to the size of particles suspended in the Earth's atmosphere. At common radar RFs, however, wavelengths are much greater than atmosphere particle sizes, so scattering effects are negligible. Radar atmospheric attenuation, then, is dominated by extinction.

The principal extinction mechanism depends on the EM frequency/wavelength. Photons at very short wavelengths, such as X-rays and ultraviolet light, have sufficient energy to ionize atoms in the atmosphere, while visible and infrared light can transition electrons to higher-energy bands. Ionization and electron band changes remove propagating EM energy. RF photons have too little energy to ionize gases or cause electrons to jump bandgaps. Instead, RF excites vibrational modes in atmospheric nitrogen, oxygen, and water molecules. The vibrations convert some portion of the radar pulses and echoes from RF to thermal energy. This mild heating of the atmosphere results in losses to radar signal power along the transmit and receive paths.

Atmospheric attenuation follows Beer-Lambert's law: The amount of energy lost per unit propagation distance is proportional to the initial intensity of the radiation. For example, if 1 mW is lost per meter for an initial 1W of radiation, 2 mW will be lost to 2W. This relationship can be formulated as a first-order differential equation whose solution takes an exponential form. Exponential loss is conveniently represented by specific attenuation, which is reported as decibels of loss per unit distance (i.e., decibels per meter or decibels per kilometer). For a constant attenuation rate, one need only multiply the specific attenuation by the propagation distance to find the total loss; if the specific attenuation is 1 dB/km and the propagation distance is 10 km, the total loss is 10 km × 1 dB/km = 10 dB. Note that, for a monostatic radar, this 10 dB of loss would occur twice, out to the target and then again back to the receiver, for a total round-trip loss of 20 dB.

Contrast the specific attenuation loss with the R-to-the-4th geometric loss in Chapter 2. Geometric loss tends to dominate at shorter ranges while attenuation losses become significant at longer ranges. For example, consider a case where specific attenuation for atmospheric loss is 0.2 dB/km. Doubling the range of a target from 10 to 20 km increases geometric losses by 12 dB, but attenuation loss by only 10 km × 0.2 dB/km = 2 dB. Doubling the range of a target from 100 km to 200 km again increases geometric losses by 12 dB, but attenuation loss by a much more significant 10 km × 0.2 dB/km = 20 dB.

Specific attenuation of RF in the Earth's atmosphere is a function of radar frequency and three atmospheric parameters: Pressure, temperature, and water vapor density. The International Telecommunications Union (ITU) provides a model, ITU-R P.835-5 [3], of these properties as a function of altitude for several standard atmospheres. An example of these functions appears in Figure 4.4 for a midlatitude (22° to 45°) region during winter. The temperature drop with altitude at top is approximately linear through the troposphere at around 5.5 K/km. Temperature achieves a minimum at the tropopause and is approximately constant in the stratosphere. The pressure drop in the middle is exponential. The water vapor drop at bottom is very rapid: The air at 5 km has one-tenth the water vapor as air at the surface. For this reason, any attenuation due to water vapor tends to be most severe at lower altitudes.

The ITU also provides a model, ITU-R P.676-10 [4], for RF specific attenuation, and another model, ITU-R P.453-10 [5], to calculate RF specific attenuation as a function of atmospheric conditions. These models were used to generate specific attenuation as a function of RF at ground level (zero altitude) for a standard atmosphere with temperature 288K and pressure 102.3 kPa as modeled in ITU-R P.835-5. Results appear in Figure 4.5(a). Two water vapor density levels are shown: a low humidity of 7.5 g/m^3, standard for continental regions with mild climates, and a high humidity of 20 g/m^3, standard for tropical zones over the ocean. Note that both frequency (in gigahertz) and specific attenuation (in decibels per kilometer) are presented in a log-log form. Gray-shaded bars highlight typical radar bands. A coarse trend of a linear increase in specific attenuation with RF (again, both plotted logarithmically) appears. Two spikes in specific attenuation are significant at radar RFs; the first is a water vapor absorption line at 22.24 GHz, and the second is an oxygen molecule absorption line at around 60 GHz. Higher humidity gives rise to a specific attenuation almost an order of magnitude higher than lower humidity over much of the RF. Specific attenuation is low—0.01 dB/km or less—for RFs in X-band or below, even under humid conditions.

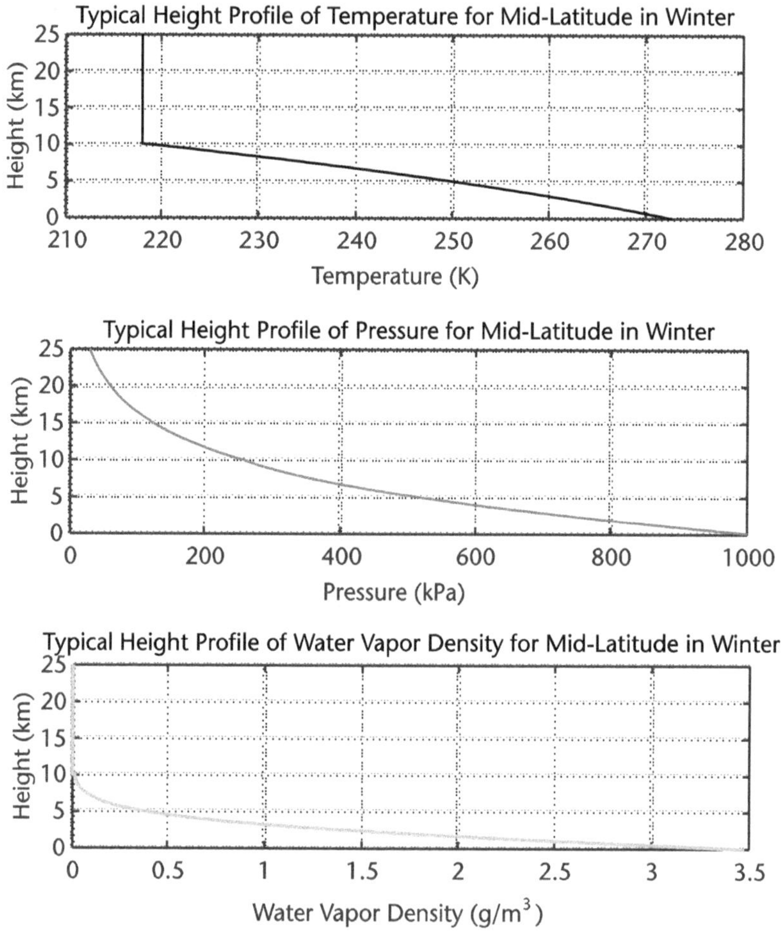

Figure 4.4 Typical atmospheric conditions versus altitude.

Figure 4.5(a) is for zero-height only. Radars and their targets are commonly at altitude, so specific attenuation as a function of altitude is needed. Combining the ITU models for specific attenuation and standard atmospheres and specific attenuation allows specific attenuation to be calculated as a function of altitude for select bands. Figure 4.5(b) shows altitude versus specific attenuation at 1 GHz and 10 GHz given the atmospheric conditions in Figure 4.4. Specific attenuation is comparable for the two bands in the stratosphere (above 10 km). The difference in specific attenuation between the two bands is most pronounced at the Earth's surface.

There are two ways to use the information in Figure 4.5(b). The first provides a rough estimate of attenuation loss given radar and target altitudes and the slant range between the two and presumes the following two approximations:

- The Earth is flat, as depicted on the left side of Figure 4.6(a).
- The specific attenuation variation with altitude is linear, as suggested in the right side of Figure 4.6(a), which contains a replica of Figure 4.5(b).

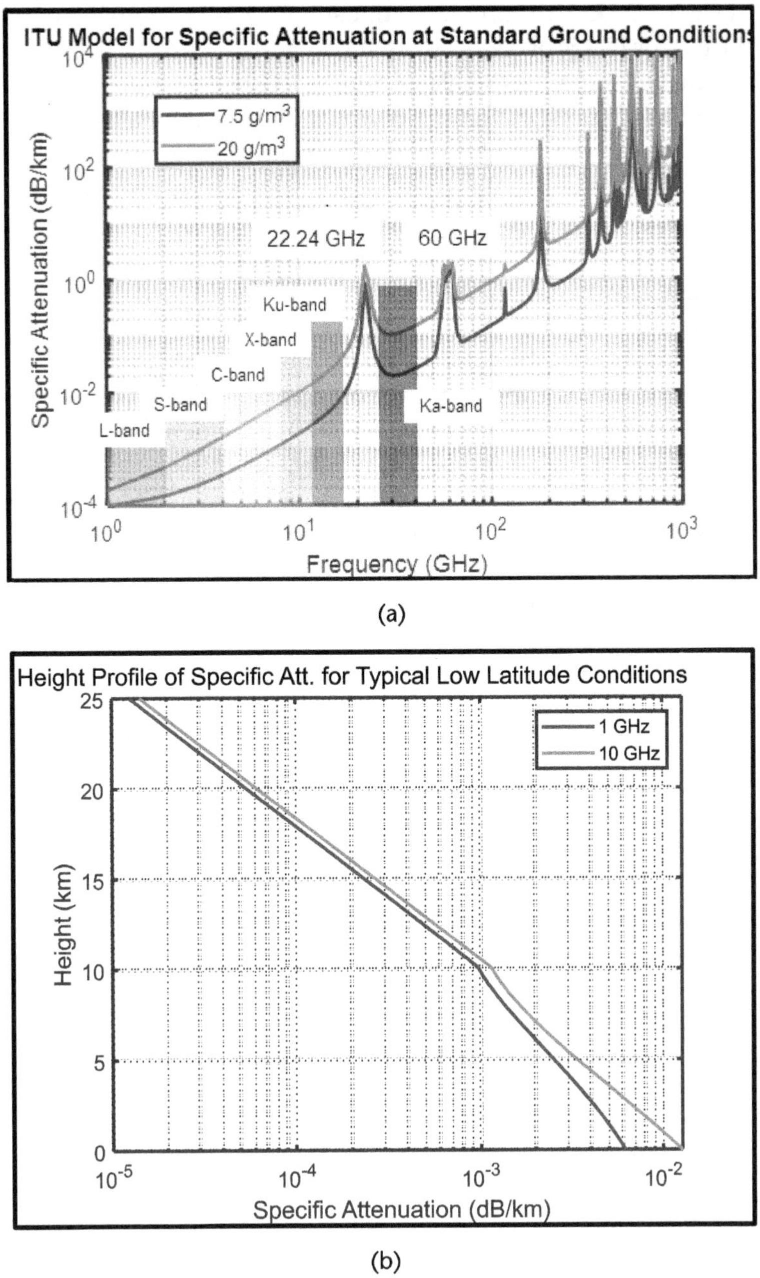

Figure 4.5 (a) ITU specific attenuation and (b) specific attenuation versus altitude.

If these approximations are reasonable, the average altitude between the radar and the target may be used to extract a single specific attenuation from Figure 4.5(b). Multiplying this specific attenuation by slant range yields total one-way atmospheric loss. For example, if the radar is at 9 km altitude and the target at 1 km altitude, the average altitude is 5 km. From Figure 4.5(b), in X-band the specific attenuation at 5 km is 3.5×10^{-3} dB/km. If the slant range between the radar and the target is 50 km, the one-way atmospheric loss is 0.175 dB, so the total two-loss is 0.35 dB. This loss is of course unchanged if the radar and target exchange altitudes.

A more accurate estimate of total atmospheric loss is achieved by numerical integration of specific attenuation from Figure 4.5(b) along the path between the target and the radar. A spherical Earth can be represented explicitly in such a model and is critical under some conditions. For example, given a radar and target at 10 km altitude, the specific attenuation at that altitude is about 10^{-3} dB/km, and a flat-Earth model would yield a two-way atmospheric loss of only 1.0 dB at a 500-km range. However, 500 km is far enough for curved Earth effects to come into play, as the propagation path will dip down into the thicker atmosphere, incurring losses much higher than the flat-Earth 1.0 dB would suggest.

To establish an upper bound on atmospheric loss, the geometry depicted in Figure 4.6(b) was evaluated using the aforementioned numerical integration method. Here, a surface-based radar (zero-height) operates against a target at zero elevation angle and a very high altitude (effectively outside the atmosphere). Table 4.1 presents the total two-way attenuation loss as a function of RF band and Earth conditions. Losses increase as frequency increases for all conditions.

Trends across conditions are more difficult to discern and appear to be frequency-dependent. For example, in L-band, losses are higher in winter and at higher latitudes, while the reverse holds in X-band. These conflicting trends may be resolved by noting that temperature variations are more important at low frequencies (colder is more lossy), and water vapor is more important at high frequencies (moister is more lossy).

A still higher bound may be established by placing the radar and target at very high altitudes and very long ranges so that the LOS vector between them is tangent to the Earth's surface. This path is simply twice the path in Figure 4.6(b), and the losses in decibels are double those in Table 4.1.

Figure 4.7 contains a plot for a more common geometry: radar and target coaltitude, with altitude in kilometers on a logarithmic ordinate and two-way loss on the abscissa. Each line is a line of constant slant range in kilometers, as noted in the legend. Each line terminates due to radar horizon limits. Conditions were 25° of latitude in winter. X-band was the RF of choice here, as that is the usual band for air-to-air radars on tactical aircraft such as fighters. Numerical integration was performed on a curved Earth, so the longer slant ranges include significant propagation through lower, lossier atmosphere. Two-way losses exceed 4 dB only at very long ranges and medium altitudes. If one were to add still higher altitudes and longer ranges to the graph, total two-way loss would asymptotically approach 11.6 dB, twice the 5.8 dB in the 25°-Winter/X-Band entry in Table 4.1.

Meteorological precipitation like rain and snow lead to greater attenuation losses beyond the clear air levels discussed thus far. Table 4.2 shows specific attenuation for rain over a range of RFs and rain intensities using a common model [6]. Spot checks against specific atmospheric attenuations levels in Figure 4.5(a) suggest light rain losses are a little higher than quiescent air losses at ground level. While losses in heavy and violent rain are severe, these tend to be localized into cells and so apply over only limited path lengths.

Specific attenuation dependence on rain rate is modest, approximately linear over a wide range of RFs and rain rates. Dependence on RF is much more dramatic,

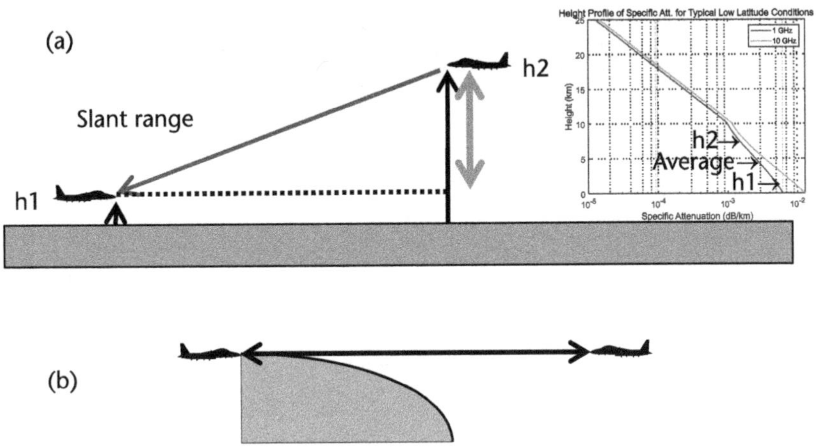

Figure 4.6 Model options: (a) average altitude with flat-Earth assumption and (b) numerical integration over spherical Earth.

following a squared or even cubic trend, and highlights the importance of RF band selection in radar design when it is vital to maintain performance under rainy conditions.

While atmospheric losses can be significant under extreme conditions, typical radar-target geometries give rise to more moderate losses of a few decibels, just one more loss in a long list of RF, ohmic, and signal processing losses. Indeed, it is not unusual to account for atmospheric losses with a fixed value, regardless of engagement geometry, in high-level radar trade studies.

4.4 Refraction

Refraction was discussed briefly in the context of the 4/3-Earth model in Section 4.2. In this section, atmospheric refraction is described in more detail, and unusual refraction conditions and their impact on RF propagation and radar performance is discussed.

As mentioned in Section 4.2, the speed of light in a vacuum is represented with the fixed variable c. Let the speed of light in any medium be denoted v. The index of refraction n in that medium is defined as

$$n \equiv \frac{c}{v} \tag{4.8}$$

Table 4.1 Selected Limits on Two-Way Attenuation Loss*

RF Latitude-Season	*L-Band (1.5 GHz)*	*S-Band (3.0 GHz)*	*C-Band (6.0 GHz)*	*X-Band (10 GHz)*
25°-Summer	3.3 dB	3.8 dB	5.0 dB	8.2 dB
65°-Summer	3.5 dB	3.9 dB	4.8 dB	7.0 dB
25°-Winter	3.8 dB	4.2 dB	4.7 dB	5.8 dB
65°-Winter	4.1 dB	4.6 dB	4.9 dB	5.7 dB

* Numerical integration results courtesy of code written by Dr. Jenny Reed, GTRI.

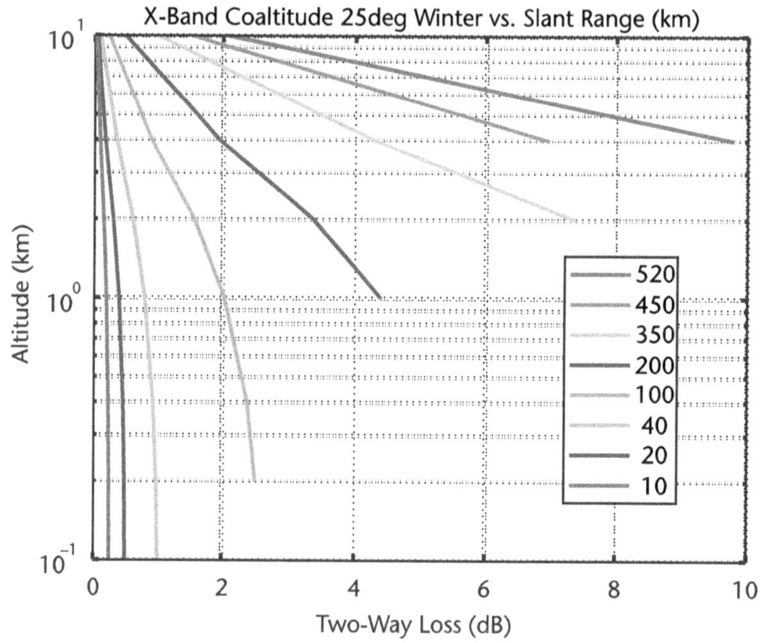

Figure 4.7 Selected coaltitude attenuation values. (Numerical integration results courtesy of code written by Dr. Jenny Reed, GTRI.)

As the speed of light is slower in any medium other than a vacuum, the index of refraction is always greater than or equal to 1.

Figure 4.8(a) depicts an EM ray starting at point P, traversing a planar interface between two media at point 0, and ending at point Q. The speed of light and index of refraction of media 1 and 2 are v_1 and n_1 and v_2 and n_2, respectively. The angle of incidence (the angle away from the normal to the planar interface) of the ray is θ_1 and θ_2 in media 1 and 2. Snell's law describes the relationship between these angles of incidence as

$$n_1 \sin\theta_1 = n_2 \sin\theta_2 \tag{4.9}$$

Transitioning to a higher index of refraction (slower speed of light) causes a smaller angle of incidence (i.e., the ray bends toward the normal).

Table 4.3 contains approximate indexes of refraction for select media. A high index of refraction glass may be found in strong reading glasses. Note the very small difference between index of refraction for a vacuum and the standard

Table 4.2 Selected Specific Attenuation Versus RF and Rain Rate

RF *Rain Rate*	*L-Band* *(1.5 GHz)*	*S-Band* *(3.0 GHz)*	*C-Band* *(6.0 GHz)*	*X-Band* *(10 GHz)*
2 mm/hr-Light	0.0002 dB/km	0.0014 dB/km	0.004 dB/km	0.024 dB/km
5 mm/hr-Moderate	0.0004 dB/km	0.0039 dB/km	0.014 dB/km	0.079 dB/km
20 mm/hr-Heavy	0.0016 dB/km	0.0187 dB/km	0.088 dB/km	0.46 dB/km
60 mm/hr-Violent	0.0045 dB/km	0.064 dB/km	0.37 dB/km	1.88 dB/km

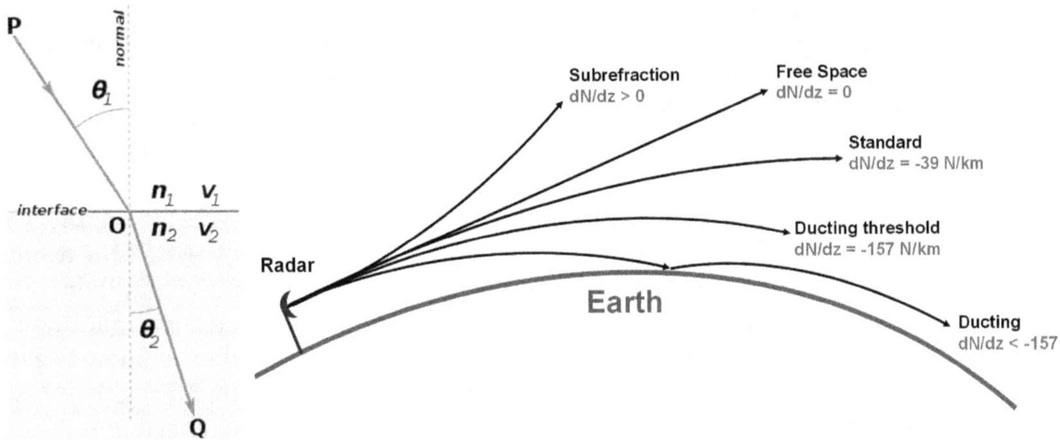

Figure 4.8 (a) Atmospheric refraction and (b) refraction conditions and possible RF propagation paths. (Figure 4.8(b) is reprinted with permission from the TEMPER User Guide courtesy of Jonathan Z. Gehman and The Johns Hopkins University Applied Physics Laboratory LLC.)

atmosphere at sea level in Table 4.3. When examining RF atmospheric propagation, it is convenient to work with RF refractivity N, defined as

$$N = (n - 1) \times 10^6 \tag{4.10}$$

From Table 4.3, the RF refractivity of a standard atmosphere and a vacuum are 293 and 0.

RF propagation through the atmosphere is not a series of passages through discontinuities, like the interface shown in Figure 4.8(a). Instead, the change in RF refractivity is continuous and gradual but may be modeled as a large number of thin horizontal layers having small RF refractivity changes between them. Under standard conditions, the speed of light increases with altitude, reaching a maximum in the vacuum of space. Therefore, RF refractivity decreases from around 300 to zero with increasing altitude. From Snell's law, an RF ray propagating upward will tend to bend away from the vertical direction. Conversely, an RF ray propagating downward will tend to bend toward the normal. In both cases, the RF path is not a straight line but rather a downward arc.

Table 4.3 Selected Indexes of Refraction

Material	*Approximate Index of Refraction*
Diamond	2.4
High index of refraction glass	1.5–1.8
Glass	1.5–1.6
Water	1.33
Air (standard atmosphere at sea level)	1.000293 ($1 + 293 \times 10^{-6}$)
Vacuum	1.0

The deviation from straight-line propagation is driven not by the absolute RF refractivity *N*, but by the rate of change of RF refractivity *dN/dz* with altitude *z*. If *dN/dz* is zero, there is no variation in RF refractivity with altitude, and RF propagation follows a straight line. Under standard atmospheric conditions, *N* decreases with altitude, so *dN/dz* is negative, and RF rays bend downward. Under unusual atmospheric conditions, *dN/dz* can be greater than zero, causing RF paths to arc upward. Under standard atmospheric conditions, *dN/dz* = –39 N/km; this value leads to the 4/3-Earth model in Section 4.2. Should *dN/dz* equal –157, the resulting RF arc has a radius equal to the radius of the Earth. Under these conditions, an RF ray launched along the local radar horizontal will neither gain nor lose altitude, but will follow the curvature of the Earth at a constant altitude. Lower (more negative) *dN/dz* cause greater arcing still, to the degree that propagating RF rays bend down to strike the Earth's surface. Should that surface support RF reflection—a smooth ocean surface is good reflector—the reflected ray starts upward but eventually bends back downward, to potentially reflect again off the Earth's surface. The ray may skip several times between the refraction layer above and the reflecting Earth below. This ducting condition may support several skips out and back. To support ducting, *dN/dz* must be less than –157 N/km, so that value is referred to as the ducting threshold. Figure 4.8(b) depicts these various propagating modes and the associated *dN/dz*.

Ordinary refraction profiles have *N* that monotonically decreases with altitude and at a moderate rate, one not high enough to support ducting. Some consequences of ordinary refraction profiles include the following:

- *Extended radar horizon:* The radar horizon is farther away than a geometric calculation would suggest. The 4/3-Earth model is one example.
- *Estimated elevation angle errors:* Because the RF ray bends downward to and from the target, the elevation angle measured at the radar receive antenna will be higher than the true geometric elevation angle to the target. At long ranges through low altitudes, this error may be several tenths of a degree.
- *Estimated slant range errors:* In general, the path between the radar and the target covers variations in index of refraction. If a constant speed of light is used when deriving slant range from detection time delay, the resulting estimate may be high or low. At long ranges through low altitudes, this error may be on the order of several tens of meters.
- *Lensing loss:* A gradual decrease in the index of refraction with altitude introduces a little more spreading of RF rays in elevation than geometric calculations would predict. This results in a decrease in energy out to the target and back again. The optical analog to this effect is a concave, diverging lens. This lensing loss is always lossy under standard conditions and may be several decibels at long ranges through low altitudes.

Extraordinary refraction profiles possess large variations in the rate of change of the index of refraction with altitude, or unusually high or low rates of change. Examples include the following:

- *Subrefraction: dN/dz* is greater than zero, so RF rays bend upward instead of downward. The radar horizon will be closer than geometric calculations would predict.
- *Ducting:* As described above, RF rays are trapped between a reflective Earth surface below and a refractive layer above, so that rays may bounce back and forth between these boundaries as they propagate.

A final measure of the influence of refraction on RF propagation is the modified refractivity *M*, defined as

$$M = N + 157 \times z \tag{4.11}$$

where z is again altitude in km. Recall that the ducting threshold corresponds to a *dN/dz* of −157 N/km. Therefore, under ducting threshold conditions *M* will assume a constant value. If altitude is plotting as a function of *M* for the ducting threshold condition, as in the plot on the far left of Figure 4.9, the result is a vertical line. The slope of the line on such a plot indicates the radius of curvature of an RF ray passing through that altitude. If the slope is infinite (vertical line), the curvature is equal to the Earth's radius. If negative, ducting may be supported. If positive, curvature is less than the Earth's surface. Figure 4.9 shows a sequence of *M* profiles at the top, and consequent propagation effects in the images below. The image colormap denotes RF intensity, where blue is low and red is high. These profiles and intensity images are taken from the user guide for TEMPER, an RF propagation tool developed by the U.S. Navy [7]. Under the right conditions, ducting layers can support RF propagation well beyond the usual 4/3-Earth radar horizon.

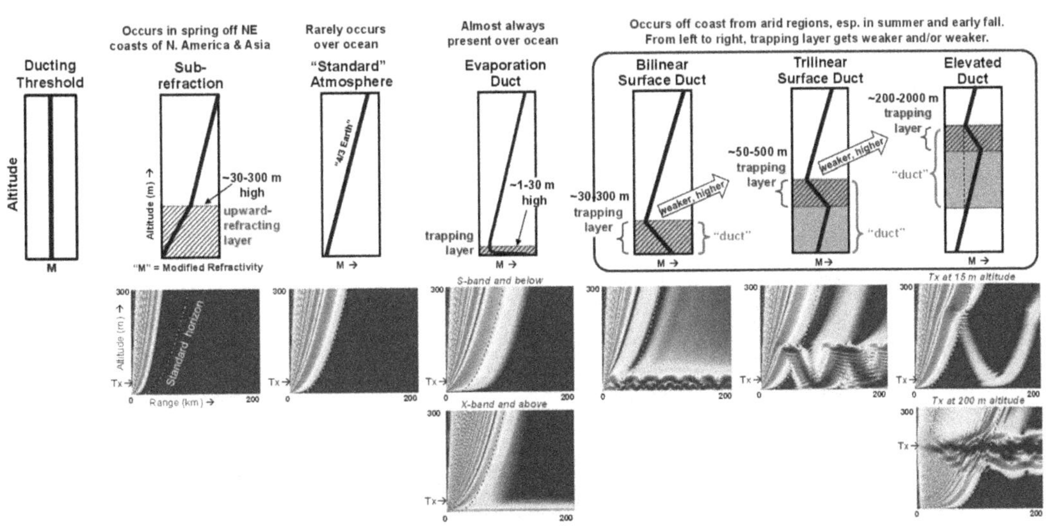

Figure 4.9 Categories of possible propagation conditions. (Reprinted with permission from the TEMPER User Guide courtesy of Jonathan Z. Gehman and The Johns Hopkins University Applied Physics Laboratory LLC.)

4.5 Multipath

Thus far, this chapter has focused on propagation phenomena confined to the free-space path between the radar and the target. Examples of such direct path effects include refraction and attenuation. RF reflecting off objects near the direct path can introduce additional propagation paths to and from the target, leading to multipath effects. In radar applications, the Earth's surface is frequently the principal source of secondary propagation paths. Figure 4.10 depicts multipath arising from the Earth's surface target, leading to multiple paths, or "multipath" effects. The upper vector represents the usual direct path, while the lower pair of vectors suggests a bounce or reflection off the Earth's surface. Thus, RF energy has two routes to the target, and two routes back to the radar.

In Figure 4.10 there are four possible round-trip trajectories, named for the constituent outbound and return paths; they are listed as follows.

- *Direct-direct* (DD): The usual free-space path.
- *Direct-indirect* (DI): Illuminate via the free-space path, and ground bounce on the return path.
- *Indirect-direct* (ID): Bounce off the ground on illumination, and free-space path on receive.
- *Indirect-indirect* (II): Ground bounce in both directions.

For a monostatic radar, the amplitude gain or loss, phase shift, and time delay along any given path are equivalent for EM energy flowing in either direction. Such reciprocity means that the DI and ID contributions are equivalent in gain, phase, and delay, and so coherently reinforce and are often combined into one single-bounce term.

The ellipse in Figure 4.10 signifies the transmit/receive antenna main beam. Note that both the target and the ground bounce point fall into the main beam. Multipath effects are present when the target or bounce point fall into the antenna sidelobes but are much more significant when all paths are confined within the main beam.

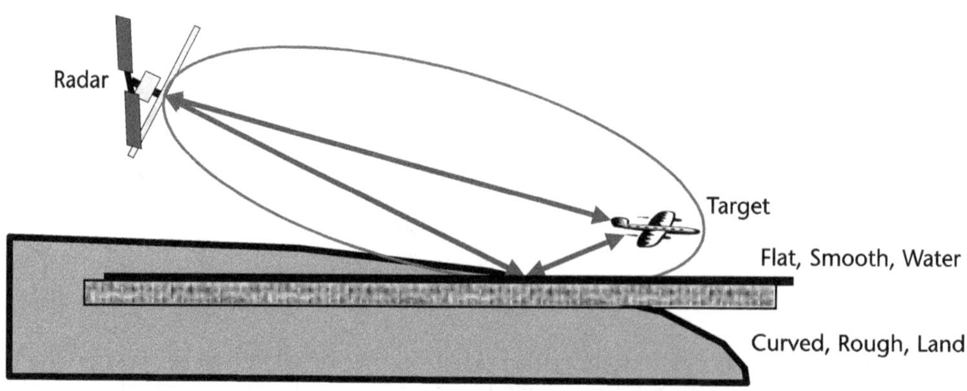

Figure 4.10 Multipath and conditions that facilitate it.

Note that the incident and scattered angles at the target are the same for DD, so this path excites a monostatic RCS response at the target. The same is true for the II path, although the monostatic elevation is lower than for DD, so the target may present a very different RCS amplitude and phase to these two paths. Meanwhile, the RCSs for the DI and ID paths are mildly bistatic, with an effective monostatic bisector falling hallway between the DD and II elevation angles.

The DD path yields the shortest time delay to the target, while the II path the longest, and DI and ID fall halfway between the two delays. The ground bounce always involves some EM absorption and scattering, so DD—no bounce—tends to produce the strongest target return, II—two bounces—the weakest, and DI and ID—one bounce—an intermediate power. In practice, multipath target returns may be stronger than the direct-path return due to target RCS variations with elevation, as well as radar main beam pointing angles that present higher gains to the bounce point than to the target. Another important consideration is the aforementioned DI and ID summation. The radar receives and processes the coherently combined DI and ID returns; it has no way to tease apart the individual path contributions. Under highly reflective bounce conditions, the DI+ID signal may have twice the voltage of the DD signal, for a gain of four, up to 6 dB over the direct path.

Strong multipath effects require a highly reflective bounce point. Figure 4.10 also offers contrasting depictions of strong and weak multipath conditions, described as follows:

- Surfaces having a high coefficient of reflection obviously accentuate multipath effects. For this reason, multipath returns are observed far more frequently over water than over land.
- Smooth surfaces reflect more power to the target on transmit (and back to the radar from the target). Rough surfaces scatter a larger fraction of the energy away from the target and radar and also serve to degrade RF wavefront coherence at the target and at the radar receive antenna. A related influence is grazing angle: A rough surface appears smoother when reflection occur at low incident and reflected grazing angles.
- Convex surfaces introduce additional geometric spreading on bounce paths. The Earth's surface is always curved, of course, but the local curvature at the bounce point is more pronounced when the radar and target are at long ranges and high altitudes. At low altitudes and short ranges, the bounce region is approximately flat.

By way of summarizing, one could imagine a scenario combining all of the attributes favorable to multipath: a shipborne radar with a wide elevation beamwidth, mounted low on the mast and not far above the waterline, operating against a low altitude antiship cruise missile coming over the radar horizon, under calm ocean conditions.

The time-delay differences between the various paths is key to understanding how multipath is manifested at the radar and predicting its impact on target detection. Time delays are found by calculating the slant ranges between the radar and the target for the different paths and scaling by the speed of light. Determining accurate slant ranges requires finding the bounce point on a spherical Earth, a rather involved geometry exercise. In Figure 4.11(a), the Earth's surface is approximated as flat, with a radar height h_r, target height h_t, and ground range R_g between the two. The flat reflecting surface allows the reflection point to be easily found by representing the target reflection with a virtual target underneath the true target at an altitude of $-h_t$. One can show that the distance along the bounce path from the radar to the true target is equal to the straight-line path length from the radar to the image target. The DD path is the hypotenuse of a right triangle of base R_g and height $(h_r\text{-}h_t)$, and the II path is the hypotenuse of a right triangle of base R_g and height $(h_r\text{+}h_t)$. The DD and II slant range R_{DD} and R_{II} are found using Pythagorean's theorem

$$R_{\mathrm{DD}} = \sqrt{R_g^2 + \left(h_r - h_t\right)^2} \tag{4.12}$$

$$R_{\mathrm{II}} = \sqrt{R_g^2 + \left(h_r + h_t\right)^2} \tag{4.13}$$

As a prelude to finding an approximation for the slant range difference between the DD and II paths, rewrite (4.12) and (4.13) as

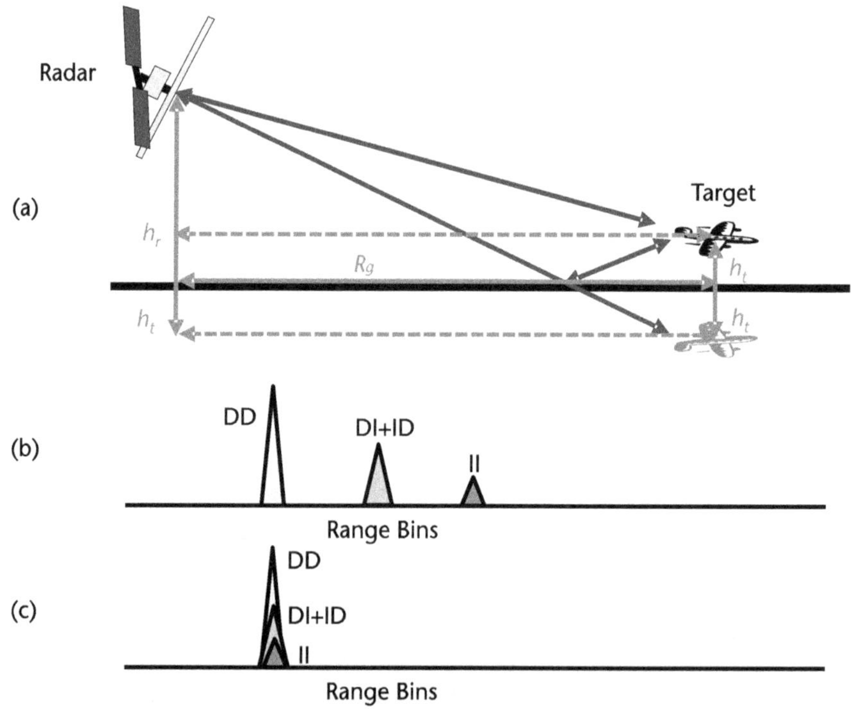

Figure 4.11 (a) Multipath with flat Earth ranges; (b) multipath returns at high altitudes and fine range resolution; and (c) returns at low altitudes and coarse range resolution.

$$R_{\text{DD}} = R_g\sqrt{1+\left(\frac{(h_r - h_t)}{R_g}\right)^2} \tag{4.14}$$

$$R_{\text{II}} = R_g\sqrt{1+\left(\frac{(h_r + h_t)}{R_g}\right)^2} \tag{4.15}$$

As stated earlier, multipath is more readily supported at low grazing angles, which holds at low altitudes and long ranges. The scenario described above with a shipborne radar operating against a cruise missile is illustrative: The radar and target altitudes are tens of meters, often less, while the range to the radar horizon is many kilometers. In short, one can often assume that $h_r << R_g$ and $h_t << R_g$. Then, the second term under the radicals in (4.14) and (4.15) is much less than 1, allowing these formulas to be simplified through the binomial approximation

$$(1+x)^{\alpha} \cong 1+\alpha x \text{ for } |x| << 1 \tag{4.16}$$

Invoking (4.16), (4.14) and (4.15) become

$$R_{\text{DD}} \cong R_g\left(1+\frac{1}{2}\left(\frac{(h_r - h_t)}{R_g}\right)^2\right) = R_g + \frac{(h_r - h_t)^2}{2R_g} \tag{4.17}$$

$$R_{\text{II}} \cong R_g\left(1+\frac{1}{2}\left(\frac{(h_r + h_t)}{R_g}\right)^2\right) = R_g + \frac{(h_r + h_t)^2}{2R_g} \tag{4.18}$$

The slant range difference between R_{DD} and R_{II} is found by taking the difference between (4.17) and (4.18), expanding the squared terms, and simplifying

$$R_{\text{II}} - R_{\text{DD}} \cong \frac{2h_r h_t}{R_g} \tag{4.19}$$

Because the DI and ID returns fall halfway between the DD and II returns

$$R_{\text{DI}} - R_{\text{DD}} = R_{\text{ID}} - R_{\text{DD}} \cong \frac{h_r h_t}{R_g} \tag{4.20}$$

Target multipath returns will appear downrange from the direct skin by a range that increases with radar and target heights and decreases with the radar-target ground range.

For example, if the radar and target are both at an altitude of 1 km, and the ground range between the two is 10 km, the DI+ID return will appear 100m downrange from the DD return, and II 100m downrange farther still. If the radar range resolution is 30m, consistent with a 5-MHz waveform bandwidth, the three detections (DD, DI+ID, II) will fall into distinct range bins with several bins between them. As another example, if the radar and target are both at an altitude of 10m and the ground range between them is 1 km, the DI+ID return will appear 0.1m downrange from the DD returns, and the II another 0.1 ms downrange. All three returns will appear in the same 30-m range bin. However, if the radar operates at 10 GHz for a wavelength of 0.03m, the three returns will be separated by several cycles of phase. They may constructively reinforce or destructively interfere, depending on the exact ground range, altitude, wavelength, and ground-bounce conditions.

These two cases are illustrated at the bottom of Figure 4.11. In 4.11(b), the radar and target altitudes are high enough, and the radar range resolution is fine enough, that the three range-compressed returns fall into separate bins. In 4.11(c), however, radar and target altitudes are low enough and the radar range resolution coarse enough that all three returns fall into the same bin. When the returns are resolved, multipath offers more than one opportunity for target detection. Recall that the DI+ID return may be significantly stronger than the DD return; as a result, multipath return may support target detection at longer ranges than the direct path alone. Finally, the range differences between the multipath returns may be applied to equations like those in (4.19) and (4.20) to generate a coarse estimate of the target altitude.

The unresolved case at the bottom of Figure 4.11 has long been of interest to radar engineers. The composite return may be significantly stronger or weaker than the direct-path skin-only return, depending upon the relative amplitude and phases of the three ground-bounce returns. The one-way propagation factor F is defined as the ratio of the multipath-induced signal complex voltage to its free-space complex voltage. The magnitude of F is therefore the ratio of the multipath and free-space amplitudes. This ratio may be as low as zero (when the direct and ground-bounce paths have the same amplitude but are out of phase and so cancel) or as high as 2 (when the two paths have equal amplitude and are in phase and so reinforce). The one-way power change due to multipath is F^2, the amplitude ratio squared. For monostatic radar the outbound and return path geometries are reciprocal, so the propagation factor squared applies at the target on transmit and again at the receive antenna, for an aggregate change to the target power of F^4. This term appears in RREs that represent multipath effects, with an upper limit of 16 (for a gain of 12 dB), and a lower limit of zero.

The propagation factor depends on radar height and target altitude, operating frequency, Earth surface type and conditions, and range between the radar and target. Furthermore, in an unfolding engagement, this factor will vary over time as the relative geometry between the radar and target evolves. Historically, a vertical coverage diagram (VCD) was used to illustrate the impact of unresolved multipath on target detectability for stationary, ground-based radars. For a detailed discussion on VCDs, the reader is referred to [8].

4.6 Summary

This chapter introduced and quantified four propagation phenomena, summarized as follows.

- The radar horizon presents a hard limit on radar-target visibility. For ground and air radars and targets, the slant range to the radar horizon scales approximately with the square root of altitude. For a target and radar at fixed altitudes, the maximum slant range for visibility is the sum of the slant ranges to the radar horizon for the radar and for the target. Under standard conditions, atmospheric refraction causes a downward bending of RF rays, which serves to increase the slant range to the radar horizon.
- Atmospheric attenuation imparts losses to radar signals due to slight RF heating of atmospheric gases and water vapor. Loss severity is quantified by specific attenuation in decibels per kilometer and depends on radar frequency and atmospheric pressure, temperature, and water vapor density. Losses tend to be higher at lower altitudes, in more humid air, and at higher RFs.
- The index of refraction is defined as the speed of light in a vacuum divided by the speed of light in the medium of interest. Within the Earth's atmosphere, index-of-refraction variations over altitude induce refraction, causing RF rays to bend away from straight-line paths. Unusual index-of-refraction profiles in the atmosphere give rise to anomalous propagation conditions such as ducting.
- Given low grazing angles and a relatively smooth and conductive surface, the Earth will coherently reflect RF energy and support a secondary propagation path between the radar and target. The addition of this indirect path to the usual free-space direct path yields four possible two-way RF trajectories between the radar and the target. The time-delay differences between these paths is proportional to the product of the radar and target altitudes and inversely proportional to the range between the two. With long time delays, additional multipath target returns will appear downrange from the direct-path return. With short time delays, direct and multipath returns will appear in the same range bin; then the relative phases between the paths is key. When the direct and multipath returns are in-phase, constructive reinforcement of the target signal occurs. Conversely, constructive cancelation occurs when the returns are out-of-phase. Multipath, then, may increase or decrease target signal over free-space conditions, depending on the instantaneous radar and target geometries.

References

[1] Skolnik, M. I., *Radar Handbook*, Third Edition, New York: McGraw-Hill Book Company, 2008.

[2] Doerry, A. W., *Earth Curvature and Atmospheric Refraction Effects on Radar Signal Propagation*, Sandia Report SAND2012-10690, January 2013.

[3] "Reference Standard Atmospheres," ITU-R P.835-5.

[4] "Attenuation by Atmospheric Gases," ITU-R P.676-10.

[5] "The Radio Refractive Index: Its Formula and Refractivity Data," ITU-R P.453-10.

[6] Richards, M. A., J. A. Scheer, and W. A. Holm, *Principles of Modern Radar: Basic Principles*, Raleigh, NC: SciTech Publishing, Inc., 2010.

[7] Gehman, J. Z., "TEMPER User Guide 3.2," Johns Hopkins University Applied Physics Laboratory LLC, 2020.

[8] Blake, L. V., *Radar Range-Performance Analysis*, Second Edition, Norwood, MA: Artech House, 1986.

CHAPTER 5

Radar Cross-Section

5.1 Introduction

Other than the radar-to-target range, the radar cross-section (RCS) is the only parameter in the RRE that is target-dependent, and therefore outside of the full control of the radar designer. We do say "full control," because there are certain radar parameters (e.g., frequency and polarization) that may be adjusted to enhance the target RCS and therefore improve detectability. The radar influence is very limited, however; once those parameters are selected, the target RCS is what it is. Thus, an accurate representation of RCS behavior in our simulation environment is key to obtaining realistic results. This chapter starts with an overview of basic RCS concepts, including a formal definition. It then presents several modeling methodologies, starting with "brute-force" approaches that range from simple summation of point scatterers and/or scattering shapes, to mathematically complex methods known as computational electromagnetics (CEM). Statistical modeling of RCS fluctuation, embodied by the well-known Swerling models, is then discussed in detail. Finally, Section 5.7 presents look-up table models as an efficient approach for incorporating measured data into the simulation environment.

5.2 The Concept of RCS

Once the EM energy emitted by the radar reaches an object of interest (which we will refer to as a "target"), it bounces off it and scatters in all directions. We are concerned with the amount of energy that is specifically scattered back in the direction of the radar receiver; this is quantified by what we call the target RCS. The IEEE [1] defines the RCS as "a measure of the reflective strength of a radar target; usually represented by the symbol σ and measured in square meters. RCS is defined as 4π times the ratio of the power per unit solid angle scattered in a specified direction to the power per unit area in a plane wave incident on the scatterer from a specified direction. More precisely, it is the limit of that ratio as the distance from the scatterer to the point where the scattered power is measured approaches infinity."

This definition can be expressed in the following equation form:

$$\sigma = \lim_{R\to\infty}\left(4\pi R^2 \frac{PD_r}{PD_i}\right) \tag{5.1}$$

Without delving too much into the details, (5.1) helps to illustrate the following two major points related to RCS:

- The RCS of a target is independent of the range between said target and the radar;
- RCS has units of area.

From the second point, the RCS is then considered an "effective area." It is generally a complex function of many factors, some related to the radar and some related to the target itself. Radar-related factors include frequency and polarization of the transmit signal, as well as the viewing angle of the radar with respect to the target. A bistatic radar configuration (radar receiver at a different location than the transmitter) will also result in a different observed RCS than for a monostatic radar. On the target side, characteristics such as its shape, material(s), size (relative to the radar wavelength), and orientation relative to the radar look direction will all impact the RCS seen by the radar. Thus, although the RCS value could be relatively constant under certain conditions, the norm is that its amplitude will fluctuate around some mean value. RCS fluctuations arise due to the physics of EM scattering. A rigorous mathematical treatment is outside the scope of this book, but the interaction between the incident wave and the target can be described qualitatively. Electrical currents are induced on the target surface due to boundary conditions. These surface currents propagate through the body of the target and reradiate back to the atmosphere not unlike an antenna. The intensity and phase of these complex currents can be calculated via a surface integral. For a given scattering direction, the integral sums up all contributions from the surface currents, and large returns occur when the currents are in phase. As the target orientation and the viewing angle from the radar changes over time, the current amplitudes and phases change to form a scattering pattern of seemingly random peaks and nulls corresponding to constructive and destructive interference. This concept is notionally illustrated in Figure 5.1 for a monostatic radar configuration. We only consider monostatic radar in this discussion.

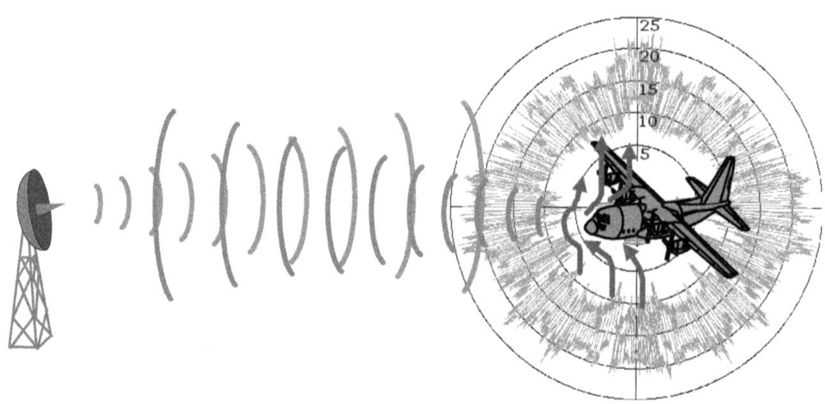

Figure 5.1 RCS scattering results from the radar EM wave reflecting off the target.

Recall that RCS has units of area [e.g., square meters (sqm) or m^2 in the SI system]. Due to the significant range in values arising from RCS fluctuations, it is also common to express RCS in decibels relative to one square meter (dBsm), defined as (using MATLAB notation)

```
RCS_dBsm = 10*log10(RCS_sqm)                                      (5.2)
```

Thus, a 1-m^2 RCS is expressed as 0 dBsm, and so on. A typical commercial aircraft may be in the 3–20-dBsm range, while military aircraft featuring stealth technology may be "much smaller" than 0 dBsm.

5.3 Scattering Surfaces

Recall that as the geometry view angle (or aspect angle) to the target changes, the total RCS response will tend to change as well. One simplified point of view that helps understand this phenomenon is when the total RCS is viewed as composed of reflections from discrete, localized regions in the target body. These localized regions are referred to by many terms in the literature, such as "scatterers," "point scatterers," "scattering centers," "scattering points," "reflectors," and "facets." These multiple scattering points will add or subtract from each other depending on their electrical phase (i.e., the path difference between each scatterer and the radar in units of wavelength). This gives rise to a phasor (or vector) summation of returns from many scattering centers, expressed as

$$\sigma_{\text{total}} = \left| \sum_{n=1}^{N} \sqrt{\sigma_n} e^{j\varphi_n} \right|^2 \tag{5.3}$$

where σ_n is the RCS corresponding to the nth scatterer located at a range R_n from the radar, and $\varphi_n = 4\pi R_n/\lambda$ is its range-dependent phase.

These scattering centers represent physical features on the target that cause significant returns; examples are specular points (such as flat surfaces on the target), cavities (such as jet engines), and dihedrals and trihedrals (corner reflectors). The relative strength of the reflections from these scattering shapes is dependent on their electrical size (dimensions relative to radar wavelength) and their orientation relative to the incident wavefront. Table 5.1 illustrates selected examples of simple shapes and their dependencies.

Models of relatively complex targets can be constructed by combining some of these simple shapes. For example, a cylinder may be used as the body of a missile or aircraft. A cone (not shown in Table 5.1) may be used as the nose or radome; flat plates and dihedrals may represent fins and tails.

One such target built using this approach is the "stovepipe airplane" built at GTRI and shown in Figure 5.2. The actual airplane was about three feet long, and it was taken to a measurement chamber where RCS data was collected as a function

Table 5.1 Simple Scattering Shapes*

Geometry	*Type*	*Frequency Dependency*	*Size Dependency*	*Formula*	*Remarks*
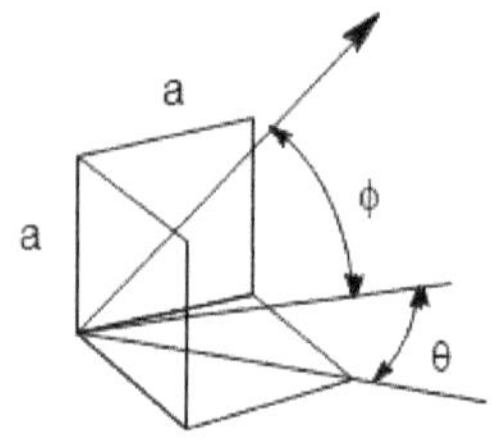	Square trihedral corner retro-reflector	f^2	L^4	Maximum $\sigma = 12\pi a^4/\lambda^2$	Strongest return; high RCS due to triple reflection
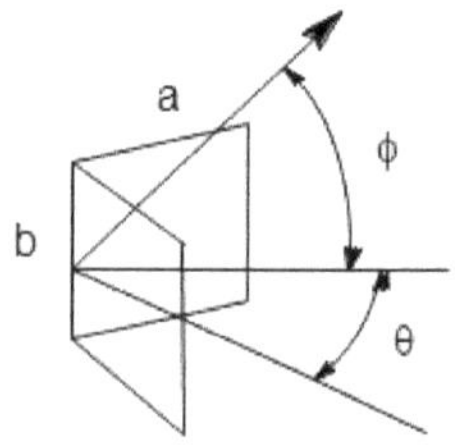	Right dihedral corner reflector	f^2	L^4	Maximum $\sigma = 8\pi a^2 b^2/\lambda^2$	Second strongest; high RCS due to double reflection, tapers off gradually from the maximum with changing θ and sharply with changing ϕ
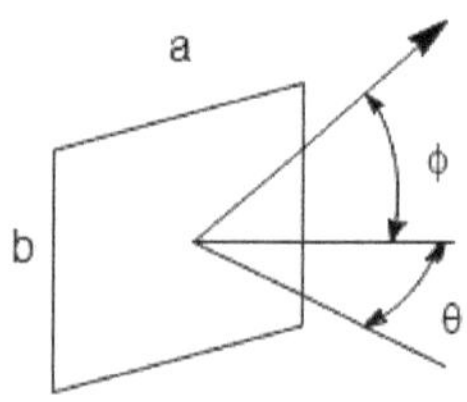	Flat plate	f^2	L^4	Maximum $\sigma = 4\pi a^2 b^2/\lambda^2$ Normal incidence	Third strongest; high RCS due to direct reflection, drops off sharply as incidence changes from normal
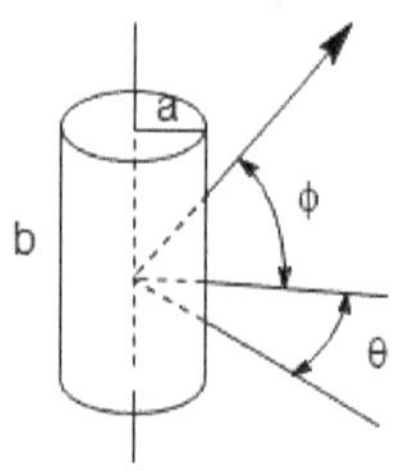	Cylinder	f^1	L^3	Maximum $\sigma = 2\pi a b^2/\lambda^2$ Normal incidence	Prevalent cause of strong, broad RCS over varying θ. Drops off sharply as ϕ changes from normal; can combine with flat plate to form dihedral corner detector
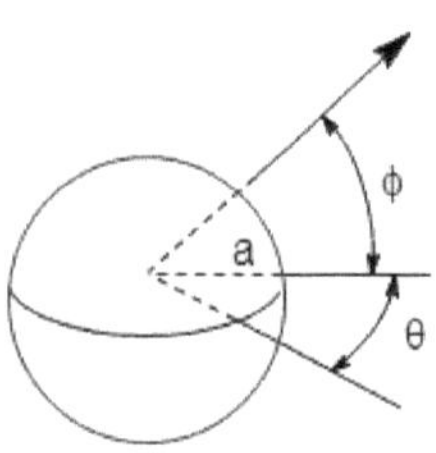	Sphere	f^0	L^2	High frequency asymptote $\sigma = \pi a^2$ Normal incidence	Prevalent cause of strong, broad RCS peaks other than those due to large openings in target body; energy defocused in two directions

*Reproduced, with permission, from [2].

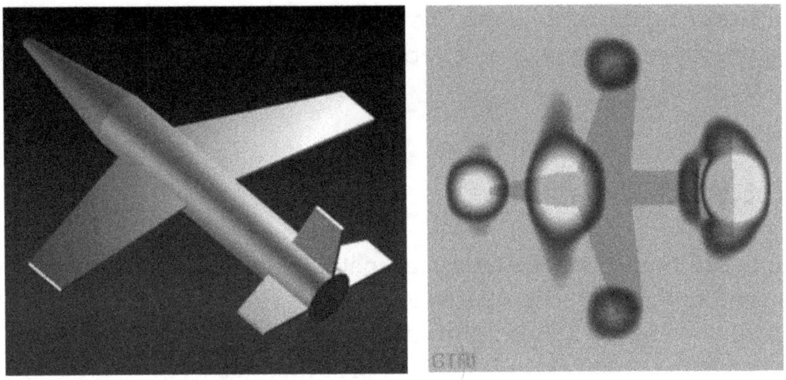

(a) (b)

Figure 5.2 (a) Image renderings of stovepipe aircraft and (b) main scattering contributions at nose aspect.

of azimuth angle. This experiment demonstrated that even a relatively smooth target is capable of showing the majority of scattering mechanisms observed in more realistic and complex targets: strong specular returns from flat surfaces, double and triple bounce reflections from corners, and diffraction effects from edges and other discontinuities.

Figure 5.2(b) illustrates the main sources of reflections and scattering for a single aspect angle—specifically, when the radar wave hits the target right on the nose (i.e., the emitter is to the left of the figure). Figure 5.3 shows RCS measurements from the stovepipe aircraft taken at 6 GHz and shown in polar format for both horizontal and vertical polarizations. The airplane outline is overlaid for

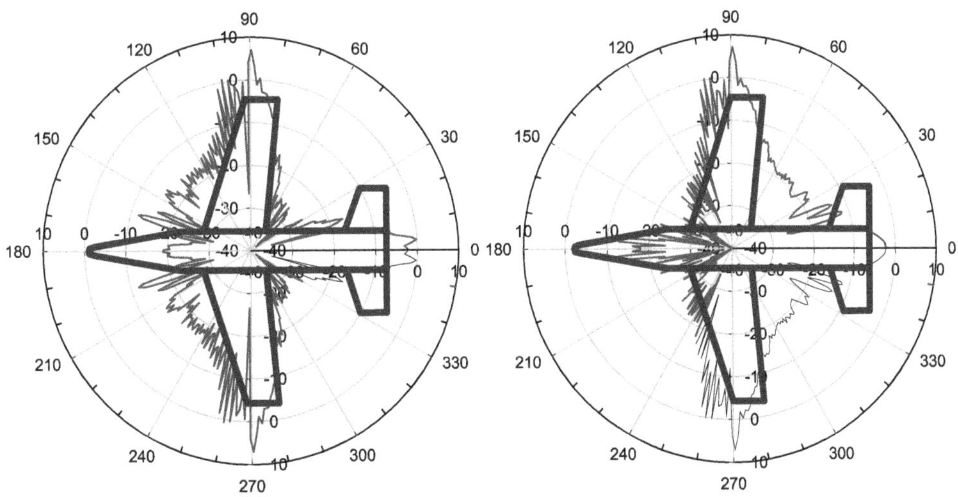

(a) (b)

Figure 5.3 Stovepipe airplane RCS at 6 GHz: (a) H-polarization and (b) V-polarization.

reference on the orientation. Note that the aspect convention adopted here is 180 degrees for nose aspect; an alternate common convention is to define the nose as 0-degree aspect.

5.4 Scatterer Integration

A closer look at the scatterer summation expression in (5.3) provides additional insight to the interaction between point scatterers in a target. We start by considering the case of a single point scatterer, the $N = 1$ case in (5.3). This is the simplest target there is: an isotropic radiator of scattered energy. The closest real-life object that approximates this target is a perfect conducting sphere, which is independent of its own orientation as well as the relative location of the radar; this is because the physical cross-sectional area facing the incident field is always the same, given by the area of the circle πr^2, where r is the radius of the sphere. The electrical size of the sphere can be expressed as kr, where k is the wavenumber $2\pi/\lambda$. The product kr is therefore the circumference of the sphere normalized by the wavelength.

The RCS for a sphere exhibits an oscillatory behavior as a function of electrical size, as illustrated in Figure 5.4. The RCS axis is scaled to the cross-sectional area. The electrical size axis may be interpreted as either increasing sphere size at a fixed frequency, or as increasing frequency for a fixed size sphere. In the Rayleigh region, the RCS increases monotonically until it reaches a peak value at $kr = 1$. At that point, induced currents that traverse the sphere circumference are in phase with the specular point, and therefore add constructively (their phase difference is exactly 2π). In the resonant region (or Mie region, named after German physicist Gustav Mie), the RCS response is a decaying exponential that converges asymptotically to a value equal to the sphere cross-sectional area. In the optical region ($kr > 10$), the oscillations have decayed to the point where the RCS is assumed to be constant; therefore $\sigma = \pi r^2$ at that point, independent of viewing angle and also

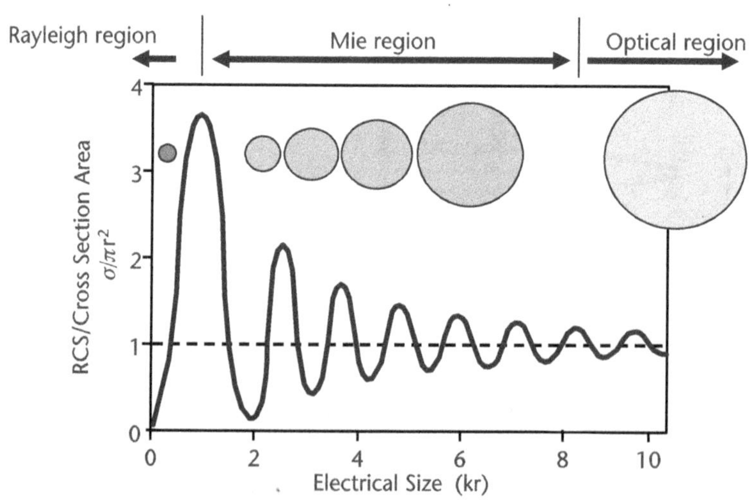

Figure 5.4 RCS of a perfect conducting sphere.

assumed to be independent of frequency. This is the basis for the point scatterer model of a fixed RCS.

The interaction between point scatterers may be examined by extending the single case to that of two scatterers [i.e., $N = 2$ in (5.3)]. For simplicity, assume that the two scatterers are identical (i.e., their complex reflectivity has the form $\sqrt{\sigma} = Ae^{j\phi}$). We will assume $A = 1$ without any loss of generality. Furthermore, assume that they are separated by a distance d, as illustrated in Figure 5.5. The radar is assumed to be in the far field, such that a planar wave front arrives at the scatterer array. This plane wave must travel an additional distance Δx from the first to the second scatterer, and this additional distance is a function of the angle between the scatterer axis and LOS between the radar and the scatterers. This angle α is referred to as the aspect angle.

Based on the stated conditions, it can be seen that the total normalized RCS is

$$\sigma = \left| 1 + e^{j\left(\frac{2\pi}{\lambda}2\Delta x\right)} \right|^2 \tag{5.4}$$

The additional factor of 2 in the exponent of the complex exponential in (5.4) is due to the round-trip distance from the radar to the scatterers and back. As the aspect angle is varied from 0 to 360 degrees, so does the distance Δx (and therefore the relative phase), such that the two scatterers interfere with each other constructively and destructively. Results from a MATLAB model of this two-scatterer example are shown in Figure 5.6. The scatterers are separated by one full wavelength ($d = \lambda$).

The RCS lobe pattern observed in Figure 5.6 is highly symmetric and looks very similar to the antenna pattern of a uniform linear array (ULA) of two elements. This is no coincidence, as in fact the math is almost identical, except for an extra factor of 2 in (5.4) due to the two-way propagation. (Transmit-and-receive antenna patterns are concerned with one-way propagation only.)

This approach may be generalized to any number of scatterers organized in arbitrary configurations. A second MATLAB example features nine identical scatterers in a square grid, as shown in Figure 5.7. In this example, the distances are not normalized; the units are in meters ($d = 0.1$m). This allows us to easily illustrate the RCS dependency in frequency.

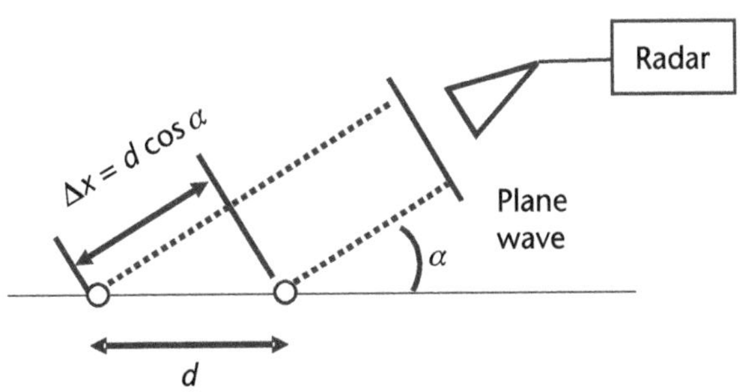

Figure 5.5 Setup for RCS of two identical scatterers.

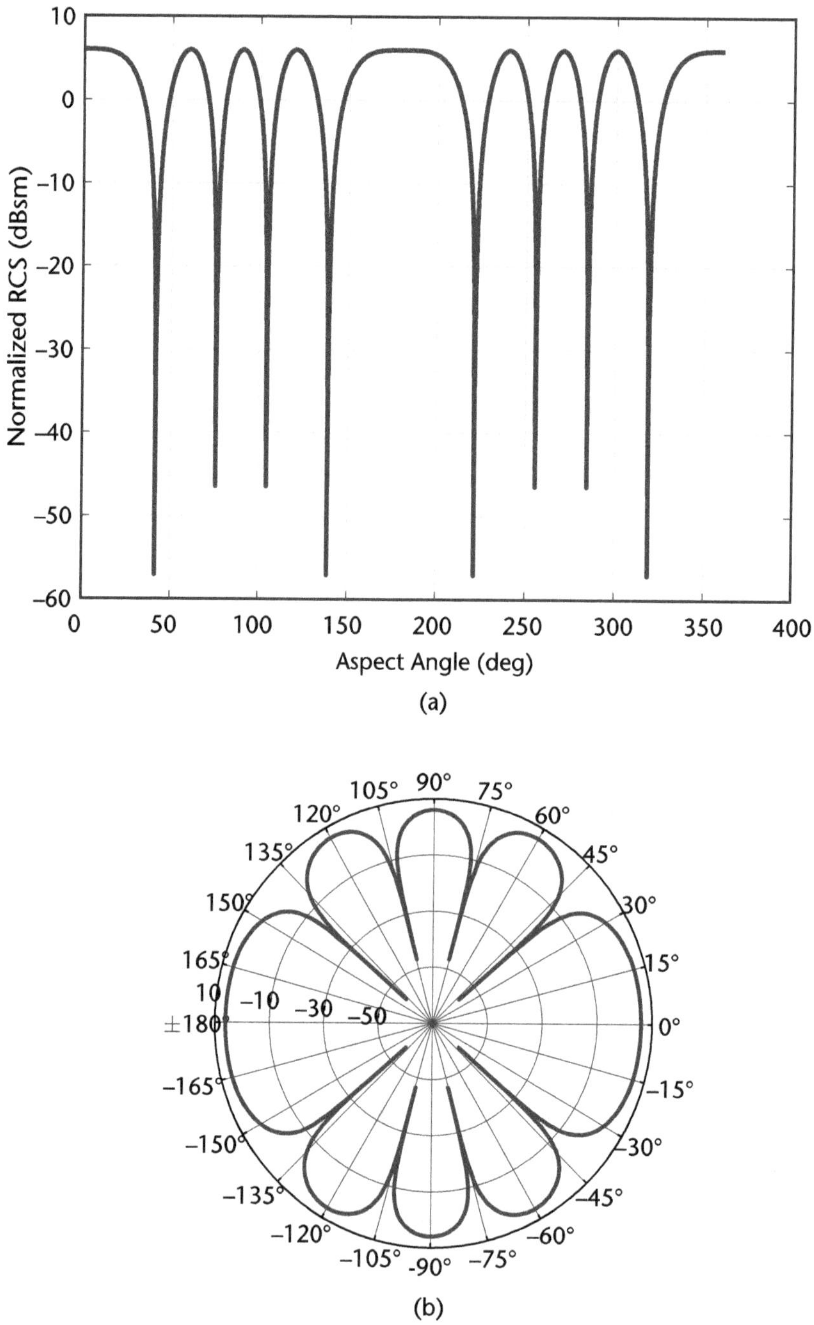

Figure 5.6 Total RCS of two-scatterer example in (a) Cartesian and (b) polar coordinates.

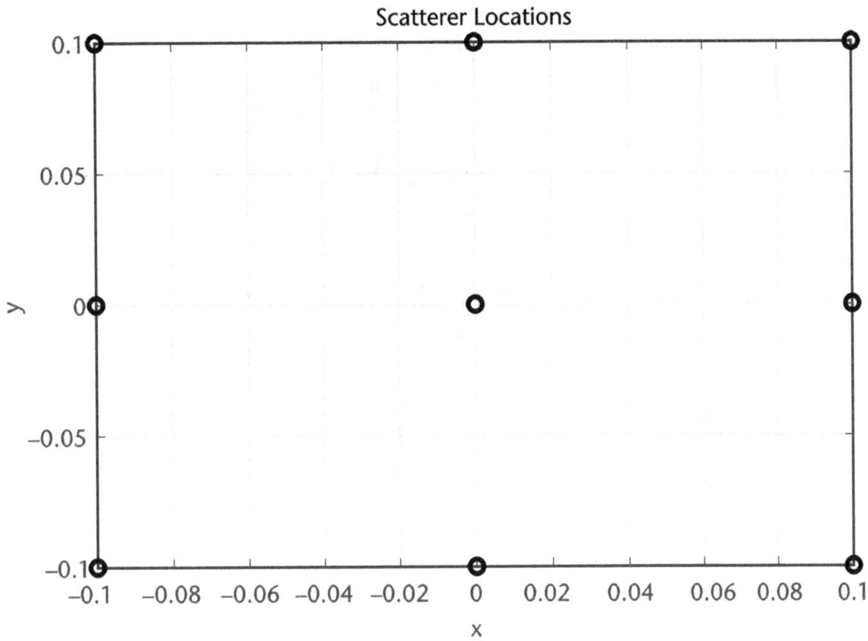

Figure 5.7 Scatterer layout for second MATLAB example ($N = 9$).

Figure 5.8 displays the RCS polar patterns at 3 GHz and 10 GHz for the same nine-scatterer array, showing the vastly different results that arise from the coherent combination of scatterer phases as their relative differences change due to the change in wavelength.

This simple modeling approach may be expanded to obtain more complex RCS behavior. The different scatterers may be assigned different amplitudes, for example. The scatterers may be arranged in more complex (3-D, nonsymmetrical) configurations. Last, shadowing effects may be added, such that scatterers on the "far side" of the radar viewpoint are obstructed and do not contribute to the total return for that particular aspect angle.

Many clutter simulations follow this same approach when calculating ground or sea clutter returns. This "brute-force" approach of calculating total RCS, although relatively straightforward in principle, can become computationally intensive as the number of scatterers increases, and they are made more heterogeneous to approximate more realistic targets.

This need to represent more complex targets takes the modeling efforts in two different directions. Section 5.5 discusses CEM, where physics-based RCS modeling and corresponding computational burden is fully embraced. A completely different approach follows in Section 5.6 on Swerling RCS models, where RCS is viewed as a statistical process.

5.5 Computational Electromagnetics

CEM consist of methods for modeling the interaction of EM fields with physical objects. This is achieved by obtaining numerical solutions to Maxwell's equations.

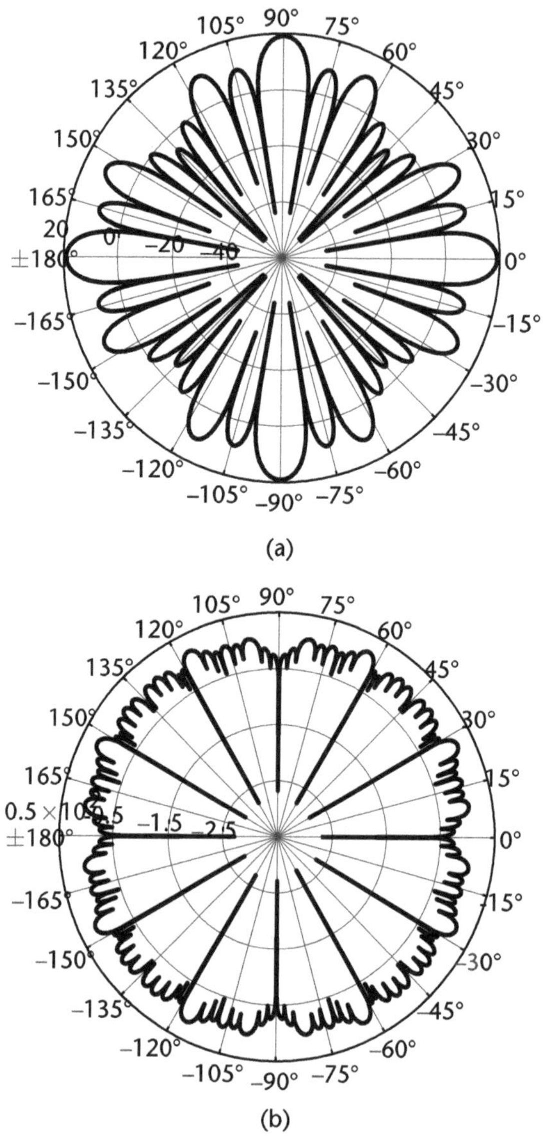

Figure 5.8 Total RCS of nine-scatterer example; results shown for (a) 3 GHz and (b) 10 GHz.

In addition to RCS modeling, these software tools are applied to a broad class of EM problems such as antenna modeling and other wave propagation phenomena.

Typically, the process starts by building a "wireframe" model of the object of interest, where the continuous surfaces are discretized into small facets. The most sophisticated tools will also incorporate additional information such as material composition, roughness, and even surface coatings and paint. The electrical size of the object (dimensions relative to the wavelength) and the number of facets are the main drivers of the execution time of the model, in addition to the actual numerical method selected.

There are two general categories of numerical algorithms in CEM. The low-frequency (LF) methods attempt to find an exact solution to Maxwell's equations; they are therefore believed to be the most accurate, but are also computationally intensive. They generally require long runtimes and large amounts of computer memory. The high-frequency (HF) methods, on the other hand, apply a number of simplifying assumptions to obtain approximate solutions, resulting in shorter runtimes. There are also hybrid methods that combine the two approaches to various degrees, in an attempt to strike a balance between run time and accuracy.

LF methods are typically applied to electrically small targets, meaning that the longest dimension is in the order of tens of a wavelength (this is where the low-frequency terminology comes from). Perhaps the best known LF example is the MoM algorithm, which attempts to solve the integral form of Maxwell's equations. Another popular LF approach is the FDTD algorithm, which tackles the differential equation form of Maxwell's equations.

HF methods make use of asymptotic approximations and are therefore significantly faster than LF methods. They are able to calculate the RCS of "large" objects (i.e., in the order of hundreds of wavelengths). Secondary effects that may be significant are seldom accounted for, however, so results may not be as accurate as those of LF methods. A good example of an HF method is the geometric optics (GO) approach, which describes the wave propagation in terms of rays and mostly models specular returns. It can be extremely fast, but it does not account for important effects such as diffraction and interference.

The next step up in complexity is the PO approach, where the EM field is now interpreted as a wave, and reflections from an object are determined by approximating surface currents induced by the field and integrating them over the object surface. A hybrid method that combines the previous two is the shooting-and-bouncing rays (SBR) approach. SBR uses GO to determine the field on the object surface, then employs PO integration to calculate the scattered field. None of these approaches model diffraction effects from edges and other boundaries. Separate approaches to handle diffraction have been developed, including the geometrical theory of diffraction (GTD), which is an extension of GO, and the UTD, which is itself an extension of GTD.

There are many commercial and government-owned CEM software suites available. Most of them implement multiple variants of the methods noted here, which the user may select as numerical solvers for a particular problem. The following are selected examples included for the benefit of the reader, and their mention is not to be construed as an endorsement.

One commercial example is Altair's Feko® [3], which features multiple solvers, including MoM and some HF techniques, as well as hybridized options. A government-owned example is Leidos' Xpatch®, developed for the US Air Force and available only to DoD contractors [4]. Xpatch uses the SBR method as its main solver, with some additional options. Another suite similar to Xpatch is Lucernhammer MT by Tripoint Industries, Inc. [5].

Many DoD contractors have developed their own proprietary software, such as GTRI's own GTS/TRACK [6], which primarily uses PO to model the RCS of targets composed of primitive geometric shapes, such as triangular facets, quadrilaterals, ellipsoids, dihedrals, and trihedrals.

5.6 Swerling Models

Statistical methods for characterizing RCS behavior are attractive for a couple reasons. The RCS amplitude fluctuations over time of many targets of interest can be modeled as a random process, due to the inherent complexity of realistic targets. In addition, a probabilistic characterization provides a mathematically rigorous framework for both evaluation and prediction of performance.

The most common statistical models for RCS belong to the family of Swerling target models, named after Peter Swerling (1929–2000). A renowned mathematician, Swerling is recognized for his many contributions to the field of radar engineering. He developed his fluctuating target models in the 1950s while working for the RAND Corporation [7]. He extended the work of another brilliant RAND mathematician, Jess Ira Marcum (1919–1992), who developed a theory for target detection by integrating multiple radar pulses in the late 1940s. In Marcum's formulation, the target RCS returns were assumed to have constant amplitude, similar to the metal sphere (point target model) discussed in Section 5.4.

This nonfluctuating target type was initially dubbed the "Marcum target," and later became known as a "Swerling type 0" target in Swerling's work. The remaining four models are numbered Swerling 1 through 4, and are defined by a probability density function (PDF) and a fluctuation rate. As a side note, the Swerling 0 target is listed as Swerling 5 in some references, but Swerling 0 is by far the most common terminology.

All fluctuating Swerling model PDFs are variations of the chi-square distribution with n degrees-of-freedom (DOF), given by

$$f(\sigma) = \frac{\left(\frac{n}{2}\right)}{\Gamma\left(\frac{n}{2}\right)\sigma_{\text{avg}}} \left(\frac{n\sigma}{2\sigma_{\text{avg}}}\right)^{\frac{n}{2}-1} e^{-\left(\frac{n\sigma}{2\sigma_{\text{avg}}}\right)}, \quad \sigma \geq 0 \tag{5.5}$$

where σ_{avg} is the average (statistical mean) RCS and $\Gamma(\cdot)$ is the gamma function.

Swerling models 1 and 2 correspond to a chi-square PDF with $n = 2$ DOFs. Setting $n = 2$ in (5.5) results in

$$f(\sigma) = \frac{1}{\sigma_{\text{avg}}} e^{-\left(\frac{\sigma}{\sigma_{\text{avg}}}\right)}, \quad \sigma \geq 0 \tag{5.6}$$

which is recognized as the exponential distribution. Thus, if all other terms in the RRE are constant, the received power from a Swerling 1 or 2 target will be exponentially distributed, which implies that the signal amplitude (magnitude) will be Rayleigh-distributed. This model is closely related to the "Rayleigh-fading" model used in the propagation of radio waves through a transmission medium.

The distinction between Swerling 1 and 2 is the rate of fluctuation; Swerling 1 is known as a "slow-fluctuation" model, while Swerling 2 has a "fast fluctuation." Historically, their fluctuation rates have also been described as "scan-to-scan" and "pulse-to-pulse," respectively, although this terminology can lead to confusion (in particular the pulse-to-pulse description) and is therefore discouraged. This point will be revisited and discussed in the context of the radar processing timeline in Chapter 12.

Swerling models 3 and 4 can be similarly characterized, as they both follow the chi-square PDF, this time with $n = 4$ DOFs. Replacing $n = 4$ in (5.5) results in

$$f(\sigma) = \frac{4\sigma}{\sigma_{\text{avg}}^2} e^{-\left(\frac{2\sigma}{\sigma_{\text{avg}}}\right)}, \quad \sigma \geq 0 \tag{5.7}$$

Just as with Swerling models 1 and 2, Swerling models 3 and 4 have "slow-fluctuation" and "fast-fluctuation" rates, respectively. Table 5.2 summarizes the definition of Swerling types.

The two questions of most interest for the radar analyst regarding Swerling models are listed as follows:

- How do I simulate them?
- How do I characterize or predict radar performance against them?

The second question is addressed in Chapter 11, and RCS modeling is discussed in this section; before we do, however, there is one additional question that is of practical interest—namely, do these models correspond to, or approximate, any real-life target types?

Recall that a spherical target is a good approximation to a Swerling 0 type, provided that it is electrically large [i.e., the radar is operating in the optical region (radius > ~1.6λ)]. For Swerling 1 and 2 (exponential) models, Swerling stated [7] that they represented targets composed of several independent scatterers of roughly the same size. This description is understood to include targets of complex shapes, such as aircraft, ships, and ground vehicles (i.e., most targets of interest to military radars). With a sufficiently large number of independent scatterers, the central limit theorem (CLT) states that the distribution of their sum approximates the Gaussian distribution [8]. Thus, we have a Gaussian-distributed voltage, which results in an exponentially distributed power.

On the other hand, Swerling 3 and 4 models are thought to apply to targets composed of one dominant scatterer in addition to many smaller scatterers of roughly the same size. Real targets that may fit this description tend to have simpler,

Table 5.2 Swerling Type Characteristics

Swerling Type	*Fluctuation Type/PDF*	*Fluctuation Rate*
0 (or 5)	Constant RCS	Nonfluctuating
1	Chi-Square w/2 DOF (Exponential)	Slow
2	Chi-Square w/2 DOF (Exponential)	Fast
3	Chi-Square w/4 DOF	Slow
4	Chi-Square w/4 DOF	Fast

smoother shapes, such as bullets, artillery shells, and reentry vehicles. There is much less confidence in this depiction, however, than in that for Swerling 1 and 2 types, based on experimental data. Partly because of this reason, the Swerling 1 and 2 models tend to be used much more often than 3 and 4 in radar performance analysis, modeling, and simulation efforts.

A Swerling 0 target is modeled as a constant RCS value (e.g., $\sigma = \sigma_{avg}$ where σ_{avg} is the mean RCS). Modeling fluctuating Swerling RCS statistics, on the other hand, consists of generating independent random variables drawn from the right PDF, at the appropriate fluctuation rate. This is illustrated with a Simulink example that generates Swerling 1 or 2 (i.e., exponential) RCS values.

All that is needed to simulate exponential random variables is a Gaussian-distributed random-number generator (RNG), which is included in most function libraries in high-level programming languages. If a Gaussian RNG is not available but a uniformly distributed RNG is, Gaussian numbers can be synthesized from uniform numbers by following the Box-Muller transformation. Box and Muller proved [9] that given two uniformly [0,1] distributed numbers U_1 and U_2, the variables X_1 and X_2 defined as

$$\begin{aligned} X_1 &= \sqrt{-2\ln(U_1)}\cos(2\pi U_2) \\ X_2 &= \sqrt{-2\ln(U_1)}\sin(2\pi U_2) \end{aligned} \tag{5.8}$$

are Gaussian-distributed with zero mean and unit variance.

It is the case that chi-square random variables are obtained by squaring Gaussian variables. The number of DOFs is simply the number of independent Gaussian variables that are added after squaring. Thus, a chi-square variable z with N DOFs is given by

$$z = \sum_{n=1}^{N} x_n^2 \tag{5.9}$$

where the variables x_n are independent, identically distributed (i.i.d.) Gaussian-distributed with zero-mean and variance σ_G^2. Specifically for the case of Swerling 1 and 2 RCS, which are exponential variables, we have $N = 2$ such that

$$z = x_1^2 + x_2^2 \tag{5.10}$$

Notice that the statistical mean value (expectation) of z is

$$\mu_z = E[z] = E\left[x_1^2\right] + E\left[x_2^2\right] = 2\sigma_G^2 \tag{5.11}$$

since the Gaussian variables are zero-mean. To obtain an arbitrary σ_{avg}, we should therefore set the variance of each of the Gaussian variables to $\sigma_G^2 = \sigma_{avg}/2$.

Depending on the particular RNG used, the variance may be set directly, or the user may need to scale the values after they are generated. MATLAB is an example of the latter, where the function `randn` will generate standard (zero-mean, unit-variance) Gaussian numbers. To force an average RCS value of "`sigmat`," a scaling such as the following is required.

```
z = (sigmat/2)*sum(randn(2,1).^2)
```
(5.12)

This MATLAB expression, sans the scaling, is functionally equivalent to (5.10); therefore `z` in (5.12) is exponentially distributed. If `z` represents RCS values, these are then Swerling 1 or 2. Adding the scaling results in a mean value of `sigmat`.

Figure 5.9 shows an equivalent Simulink subsystem for generating Swerling 1 or 2 RCS values. The random-number block is a Gaussian RNG, and its corresponding block parameters can be set as vectors, as shown in Figure 5.9(a). Two Gaussian variables are created at each sample time; the means are set to zero, and the variances are set to `RCS/2`, where `RCS` is the desired average cross-section. The initial seeds are set to different values to ensure independent realizations.

The Gaussian variables exit the RNG together in a single wire but are then separated into individual signals by the Demux block. The approach taken in the example is to create a complex quantity using the Gaussian variables as the real and imaginary parts. The magnitude squared of this complex signal is taken, and it is now exponentially distributed. As an alternate approach, each Gaussian sample could have been squared individually, then added together; this would have required three blocks instead of the two used.

The distinction between whether the RCS values are Swerling 1 or 2 (slow or fast fluctuating) is based on the choice of the scintillation or fluctuation rate `Tscint`, which is set as the sample time for the RNG.

This model was tested by generating a large number of histograms using groups of 1,024 scalar RCS values at a time, then averaging the histograms as an approximation to the continuous PDF. As an additional feature, square roots of the RCS values are collected and processed in a similar way to verify that they follow Rayleigh statistics. The model infrastructure is not intended as a rigorous verification of the model; it is only meant for visual "sanity checks." The two plots created at the end of the run—shown in Figure 5.9(b, c)—display the averaged histograms overlaid with the expected PDF functions, showing a good match. In addition, empirical values for the RCS mean, variance, and standard deviation are calculated and shown to be close to the desired value of 1 sqm.

The simulation of Swerling 3 and 4 RCS values follows in a straightforward manner. Since these chi-square variables have four DOF, we just need to generate four i.i.d. Gaussian variables and scale them properly, as shown in (5.13). This assumes the Gaussian RVs have unit variance.

$$z = \left(\frac{\sigma_{\text{avg}}}{4}\right)\left(x_1^2 + x_2^2 + x_3^2 + x_4^2\right) \tag{5.13}$$

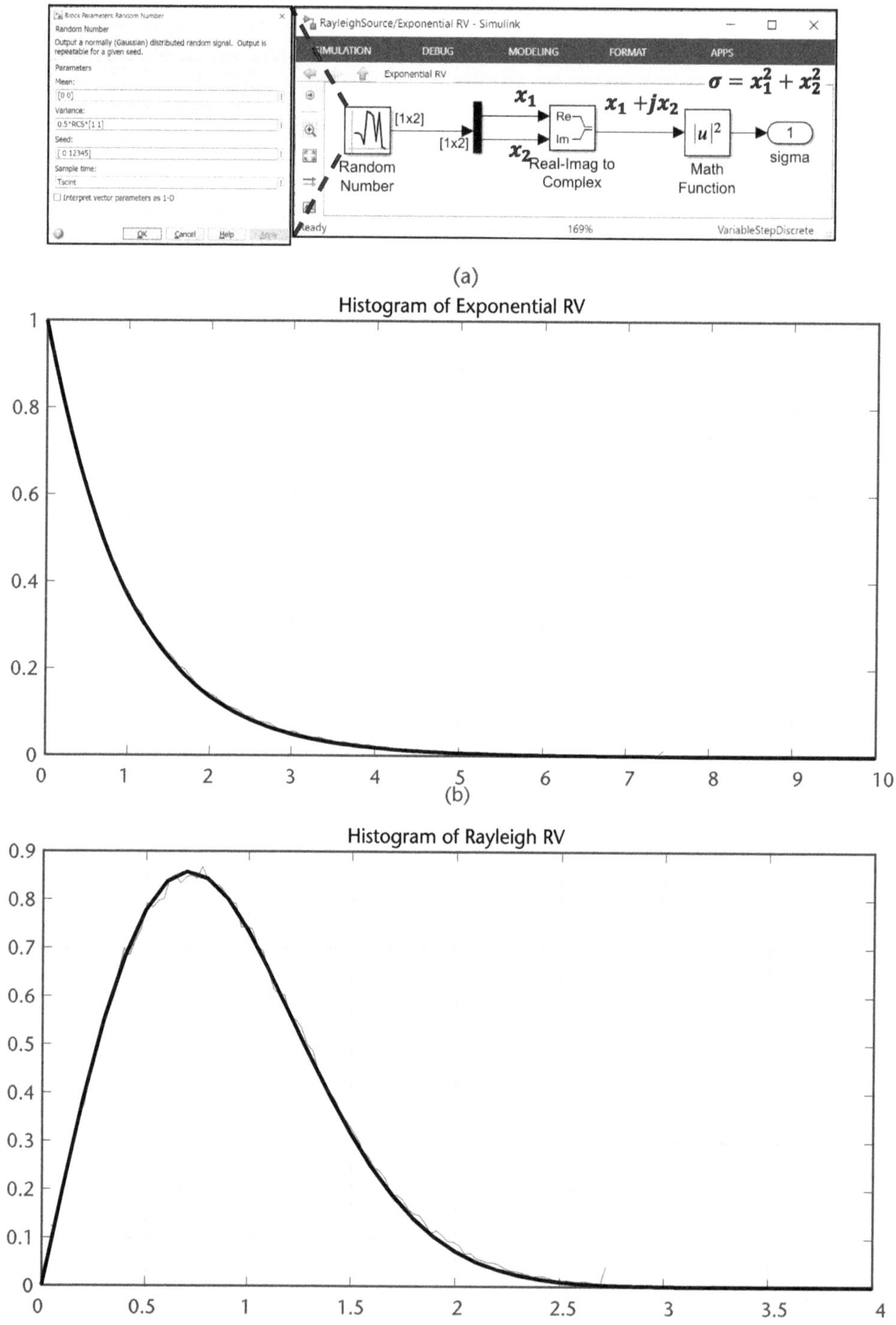

Figure 5.9 Simulink generation of Swerling 1 or 2 RCS values: (a) block diagram, (b) histogram plots for exponential RCS, and (c) Rayleigh magnitude.

5.7 RCS Table Lookup

One additional modeling approach that is often used is the importing of RCS data generated from sources outside of the simulation framework that will employ it. RCS data sets may be obtained from laboratory measurements, or created offline from high-fidelity CEM tools such as those discussed in Section 5.5. Whichever the source, a convenient way to utilize this data is to set them up as lookup tables (LUTs) that return an RCS value based on the input(s) entered, and interpolate between values when an exact match is not available in the data.

The dimensionality of the LUT will match that of the data. RCS values may be available as a function of some combination of azimuth angle, elevation angle, frequency, or polarization, for example. Both MATLAB and Simulink offer utilities that can easily organize the data in LUT for the user benefit. A simple 1-D example of RCS versus azimuth is illustrated both in Cartesian and polar formats in Figure 5.10.

The first step is to load the data into your model. The manner in which this will be accomplished depends entirely on the format of the data file. In this example, the 1-D RCS data comes from an Excel spreadsheet and is divided in two columns, one for the angle values and one for the corresponding RCS values in decibels relative to one square meter. Reading an Excel spreadsheet into MATLAB is straightforward, as there are several utility functions available for this action. The function `xlsread` is used here, in the following format:

```
data = xlsread('Complex RCS Data.xls');
```
(5.14)

The argument is simply the filename, and the output argument `data` is a two-column array of angles and RCS values. We note that although this function is available and still works with recent MATLAB versions, it is not recommended as of version R2019a on, and other options such as `readtable`, `readmatrix`, or `readcell` are suggested instead.

Once the data is loaded in the MATLAB workspace, it is available for use in either a MATLAB or a Simulink simulation. A straightforward MATLAB implementation option is

```
RCSdBsm = interp1(azaspectdata, rcsdBdata, angle)
```
(5.15)

where the first two input arguments correspond to the first and second columns of array `data` in (5.14), respectively. The default interpolation method used in `interp1` is linear, but other options are available. Similarly, there are options for handling input values that lie outside the table.

Figure 5.11 shows an equivalent Simulink implementation using a LUT block. It takes a single input signal, although it may be multidimensional; thus, multiple angles can be queried if packed as a vector. The output dimension will match that of the input.

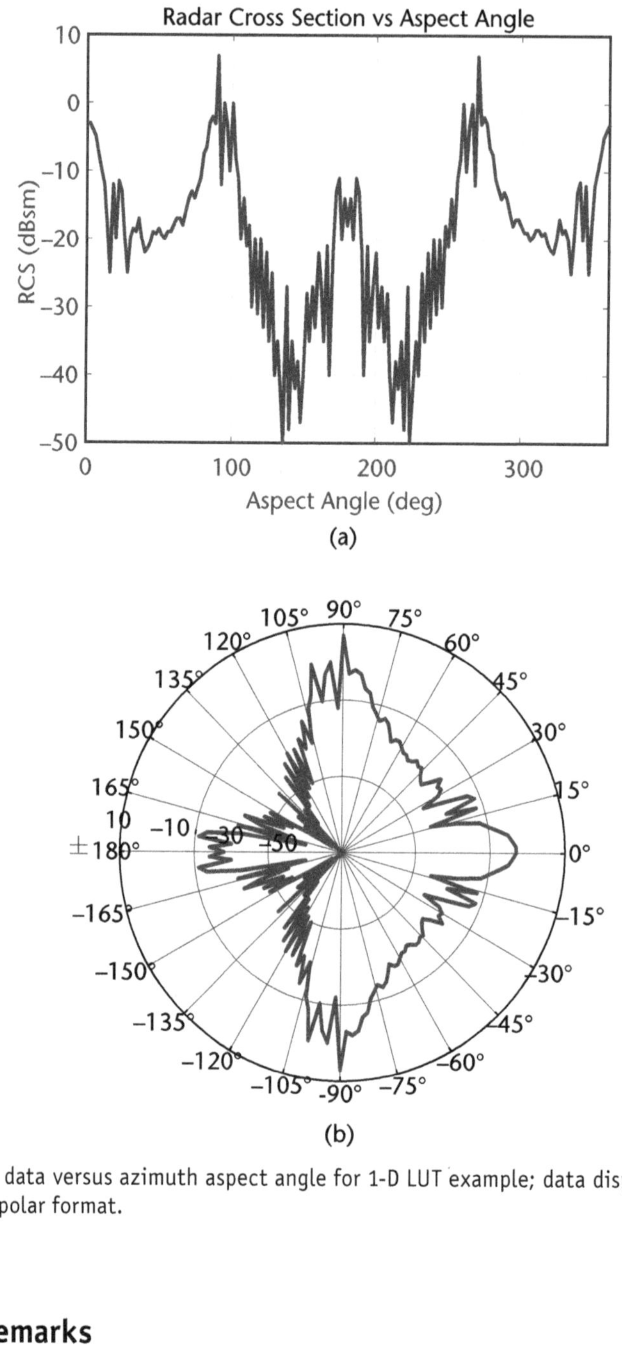

Figure 5.10 RCS data versus azimuth aspect angle for 1-D LUT example; data displayed in (a) Cartesian and (b) polar format.

5.8 Concluding Remarks

This chapter presented a number of approaches to RCS modeling. The most widely known and used by far are the Swerling models. They are the de facto standard RCS models used by the radar community to analyze and predict performance. As discussed in Chapter 11, analysis of radar detection is so intertwined with the Swerling models that the performance equations for probability of detection (the ultimate figure of merit for radar detection) are categorized based on the Swerling target type.

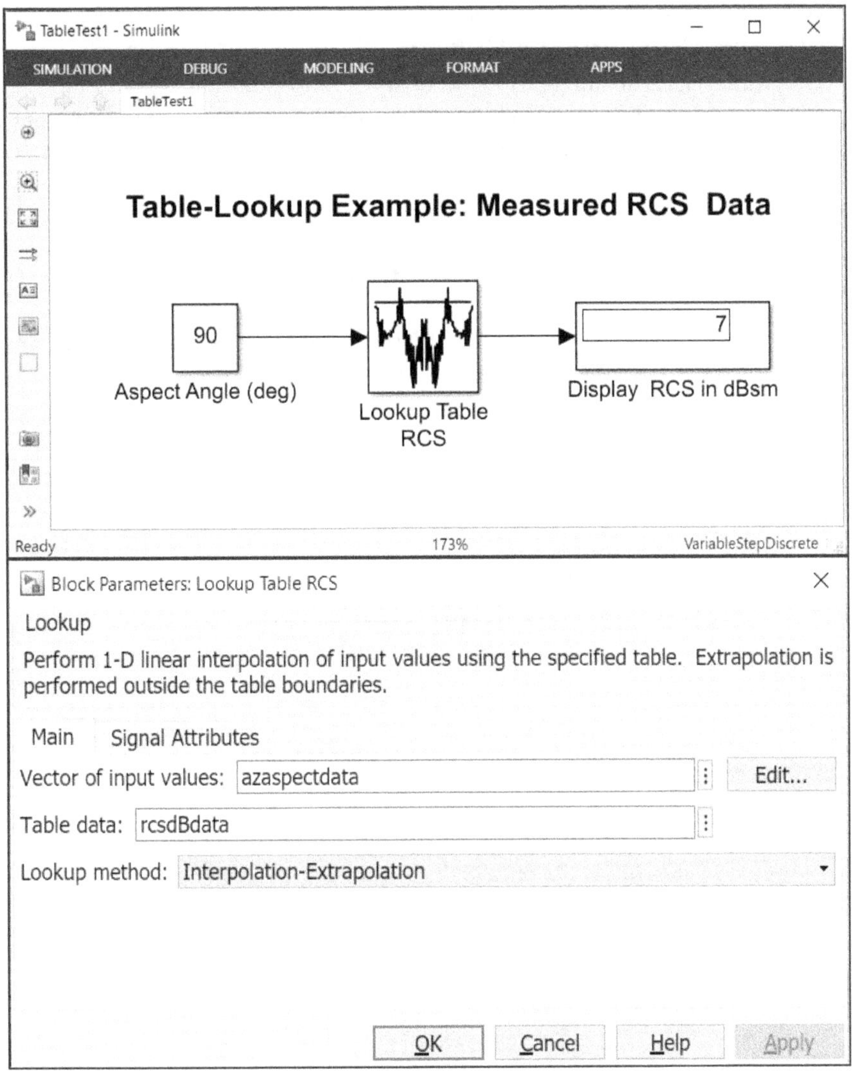

Figure 5.11 Simulink example of RCS LUT: block diagram (top) and LUT block parameter window (bottom).

The software models presented here for Swerling RCS implementation (as well as other options, such as the LUT) are integrated into the radar engagement simulation presented in Chapter 15.

References

[1] IEEE Aerospace and Electronic Systems Society, sponsored by the Radar Systems Panel, "IEEE Standard Radar Definitions," *IEEE Std. 686™-2008*, May 21, 2008.

[2] Knott, E., J. Shaeffer, and M. Tuley, *Radar Cross Section*, Norwood, MA: Artech House, 1985.

[3] https://www.altair.com/feko/.

[4] https://www.afrl.hpc.mil/software/description.html?app=xpatch.

[5] http://lucernhammer.tripointindustries.com/index.html.

[6] Hines, N. R., and Mavris, D. N., "A Parametric Design Environment for Including Signatures Analysis in Conceptual Design," *5th World Aviation Congress and Exposition*, San Diego, CA, October 10–12, 2000.

[7] Swerling, P., "Probability of Detection for Fluctuating Targets," *RAND Memorandum RM-1217*, March 17, 1954.

[8] DiFranco, J. V., and Rubin, W. L., *Radar Detection*, Norwood, MA: Artech House, 1980, pp. 96–98, 104.

[9] Box, G. E. P., and M. E. Muller, "A Note on the Generation of Random Normal Deviates," *Annals of Mathematical Statistics*, Vol. 29, No. 2, June 1958, pp. 610–611.

CHAPTER 6

Clutter

6.1 Introduction

In radar vernacular, clutter refers to distributed, passive, and undesirable radar returns. Moving from right to left through that description, the contributing terms are defined as follows:

- Returns are echoes from objects collected by the radar receive antenna.
- Radar indicates the echoes are the result of RF transmissions.
- Undesirable suggests these returns degrade the performance of the radar in some way. For example, clutter often masks the returns from moving ground vehicles, underway ships, and flying aircraft, lowering the probability of detection and degrading the accuracy of measurements on these targets of interest.
- Passive excludes objects employing electronic amplification. Noise jammers and coherent repeaters may generate undesirable returns that degrade performance but are distinct from clutter.
- Distributed indicates there are multiple objects generating returns dispersed over 2-D or 3-D space. Two-dimensional clutter tends to originate from the Earth's surface, while 3-D clutter is usually meteorological in nature, frequently rain and sometimes hail or snow.

While the Earth's surface is an important source of clutter, its presence does not always equate to an undesirable return. For example, in ground mapping and synthetic aperture radar (SAR) imaging modes, the Earth's surface is the target, and ground and ocean returns are not considered clutter, per se. Similarly, the Earth's surface may modify target returns through multipath effects, as described in Chapter 4, but multipath is not considered a clutter effect; should the target vanish, multipath returns disappear, but direct clutter returns from the Earth's surface persist. Finally, jammer returns may scatter off the Earth's surface before arriving at the radar-receive antenna in an electronic attack known as terrain-scattered interference or terrain-bounce jamming. These so-called hot clutter returns go away if the jammer turns off. The development below is dedicated instead to cold clutter, radar returns from the Earths' surface and atmosphere.

This chapter eschews a detailed recounting of clutter physics and phenomenology (found in sources such as [1–4]), instead focusing on methods for modeling and simulation (M&S) of clutter, with an emphasis on 2-D clutter. There are several reasons for constructing and operating high-fidelity M&S representations of

2-D (Earth-surface) clutter. First, 2-D clutter, particularly that coming from land, is often stronger than target returns by many orders of magnitude. The prospect of strong ground clutter drives the dynamic range requirements of many radars, and its presence decreases target-detection probability and parameter estimation accuracies to levels well below those predicted for thermal noise interference alone. Second, many modern radar waveform and processing schemes such as sensitivity time control (STC), sidelobe blanking (SLB), moving target indication (MTI) cancelation, pulse-Doppler filtering, and space-time adaptive processing (STAP) are direct responses to challenging 2-D clutter conditions. Finally, M&S for clutter is much more complicated than for ground and air targets. Meanwhile, 3-D clutter is usually a secondary concern (with the notable exception of weather radar).

A final note on terminology: While we may imagine RF energy reflecting off a tree or a drop of rain, reflection harkens back to Snell's Law—the angle of reflection equals the angle of incidence—and suggests energy continuing away from the radar transmitter. This book focuses on monostatic radar operation, whereby transmit-and-receive antennas are colocated, and indeed are usually time-shared by the same aperture. Instead of energy reflected away from the radar, we are interested in energy scattered back to receive antenna. Therefore, this chapter employs the term back-scattered in lieu of reflected, as it is serves as a reminder of the monostatic geometry.

6.2 From Target Models to Clutter Models

While clutter lives in a 2-D or 3-D space, targets are as treated as zero-dimensional (0-D) entities by search-and-track radars. These sensors achieve best detection and parameter estimation performance not by over-resolving and imaging targets. Instead, they employ range/angle/Doppler resolution cell sizes comparable to or larger than the physical size of the expected targets. Aircraft have wings and tanks turrets, but search-and-track radars collapse these features down to an infinitesimally small point in space with time-varying locations and velocities and accelerations. The actual physical size, shape, location, and makeup of these constituent target components is unimportant beyond their tendency to induce RCS fluctuations on a 0-D target. In Chapter 5, complex RF scattering off the wings, tail, and fuselage of an aircraft, or turret, hull, tracks, and barrel of a tank, was rolled into one of the Swerling models. The four Swerling models include two levels of RCS statistical fluctuation, as given by the chi-squared (with 2 or 4 degrees of freedom) PDFs, and two rates of RCS decorrelation, either scan-to-scan or pulse-to-pulse (CPI-to-CPI for modern pulse-Doppler radars).

Surface clutter may be modeled as a large (approaching infinite) number of 0-D targets confined to a 2-D surface. Similarly, volume clutter may be modeled as a large number of point targets distributed in 3-D space. Land and sea possess very different clutter properties and are treated separately. (In the radar community, sea clutter models may be applied to freshwater bodies of water as well, so these models are not so much sea-specific, but rather are more general water clutter models.)

Distributed land clutter is modeled as a large number of small surface patches, with each patch containing a 0-D target. Patch statistics are locally homogeneous, but the larger landscape may consist of a heterogeneous collection of these homogeneous areas. This portion of the model captures distributed returns from terrain such as forest, grassland, and cropland. The distributed clutter model may be augmented with discrete returns using a number of strong, point-like scatterers placed randomly upon the Earth's surface. Each discrete is a 0-D target and represents impulsive returns from buildings and similarly large, high-RCS man-made objects.

Like land clutter, distributed sea clutter may be modeled by a large number of small surface patches, each containing a 0-D target. Unlike land, the sea is homogeneous over a wide region. While land clutter is mostly static (with windblown vegetation an important exception), the sea surface is always in motion, so sea clutter models are dynamic. In addition, observed sea clutter is highly dependent on the direction (with respect to the radar location) of the wind over the ocean surface, so that having the radar upwind or downwind from the surface will give rise to very different cutter levels at the radar. At fine radar resolutions, impulsive clutter returns known as sea spikes become noticeable. Sea spikes may be modeled with a number of large 0-D targets placed onto the ocean surface, like discretes for land clutter.

Atmospheric clutter includes rain, snow, hail, and similar metrological precipitation distributed over a 3-D space. Distributed 3-D clutter is modeled with a volume of 0-D point targets in small volume pixels (voxels), analogous to the 2-D patches for land and sea clutter. Like sea clutter, volume clutter is ever-changing due to the internal motion of raindrops, snowflakes, and so on, so these models are dynamic, like sea clutter models.

A common theme for clutter modeling is the summation over returns from a large number of 0-D targets in 2-D and 3-D space. Revisiting 0-D targets, recall that these models have a mean RCS σ with units of area (square-meters). A 2-D clutter patch is given an analogous mean backscatter coefficient σ^0 with RCS per unit area (and so is unitless), and a 3-D clutter voxel has a mean backscatter coefficient η with RCS per unit volume (for units of 1/meter). The Swerling target models acknowledge that the instantaneous RCS of a target is defined by a PDF with the given mean. Similarly, the RCS of 2-D patches and 3-D voxels varies from patch to patch and from voxel to voxel, consistent with some underlying PDF. The Swerling models capture time variations in RCS in a decorrelation parameter, and clutter patches and voxels have a related decorrelation time or rate. In summary, clutter modeling begins with point-target RCS modeling and extends those models to two and three dimensions.

6.3 Principles of Area Clutter Modeling

Models for 2-D clutter, be it land or sea, begin by partitioning an area over the Earth's surface into a large number of patches, or facets [5]. Figure 6.1 depicts a clutter M&S Cartesian coordinate system with receive aperture at the origin and mounted upon an airborne platform. It is convenient to inscribe a polar grid onto the Earth's surface, with the origin of the polar coordinate system at the nadir point below the radar. In this way, a circle of constant radius on this polar grid has constant slant range from the radar. Patches are defined by circles of inner

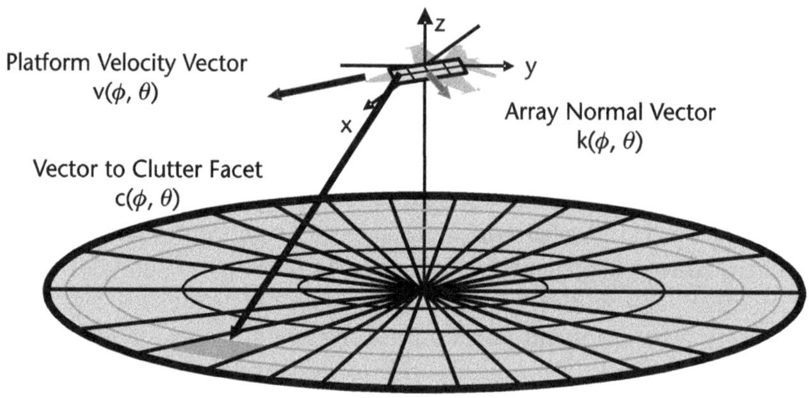

Figure 6.1 Area clutter coordinate system.

and outer radii, as shown in Figure 6.1, and a user-defined azimuth extent. If, for example, a user were interested in modeling clutter in a radar range bin centered at 100 km and having a 1-km width, a ring of patches having inner and outer radii corresponding to slant ranges of 99.5 km and 101.5 km, respectively, would be established. The azimuth extent of the patches is driven by antenna beamwidth and Doppler bandwidth considerations. For the antenna, patches should be smaller than the projection of the main beam onto the Earth's surface. Similarly, patches should be smaller than the projection of a Doppler filter bandwidth onto the Earth's surface, found through the usual mapping between Doppler resolution and cone angle. In general, smaller patches in radius and azimuth support higher fidelity and finer range-angle-Doppler resolutions but require more computational processing and memory.

While a rectangular grid of clutter patches seems like a reasonable alternative to the polar grid, the latter possesses a number of advantages. For example, slant ranges to facets on a rectangular grid generally give rise to returns that straddle radar sample times and range bins. In contrast, the clutter polar grid is set up explicitly so that range rings on the Earth's surface align with expected radar range bins. Furthermore, the polar grid allows for a limited slant range swath of clutter to be modeled. In the extreme, clutter in just one range bin—perhaps corresponding to a target bin—can be modeled with a single clutter range ring. Moreover, the radius and azimuth form of the polar grid allow for independent control of the fidelity and resolution of clutter in slant range and angle/Doppler, respectively. For example, finer radar range resolution achieved using wider waveform bandwidths can be supported with smaller range rings, and finer Doppler resolution (the product of longer CPIs) and narrow antenna beamwidths (a consequence of larger apertures) call for smaller facet azimuth extents. Finally, the polar grid may be readily extended from a planar Earth to a spherical Earth, as a line of constant slant range from the radar intersects the Earth spheroid as a circle centered at the radar nadir point.

Referring back to Figure 6.1, given a platform velocity vector $\mathbf{v}$, a receive aperture unit normal vector $\mathbf{k}$, and a vector from the receive antenna to the clutter patch at the lth azimuth angle in the kth range bin $\mathbf{c}_{kl}$, all a function of their azimuth and elevation pointing angles ϕ and θ, each patch contributes the following:

- A slant range, given by the norm $||\mathbf{c}||$.
- A Doppler shift, a function of the dot product between **c** and **v** and the operating RF of the radar.
- A receive-antenna gain, a function of the angle between **c** and **k**, and between **c** and a second aperture vector (not shown) that defines the orientation of the aperture, and the gain pattern of the aperture. This single-channel development can be extended to multiple channels by using an additional set of vectors $\{\mathbf{d}_m\}$ to define the locations of M subarrays. The amplitude and phase progressions of the clutter patch across channels can be found by applying **c** and $\{\mathbf{d}_m\}$ to standard matrix-vector operations.
- A phase, $\phi(k,l)$.
- An amplitude, $a(k,l)$.

Mathematically, the clutter signal x_c from the facet at the kth range bin and the lth azimuth angle in the mth channel and the nth pulse is

$$x_c(k,l,m,n) = a(k,l)e^{j\phi(k,l)}g(k,l,m)e^{j\varphi(k,l,m)}e^{j2\pi f_d(k,l)nT} \tag{6.1}$$

where $g(k,l,m)$ is the complex gain of the mth channel in the direction of the facet, $\varphi(k,l,m)$ is the additional phase induced by the displacement of the mth channel phase center from the reference phase center of the receive antenna, and $f_d(k,l)$ is the Doppler shift to the facet, with T the time between pulses. The facet backscatters RF to the radar as an RCS, so the signal x_c is a complex RCS with units of meters. Therefore, the amplitude $a(k,l)$, in turn, has units of square-root-RCS, or meters.

Simulating clutter returns is an exercise in numerical integration over the contributions of all modeled facets. For example, the total clutter in the kth fast-time sample or range bin, mth channel, and nth pulse is simply the coherent sum of the contributions from all facets in that ring

$$x_c(k,m,n) = \sum_{l=1}^{L} x_c(k,l,m,n) \tag{6.2}$$

In the list of five contributions from each clutter patch, 1 through 3 (slant range, Doppler, and complex channel gain) are deterministic in nature, whereas 4 and 5 (amplitude and phase) are stochastic. The remainder of this section develops methods for assigning amplitude and phase to each facet.

In this facetized clutter model, each patch is assumed to consist of a single 0-D target. These targets, in turn, consist of a large number of small, independent scatterers. For clutter, these scatterers may be leaves on a tree for forest clutter, blades of grass for pasture clutter, rocks and pebbles for soil clutter, or tiny wavelets on the ocean surface. The CLT says the aggregate phase $\phi(k,l)$ from a large number of independent and identically distributed scatterers is uniformly distributed from 0 to 2π.

The amplitude (in meters) $a(k,l)$ is drawn randomly from the RCS PDF of the 0-D target in the (k,l) facet. To define the RCS PDF, first the mean facet RCS σ_{Facet} must be determined. The mean RCS is simply the product of the facet area and backscatter coefficient

$$\sigma_{\text{Facet}} = A_{\text{Facet}}\sigma^0 \tag{6.3}$$

For example, if the backscatter coefficient is −20 dB, or 0.01, and the facet area is 10 m^2, the average facet RCS is 0.1 m^2, or −10 dBsm.

Calculating the area of a facet is a relatively straightforward exercise and is shown for a flat Earth in Figure 6.2. In the side view, the radar is at an altitude h and ground range equal to zero. The facet under consideration is at an altitude of zero and a slant range of R_S. The grazing angle d at the facet is

$$\delta = \sin^{-1}\left(\frac{h}{R_S}\right) \tag{6.4}$$

and the ground range to the facet R_G is

$$R_G = R_S \cos\delta \tag{6.5}$$

In this example, the extent of the facet is set by the range bin width ΔR, which in turn is set by the pulse width τ and speed of light c as $c\tau/2$.

However, this relationship is developed in the slant plane. When projected onto the ground plane, the extent increases by an amount equal to $1/\cos\delta$, for a downrange extent of $\Delta R/\cos\delta$.

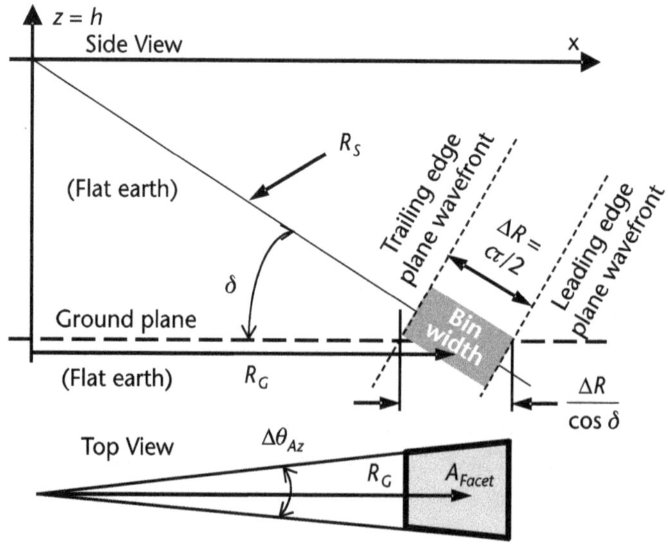

Figure 6.2 Area at a facet.

The top view suggests that the cross-range extent of the facet is approximately equal to the arc-length given by the product of the ground range to the facet and the azimuth extent of the facet $\Delta\theta_{Az}$, or $R_G\Delta\theta_{Az}$.

The area of the facet is equal to the product of the downrange and cross-range extents

$$A_{\text{Facet}} = \left(\frac{\Delta R}{\cos\delta}\right)\left(R_G \Delta\theta_{Az}\right) \tag{6.6}$$

Applying (6.5) to (6.6) yields

$$A_{\text{Facet}} = \Delta R\left(R_S \Delta\theta_{Az}\right) \tag{6.7}$$

All three contributors to facet area are defined at the beginning of the simulation. Recall that the facet range and angle extents are set by model fidelity and radar resolution concerns, and slant range(s) by operational considerations such as fast-time sampling instances or the distance to targets.

Referring back to (6.3), once facet area is calculated, all that remains to finding average facet RCS is a value for the backscatter coefficient.

6.4 Land Clutter Backscatter Coefficients

The particular value assigned to backscatter coefficient σ^0 is a function of the following.

- *RF band:* Usually documented by radar band (e.g., L-band and X-band). There is a general trend of increasing backscatter coefficients at higher RFs.
- *Polarization:* Examples include vertical and horizontal and may be broken out by copolarized and cross-polarized returns. For land clutter, copolarized horizontal and vertical backscatter coefficients tend to be comparable.
- *Terrain type:* Examples included farmland, and urban. The backscatter coefficient grows as surface roughness increases, where roughness means deviations from the local horizontal plane and includes the contributions from all RF scatterers, so that a forest of trees is quite rough with a high backscatter coefficient, while a lake on a windless day is very smooth. When roughness is measured in wavelengths, the trend of higher backscatter with increasing roughness is tied to the first bullet, the trend of higher backscatter with increasing RF.
- *Grazing angle at the Earth's surface:* Grazing angle is equal to 90° minus incidence angle. Backscatter coefficients always increase with increasing grazing angle.

σ^0 is commonly plotted as a function of grazing angle for a given operating frequency, polarization, and terrain type. Figure 6.3 is a classic example [6] showing backscatter coefficient versus grazing angle for four terrain types in X-band for

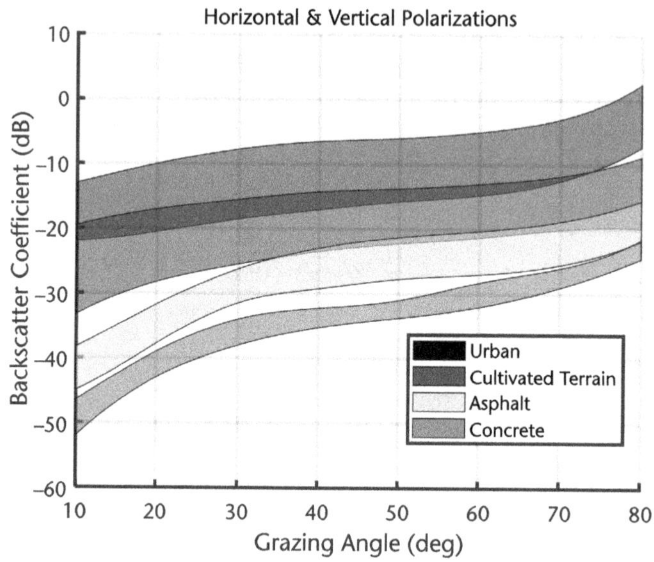

Figure 6.3 Terrain and grazing angle dependencies of backscatter coefficients. (After Figure 13.8 in [6].)

copolarized horizontal and vertical polarizations. σ^0 is shown as a range of values for each terrain type, as these categories are quite broad. For example, concrete may be wet or dry, laid rough or finished smooth, and so on. Note the increase in σ^0 as grazing angle and terrain type roughness increase.

Within a clutter simulation, σ^0 is stored as LUTs or closed-form fits to σ^0 versus grazing angle. A simulation of cultivated terrain clutter using the data in Figure 6.3 might employ a σ^0 of –20 dB for range rings at short ranges, closer to the nadir point and so at steeper grazing angle, and decreasing to –30 dB at longer ranges near the horizon and so at grazing angles of a few degrees.

It is often convenient to partition gazing angles into the three regimes depicted in Figure 6.4:

- *Plateau:* Intermediate grazing angles with a modest impact on σ^0. The physics and phenomenology is consistent with rough surface scattering. The plateau region extends down to 5–10° and up to 60–70°.
- *Specular:* Steep grazing angles exhibit a more rapid rise in σ^0 with grazing angle than rough surface scattering physics would predict. Instead of constructively reinforcing and destructively interfering independently, local collections of scatterers begin to backscatter quasi-coherently, leading to an RCS augmentation and a rapid rise in σ^0.
- *Microshadowing:* Low grazing angles exhibit a more rapid falloff in σ^0 with grazing angle than rough surface scattering physics would predict. While ideal objects scatter independently, real clutter (e.g., houses, trees, and undulating terrain) possess nonzero height and cast shadows onto objects further down range into shadow. Shadows lengthen as grazing angle approaches zero, plunging a large fraction of the clutter scene into shadow and so inducing a dramatic decline in σ^0.

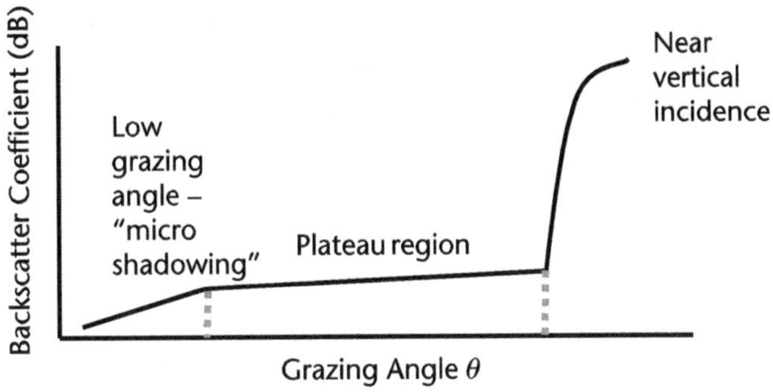

Figure 6.4 Grazing angle regions.

The plateau region is commonly represented with the constant-gamma model, which has the form

$$\sigma^0 = \gamma \sin\delta \tag{6.8}$$

with γ taking on a constant value for a given RF, polarization, and terrain type. The constant-gamma model is especially useful for modeling clutter for airborne and space-based radar, as clutter is often dominated by returns from grazing angles within the plateau region. Note that the constant-gamma model will yield backscatter coefficients that are too high in the microshadowing regions and too low for specular returns. Table 6.1 contains typical values of gamma for various terrain types in X-band [7].

One model for specular clutter takes the form

$$\sigma^0 = \left(\frac{\rho}{\tan(\beta_0)}\right)^2 \exp\left(-\left(\frac{\tan(\beta)}{\tan(\beta_0)}\right)^2\right) \tag{6.9}$$

where β is the angle of incidence, so that $\beta = 90° - \delta$, ρ is the Fresnel coefficient at vertical incidence, and $\tan(\beta_0)$ is the RMS slope of the surface. The backscatter coefficient achieves a maximum of $(\rho/\tan(\beta_0))^2$ at 90°. Representative values for ρ and $\tan(\beta_0)$ may be found in Table 6.2 [1].

Table 6.1 Gamma for Various Terrain Types

Terrain Type	*Values of Gamma*
Mountains	–5 dB
Urban	–5 dB
Wooded hills	–10 dB
Rolling hills	–12 dB
Farmland	–15 dB
Desert	–20 dB
Smooth water	–45 dB

There are no standard formulae for the microshadowing regions, and models tend to be handcrafted from empirical observations.

An example backscatter model for grass appears in Figure 6.5. Constant-gamma provides the model for the plateau region, with support for grazing angles between 5° and 60°. The adjacent specular and microshadowing regions are clearly visible. There is a fourth region at very low grazing angles that takes on a constant backscatter coefficient of –37 dB. This model may be justified by noting that some of the objects having height that give rise to microshadowing will provide nonzero backscatter to the radar even down to a grazing angle of zero.

From (6.3) the average RCS of a facet is determined by the area of the facet given by (6.7) and the average backscatter coefficient provided by functions such as (6.8) and (6.9) incorporated into a larger lookup function such as the one in Figure 6.5. Clutter backscatter coefficient tables are widely available [2, 3]. The next task is to define the RCS statistics, as done for the Swerling targets, from which random draws of actual facet RCS amplitude and phase are made.

6.5 Land Clutter Backscatter Statistics

The clutter in a facet often consists of a large number of small scatters such as blades of grass, leaves on a tree, and rocks in soil, scattering independently and superimposing their returns randomly. This model is consistent with the assumptions behind the Swerling models 1 and 2. To review, the relevant PDFs are listed as follows:

- Real and imaginary parts of the facet RCS are zero-mean and uncorrelated Gaussian with same variance.
- Amplitude (square-root of RCS) follows a Rayleigh distribution.
- Power (RCS) exhibits an exponential PDF.
- Phase is uniformly distributed over 2π.

Table 6.2 Specular Model Parameter Values

Terrain Type	*Typical Values of* ρ
Land	0.5
Water	0.8
	Typical Values of $Tan(\beta_0)$
Calm	0.00
Smooth	0.05
Slight	0.12
Moderate	0.14
Rough	0.15
Very rough	0.16
High	0.18
Very high	0.22

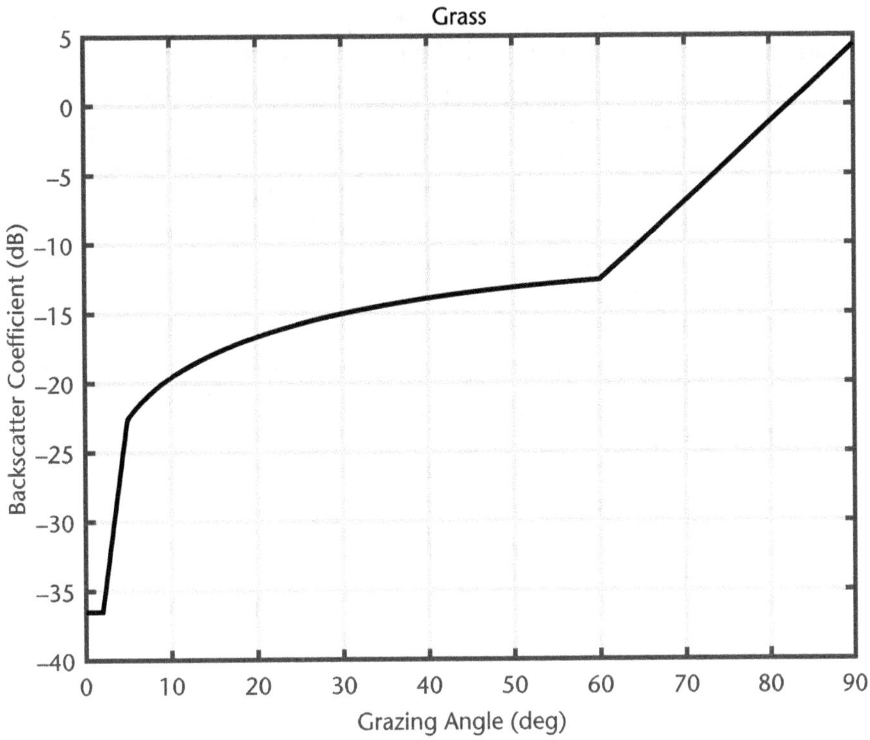

Figure 6.5 Representative model for backscatter coefficient for grass.

In practice, each facet is defined by two Gaussian random draws—one for the real part and the other for the imaginary part—and scaled by the square root of the mean facet RCS $\bar{\sigma}$. In MATLAB, with `sigma _ avg` as the mean RCS,

```
facet_complex_rcs(k,l) = sqrt(sigma_avg/2)*(randn(1,1)+1j*randn(1,1));
```
(6.10)

Alternatively, amplitude and phase may be drawn explicitly from exponential and uniform PDFs, respectively

```
a(k,l) = sqrt(-sigma_avg*log(rand(1,1)));
```
(6.11a)

```
phi(k,l) = 2*pi*rand(1,1);
```
(6.11b)

The exponential PDF for RCS σ

$$f(\sigma) = \frac{1}{\bar{\sigma}}\exp\left(-\frac{\sigma}{\bar{\sigma}}\right) \tag{6.12}$$

is convenient for radar signal processing as the function is defined by a single parameter, $\bar{\sigma}$. As it happens, that parameter corresponds to the mean of the PDF, determines the mean, and the PDF is defined completely. Moreover, the mean is readily estimated from typical radar data products, best exemplified by the well-known cell-averaging constant false alarm rate (CA-CFAR) algorithm for automatically setting a target detection threshold. With an estimate of the mean from the CA-CFAR, a threshold achieving some desired probability of false alarm may be set with confidence given exponentially distributed interference like thermal noise, and some kinds of clutter.

However, decades of calibrated collections of land clutter suggest deviation from the simple exponential PDF is common [4]. The observed PDFs have longer tails than the reference exponential, so that large values appear more frequently than predicted by the exponential distribution with the same mean value. Under these conditions, the CA-CFAR implementation will suffer from higher false-alarm rates. To evaluate the performance impact on a CA-CFAR implementation and to design and mature more robust CFAR solutions, several alternatives to the exponential PDF have been proposed.

Perhaps the oldest nonexponential PDF for clutter is the log-normal [8]

$$f(\sigma) = \frac{1}{\sigma s\sqrt{2\pi}} \exp\left(-\frac{(\ln\sigma - \mu)^2}{2s^2}\right) \tag{6.13}$$

where μ is the mean of $\ln\sigma$, and s is the standard deviation of $\ln\sigma$. A log-normal PDF for clutter suggests clutter power in decibels, or RCS in decibels relative to one square meter, exhibits a normal distribution. While there is no physical justification for application of the log-normal PDF to radar clutter, historical clutter collections are often presented as histograms of decibel measurements, so constructing a log-normal PDF from such records is straightforward. Well-chosen values for μ and s will yield a PDF having higher and longer tails than an exponential for the same mean level. In a MATLAB clutter simulation, (6.11) is used but with a random draw for a log-normal PDF for $a(k,l)$

```
a(k,l) = 10^((mu_dBsm+s_dBsm*randn(1,1))/20);
```
(6.14)

with the mean and standard deviation in units of decibels relative to one square meter.

Another oft-invoked two-parameter PDF for modeling of land clutter is the Weibull distribution

$$f(\sigma) = \frac{b\ln(2)}{\sigma_m}\left(\frac{\sigma}{\sigma_m}\right)^{b-1} \exp\left(-\ln(2)\left(\frac{\sigma}{\sigma_m}\right)^{b}\right) \tag{6.15}$$

where σ_m is the median of σ. Like the log-normal PDF, it has no physical justification, but the tails of the PDF can be augmented or suppressed by judicious selection of the shape parameter b, and a similarly modified form of (6.11) is employed in clutter simulations. Weibull has the attractive feature that both exponential and log-normal PDFs are special cases. A common application is simulating clutter in the microshadowing regions, where the modulation of otherwise exponentially distributed clutter by shadows gives rise to heavy-tailed PDFs that match well to Weibull functions.

Figure 6.6 is an exceedance plot of two distributions, exponential and Weibull, for the same mean backscatter coefficient of –25 dB. Exceedance is equal to one minus the cumulative density, and the ordinate of an exceedance plot shows which fraction of the data samples exceeds the corresponding value on the abscissa. With $b = 3.4$, the Weibull PDF is much more likely than the exponential to produce high backscatter values. For example, at a backscatter coefficient of –10 dB (15 dB higher than the mean), the exponential PDF is around 10^{-10}, while the Weibull is close to 10^{-2}, about 100 million times higher.

A final PDF often applied to simulating clutter is the K-distribution. It begins with the exponential PDF in (6.12) but allows the mean backscatter coefficient to fluctuate from facet to facet. Without fluctuations, the K-distribution reduces to the exponential; with fluctuations it exhibits a heavier tail for the same mean value. The PDF for the mean backscatter coefficient is the gamma distribution

$$f(x;\alpha,\beta) = \frac{\beta^{\alpha}}{\Gamma(\alpha)} x^{\alpha-1} e^{-\beta x} \tag{6.16}$$

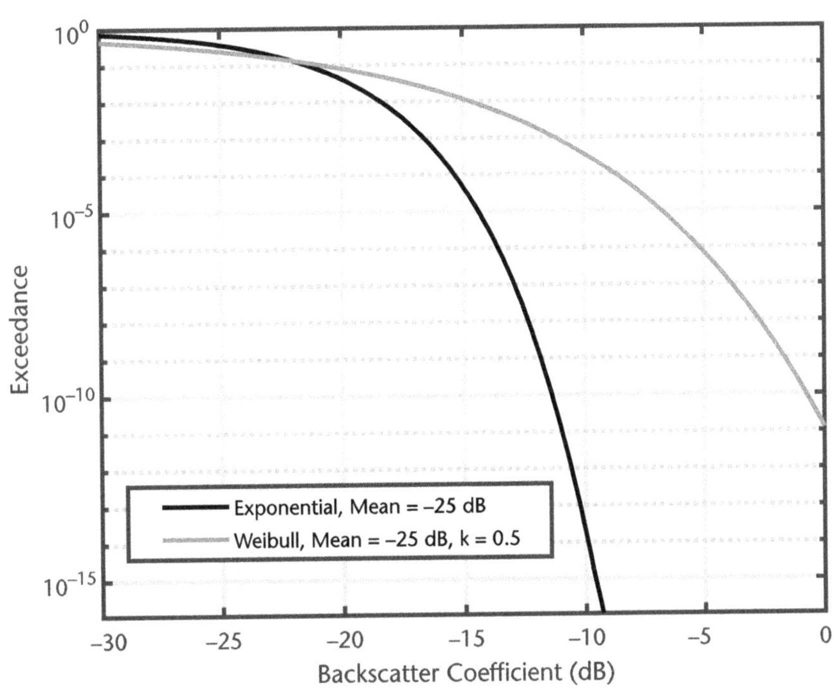

Figure 6.6 Probability densities of comparable exponential and Weibull PDFs.

where $\Gamma(\alpha)$ is the gamma function. Gamma is another two-parameter PDF with the mean equal to α/β and variance equal to α/β^2. By setting the mean to the desired mean backscatter coefficient $\bar{\sigma}$, the variance becomes $\bar{\sigma}/\beta$ and so depends on a single parameter β. As β approaches infinity, the gamma variance approaches zero, and the K-distribution reduces to the exponential. A finite β yields a K-distribution with a heavier tail than the exponential. The physical model behind the K-distribution—in which individual facets are exponentially distributed, but mean clutter power can vary from location to location—has some basis in reality. The K-distribution has been applied to sea clutter simulations as well as land clutter, but with different gamma distribution parameters.

6.6 Land Clutter Discretes

Another approach to generating stressing, heavy-tailed clutter is to start with the facetized distributed clutter model of Section 6.3, employing the exponential PDF in Section 6.4 for random draws of facet backscatter coefficients, but then adding a laydown of discretes to the Earth's surface. In this context, a discrete is a stationary, nonfluctuating, high-RCS, 0-D target, intended to represent a building or similar impulsive returns.

The density, or number of discretes per unit area, are derived from empirical models. Density decreases as the RCS of the discrete increases so that moderate strength discretes are common and very strong discretes exceedingly rare. One discrete model was created by exploiting SAR imagery collected by the Forward-Looking Advanced Multimode Radar (FLAMR) [9]; the resulting areal density as a function of discrete RCS appears in Table 6.3. Density takes on a power-law trend, so that discretes that are 10 times stronger are also 10 times sparser.

To add discretes to a clutter simulation, first the total area over which distributed (faceted) clutter is being modeled is found. Next, this area and the densities in Table 6.3 are applied to a Poisson probability mass function (PMF) generator, so that the total number of discretes of a given RCS may vary from run to run. Then, the number generated are distributed randomly over the Earth's surface active in the clutter simulation. Figure 6.7 shows one FLAMR realization, with discrete RCS versus downrange and cross-range location. There were no 60 dBsm discretes in this example.

Figure 6.8 contains simulated range-Doppler maps of distributed-clutter-only, discretes-only, and discretes plus distributed clutter. The presence of discretes

Table 6.3 FLAMR Discrete Densities

Discrete RCS	*Discrete Density*
60 dBsm	$0.01/\text{NM}^2$
50 dBsm	$0.1/\text{NM}^2$
40 dBsm	$1/\text{NM}^2$
30 dBsm	$10/\text{NM}^2$
20 dBsm	$100/\text{NM}^2$

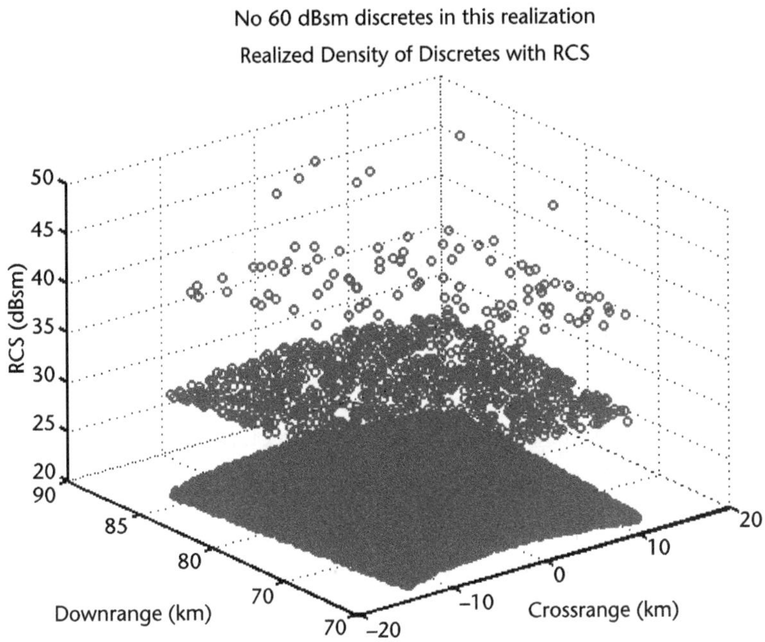

Figure 6.7 Example discrete laydown.

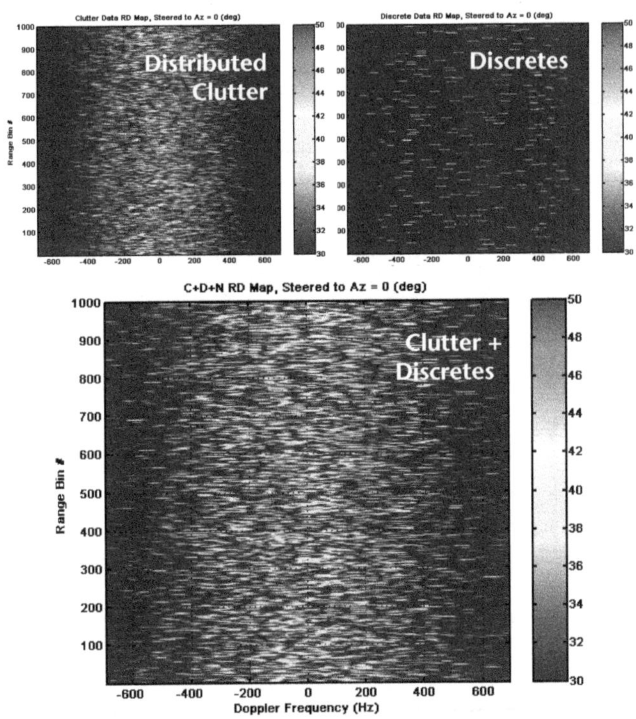

Figure 6.8 Example of distributed clutter and discrete range-Doppler maps.

serves to add many high-powered, concentrated spikes to the radar range-Doppler map.

Range-Doppler maps containing thermal noise only, distributed clutter plus thermal noise, and discretes plus distributed clutter plus thermal noise were applied to an ideal CFAR that estimated background interference levels based on the contributions of distributed clutter and thermal noise only. Each range-Doppler pixel power was normalized by the corresponding estimate of the background interference power, and the ratio sorted and plotted as an exceedance plot in Figure 6.9. Here, the thermal noise only and distributed clutter plus thermal noise exceedances take the form of an exponential PDF. In both cases, applying the normalized range-Doppler pixels to a threshold of 11 dB would serve to maintain a probability of false alarm of 10^{-5}. To maintain that performance against the discrete data requires a threshold of 27 dB, effectively desensitizing the radar against targets by 16 dB.

6.7 Land-Clutter Temporal Correlation

Recall a key distinguishing feature of the Swerling target models was RCS correlation time, with Swerling models 1 and 3 decorrelating slowly (scan-to-scan) and 2 and 4 decorrelating rapidly (pulse-to-pulse, or CPI-to-CPI). Land-clutter backscatter may decorrelate in a similar fashion, due primarily to windblown vegetation. Random motion of blades of grass and leaves on trees cause variations in backscattered

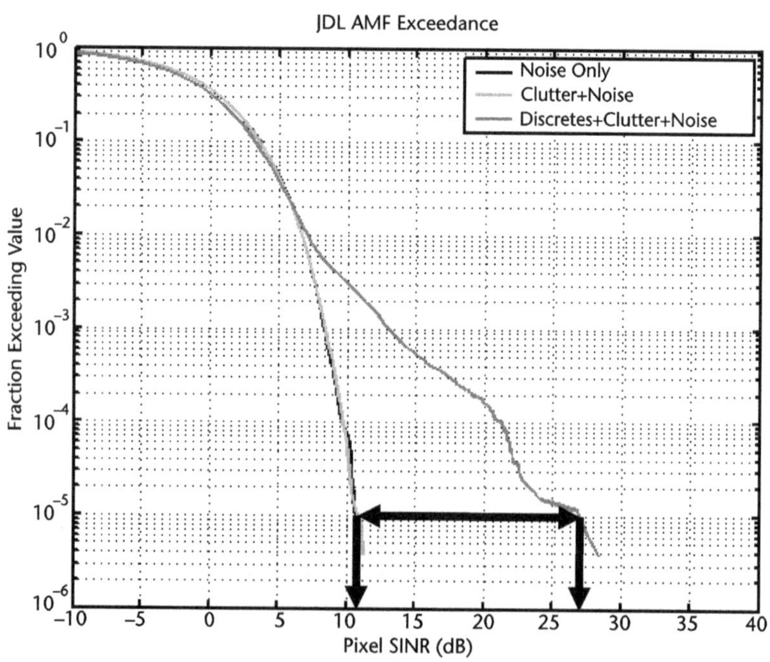

Figure 6.9 Impact of discretes on detection exceedance.

amplitude and phase over time and lead to spreading of clutter in Doppler. This phenomenology is sometimes referred to as intrinsic clutter motion (ICM).

Unlike the Swerling models, clutter decorrelation is not binary (perfectly correlated or one hundred percent–decorrelated) but is a gradual process. Table 6.4 contains approximate clutter decorrelation times as a function of wind speed in miles per hour (MPH) and RF. Not surprisingly, higher wind speeds cause more dramatic foliage motion and thus shorter decorrelation times. Higher RFs have shorter wavelengths, so that the same foliage motion gives rise to more pronounced phase changes and, hence, backscatter decorrelation.

Figure 6.10 shows the effects on the clutter Doppler spectrum of windblown trees [10]. Clutter power increases significantly between 0.5 and 2.0 m/s for breezy and windy conditions over light air. Dismounts—people running or walking—present with range rates in this regime, so ICM is an important consideration for radars designed to detect and characterize such targets.

Random decorrelation vectors may be generated in MATLAB via the following procedure [11]. First, the temporal autocorrelation function $R(\tau)$ is defined, where τ is the time lag, $R(0) = 1$, and $R(\neq 0) \leq 1$. A special case is $R = 1$ for all time lags, which corresponds to the condition of no clutter decorrelation. Common decorrelation functions for land clutter include exponential (really Laplacian, as the function is symmetric about $\tau = 0$) and Cauchy (also known as Lorentzian). Next, the autocorrelation function is used to populate a temporal (slow-time, pulse-to-pulse) decorrelation covariance matrix $\mathbf{R}_{\text{Decorr}}$

$$\mathbf{R}_{\text{Decorr}} = \begin{bmatrix} R(0) & R(T) & \cdots & R((N-1)T) \\ R(T) & R(0) & \cdots & R((N-2)T) \\ \vdots & \vdots & \ddots & \vdots \\ R((N-1)T) & R((N-2)T) & \cdots & R(0) \end{bmatrix} \tag{6.17}$$

where T is the pulse repetition interval. The matrix is N by N, where N is the number of slow-time pulses over the CPI. For no decorrelation, the decorrelation matrix is equal to the all-ones matrix, which may be expressed as $\mathbf{1}\mathbf{1}^T$, where $\mathbf{1}$ is the all-ones vector and T is the vector transpose operation.

Land ICM models acknowledge that only a fraction of the clutter decorrelates: Wind may blow leaves and limbs, but tree trunks and the ground remain unmoved. Therefore, the final land ICM covariance matrix $\mathbf{R}_{\text{ICM}}$ has AC (decorrelating)

Table 6.4 Temporal Decorrelation Times for Trees Versus RF and Wind Speed

Wind Speed (MPH)	*10 GHz*	*16 GHz*	*35 GHz*	*95 GHz*
2	210 ms	150 ms	65 ms	30 ms
3–5	92 ms	55 ms	27 ms	12 ms
6–10	45 ms	28 ms	23 ms	8 ms
11–14	21 ms	11 ms	2.8 ms	0.8 ms

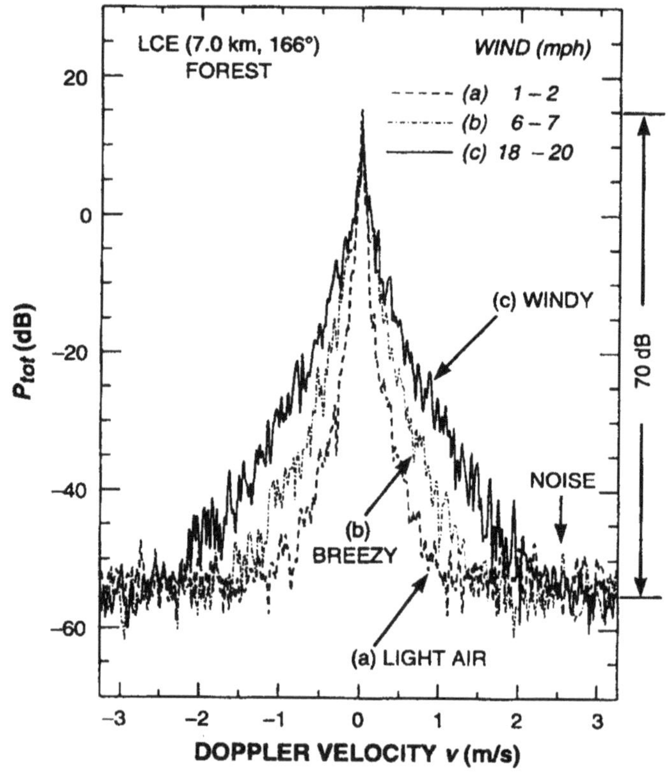

Figure 6.10 Doppler spread of windblown trees.

and DC (correlated) components, with k_{ACDC} setting the relative strength of the two constituents

$$\mathbf{R}_{\text{ICM}} = \left(1 - k_{\text{ACDC}}\right)\mathbf{1}\mathbf{1}^T + k_{\text{ACDC}}\mathbf{R}_{\text{Decorr}} \tag{6.18}$$

Next, a matrix square-root (`sqrtm()` in MATLAB) is applied to the ICM covariance matrix to produce a transformation matrix **T**

$$\mathbf{T} = \left(\mathbf{R}_{\text{ICM}}\right)^{1/2} \tag{6.19}$$

To decorrelate the slow-time response from a clutter facet, a unit-power white complex Gaussian noise vector **n** of length-N is formed; in MATLAB

```
n = sqrt(1/2)*(randn(N,1)+1j*randn(N,1));          (6.20)
```

and passed through **T** to generate a decorrelation vector $\mathbf{x}_{\text{Decorr}}$

$$\mathbf{x}_{\text{Decorr}} = \mathbf{Tn} \tag{6.21}$$

This vector has the desired ICM decorrelation properties

$$\begin{aligned} E\left[\mathbf{x}_{\text{Decorr}}\mathbf{x}_{\text{Decorr}}^{H}\right] &= E\left[\mathbf{Tnn}^{H}\mathbf{T}^{H}\right] = \mathbf{T}E\left[\mathbf{nn}^{H}\right]\mathbf{T}^{H} \\ &= \mathbf{TIT}^{H} = \mathbf{R}_{\text{ICM}}^{1/2}\left(\mathbf{R}_{\text{ICM}}^{1/2}\right)^{H} = \mathbf{R}_{\text{ICM}} \end{aligned} \tag{6.22}$$

where H is the Hermitian, or conjugate transpose, operation. The slow-time facet signal in (6.1) is modulated by this decorrelation vector along the n dimension. Note that each facet receives a different decorrelation realization from (6.20) and (6.21).

One final note: This decorrelation process should be applied to only certain types of land clutter, mostly forested areas, and perhaps tall grass and crops. There is negligible windblown decorrelation of rocky soils, fallow ground, and suburban and urban areas.

6.8 Site-Specific Clutter

The clutter models and simulations discussed thus far have been for generic sites: The radar is located on or above a bland Earth, spherical perhaps, but with no terrain height variations, and the terrain itself has been homogeneous, without pronounced variation in type or backscatter coefficient. Site-specific clutter models attempt to generate clutter as observed by a real radar in an actual physical location. Site-specific models capture terrain height and terrain type variations. One obvious reason for building site-specific clutter M&S environments is a desire to predict the effectiveness of a specific radar in an exact location for test and evaluation purposes, or for quantifying the operational strengths and weaknesses of a radar. A more subtle, yet common, motivation is the desire to stress a radar design, including its notional signal processing chain, with challenging clutter conditions. For example, a simple CA-CFAR threshold-setting algorithm performs well when presented with clutter from a region exhibiting constant-gamma backscatter but will struggle near the interface between high backscatter land and low backscatter water and the discontinuity between strong clutter on the nearside of a mountain and the shadowed region beyond. Under such conditions, the resulting target detection threshold will be too high on the low-clutter side, leading to missed target detections, and too low on the high-clutter side, producing false alarms.

The following two types of terrain information are required for a site-specific model:

- Terrain height (elevation) information as a function of absolute location;
- Terrain type as a function of absolute location.

Interestingly, terrain elevation does not affect radar clutter directly. Instead, it gives rise to the following two indirect consequences:

- Terrain height variations cause the local slope of the terrain to deviate from its flat-Earth value of zero, which in turn influences the local grazing angle. Moving away from the radar, a rise in terrain height leads to increased grazing angles over flat terrain, while falling terrain causes decreased grazing angles. Section 6.4 made clear the dependence of backscatter coefficient on grazing angle.
- Elevated terrain may cast shadows onto the terrain further downrange.

Terrain type directly affects the backscatter coefficient, which was detailed in Section 6.4.

Terrain height is provided by Digital Elevation Models (DEMs), which are available from entities such as the United States Geological Survey (USGS). DEMs usually report height as a function of latitude and longitude. A DEM that includes the state of Tennessee appears in Figure 6.11, with elevation in the color bar and longitude and latitude along the horizontal and vertical axes.

Terrain type is extracted from Land Cover Classification (LCC) data sets, also available from the USGS. LCC information is often reported using equal-area map projections, such as the Albers projection. Figure 6.12 contains a color-coded land cover for Tennessee and a legend-mapping color to land cover numerical types and descriptions.

A number of coordinate and data transformations are required to realize site-specific clutter. The order may vary across implementations; what follows is a description of the order used by the author to build a site-specific clutter simulation.

First, the LCC data in Albers coordinates is interpolated to latitude-longitude. Now the DEM and LCC share a common 2-D sampling.

Next, the local 2-D slope (vector normal) is found at each DEM location by using nearby DEM samples to construct facet vertices, which in turn define the vector normal at that sample point.

Then, the LCC terrain types are mapped to terrain types for which backscatter coefficient information is available. For example, types 41, 42, and 43 in Figure 6.12 (deciduous, evergreen, and mixed forest) might be collapsed to simply "forest" if only forest backscatter data is available. Everything thus far may be considered preprocessing, as nothing site-specific has yet occurred.

DEM data available (for a fee) from the USGS http://edc.usgs.gov/

Figure 6.11 DEM of Tennessee.

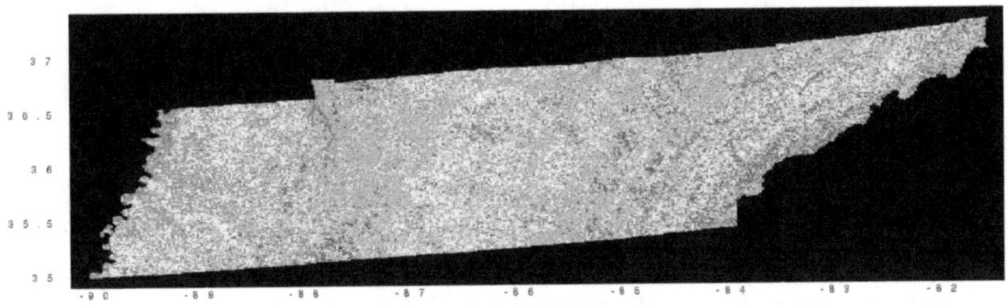

National Land Cover Dataset Classification System Legend

Color Key	RGB Value	Class Number and Name
	102, 140, 190	11 - Open Water
	255,255,255	12 - Perennial Ice/Snow
	253, 229, 228	21 - Low Intensity Residential
	247, 178, 159	22 - High Intensity Residential
	231, 86, 78	23 - Commerical/Industrial/Transportation
	210, 205, 192	31 - Bare Rock/Sand/Clay
	175, 175, 177	32 - Quarries/Strip Mines, Gravel Pits
	83, 62, 118	33 - Transitional
	134, 200, 127	41 - Deciduous Forest
	26, 129, 78	42 - Evergreen Forest
	212, 231, 177	43 - Mixed Forest
	220, 202, 143	51 - Shrubland
	187, 174, 118	61 - Orchards/Vineyards
	253, 233, 170	71 - Grasslands/Herbaceous
	252, 246, 93	81 - Pasture/Hay
	202, 145, 71	82 - Row Crops
	121, 108, 75	83 - Small Grains
	244, 238, 203	84 - Fallow
	240, 156, 054	85 - Urban/Recreational Grasses
	201, 230, 249	91 - Woody Wetlands
	144, 192, 217	92 - Emergent Herbaceous Wetlands

To obtain land cover classification data: http://landcover.usgs.gov/natllandcover.asp

Figure 6.12 LLC of Tennessee.

For a particular clutter run, the platform is assigned to an absolute location in latitude, longitude, and height, and the radar antenna orientation set in roll, pitch, and yaw. For each latitude-longitude sample point, the following parameters are calculated:

- The slant range from the radar;
- The azimuth angle from the radar;
- The elevation (or depression) angle from the radar;
- The grazing angle from the radar (using the local slope calculated previously);
- The clutter type;
- Whether or not the location falls into shadow due to intervening terrain.

Actual site-specific modeling begins with a faceted Earth on a polar grid as depicted in Figure 6.1 and described in Section 6.3. As before, each facet has a slant range and azimuth angle. These are interpolated into the available slant ranges and azimuth angles versus latitude and longitude (the first and second bullets above) to derive the following:

- The elevation (or depression) angle to the facet;
- The grazing angle at the facet;

- The clutter type for the facet;
- Whether or not the facet falls into shadow.

If the facet falls into shadow, it is assigned a backscatter coefficient of zero. If not, the clutter type and grazing angle are used to look up the backscatter coefficient. As before, the azimuth and elevation angles serve to derive transmit and receive antenna gains on the facet. To add to the realism, facet RCS amplitude may be drawn from a heavy-tailed PDF as described in Section 6.5, the slow-time response decorrelated with ICM using the procedure from Section 6.7, and discretes added by means of Section 6.6.

In the site-specific flow diagram in Figure 6.13, a DEM for the Delaware-Maryland-Virginia region appears in the upper left, with a plot inserted of backscatter coefficient versus grazing angle. LCC information for the same region appears in the lower left, with a table showing the mapping from terrain type to gamma (for the constant-gamma model). The right side of Figure 6.13 is an intermediate data product showing patch RCS in an east-north coordinate system with the radar at the origin and at altitude. (This product is not in the critical path of the site-specific clutter simulation procedure presented in this section, but it provides useful diagnostic information.) In this east-north coordinate system, all patches have the same area, so the image also suggests backscatter coefficient trends. As expected,

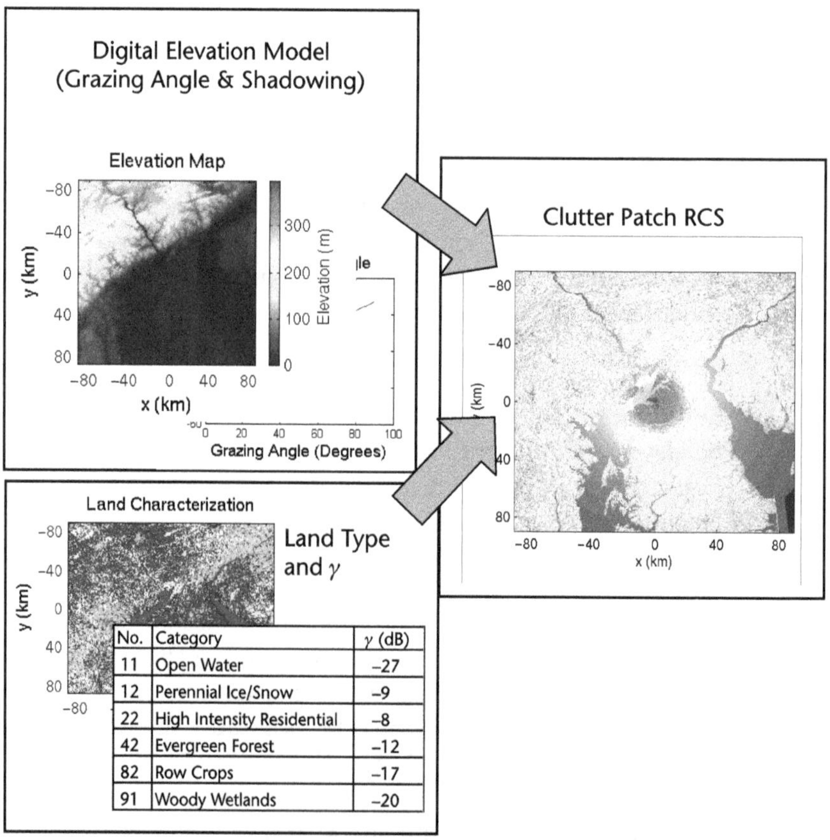

No.	Category	γ (dB)
11	Open Water	–27
12	Perennial Ice/Snow	–9
22	High Intensity Residential	–8
42	Evergreen Forest	–12
82	Row Crops	–17
91	Woody Wetlands	–20

Figure 6.13 Site-specific clutter example. (*From:* [12].)

clutter RCS is higher at steep look-down angles and on land. To the north is gently rolling terrain, with the near side of ridges highlighted by high RCS in dark gray.

6.9 Sea Clutter

In light of the extensive and extended attention given to land clutter thus far, it is worthwhile to begin this section by comparing and contrasting land and sea clutter [4]. Like land clutter, the backscatter coefficient for sea clutter increases with increasing grazing angle, RF, and surface roughness. While horizontal and vertical copolarized backscatter coefficients are comparable for land, vertical tends to be significantly higher than horizontal for the sea. Because water has higher RF reflectivity than land, and the sea is smoother under most conditions, at sea more RF is reflected away from the radar at low grazing angles, but more is reflected back at high grazing angles. So, backscatter coefficients tend to be lower for sea than land except in the specular region, at very steep grazing angles. Ocean clutter is much more homogeneous than land; away from land, the same sea is presented from horizon to horizon, while over land a great variety of terrain types may contribute to clutter returns. Ocean clutter exhibits far greater temporal variations, both in the short term, as the surface is in constant motion, and in the long term, as storms come and go.

Despite significant phenomenological differences, sea clutter M&S employs the same numerical integration over a polar grid of facets, depicted in Figure 6.1 and 6.2, as land clutter.

Figure 6.14 is a plot of ocean backscatter coefficient versus grazing angle for four surface roughness conditions given by RMS roughness in meters. Comparing to land in Figure 6.3, backscatter coefficients for 0.1m are like those for concrete or asphalt, depending on the grazing angle regime, while 1m, 2m, 10m, fall into asphalt and cultivated terrain.

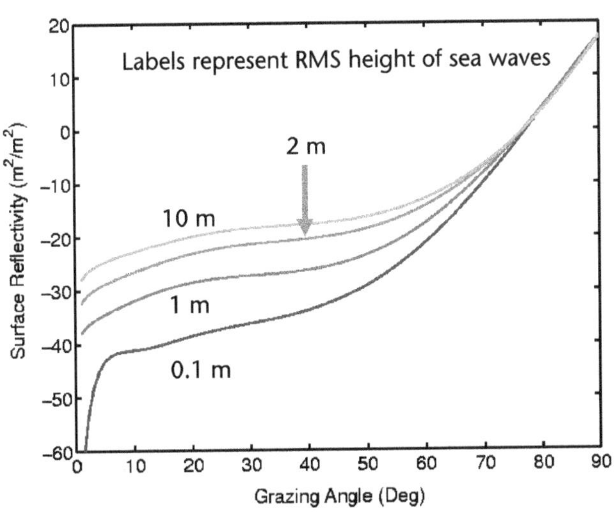

Figure 6.14 Sea-state backscatter-coefficient dependencies on surface roughness and grazing angle.

Long-term variations in ocean surface roughness are commonly reported as sea state, a natural number from 1 to 8, with 8 being the roughest. The chart in Figure 6.15 shows how sea state (row number 6) maps to other measures of ocean conditions. Noting that the roughness values in Figure 6.14 correspond to 0.3, 3, 6, and 30 feet, converting these from RMS to crest-to-trough with a factor of $2\sqrt{2}$ (derived using a sinusoidal wave model) to 1, 10, 20, and 100 feet, and examining row number 5 in Figure 6.15, the plots in Figure 6.13 correspond to sea state 1–2, 5, 6–7, and 8. Sea states above 5 are unusual, so the lower two backscatter coefficient plots in Figure 6.14 may be considered more typical.

Short-term changes to the ocean surface are captured by the ICM decorrelation procedure in Section 6.7. The entire surface moves over time, so there is no DC component in (6.18), and $k_{\text{ACDC}} = 1$. For the sea, Gaussian is a good autocorrelation function with which to populate the decorrelation covariance matrix $\mathbf{R}_{\text{Decorr}}$ in (6.17).

Under coarse resolution conditions with relatively large facet sizes, the CLT readily applies to sea clutter, and backscatter coefficient realizations may be drawn from an exponential PDF. At fine resolutions employing small facets, the sea begins to appear heterogeneous (one facet may be on a wave and the next between waves) and backscatter coefficient PDFs grow heavy-tailed. Under these conditions, modelers often resort to the K-distribution [13]. At low grazing angles, wave-induced microshadowing can occur, a job for the Weibull distribution. Finally, at very fine resolutions, sea spikes—transient, high-RCS, point-like returns—become pronounced. These may be modeled with a laydown of short-lived discretes, as described in Section 6.6.

6.10 Volume Clutter

While our clutter originates from a large number of scatterers over the Earth's surface, volume clutter arises from a large number of scatterers suspended throughout an enclosure in air. Models and simulations for volume clutter expand the polar grid of ground facets in Figure 6.1 to a spherical coordinate system of voxels. Here, a voxel is a discrete unit of volume, analogous to a facet as a discrete unit of area. The coordinate system consists of spheres of constant radius, centered on the radar and so with slant range R_S, and angle planes of constant azimuth angle θ_{Az} and elevation angle θ_{El}. Like (6.3) for the average RCS of a facet, the average RCS in a voxel is the product of the volume of the voxel and the backscatter coefficient η

$$\sigma_{\text{Voxel}} = V_{\text{Voxel}}\eta \tag{6.23}$$

and η having units of RCS per unit volume, or 1/m. A voxel has a slant range extent ΔR, set by the radar waveform range resolution or simulation fidelity, and azimuth and elevation extents $\Delta\theta_{\text{Az}}\,\Delta\theta_{\text{El}}$, set by transmit and receive antenna azimuth and elevation beamwidths. Reminiscent of the derivation of the area of a facet given in (6.7), the volume of a voxel is the product of the slant range extent and cross-range extent and vertical extent [14]

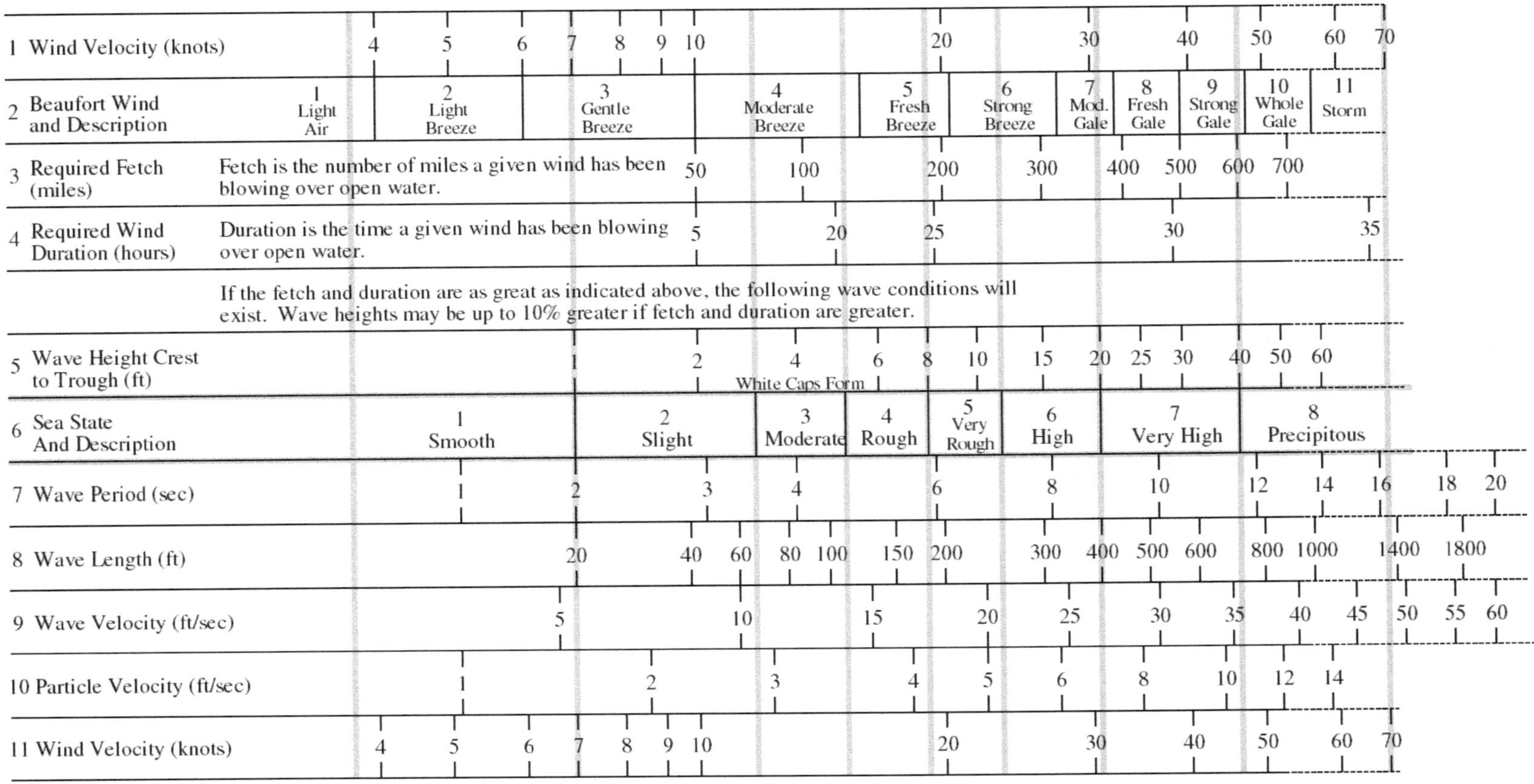

This table applies only to waves generated by the local wind and does not apply to swell originating elsewhere. (From: Undersea Technology, May 1964, p.34)

Warning: Presence of swell makes accurate wave observations exceedingly difficult.

Notes: (a) The height of waves is arbitrarily chosen as the height of the highest one-third of the waves.
(b) Only lines 7, 8, and 9 are applicable to swell as well as waves.
(c) The above values are only approximate due both to lack of precise data and to the difficulty in expressing data.

Figure 6.15 Sea state chart.

$$V_{\text{Voxel}} = \Delta R\left(R_S \Delta\theta_{\text{Az}} \cos\theta_{\text{El}}\right)\left(R_S \Delta\theta_{\text{El}}\right) \tag{6.24}$$

From extensive measurements and regression analyses, a closed-form expression for the backscatter coefficient of rain is

$$\eta = \frac{ar^b}{\lambda^4} \times 10^{-15} \tag{6.25}$$

where r is rain rate in mm/hour, λ is RF wavelength in m, and η has units 1/m. Table 6.5 contains representative values for the fit parameters a and b.

Random relative raindrop motion causes temporal decorrelation. Like sea clutter, all scatterers move, so there is no DC component to the ICM matrix, and a Gaussian PDF should be used to build up the decorrelation matrix.

One key point about volume clutter is that unlike surface clutter, which presents from horizon to horizon, volume clutter is localized, and severe volume clutter with strong backscatter is highly localized. For example, while light rain of around 2 mm/hour may manifest over several tens of kilometers, heavy rain of perhaps 20 mm/hour will be localized to cells a few kilometers across.

6.11 Clutter Model Results

Figures 6.16–6.21 show example results from the MATLAB clutter model *Fast_Clutter*. Each simulation run creates raw I/Q data using the construct described in this chapter, then pulse-compresses and Doppler-filters the data to form a range-Doppler map of the scene. The baseline scenario simulates the radar flying due north at 300 m/s at an altitude of 1,450m. The antenna look angles θ_{AZ} and θ_{EL} relative to the velocity vector are +20° (positive to the right), and −10° (negative down), respectively. The antenna pattern is modeled as a 2-D, symmetric sinc function with a one-sided, half-power beamwidth (HPBW) of 2°. Other relevant parameters may be examined directly from the MATLAB script.

The model features several parameters to control the granularity and extent of the terrain sampling in azimuth and range, including the number of range ambiguities; this provides the user with the ability to trade off runtime speed with model fidelity. Figure 6.16(a) shows the clutter range-Doppler spectrum under nominal conditions, which are listed within the figure. The effect of varying these parameters is shown next.

The clutter Doppler bandwidth is

Table 6.5 Parameter Values for Rain Rate-to-Backscatter Coefficient Fit

Conditions	*a*	*b*
Thunderstorm	138	1.37
Stratiform	57	1.60

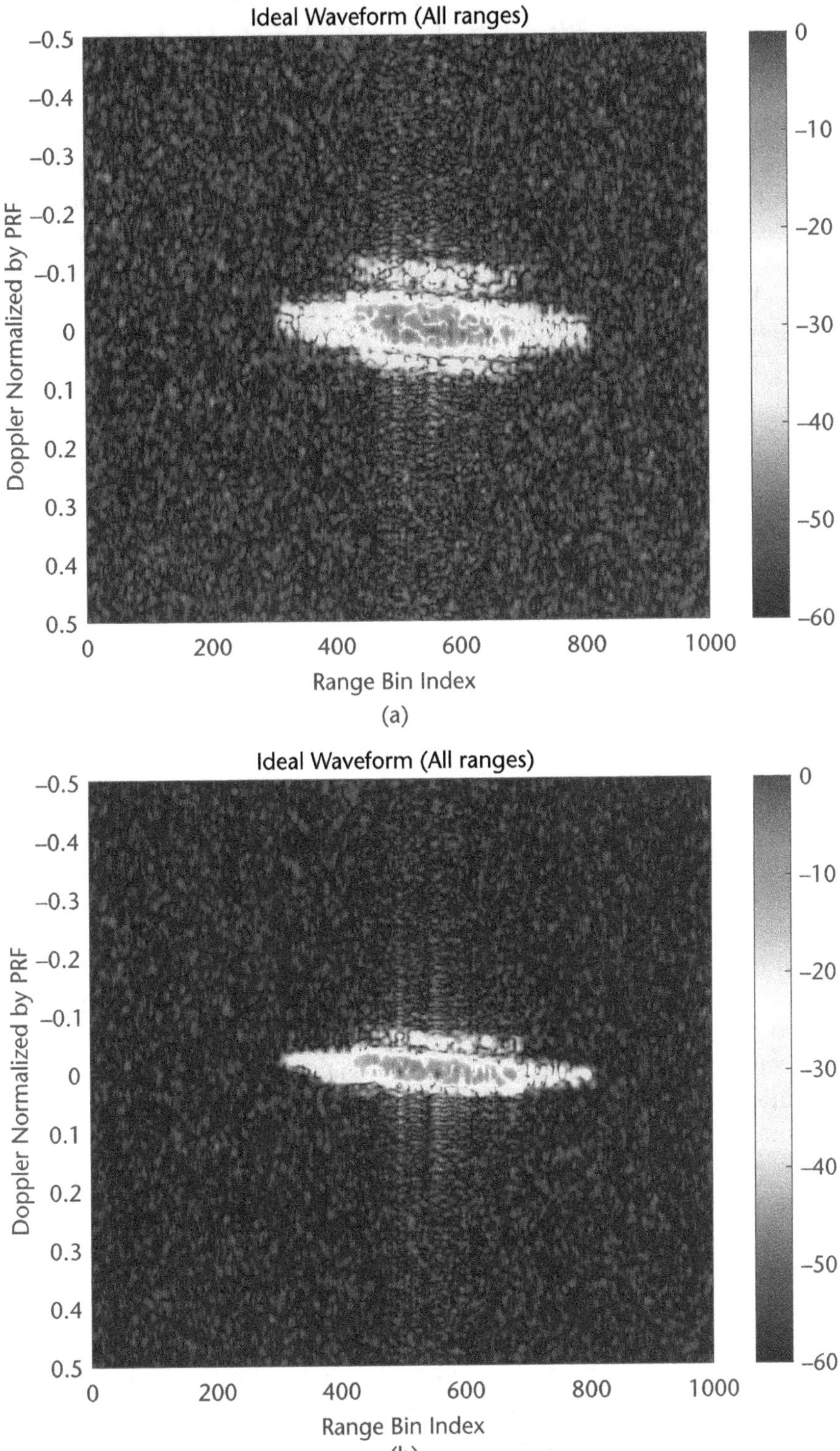

Figure 6.16 (a) Baseline clutter spectrum and (b) azimuth angle changed from 20° to 10°.

$$BW_{\text{Dop}} = \frac{4}{\lambda} V_R \cos\left(\theta_{\text{EL}}\right)\sin\left(\theta_{\text{AZ}}\right)\sin\left(\varphi_{\text{AZ}}\right) \tag{6.26}$$

where V_R is the radar speed, θ_{AZ} and θ_{EL} are the look angles, and φ_{AZ} is the one-sided HPBW in azimuth. The clutter bandwidth is proportional to the azimuth angle, as illustrated by Figure 6.16(b), which shows a much narrower spectrum when the azimuth angle is halved to 10°. Figure 6.17 shows the opposite effect of wider Doppler bandwidth when the azimuth angle increases.

The next set of results in Figure 6.18 show the dependency to platform velocity. As (6.26) indicates, the clutter bandwidth is also proportional to this velocity.

The simulation sampling parameters can also affect the resulting clutter spectrum significantly. Fine angular sampling ensures consistent coverage of a large number of point scatterers, at the expense of longer runtimes. A modest increase in the sampling interval from the nominal 0.25° can shorten the execution time with negligible or minor effects, as illustrated in Figure 6.19.

Further increases, however, will start showing a visibly degraded spectrum. Figure 6.20 shows how coarse sampling causes discrete clutter streaks across Doppler rather than a continuous return.

Finally, Figure 6.21 shows the effect of including all range ambiguities out to the radar horizon. This case takes the longest to run but has the most fidelity, as it includes range-ambiguous sidelobe clutter that may realistically impact radar detection.

6.12 Summary

This chapter began by defining clutter as distributed, passive, and undesirable radar returns. It then performed M&S of area and volume clutter beginning with a review of the properties of point targets, particularly the RCS fluctuation PDFs and decorrelation times of the Swerling models. Models for 0-D targets may be extended to 2-D and 3-D clutter. The analog to average RCS of a target for area clutter is the backscatter coefficient with units RCS per unit area, and for volume clutter a backscatter coefficient with units RCS per unit volume.

A 2-D clutter simulation numerically integrates over facets to produce the aggregate response of clutter over fast-time, slow-time, and angle. Two-dimensional clutter is facetized on a polar grid at the nadir point of the target. The product of facet area and backscatter coefficient yields the average RCS of a facet. The area of a facet is found using geometric constructs.

Land clutter backscatter coefficients are a function of RF, polarization, terrain type, and grazing angle. Grazing angles are conveniently partitioned into plateau, specular, and microshadowing regions. The constant-gamma model is often employed for the plateau region; an analogous expression was given for backscatter in the specular region. Facet backscatter coefficients may be drawn from an exponential distribution for rough surfaces. Complex clutter scenes generate strong returns more frequently than would be predicted by the exponential PDF, and such heavy-tailed statistics have been modeled by log-normal, Weibull, and

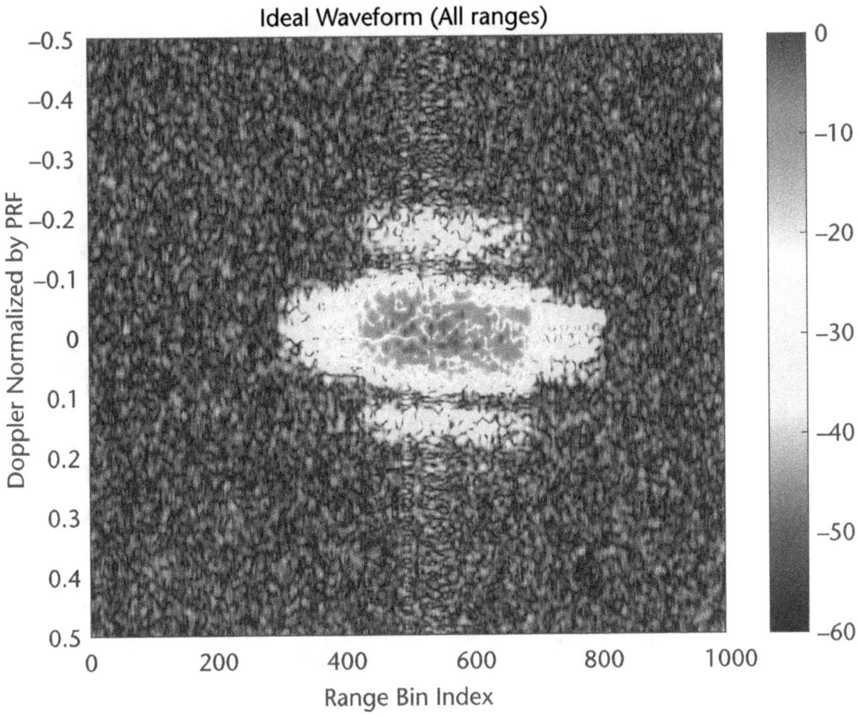

(a)

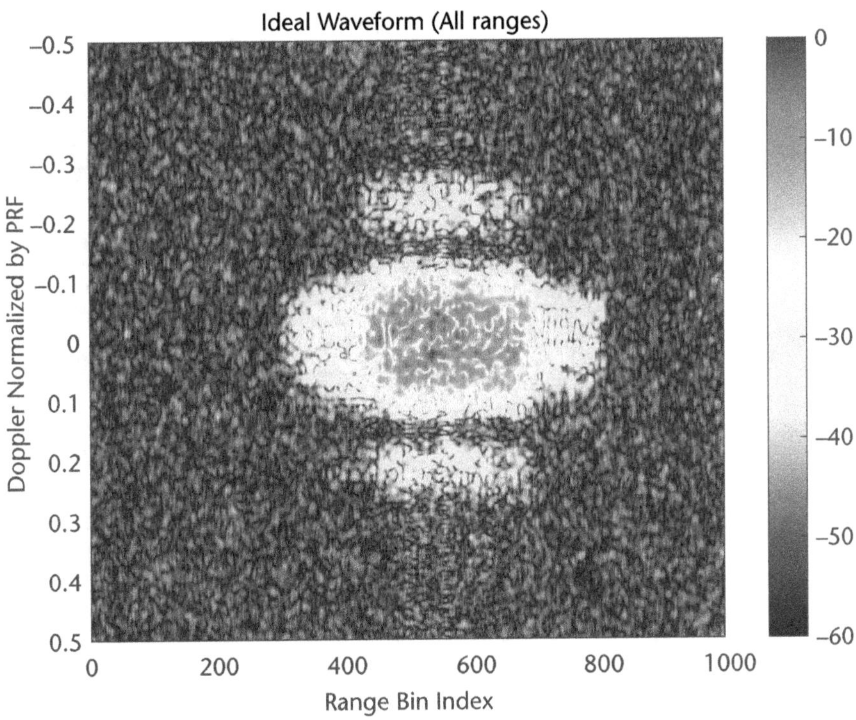

(b)

Figure 6.17 Additional azimuth angle variations: (a) increase to 45° and (b) increase to 80°.

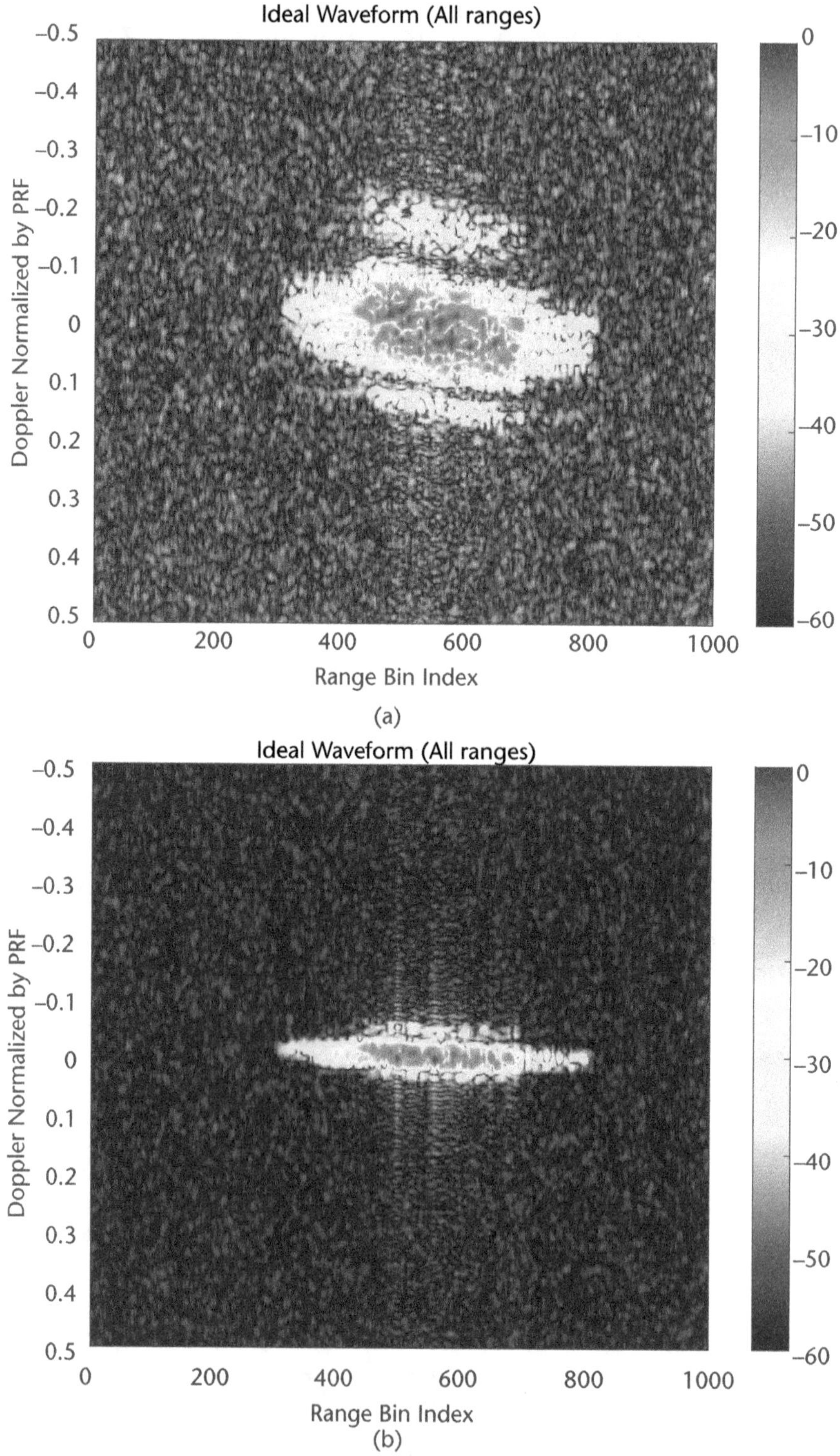

Figure 6.18 Radar velocity dependency: (a) increase to 600 m/s and (b) decrease to 150 m/s.

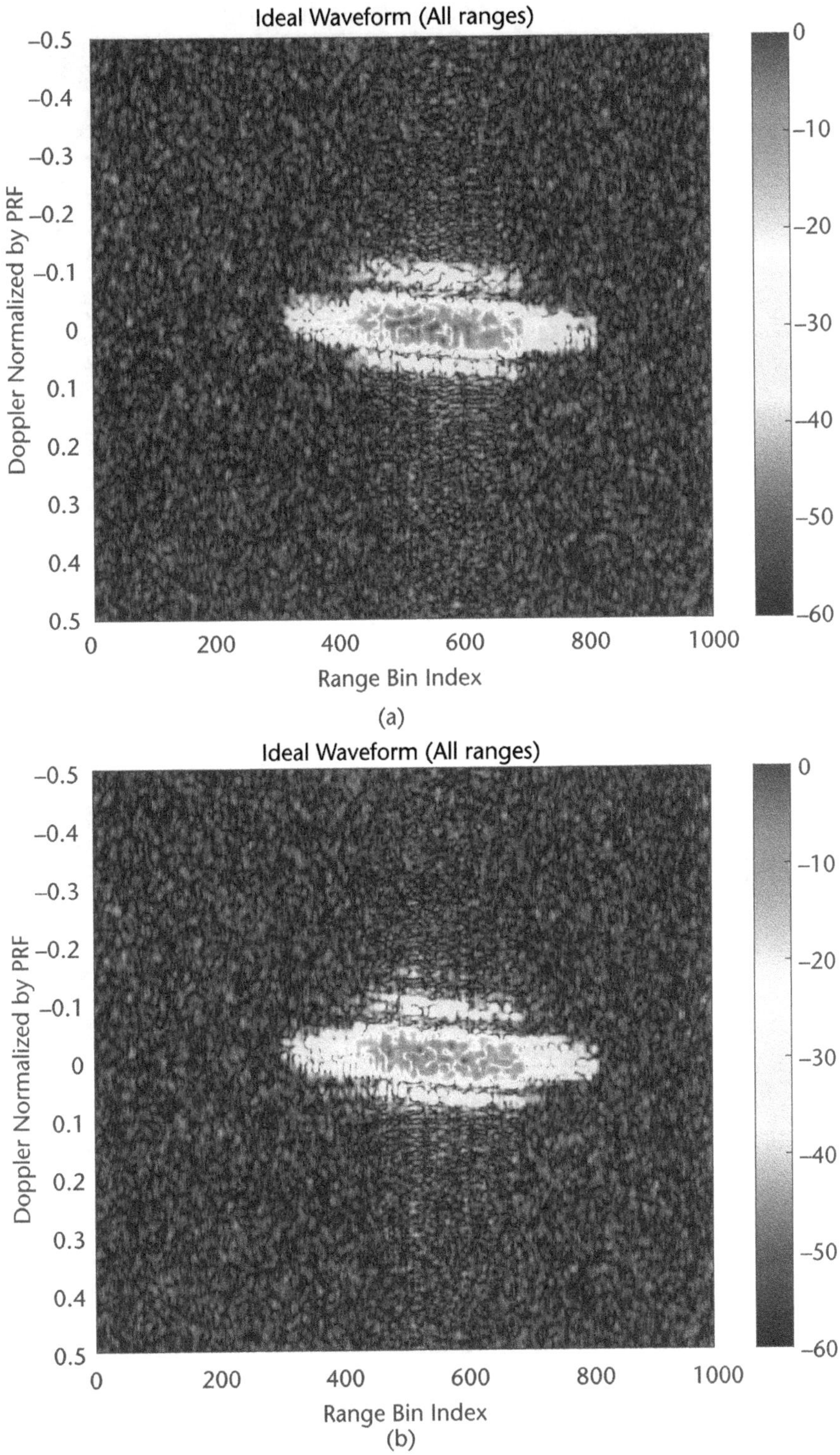

Figure 6.19 Angular sampling dependency: (a) increase to 0.5° and (b) increase to 1.0°.

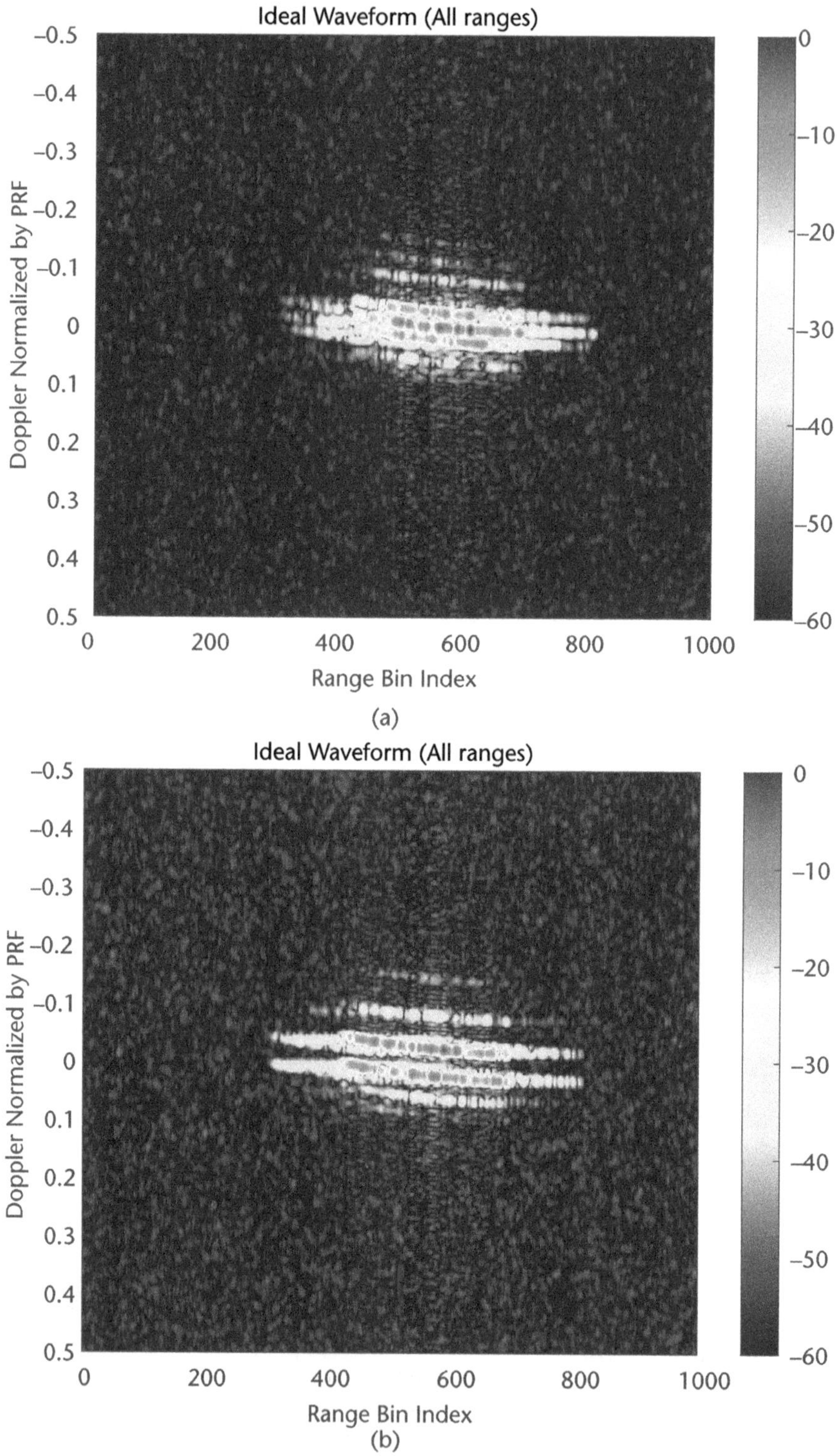

Figure 6.20 Angular sampling dependency: (a) increase to 2.0° and (b) increase to 4.0°.

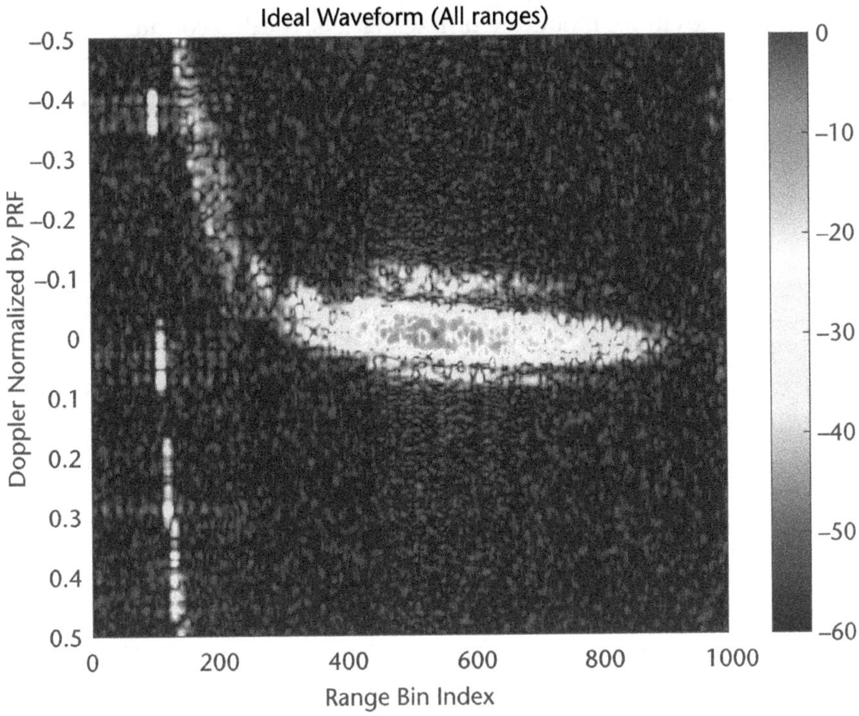

Figure 6.21 Clutter modeled over full-range ambiguity.

K-distribution PDFs. Heavy tails may also be created by distributing high-RCS point targets about the modeled scene to represent discrete returns. The decorrelation effects due to windblown foliage are captured by modulating facet returns with random decorrelation realizations created from an ICM covariance matrix, which in turn is built up from a mathematical model of clutter autocorrelation.

A site-specific clutter simulation requires access to a DEM for terrain height and LCC data for terrain type. Terrain height allows for the estimation of local terrain slope, which affects grazing angle, and prediction of shadowing effects. Terrain type is used to access relevant backscatter coefficient information. Numerical integration over facets proceeds as before, but now each facet may fall into shadow, or possess a slope that modifies grazing angle, and a terrain type that, in conjunction with grazing angle, determines the backscatter coefficient. Such a simulation requires significant transformations and interpolation between coordinate systems.

Sea-clutter M&S methods are similar to those for land clutter. Sea clutter backscatter coefficients are lower at most grazing angles but higher in the specular region. Sea state is an integer reporting of surface roughness and therefore clutter backscatter strength. Sea-clutter backscatter PDFs are exponential at low resolutions but grow heavy-tailed at fine resolutions, encouraging use of the log-normal, Weibull, and K-distributions introduced for land clutter. Ocean decorrelation is modeled the same way as land clutter, but without a DC component and employing a Gaussian autocorrelation function.

Volume clutter modeling utilizes voxels, the 3-D counterpart to 2-D facets. The procedure is similar, with the user setting voxel volume and choosing a backscatter coefficient to set the average voxel RCS. This chapter presented mathematical models that map rain rate to backscatter coefficients. Strong volume clutter, as observed for rain, tends to be highly localized.

References

[1] Beckmann P., and A. Spizzichino, *The Scattering of EM Waves from Rough Surfaces*, New York: Pergamon, 1963.

[2] Currie, N. C., et al., *Millimeter-Wave Radar Clutter*, Norwood, MA: Artech House, Boston, 1992.

[3] Ulaby, F. T., and M. C. Dobson, *Handbook of Radar Scattering Statistics for Terrain*, Norwood, MA: Artech House, 2019.

[4] Long, M. W., *Radar Reflectivity of Land and Sea*, Third Edition, Norwood, MA: Artech House, 2001.

[5] Melvin, W. L., "A STAP Overview," *IEEE Aerospace and Electronic Systems Magazine*, Vol. 19, No. 1, 2004.

[6] Skolnik, M. I., *Radar Handbook*, Second Edition, New York: McGraw-Hill Book Company, 2008.

[7] Barton, D. K., and S. A. Leonov (eds.), *Radar Technology Encyclopedia*, Norwood, MA: Artech House, 1997.

[8] Nathanson, F. E., *Radar Design Principles: Signal Processing and the Environment*, Second Edition, New York: McGraw-Hill, 1991.

[9] Pietsch, R., and W. A. Rasco, "Synthetic Aperture Radar Data Variance Analysis," AFWAL-TR-80-1058, Air Force Wright Aeronautical Laboratory, 1980.

[10] Billingsley, B., *Low-Angle Radar Land Clutter*, Norwich, NY: William Andrew Publishing, 2002.

[11] Melvin, W. L., and G. A. Showman, "An Approach to Knowledge-Aided Covariance Estimation," *IEEE Transactions on Aerospace and Electronic Systems*, Vol. 42, No. 3, 2006, pp. 1021–1042.

[12] Zywicki, D. J., and W. L. Melvin, "Application of Cultural Data Bases in Site-Specific Clutter Modeling," *Proc. 2002 KASSPER Conf.*, Washington, D.C., April 2002.

[13] Ward, K., R. Tough, and S. Watts, *Sea Clutter: Scattering, the K Distribution and Radar Performance*, IET Radar, Sonar and Navigation Series 20, London: Institution of Engineering and Technology, 2013.

[14] Richards, M. A., J. A. Scheer, and W. A. Holm, *Principles of Modern Radar: Basic Principles*, Raleigh, NC: SciTech Publishing, 2010.

CHAPTER 7

Waveforms

7.1 Introduction

The choice of radar waveform is an important design decision that bears significant impact in system performance. High-fidelity, engineering-level models are often referred to as "waveform-level," as actual signal samples are simulated and processed within this category of models.

As with many other engineering concepts, the first step in developing a computer model of a radar waveform is to define a mathematical representation, which may then guide the software implementation. A "waveform" is a description of the EM fields radiated by the radar antenna. These fields are time-harmonic, meaning that the variation over time is of sinusoidal form. Thus, a simple sine (or cosine) function serves as a suitable mathematical model for the waveforms of interest. An alternate but often convenient representation is that of a complex exponential. Both of these forms are shown in Table 7.1.

Notice that the signal amplitude A, its frequency f, and its initial phase θ are highlighted in bold. These are the parameters that a waveform designer may manipulate over the signal duration to define a modulation on the signal. More on this later.

Clearly, the expressions in Table 7.1 are related by $s_r(t) = \mathrm{Re}[s_c(t)]$. The signals that we measure in the real world (voltages, currents) are obviously real; however, the complex representation $s_c(t)$ is often used to model a signal in a quadrature receiver, where the I and Q channels can be mapped to the real and imaginary parts of the complex signal formulation. "I/Q-level model" is often used as a synonym for a waveform-level model.

Another key consideration in waveform modeling is the corresponding processing that goes along with the particular waveform choice. In general, the main processing steps are driven by the waveform type; more detailed implementation is dictated by the particular system being simulated. Even after the detailed processing steps are established, the model developer ought to consider the level of fidelity to be applied to the processor. The simplest, lowest fidelity case is to assume ideal conditions, meaning utilizing the full precision of the software modeling tool. In the case of MATLAB, all variables and operations are performed using double precision (64-bit floating point) as a default, meaning a very large dynamic range (–1.79769e308 to +1.79769e308) and very small step size (in the order of `eps`[1] = 2.22e-16) in representing signal values. For more realistic conditions, a

[1] `eps` (short for "epsilon") is a function that returns MATLAB's floating point relative accuracy.

Table 7.1 Mathematical Models for Radar Waveforms

Real Representation	*Complex Representation*
$s_r(t) = \mathbf{A}\cos(2\pi \mathbf{f}t + \boldsymbol{\theta})$	$s_c(t) = \mathbf{A}\exp(j(2\pi \mathbf{f}t + \boldsymbol{\theta}))$

receiver/processor model with hardware effects could be considered instead. These hardware effects could be representative of analog or digital components. Example analog effects include oscillator phase noise and other spurious signals, as well as amplifier saturation and other nonlinearities. These effects are explored in more detail in Chapter 10. Digital effects are those stemming from quantizing signal amplitudes into a discrete set of values based on the number of bits available in an analog-to-digital converter (ADC). Additional effects are the result of a finite bit depth when performing math operations.

In addition to amplitude quantization, an ADC also performs sampling, or temporal quantization, at a defined sampling rate. Of course, in a computer simulation, we are always dealing with sampled signal values, even if we choose not to quantize the signal amplitude. The sampling rate (or sampling frequency) *Fs* is therefore another key decision in the model design process. It is well known that to properly capture all the information within a continuous-time signal, the sampling rate must be at least twice its bandwidth *B*. This principle is known as the Nyquist (or Nyquist-Shannon) sampling theorem [1]. If there are signal components at frequencies above *Fs*/2, *aliasing* will occur, which means they will be indistinguishable from signal components at lower frequencies. This places a restriction in the choice of *Fs* used, not only to model our waveforms, but also in the ADC rates in an actual signal processor. In particular, it drives the model designer to adopt a baseband representation of radar waveforms. The reason is one of practicality; to illustrate, assume an X-band system being modeled, with a center frequency of 10 GHz. Avoiding aliasing would require a sampling rate of at least 20 GHz [or more accurately, gigasamples-per-second (Gsps)]. In general, this rate would result in exceedingly long data records. Simulating a radar dwell of 64 pulses at a pulse repetition frequency (PRF) of 32 kHz (6.4-ms duration) would take a vector of 64 million samples! These vector lengths are unnecessary and will cause slow runtimes and memory issues.

If the signal samples are complex, the sampling requirement may be reduced by one-half, as one complex sample is equivalent to two real samples. In the aforementioned example, the sampling rate may then be reduced to at least 10 GHz. One complex sample equals two real samples taken at 10 GHz, as opposed to taking one real sample at a 20-GHz rate. The memory requirements remain the same, however, as 32 million complex samples take as much space as 64 million real samples.

Rather than creating the signal samples directly at RF, a more sensible approach is to realize that a high center frequency does not add significant additional information to the time samples of the simulated waveform and can therefore be maintained as a separate parameter. Using the mathematical formulations in Table 7.1, it is straightforward to show that

$$s_c(t) = z(t)\exp\left(j2\pi \mathbf{f}t\right) \tag{7.1}$$

The function $z(t) = \mathbf{A}\exp(j\boldsymbol{\theta})$ is known as the complex envelope, or complex baseband representation of the waveform. This complex envelope $z(t)$, jointly with the center frequency f, contains all the information provided by the original model $s_c(t)$. The sampling requirements for $z(t)$, however, are greatly relaxed. If the bandwidth of $z(t)$ is $B = 10$ MHz, for example, now the complex sampling rate just needs to be at least 10 MHz (recalling that $z(t)$ is complex), which is a factor of 1,000 (three orders of magnitude!) lower than before.

Another common assumption in waveform modeling is that signals are narrowband, meaning that the signal bandwidth is much smaller than the center frequency ($B << f$). This assumption makes it possible, among other things, to define the effects of relative motion as a Doppler frequency shift based on the wavelength λ corresponding to the center frequency f. Similarly, it simplifies the modeling and application of gains for antennas and other components that are frequency-dependent.

Section 7.2 presents a taxonomy of common radar waveforms, followed by more in-depth discussions of both CW and pulsed waveforms. In addition, Section 7.2 provides simple Simulink examples of both types of waveforms. The chapter concludes in Section 7.5 with a presentation of a "waveform generator" model that not only implements most waveform types in the taxonomy but is also a good illustration of MATLAB and Simulink interaction.

7.2 Taxonomy of Radar Waveforms

The two main categories of waveforms used in radar and communications are pulsed and CW, as depicted in Figure 7.1. Both of these types of waveforms may be transmitted in unmodulated form, but more often than not, some type of modulation is imposed on them.

The generic waveform equations displayed in Table 7.1 apply equally to a CW or a pulse waveform. For a pulse waveform, the distinguishing feature from CW is the finite duration of a single pulse, or pulse width. This is not explicitly represented in Table 7.1, however; thus, we can add a rectangular function of width τ, defined here as

$$rect\left(\frac{t}{\tau}\right) = U(t) - U(t-\tau) \tag{7.2}$$

where $U(t)$ is the standard unit step function, also known as the Heaviside function in honor of the British mathematician Oliver Heaviside [2]. We note here that our definition of the *rect* function is causal (starts at $t = 0$). The more common definition in the literature spans the interval ($-\tau/2$ to $+\tau/2$), which is symmetric around zero but is noncausal.

The mathematical representation for a *causal* square pulse waveform is therefore

$$p(t) = s(t)rect\left(\frac{t}{\tau}\right) \tag{7.3}$$

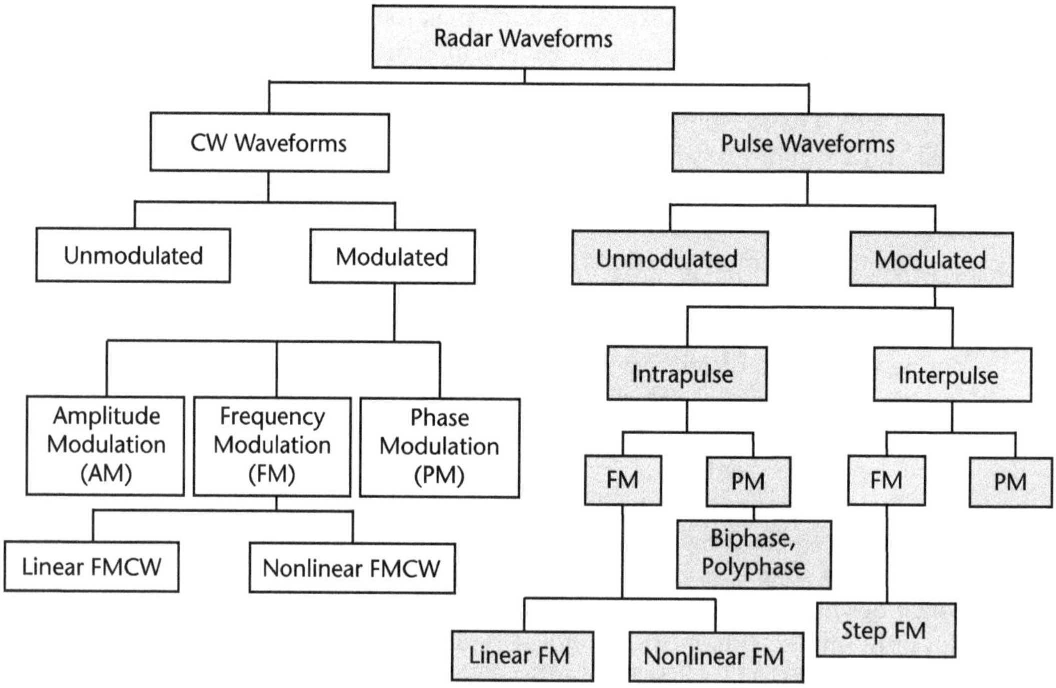

Figure 7.1 Common radar waveforms.

for either real or complex $s(t)$ [i.e., $s(t) = s_r(t)$ or $s(t) = s_c(t)$].

It was stated earlier that the signal amplitude A, its frequency f, and its initial phase θ in Table 7.1 are parameters that may be used to modulate the signal. The most common types of modulation are, in fact, amplitude modulation (AM), frequency modulation (FM), and phase modulation (PM). Specifically for radar applications, efficient transmission mandates constant power; therefore FM and PM tend to be the preferred options. FM in particular is a very popular and versatile choice, capable of high time-bandwidth products that provide high-range resolution and SNR gain through matched filtering and yet are also amenable to narrowband performance via stretch processing [3]. Linear FM (LFM) is by far the most common type encountered, although there are systems known to use nonlinear FM (NLFM) waveforms.

Due to the additional parameters involved in defining a pulse waveform, this category has additional waveform choices not available to CW. Modulation, whether FM or PM, may be applied within a pulse itself ("intrapulse modulation") or from pulse to pulse ("interpulse modulation"). Other modulation types such as pulse position modulation (PPM), although possible, are not common in radar operations and therefore not included in the taxonomy.

7.3 CW

CW is one of the two main waveform categories. As the name implies, a CW signal is always "ON." An unmodulated CW signal is the simplest waveform to model—it is fully defined by Table 7.1 with fixed amplitude, frequency, and phase. As simple as it may be, it is useful for estimating target radial velocity (velocity along the radar-target LOS) via the Doppler effect. In order to estimate target range, however, it requires some form of modulation to provide a timing reference.

From a systems perspective, one of the major design challenges of a CW radar is precisely the fact that it is always on; therefore, the system must be able to receive while it is transmitting. Receiver *isolation* then becomes a key design parameter; it is the level of rejection of the transmit signal leaking into the receiver. It may be achieved through a combination of design features that may include RF circulators as well as potentially having two separate antennas for transmit and receive. Insufficient isolation degrades performance by desensitizing the receiver (i.e., by reducing the radar's ability to detect small target reflections in the presence of stronger transmit leakage). It can also potentially damage the sensitive electronics in the receiver front end. These considerations tend to drive CW radar toward low-power, short-range applications. On the other hand, CW radars tend to be simpler and therefore relatively low-cost, compared to pulse systems. CW radar has no blind velocities or blind ranges stemming from receiver blanking that occurs on pulse systems. Moreover, the typical receiver follows a *homodyne* design with a single mixing stage for direct conversion down to baseband. The local oscillator (LO) signal stems from the transmit signal itself; this guarantees the phase coherency required for Doppler processing.

The block diagram for a generic CW radar shown in Figure 7.2 displays many of the features just discussed. A fixed-frequency waveform is assumed in this illustration. In this case a single antenna is used for transmit and receive, so a circulator is used to provide isolation. A replica of the transmit waveform is used for the local oscillator; furthermore, quadrature demodulation is used to create I/Q channels.

The RF source creates a CW signal at a fixed carrier frequency f_c, which is radiated through the antenna. Upon reflecting from a moving target, the return signal frequency changes by the Doppler frequency f_d. The homodyne I/Q receiver downconverts the received CW to baseband (i.e., removes the carrier in one step).

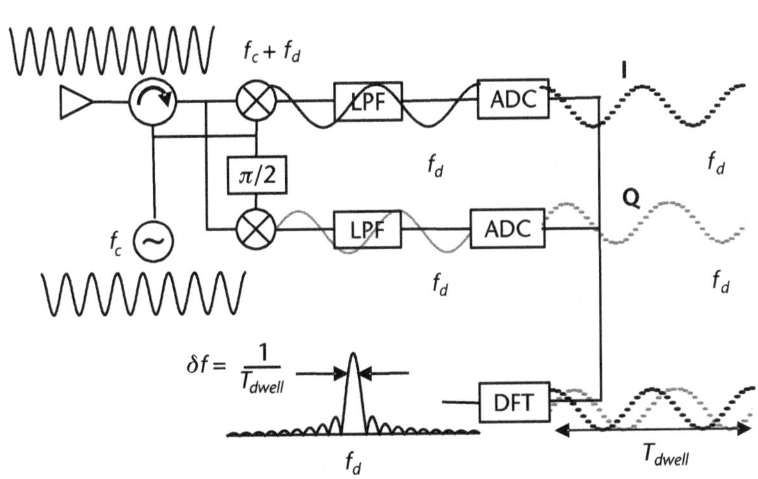

Figure 7.2 Block diagram and signal processing of fixed-frequency CW radar.

The sum component (i.e., high frequency) output of the mixer is eliminated by the low-pass filters (LPFs). The downconverted baseband CW is now at f_d, which needs to be measured. This measurement is accomplished by digitizing the signal using ADCs, then performing a discrete Fourier transform (DFT) on the sampled data. The DFT is formally introduced in Chapter 8, but it suffices to say here that the peak of the Fourier spectrum is located at the target Doppler frequency, with a frequency resolution (mainlobe width) inversely proportional to the duration of the data collection or dwell time T_{dwell}. The peak location automatically provides the Doppler estimate.

7.3.1 Simulink Example

Creating a Simulink model of a fixed-frequency CW is a trivial exercise, given that a "Sine Wave" signal block is available in the "Sources" sublibrary. Figure 7.3 shows the Simulink model and output plots in the time and frequency domain.

For this numerical example, the carrier frequency was set to 200 kHz, and the waveform is sampled at a 4-MHz rate (oversampled factor of 20×). The simulation runs for 1 ms, so only a portion of the time-domain samples are shown in the scope display. A total of 4,096 samples are collected for the spectrum analyzer, which is set to average two 2048-point FFTs. Thus, the frequency resolution of the measurement is 162.8 Hz, with a peak clearly detected at 0.2 MHz, which is the expected result.

Even a trivially simple CW model such as this can be useful to model a number of signals of interest. It could represent a transmit signal if the carrier frequency is set to the radar RF. Recall that this is not always practical; however, it is an option if there is a need to model it for a short period of time. On the receive side, the model could represent a baseband signal, where in this case the frequency value would correspond to the target Doppler.

7.3.2 Range Estimation

CW waveforms are inherently good for measuring target Doppler, and hence velocity, but they are incapable of measuring range due to their continuous nature. Some type of modulation is needed to provide a timing reference, and FM tends to be the most popular choice; common FM options are illustrated in Figure 7.4.

The sawtooth and triangle functions, also called "chirp" waveforms, are examples of linear modulations and the best choices for range estimation. The sine and square functions are nonlinear modulations; while range estimation can still be achieved with these waveforms, it is not as robust. The sawtooth LFMCW will be used here to describe range estimation via stretch processing; it may be used with either a positive or negative FM slope, and tends to be the top choice among modern systems.

Figure 7.5 illustrates the main parameters associated with the sawtooth LFM. The frequency deviation ΔF is the difference between the minimum and maximum frequency offsets; it also corresponds to the signal bandwidth. The duration of the linear ramp T_m is the modulation interval, and its inverse $f_m = 1/T_m$ is the modulation frequency. The LFM slope is therefore $\Delta F/T_m$ and may be either positive or

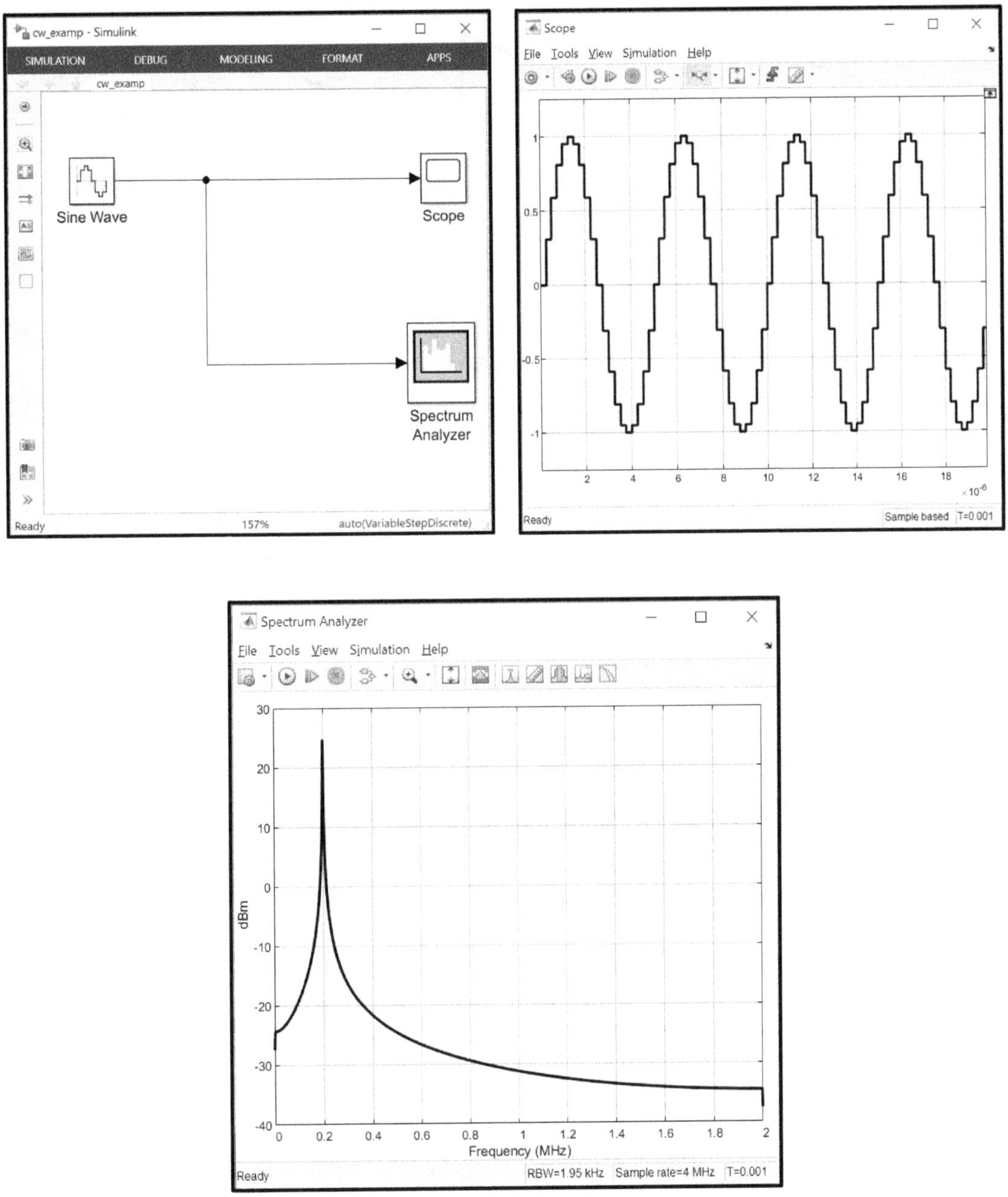

Figure 7.3 Simple Simulink model of unmodulated CW signal.

negative. In this example, $\Delta F = 150$ kHz and $T_m = 100\ \mu$s, resulting in a slope of 1.5 GHz/s. The bottom part of Figure 7.5 shows the actual LFM waveform over two modulation cycles, with a visible transition from max offset back to min offset at the end of the first cycle.

The approach for estimating target range is illustrated in Figure 7.6 using a single modulation interval of the transmit (solid line) and receive (dashed line) modulation waveforms. The received waveform is delayed relative to the transmit

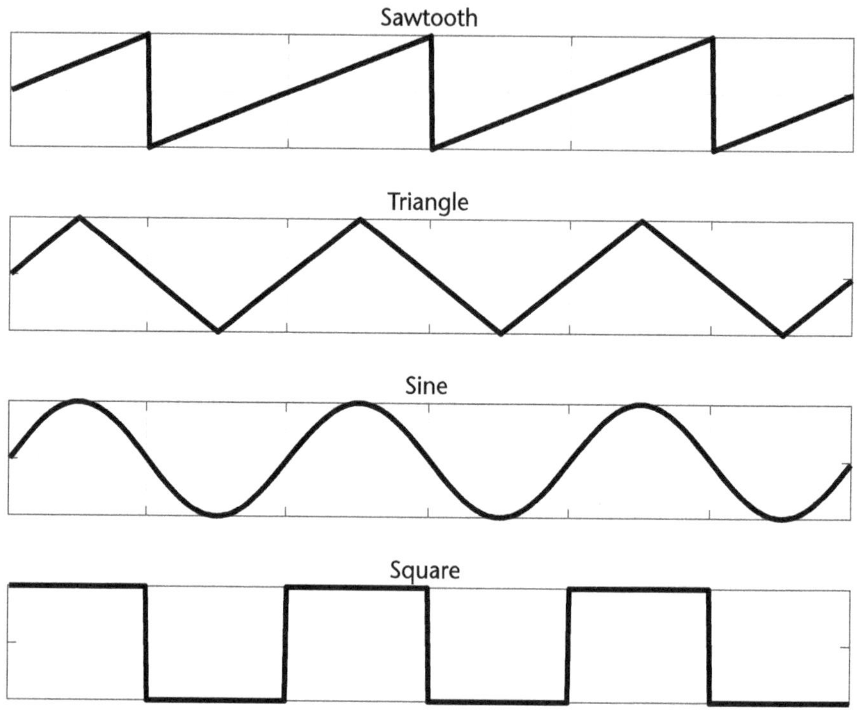

Figure 7.4 Example FM options.

by the round-trip time delay t_d, which is proportional to the target range R by the relation

$$t_d = \frac{2R}{c} \tag{7.4}$$

Mixing the receive-and-transmit waveforms (and rejecting the sum term via LPF not shown in Figure 7.6) results in an output CW signal whose frequency equals the difference between them. This difference is illustrated by the vertical separation between the two ramps; the vertical separation is constant, and therefore, so is the frequency, referred to as the *beat frequency* f_b. The beat frequency relates to t_d by the straight line equation $y = m * x$ where m is the slope. We therefore may write

$$f_b = \left(\frac{\Delta F}{T_m}\right)t_d = \left(\frac{\Delta F}{T_m}\right)\left(\frac{2R}{c}\right) = \left(\frac{2\Delta F}{cT_m}\right)R \tag{7.5}$$

which explicitly shows that the beat frequency is proportional to target range. The baseband signal can then be sampled and processed using a fast Fourier transform (FFT) to obtain the frequency spectrum. The beat frequency is measured by detecting the location of the spectrum peak, and solving for R in (7.5) results in a range estimate. Since the chirp or linear ramp is removed from the mixer output, this technique is also referred to as dechirping or deramping.

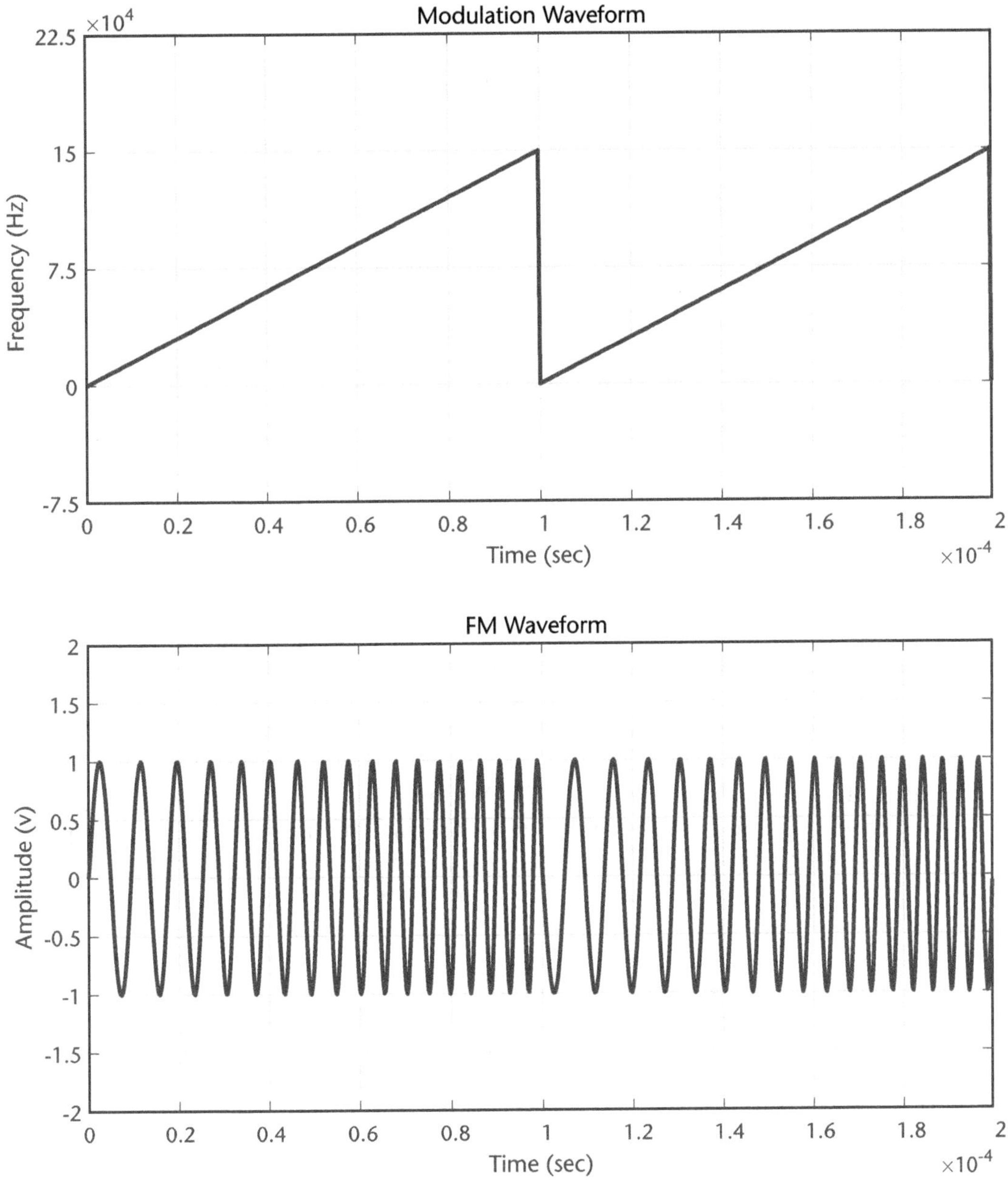

Figure 7.5 LFM parameters and example waveform.

The baseband output of the mixer is yet another example of a signal that could be modeled with the CW example from Figure 7.3, just by making the sine wave frequency equal to the beat frequency. Multiple replicas of this model at different beat frequencies could easily simulate a range-extended target composed of multiple scatterers, thereby forming a high-range resolution (HRR) profile via the FFT.

LFM and other CW modulation examples will be shown as part of the waveform generator described in Section 7.5.

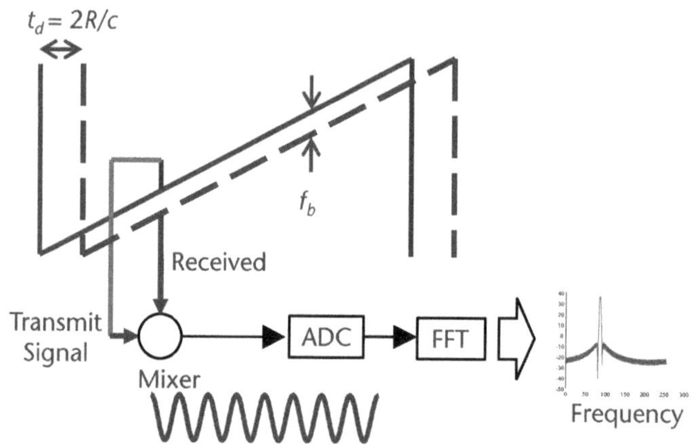

Figure 7.6 LFMCW range processing.

7.4 Pulse Waveforms

Pulse is the second major category of radar waveforms. Unlike CW, it toggles between "ON" and "OFF" states; during the "ON" state, the radar transmits EM energy for a time duration denoted as τ, defined here as the pulse length or pulse width. It then turns off over a period of time before transmitting the next pulse. The time separation between pulses (leading edge to leading edge) is the PRI, also called interpulse period (IPP). The inverse of the PRI is the PRF, which has units of hertz. The pulse width τ and the PRI (or PRF) fully define the pulse train. These quantities are illustrated in Figure 7.7. As was the case with modulated CW, range estimation is based on measuring the round-trip delay t_d of the target return given by (7.4); in this case, the leading edge of each transmitted pulse provides a timing reference for measuring t_d. The "OFF" time between pulses becomes a "listening time" where the radar looks for target returns. The easiest form of range estimation is achieved when the reflection from a given pulse arrives at the radar before the next pulse is transmitted; otherwise the estimate is ambiguous. Figure 7.7 also illustrates this phenomenon.

7.4.1 Range Ambiguities

The timeline in Figure 7.7 shows pulses being transmitted at integer multiples of the PRI, and pulse returns (echoes) from two targets. Target 1 is at a range where

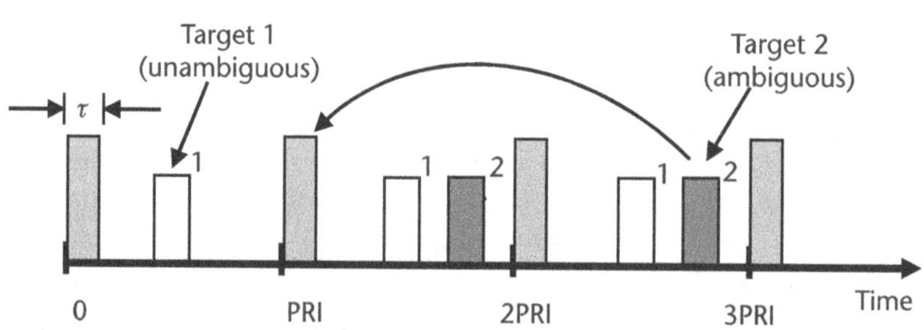

Figure 7.7 Notional pulse train and ambiguous target returns.

the round-trip delay is shorter than the PRI; therefore its returns (depicted as white pulses) belong to the transmitted pulses immediately to their left; the range estimate is said to be *unambiguous*. Target 2, however, is at a range with t_d > PRI, such that the return from transmit pulse 1 does not arrive until after pulse 2 has been transmitted, and so on. The range estimate will still be based on the transmit pulses immediately to the left, so the measurement is *ambiguous*; it will be off by a full PRI in the example.

This same discussion may be framed in the range dimension by introducing the concept of the maximum unambiguous range R_u, defined as

$$R_u = \frac{cPRI}{2} = \frac{c}{2PRF} \tag{7.6}$$

Targets at a range $R > R_u$ will have ambiguous (erroneous) range measurements. Moreover, targets at a range $R + kR_u$ (where k is an integer) will show up at the same range as a target at range R. Incidentally, modulated CW waveforms used for range measurement (such as the LFMCW example discussed in Section 7.3) suffer from the same range ambiguity limitation, where the modulation interval T_m becomes analogous to the PRI.

Another important parameter is the ratio of the pulse width to the PRI, known as the transmit duty factor or duty cycle

$$d_T = \frac{\tau}{PRI} \tag{7.7}$$

The duty factor is often described as a percentage. It is one way to categorize a pulsed waveform as either low-, medium-, or high-PRF. In a low-PRF (LPRF) waveform there is ample spacing between pulses, such that range measurements are expected to be unambiguous. LPRF duty factors tend to be < 10%. At the other end, high-PRF (HPRF) waveforms have pulses very close together so range measurements are highly ambiguous, with duty factors typically > 40%. In fact, HPRF waveforms tend not to be processed in range at all because of this reason. Duty factors that approach or surpass 50% are not suitable radar waveforms, as the radar returns will be mostly eclipsed due to receiver blanking. In between these two extremes, we have medium-PRF (MPRF) waveforms as a compromise between the two.

7.4.2 Doppler Ambiguities

A similar ambiguity phenomenon occurs in the frequency domain. The Fourier transform of a square pulse train is a "line spectrum" shaped by a "sinc" ($\sin(x)/x$) function envelope, as shown in Figure 7.8. The separation between the lines is given by the PRF, such that an HPRF waveform shows ample separation between each line. Last, the spectral "lines" are not lines at all; they are not impulses or delta functions, but rather narrow sinc functions themselves, which arise from a finite dwell or observation time (finite number of pulses). This is revisited in Chapter 8.

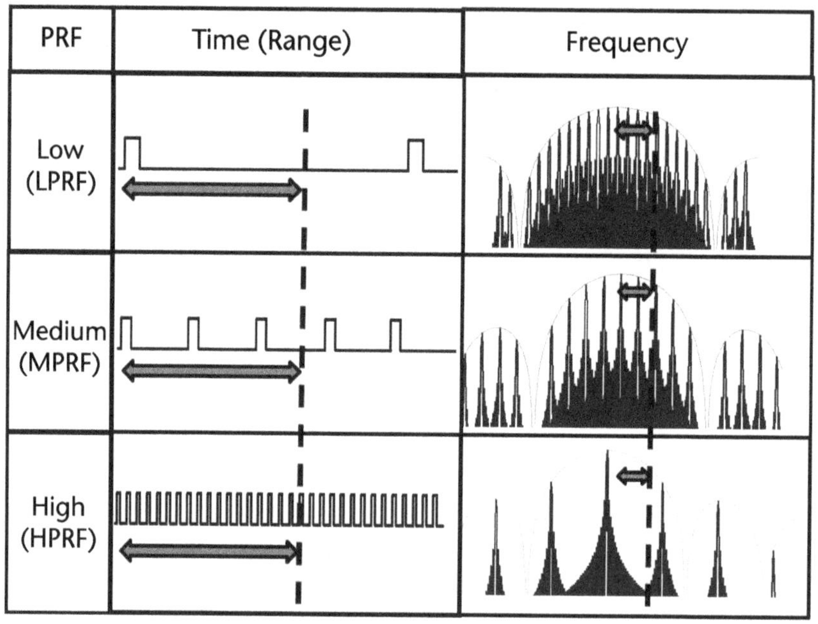

Figure 7.8 PRF modes: time and frequency.

A train of return pulses reflected from a moving target will exhibit the same spectral shape, but the spectrum will be shifted by the Doppler frequency offset. Target Doppler can therefore be estimated by measuring the frequency separation between a line in the received spectrum and its nearest line in the corresponding transmit spectrum. If the Doppler shift is greater than PRF/2, however, the adjacent transmit line will be closer and likely used as the reference, thereby introducing a measurement error. Moreover, targets with Doppler shifts equal to $f + kPRF$ (where k is an integer) will show up at the same frequency location as a target at Doppler shift f, again, a situation analogous to ambiguities in the range dimension.

The vertical dashed lines in Figure 7.8 indicate the location of some arbitrary target at some range and Doppler. As the PRF varies from low to medium to high, the reader can appreciate how the severity of ambiguity in each dimension changes and also appreciate the fact that they are inversely related. Use of LPRF waveform results in highly ambiguous Doppler (many lines between target line and its true reference, indicated by the gray horizontal arrow), yet range is largely unambiguous (i.e., very large R_{u}). The opposite case occurs at HPRF, while MPRF presents a situation where the target is likely to be ambiguous in both range and Doppler, but at a manageable level. Range and Doppler ambiguities must be resolved by changing PRF values over multiple dwells.

7.4.3 Pulse Modulations

In terms of pulsed-waveform modulation, the same common choices seen for CW waveforms (FM and PM) apply here; however, the addition of the PRF parameter to the waveform definition adds one more layer of waveform options, as the modulation may be applied to each single pulse (intrapulse modulation), applied across multiple pulses (interpulse modulation), or combination thereof. Figure 7.9 illustrates the four combinations, with biphase modulation used for the PM examples.

The two phase states (nominally 0°/180°) are denoted by the positive and negative signs; there are also polyphase modulation options available. For the intrapulse LFM, the modulation interval is the pulse width itself (i.e., $T_m = \tau$).

One last but crucial aspect of the pulse waveform is whether the pulse train is coherent or not, meaning whether the phase relationship of the underlying carrier sinusoid signal is preserved from pulse to pulse. This is best illustrated in Figure 7.10. A coherent RF source, such as the common master oscillator power amplifier (MOPA) configuration, may be modeled as a free-running oscillator followed by an ON/OFF switch that closes every PRI and stays closed for a duration equal to the pulse width. The resulting pulse train will pick up the initial phase at every subsequent pulse at the value the oscillator CW signal would have at that point it time.

A noncoherent RF source, on the other hand, can be thought of as turning the oscillator itself on and off, where each time it will start at some random phase. Phase coherency is required for Doppler measurement, and is therefore the type of pulse waveform considered herein.

7.4.4 Simulink Example

As was the case for CW, creating a coherent pulse train model is straightforward. The on-off keying (OOK) function represented by the switch in Figure 7.10 will be implemented by a Pulse Generator block. Figure 7.11 shows the model as well as the resulting waveforms.

In Figure 7.11, the coherent pulse train in the scope window in (b) is the product of the two waveforms in the bottom right. This is a simple model of the

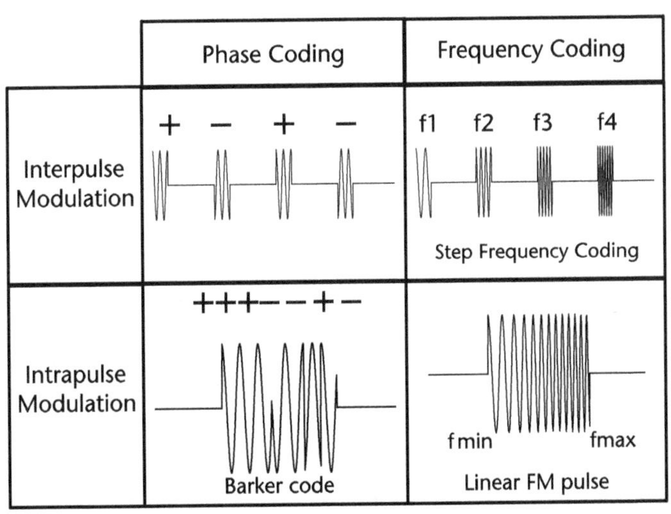

Figure 7.9 Pulse-modulation examples.

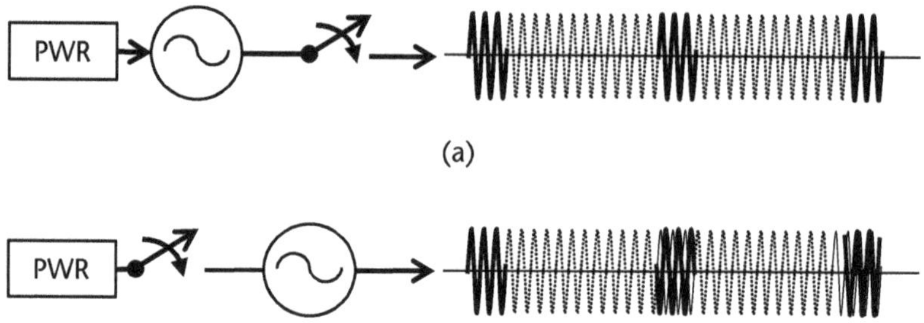

(a)

(b)

Figure 7.10 (a) Coherent and (b) noncoherent pulse trains.

aforementioned MOPA configuration; the PA portion of the configuration could be represented explicitly by a Gain block at the output of the Multiplier block; otherwise, it can be accounted for when setting the amplitude of the sine wave source. Last, it is noticed that each coherent pulse in Figure 7.11(b) has the exact same initial phase; this is a consequence of the chosen PRI being an integer multiple of the sine wave period. This is not required for coherency, however. As an exercise, the reader may change the sine wave frequency by as little as 1% and observe how the pulses will now have a different starting phase, while still maintaining coherency. Another insightful exercise would be to add a Spectrum Analyzer block to the model (a Zero-order Hold block with a *Ts* sampling interval prior to the analyzer will also be needed to make it work). This will enable the reader to observe the "line spectrum" of the coherent pulse train.

In a manner similar to the CW model discussed in Section 7.3, the simplicity of this pulsed-waveform model belies its potential usefulness. It could be used to represent a transmit signal by setting the waveform frequency to the transmit RF; or it could represent a received signal already downconverted to baseband by setting it to the target Doppler frequency. The one shortcoming this model has in its current form is the lack of any modulation; in particular, the implementation of pulse-compression options that are so common in modern systems. The "waveform generator" model presented in Section 7.5 shows examples of LFM pulses; biphase modulation is covered in Chapter 8 as an example of matched-filter processing.

7.5 Waveform Generator Model

The model presented here is a collection of waveform-generation blocks available in Simulink; it implements many of the cases featured in the waveform taxonomy discussed in Section 7.2. The model itself is a good illustration of MATLAB/ Simulink interaction, as the waveform generation takes place in Simulink but is invoked from MATLAB through a GUI. It also illustrates the use of "Enabled Subsystems," another useful Simulink feature.

The GUI is activated by typing the name "wavegen" in the command prompt:

```
>> wavegen
```

This will open the main GUI window shown on the upper left side of Figure 7.12. It features a dropdown menu of waveform choices and a "Continue" push-button. The options available in the menu are summarized in Table 7.2.

Upon selecting a waveform choice and pressing "Continue," a secondary GUI window opens, with default parameters corresponding to the waveform choice. After pressing "Start," a number of waveform samples are created and displayed in several plots in both time and frequency.

In Figure 7.12, the CW option was selected, resulting in the two simple plots shown. Figure 7.13 shows the secondary GUI and corresponding plots resulting from selecting the pulse waveform with step-FM modulation.

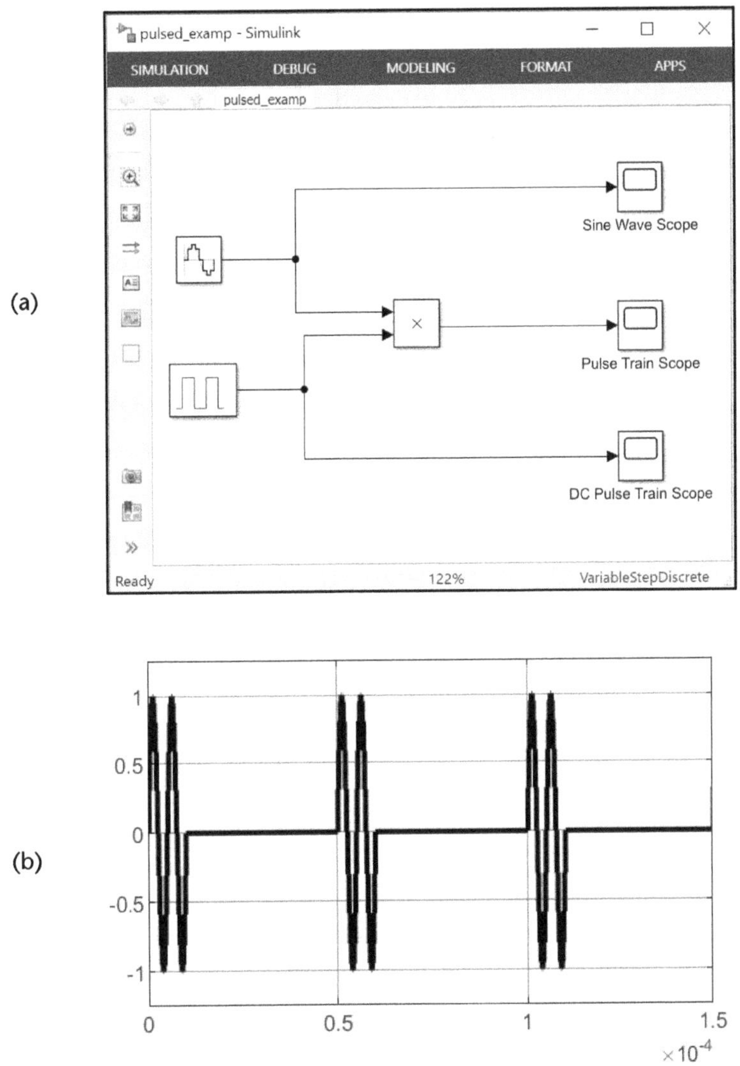

Figure 7.11 Simple Simulink model of unmodulated coherent pulse signal.

(c)

(d)

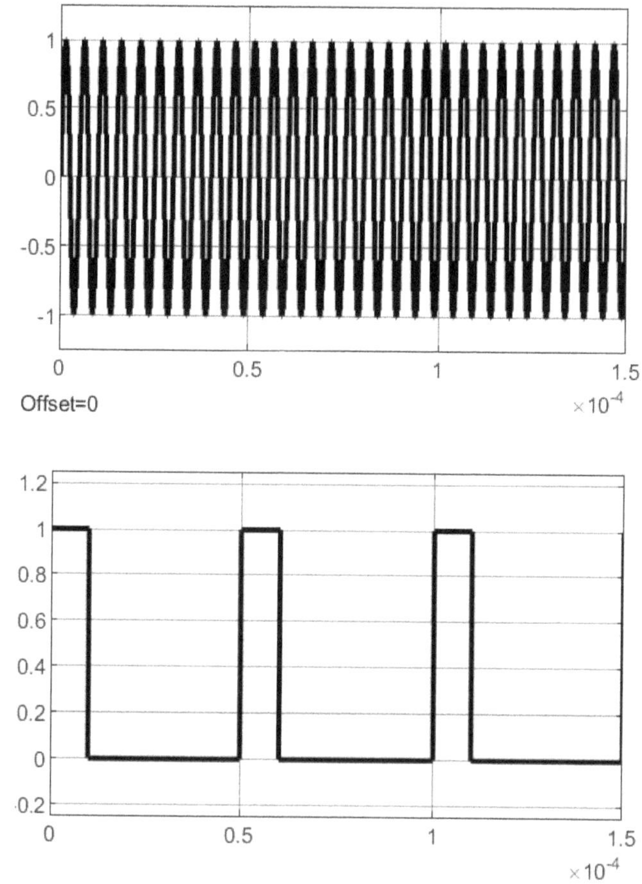

Figure 7.11 *(Continued)*

The modulation waveform for this example is a series of N discrete steps equally spaced between user-defined minimum and maximum values. Careful design of these parameters yields a variety of different waveforms. In Figure 7.13, the FM step duration was matched to the PRI (0.25 ms), resulting in a form of interpulse modulation where each pulse is at a fixed frequency, but the frequency increases from pulse to pulse. This is achieved by defining the modulation frequency f_m as

$$f_m = \frac{PRF}{N} = \frac{(4.0 \text{ kHz})}{4} = 1.0 \text{ kHz} \tag{7.8}$$

This form of step-FM is a waveform commonly used in instrumentation radars where narrowband pulses are processed together to synthesize a wideband (i.e., high-resolution) response.

As an exercise to the reader, a completely different response may be obtained by matching the modulation period T_m to the pulse width τ—in other words, by setting the modulation frequency as

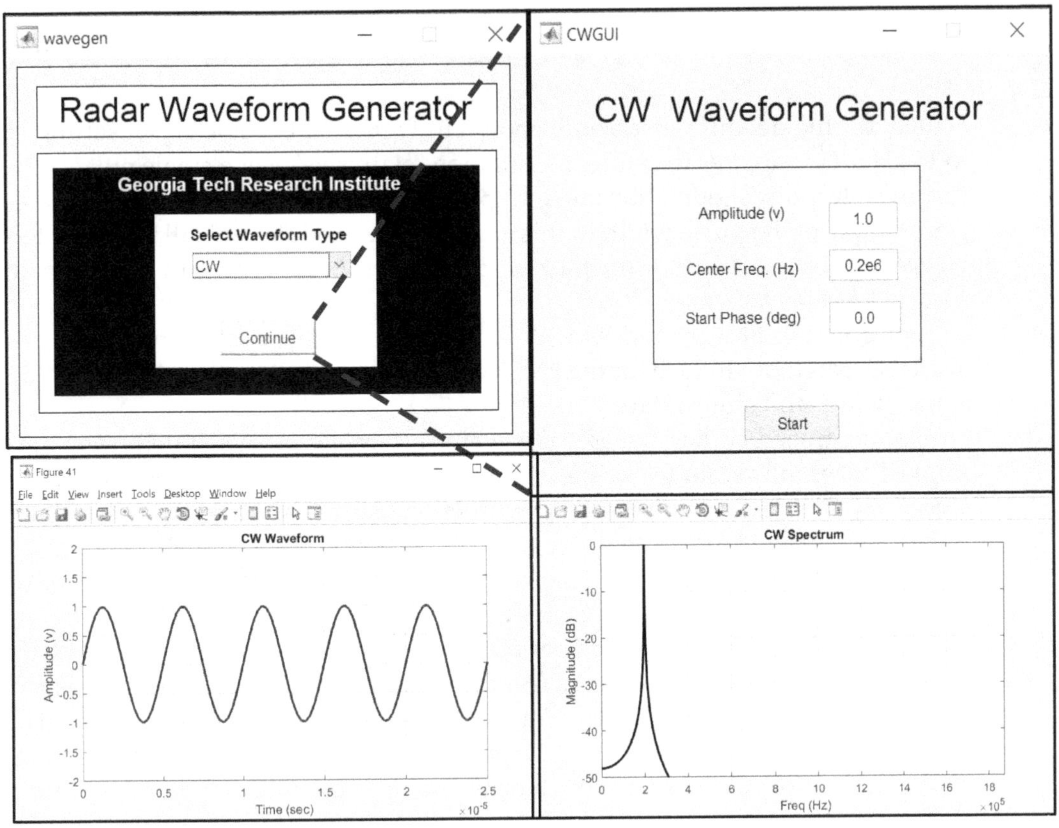

Figure 7.12 Waveform generator model GUI and CW plots.

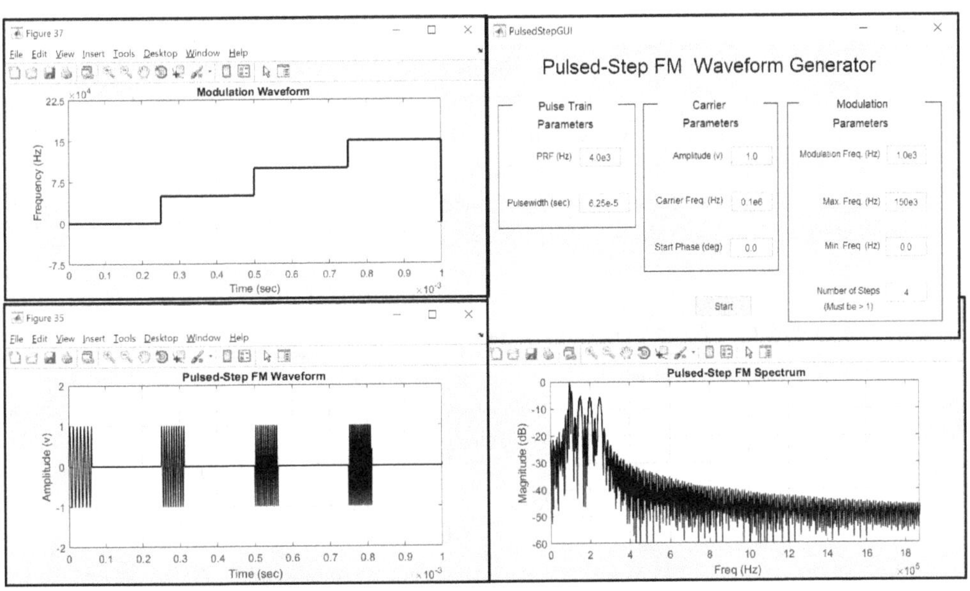

Figure 7.13 Step-FM GUI and plots.

$$f_m = \frac{1}{\tau} \tag{7.9}$$

Which, for the default pulse width shown in Figure 7.13, would mean setting f_m to 16 kHz. This applies the entire modulation "staircase" over a single pulse, thus turning interpulse modulation into intrapulse modulation. Since the PRI is an integer multiple of the pulse width in this example, the modulation will be replicated on every pulse. Last, increasing the number of steps (try $N > 20$) will approximate a pulse LFM waveform.

For both the FMCW and AMCW waveforms, three modulation options are available: Sawtooth (which for the FMCW is simply the LFM shown in Figure 7.5), square wave, and sinusoidal.

As mentioned earlier, the actual waveform samples are generated from Simulink. The Simulink model is called "waveformGen1p5" and is invoked from the secondary GUI "Start" button callback function using the `sim` command as illustrated in Figure 7.14 (upper left).

The input argument in the `sim` command is a string variable with the name of the Simulink model. The model's top-level is shown in the upper right. The constant "WType" contains the index of the waveform selected from the main GUI dropdown menu (i.e., the numbers from column 1 in Table 7.2). This index is fed to the Waveform Generator subsystem, whose contents are shown at the bottom of Figure 7.14. This subsystem can be interpreted functionally as an if-then construct. Through multiple Go To and From blocks, the selected index is compared to the six possible values, resulting in Boolean true (logical 1) or false (logical 0) signals. These Boolean signals control the six "Enabled Subsystems" that correspond to the possible waveform choices. Enabled subsystems are only executed when their control signal is high, or positive; thus, only the subsystem for the selected waveform executes in this case—all the other subsystems remain disabled. The waveform samples and corresponding plots are created by the selected subsystem, using the parameter values specified in the secondary GUI window. These parameters are passed across GUI subfunctions by declaring them `global` variables. The global variable definitions are contained in the *waveform_init* script shown in line 118 of the Start button callback code in Figure 7.14.

In addition to creating the time and frequency plots, the waveform samples exit their Enabled Subsystem. Through the Multiport Switch block on the right

Table 7.2 Waveform Generator Options

WType	*Waveform Option*	*Description*
1.	Pulsed	Unmodulated coherent pulse train
2.	Pulsed LFM	Coherent pulse train with LFM
3.	Pulsed-Step FM	Coherent pulse train with stepped FM
4.	CW	Unmodulated CW
5.	FM CW	Frequency-modulated CW
6.	AM CW	Amplitude-modulated CW

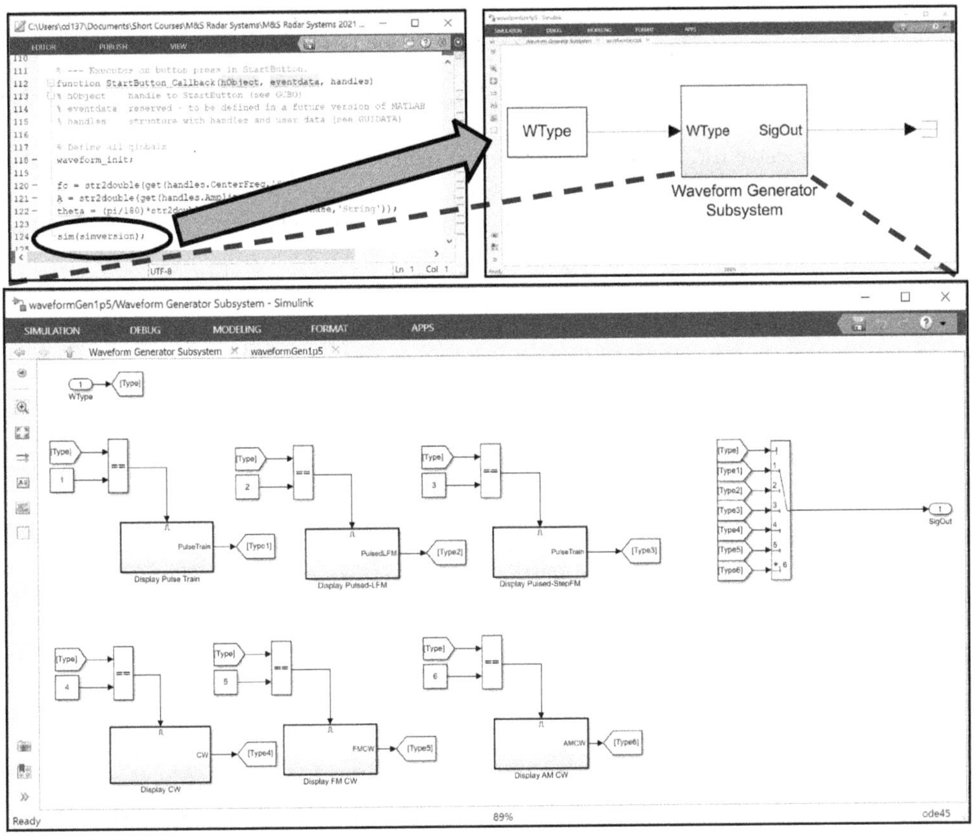

Figure 7.14 MATLAB/Simulink interaction in Waveform Generator model.

side, the samples exit the Waveform Generator subsystem and are terminated at the top level of the model. This was implemented as a future growth option in which the output of the generator could feed a larger simulation.

Last, the reader will notice that the contents of the CW and Pulsed Enabled Subsystems (options 1 and 4) are very similar to the simple examples illustrated in Figures 7.3 and 7.11. Indeed, each of the waveform options in this model is relatively straightforward to build, and the model is easily extensible to additional waveform choices.

7.6 Concluding Remarks

Waveform modeling is a key step in developing engineering-level models. Simulink makes this development easy; however the principles outlined here apply regardless of the software environment. The important family of pulse PM waveforms (binary PM in particular) is discussed in Chapter 8.

References

[1] Proakis, J. G., and D. G. Manolakis, *Introduction to Digital Signal Processing*, New York: MacMillan Publishers, 1988.

[2] Bracewell, R. N., *The Fourier Transform and Its Applications*, McGraw-Hill, New York, 2000.

[3] Keel, B. M., and J. M. Baden, "Stretch Processing," in *Principles of Modern Radar, Vol II: Advanced Techniques* (W. L. Melvin and J. A. Scheer, eds.), Raleigh, NC: Scitech Publishing, Inc., 2013, pp. 26–40.

CHAPTER 8

Range and Doppler Processing

8.1 Introduction

The two most important measurements made from target returns are range and angle, which allow the radar to estimate the target position in space (geolocation). The third most important measurement is arguably the target Doppler frequency, from which target velocity information can be derived. Angle measurement modeling is discussed in Chapter 9, which expands on the monopulse processing concepts introduced in Chapter 3. This chapter discusses range and Doppler processing, with a focus on pulse-Doppler waveforms. Pulse-Doppler processing involves sampling the received radar data at two different time scales, termed fast-time and slow-time. Fast-time sampling is associated with range processing, and is performed by the receiver ADC. Multiple samples are typically taken within an individual pulse, with sampling intervals that may be in the order of microseconds to nanoseconds ($\sim 10^{-6}$–10^{-9} sec). On the other hand, slow-time is commonly used for Doppler processing. It entails sampling between pulses, so the sampling frequency matches the waveform PRF, with sampling intervals in the order of milliseconds to microseconds ($\sim 10^{-3}$–10^{-6} sec). Since both range and Doppler processing involve sampled data, aliasing and other sampling artifacts can occur on either dimension.

The rest of the chapter is organized as follows. Section 8.2 develops a mathematical model that ties in target velocity with the Doppler frequency. This leads to a formal introduction of the FT in Section 8.3, followed by the DFT and FFT in Section 8.4. The FFT is the tool of choice for Doppler processing and velocity estimation, but it also plays a part in range processing. Section 8.5 introduces pulse compression waveforms as a key component of range processing, with detailed discussions on PM and FM waveforms. Modeling of matched filtering and pulse compression of both PM and LFM pulses is presented in Sections 8.5.1 and 8.5.2, respectively, where the LFM matched filtering is implemented via the FFT.

Last, a consolidated view of both range and Doppler processing from a systems perspective is presented in Sections 8.6 and 8.7. The result of this processing is a range-Doppler map (RDM), which is the principal data product in pulse-Doppler radar systems. Concluding comments are presented in Section 8.8.

8.2 Target Velocity and Doppler

We now consider a first-order mathematical model of a target return when there is relative motion between the target and the radar. For the transmit waveform, consider the complex exponential model first shown in Table 7.1, reproduced here for convenience:

$$x_{tx}(t) = A\exp\left(j\left(2\pi f_0 t + \theta\right)\right) \tag{8.1}$$

where f_0 is the transmit RF. Assume that the target is at a range R; furthermore, assume that this range is changing at a constant rate (no acceleration). Then range can be expressed as a function of time as

$$R(t) = R_0 + \dot{R}t \tag{8.2}$$

where R_0 is the range at time $t = 0$, and $\dot{R}$ is the rate of change of range as a function of time, dR/dt. Equation (8.2) stems from the Taylor series expansion of $R(t)$, assuming that acceleration and all other higher-order derivatives are zero.

The target reflection received at the radar is modeled as a scaled, time-delayed version of the transmit signal, given by

$$x_{rx}(t) = \hat{A}\exp\left(j\left(2\pi f_0\left(t - t_d\right) + \theta\right)\right) \tag{8.3}$$

where the round-trip time delay was defined in (7.4) as proportional to target range. Replacing (8.2) in (7.4) results in

$$t_d = \frac{2R_0}{c} + \frac{2\dot{R}t}{c} \tag{8.4}$$

Thus, we may rewrite (8.3) as

$$\begin{aligned} x_{rx}(t) &= \hat{A}\exp\left(j\left(2\pi\left(f_0 - \frac{2\dot{R}}{\lambda}\right)t - \frac{4\pi R_0}{\lambda} + \theta\right)\right) \\ &= \hat{A}\exp\left(j\left(2\pi\left(f_0 + f_{\text{Dop}}\right)t + \varphi\right)\right) \end{aligned} \tag{8.5}$$

where φ is a range-induced phase term, and f_{Dop} is the Doppler frequency shift on the RF due to target relative motion.

One key point is that the range rate $\dot{R}$ relates to the target relative velocity *along the radar-target LOS*, and not to the actual total velocity. This LOS velocity (or radial velocity) is the projection of the velocity vector onto the LOS vector, as illustrated in Figure 8.1. The target moves based on a velocity vector $\mathbf{v}_t$ pointing in any arbitrary direction. The component of that vector that lies in the direction of the LOS vector is calculated by the dot product of the two, with the LOS vector normalized to unit magnitude; that is,

$$V_{\text{LOS}} = \mathbf{v}_t \bullet \mathbf{LOS}_\mathbf{u} = V\cos(\phi), \text{ where } \mathbf{LOS}_\mathbf{u} = \frac{\mathbf{LOS}}{|\mathbf{LOS}|}, V = |\mathbf{v}_t| \tag{8.6}$$

Given the radial velocity, the Doppler frequency shift is then

$$f_{\text{Dop}} = \frac{2V_{\text{LOS}}}{\lambda} = -\frac{2\dot{R}}{\lambda} \tag{8.7}$$

The Doppler shift is positive (receive frequency higher than transmit) for positive V_{LOS}, meaning that the target is approaching the radar. Notice, however, that for an approaching target, $\dot{R} < 0$ (range is decreasing over time); therefore, if target Doppler is defined in terms of range rate, it needs the negative sign shown in (8.7). Oversight of this subtle but important detail is a common modeling error, where V_{LOS} and $\dot{R}$ are often used interchangeably. In reality, although they are equal in magnitude, they are opposite in sign.

8.3 The Fourier Transform

The Fourier transform (FT), named after the French mathematician Jean-Baptiste Joseph Fourier, is a mathematical tool for describing how a signal of interest is distributed in the frequency domain. The original signal may be defined in either time or space, and the corresponding transform domain is therefore characterized as a temporal or spatial frequency. In this context we only consider time-domain signals, and thus the temporal or standard frequency domain. For a continuous-time signal $x(t)$, its continuous-time FT (CTFT) $X(f)$ is defined as

$$X(f) = \int_{-\infty}^{\infty} x(t)e^{-j2\pi ft}dt \tag{8.8}$$

and is often referred to as the frequency *spectrum* of $x(t)$. Equation (8.8) is the *forward* transform. There is a corresponding *inverse* transform that takes $X(f)$ and returns it to the time-domain:

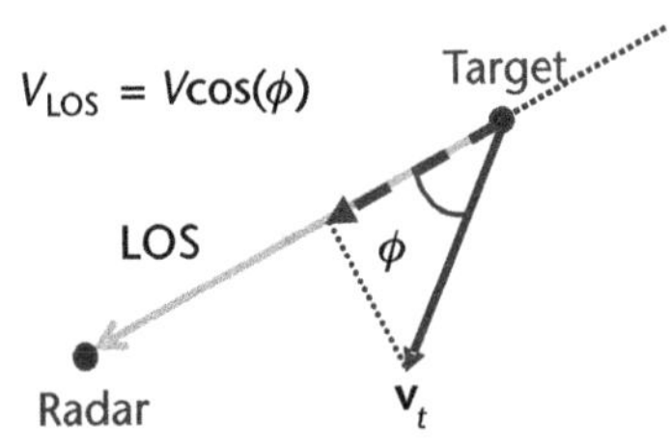

Figure 8.1 Definition of radial velocity V_{LOS}.

$$x(t) = \int_{-\infty}^{\infty} X(f)e^{+j2\pi ft}df \quad (8.9)$$

Table 8.1 shows a number of FT pairs that are relevant to radar analysis and modeling. Row 1 shows that the FT of a CW signal at a frequency f_0 of infinite duration is just an impulse function (also known as the Dirac delta function, named after English physicist Paul Dirac) located at f_0. If the function is real (sine or cosine) as shown in Table 8.1, the transform will have components at $+/- f_0$ (the one illustrated corresponds to cosine). A complex exponential such as the complex waveform representation of Table 7.1, however, will only have one component at f_0.

The second row of Table 8.1 depicts a single-square pulse at baseband (or DC pulse, as in the Simulink example of Figure 7.11). It has a well-known FT known as a sinc function, defined as

$$\text{sinc}(x) = \frac{\sin(\pi x)}{\pi x} \quad (8.10)$$

This function is symmetric with a maximum value of 1 at $x = 0$, followed by decaying oscillations with zero crossings at +/− integer values of x. These oscillations are infinite in extent; only the first few zero crossings are shown in Table 8.1's figures. The crossings nearest to the peak define the width of the sinc mainlobe, which is 2 for the nominal sinc function defined in (8.10). We often consider the absolute value of the sinc function, in which the negative lobes flip to the positive side, and the zero crossings become nulls in the pattern. The mainlobe width defined by the nearest zero crossings is thus known as the null-to-null width. Notice the analogy with an antenna of rectangular aperture and uniform illumination function as discussed in Chapter 3.

Equation (8.10) is consistent with the MATLAB definition; an alternate definition is simply $\sin(x)/x$, whose zero crossings/nulls occur at integer multiples of $+/-\pi$.

Specifically for a square pulse of arbitrary duration τ as shown in Table 8.1, the corresponding sinc results from the scaling property of FTs and is

$$p\left(\frac{t}{\tau}\right) \longleftrightarrow \tau\,\text{sinc}(f\tau) = \tau\frac{\sin(\pi f\tau)}{\pi f\tau} \quad (8.11)$$

Strictly speaking, this Fourier relationship assumes that the time-domain pulse is symmetric around zero. Therefore, $p(t/\tau)$ as defined here differs from the rectangle function $\text{rect}(t/\tau)$ from (7.2), which was defined between 0 and τ. In fact, $\text{rect}(t/\tau) = p(t/\tau - 1/2)$, and this is done to force the *rect* function to be causal. This delay in the time domain, however, induces a phase shift in the frequency domain given by $\exp(j\pi f\tau)$ (yet another important FT property!). We therefore define (8.11) using a time-symmetric pulse for clarity. The main point is that both the peak

Table 8.1 Common Fourier Transform Pairs

Waveform	*Time Domain x(t)*	*Frequency Domain X(f)*
CW (infinite duration)	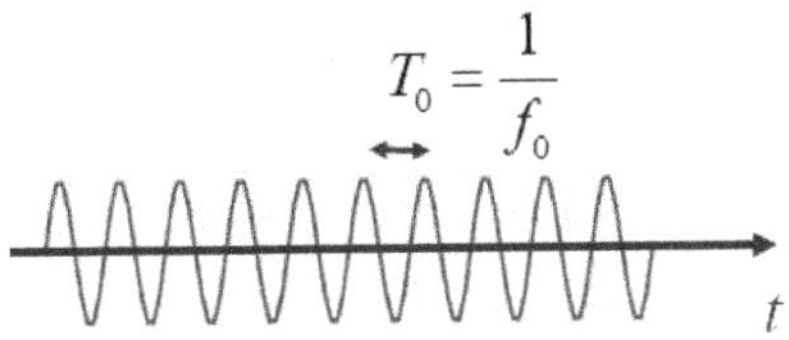	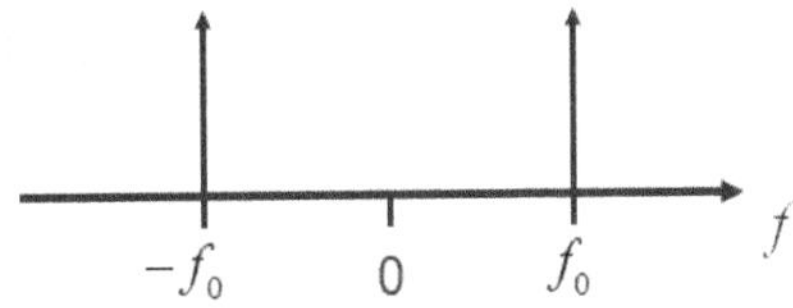
Single-square pulse (baseband)	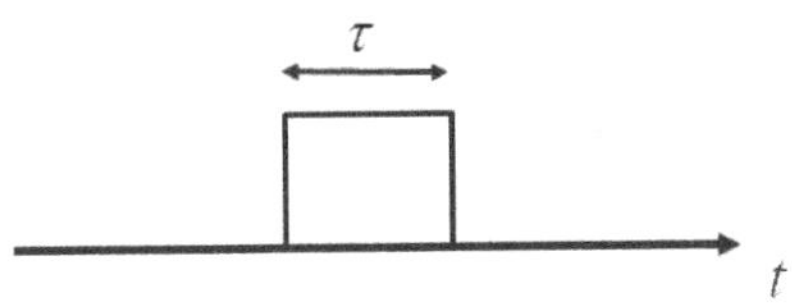	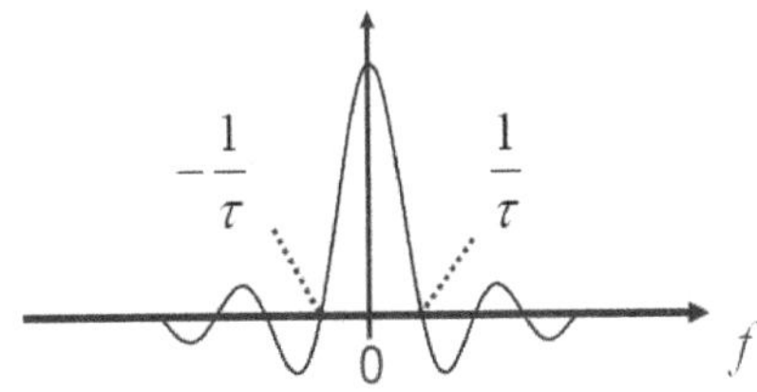
Single square pulse (RF)	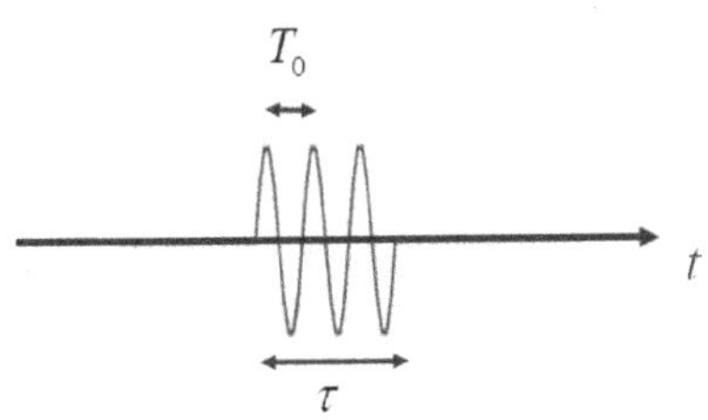	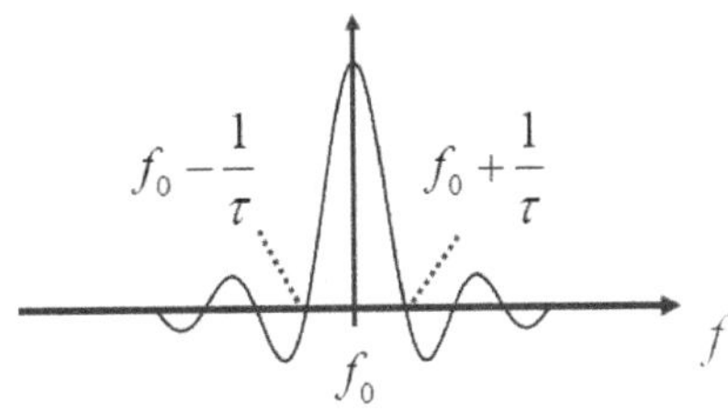
Infinite pulse train	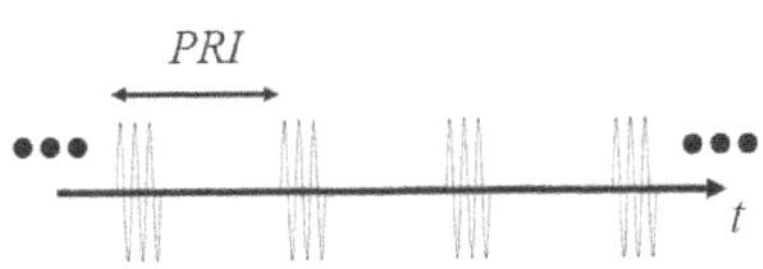	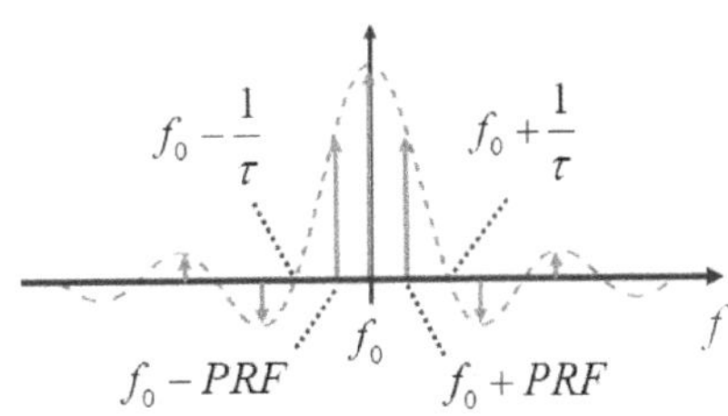
Finite pulse train	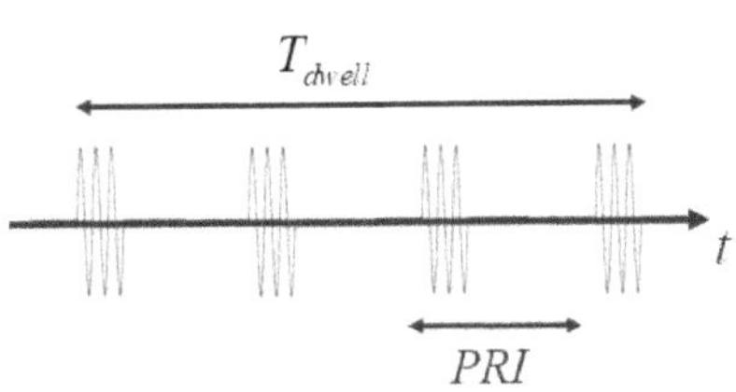	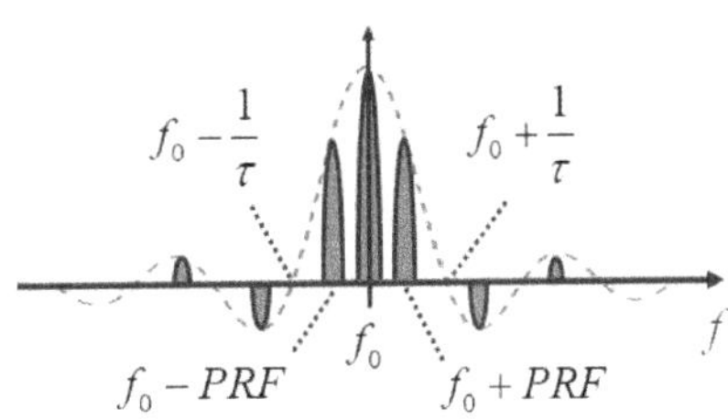

and the mainlobe width of the sinc function scale with pulse width, such that the null-to-null bandwidth is $2/\tau$ as shown in Table 8.1.

When the pulse contains a sinusoid at a frequency f_0, as seen in row 3, it has the effect of shifting the spectrum from 0 to f_0.

Rows 4 and 5 show the Fourier spectrum of a pulse train as a line spectrum, as first observed in Chapter 7. The lines are separated by the PRF, and the spectrum envelope is a sinc function corresponding to the FT of a single pulse. For an infinite length pulse train, the lines have zero width (i.e., they are delta functions). For a pulse train with a finite dwell duration T_{dwell}, the lines will be narrow sinc functions themselves, with a null-to-null bandwidth of $2/T_{\text{dwell}}$.

For M&S applications, since we deal with sampled data, we typically care about the DTFT, defined as

$$X(\omega) = \sum_{n=-\infty}^{\infty} x_n e^{-j\omega n} \tag{8.12}$$

In (8.12), the continuous time signal $x(t)$ has been sampled at multiples of the sampling interval T_s, such that

$$x_n = x(t)\big|_{t=nT_s} \text{ and } T_s = \frac{1}{F_s} \tag{8.13}$$

The radial frequency $\omega = 2\pi f/F_s$ is normalized by the sampling frequency, and therefore defined uniquely between $-\pi$ and π. Beyond these limits, the spectrum $X(\omega)$ replicates (i.e., it is periodic with period 2π). Notice also that while the time-domain signal is sampled, its spectrum is continuous in ω. The inverse DTFT is

$$x_n = \frac{1}{2\pi}\int_{-\pi}^{\pi} X(\omega)e^{+j\omega n}d\omega \tag{8.14}$$

Due to the periodicity of $X(\omega)$, the limits of integration may be any interval of duration 2π, such as $[0, 2\pi]$; however, the interval $[-\pi, \pi]$ is a common convention.

8.4 The Discrete Fourier Transform

The discrete Fourier transform (DFT) is similar to the DTFT in that it too operates on sampled time data and provides a view of the signal distribution in the frequency domain. One major difference, however, is that its calculation results in a sampled spectrum (that is, a function of a discrete frequency index k), whereas the DTFT is a function of the continuous frequency variable ω. Given a set of N time samples x_n as defined by (8.13), the N-point DFT is defined as

$$X_k = \sum_{n=0}^{N-1} x_n e^{-j\frac{2\pi}{N}kn}, \quad k = 0,\ldots,N-1 \tag{8.15}$$

for the forward transform, where X_k is the signal strength at the discrete frequency $k * F_s/N$. The collection of X_k samples therefore spans the interval $[0, F_s]$ and is often thought of as a filter bank where each filter has a spectral width of F_s/N. The

DFT X_k can be seen as a frequency sampling of the DTFT $X(\omega)$; specifically at N equally spaced points within the interval $[0, 2\pi]$, such that

$$X_k = X(\omega)\big|_{\omega=2\pi kF_s/N}, \quad k = 0,\ldots,N-1 \tag{8.16}$$

The corresponding inverse transform (IDFT) is

$$x_n = \frac{1}{N}\sum_{n=0}^{N-1} X_k e^{j\frac{2\pi}{N}kn}, \quad n = 0,\ldots,N-1 \tag{8.17}$$

In comparing (8.17) and (8.14), one key distinction is that the time sequence from the IDFT is only defined for n between 0 and $N-1$; for values of n outside of this interval, the sequence repeats—so (8.17) is in fact a periodic sampled function with a period of N samples. In general, this is not the case for (8.14).

Computationally efficient methods for computing (8.15) and (8.17) are collectively known as FFTs, and the terms DFT and FFT are often used interchangeably, since their result is numerically identical.

Two additional relevant points on DFTs/FFTs are listed as follows:

- To sample the frequency domain signal at a finer interval than F_s/N, zero-padding is often used. If the input data sequence is augmented with $(L - N)$ trailing zeros, then an L-point FFT can be performed, with an output sampling interval of F_s/L, $L > N$.
- The finite length of the input vector implies that the observation or dwell time on the signal is finite. This has the following two major effects on the frequency spectrum of the signal:
  - The frequency resolution δ_f (the ability to distinguish between two signals at closely spaced frequencies) is inversely proportional to the dwell time:

$$\delta_f = \frac{1}{T_{\text{dwell}}} \tag{8.18}$$

    We note here that zero-padding decreases the spacing between frequency samples; however, it does not improve the frequency resolution, as the observation time remains unaffected.
  - The sharp on/off transitions in the edges of the time-domain signal are discontinuities that manifest as high sidelobes in the frequency domain. The sinc function from (8.11), for instance, has near-in sidelobes that are only about 13-dB lower than the peak of the spectrum, as seen in Figure 8.2. To reduce the size of the sidelobes, the discontinuities are smoothed out by using an amplitude taper function. This amplitude taper, also known as a window function, is a set of amplitude weights w_n applied to each sample x_n in the DFT equation. The weighted DFT equation is

$$X_k = \sum_{n=0}^{N-1} w_n x_n e^{-j\frac{2\pi}{N}kn}, \quad k = 0,\ldots,N-1 \tag{8.19}$$

The sinc function displayed in Figure 8.2 is obtained by setting $w_n = 1$ in (8.19). This constant window function is also known as a uniform window since all weights have the same value. Other names include rectangular or boxcar window. Several window functions are available in the literature [1]. Both MATLAB and Simulink have several window functions available, including Blackman, Chebyshev, Hamming, Hann, Kaiser, and Taylor windows. See [1] and MATLAB's online help for their definitions and additional information.

One popular choice is the Hamming window, named after its proponent Richard Hamming. The frequency spectrum for the Hamming window is also shown in Figure 8.2, featuring sidelobes that are 40-dB below its peak. The window spectra in Figure 8.2 have been zero-padded to show the details, although they are plotted against the unpadded index k spanning the interval [–N/2, N/2 – 1], a shifted version of (8.19).

Lower sidelobes, however, come at a price. The window function has the following effects on the frequency spectrum:

- The width of the spectrum mainlobe increases. This worsens the Doppler resolution.
- The peak decreases. (There is a SNR loss associated with the window function.)

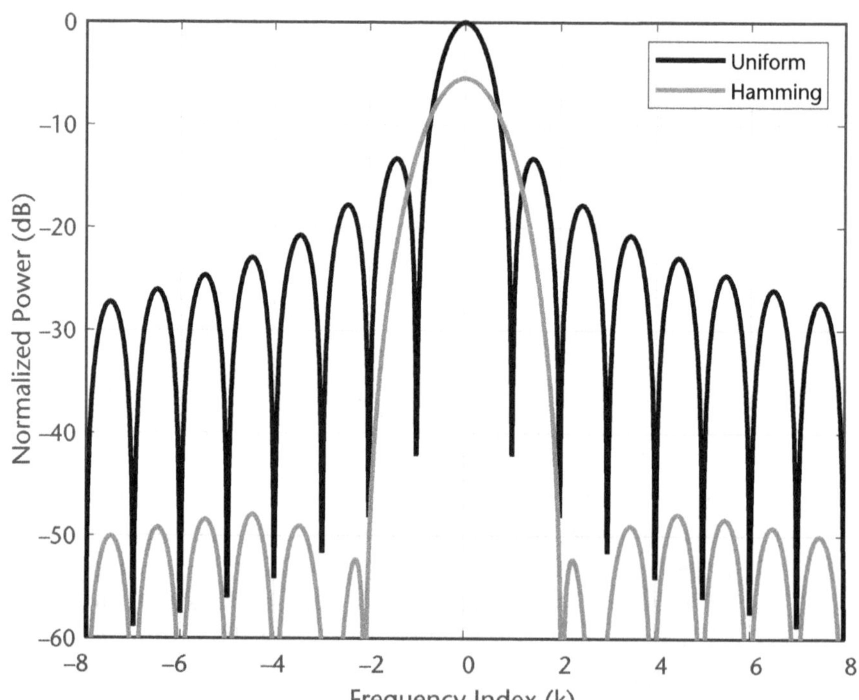

Figure 8.2 Frequency spectra of window functions: uniform (black) and hamming (gray).

Both of these effects are apparent in the Hamming window spectrum. The signal power and SNR losses due to windowing are [1]:

$$L_{\text{signal}} = \frac{1}{N^2}\left(\sum_{n=0}^{N-1} w_n\right)^2, \quad L_{\text{SNR}} = \frac{\left(\sum_{n=0}^{N-1} w_n\right)^2}{N\sum_{n=0}^{N-1} w_n^2} \tag{8.20}$$

These are the signal power and SNR processing gains, respectively, normalized by their corresponding unwindowed gains N^2 and N. Notice that L_{SNR} is mathematically identical to the antenna aperture efficiency defined in (3.25). Antenna tapers use the same type of window functions as the FFT.

In MATLAB, the FFT operation is implemented through the `fft` function. A typical function call looks like this:

```
>> X = fft(x)
```
(8.21a)

where `x` is the time-domain input vector/sequence of length N. With a single input argument as shown in (8.21a), the number of points in the FFT is equal to N. To explicitly define an FFT length `L`, use

```
>> X = fft(x,L)
```
(8.21b)

If `L` is longer than N, (8.21b) automatically implements zero-padding.

The output vector/sequence `X` spans the frequency range $[0, F_s * (N-1)/N]$, with `X(1)` centered at 0. Notice that X_k as defined in (8.15) and (8.19) is a zero-indexed sequence, while vectors in MATLAB are one-indexed. Also, since the frequency spectrum is symmetric around 0, it is often convenient to display it with 0 in the center of the sequence rather than at the beginning (as exemplified by Figure 8.2). The MATLAB function `fftshift` is used for this purpose. The resulting sequence will then span the interval $[-F_s/2, (F_s/2) * (N-2)/N]$.

Consider the following Doppler processing example, which operates on a CW signal. The example is contained in the MATLAB script `fft _ example`. The relevant parameters are shown in Table 8.2.

Table 8.2 Simulation Parameters for the Doppler/Velocity Measurement Example

Example Parameters	*Value (units)*
Target radial velocity (v)	265 m/s
RF (f_0)	10 GHz
Doppler frequency (f_{Dop})	17.7 kHz
Sampling rate (F_s)	150 kHz
Number of samples collected (N)	64
Number of points in FFT (N_{FFT})	64

The processing flow is illustrated in Figure 8.3. A continuous-time signal $x(t)$ is sampled at the rate F_s, meaning that the sample times will be at integer multiples of $T_s = 1/F_s$. A total of N samples are collected, then processed via an FFT using N_{FFT} *points*. In the figure, $N_{\text{FFT}} = N$.

Notionally, the estimated signal frequency $\hat{f}$ comes from locating the peak of the frequency spectrum; then a velocity estimate is calculated by solving (8.7) for the radial velocity:

$$\hat{V}_{\text{LOS}} = \frac{\lambda}{2}\hat{f} \tag{8.22}$$

It is often the case, however, that the observed peak is not necessarily located at the exact signal frequency; this is because, as stated earlier in this section (Section 8.4), the DFT is a sampled frequency spectrum, and the samples may very well miss in this section the absolute peak. In the example shown in Figure 8.3, the largest peak is at 18.75 kHz; the next largest to its left is at 16.41 kHz. The signal loss due to the frequency sample missing the peak is known as a scalloping loss or picket fence loss. The mainlobe broadening caused by windowing actually reduces this loss.

The frequency estimate may be refined in a number of ways; for example, with knowledge of the spectrum shape (e.g., a sinc function for nonwindowed data), curve-fitting followed by interpolation will yield a more accurate estimate. Another, arguably simpler option is to increase the sample density on the frequency domain via zero-padding.

A concept related to measurement accuracy is the resolution that results from the time duration of the observation. A short observation time produces a broad mainlobe spectrum, as shown by (8.18). This can make the peak estimation more challenging, as there are bound to be multiple samples on the mainlobe with similar amplitudes. Increasing the observation time for the same sampling rate decreases the number of mainlobe lines and thus eliminates some of the potential ambiguity. To illustrate these issues, we may try some excursions from the example in Figure 8.3. Consider, for example, the following:

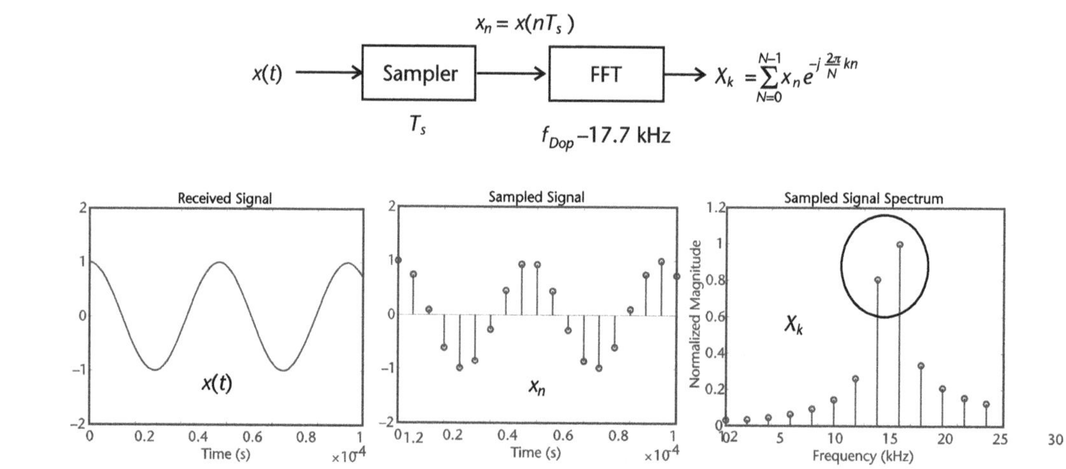

Figure 8.3 Velocity measurement via FFT.

- *Excursion #1:* On line 44 of `fft _ examp`, change the value of `Nfft` from `N` to `4*N`. This implements zero-padding. Rerun the script. What happens?
- *Excursion #2:* On line 13, set `N` to 256 (leave `Nfft` equal to `N`). This increases the observation time. Rerun the script. What happens?

The results of these excursions are shown in Figure 8.4: The left plot shows the original spectrum with short observation time and coarse frequency sampling; and the center plot shows the denser sampling stemming from zero-padding. The largest peak is now at 17.58 kHz, which is much closer to the true value of 17.7 kHz than the previous case at 18.75 kHz. The right plot of Figure 8.4 shows the effect of a longer observation time: a clear single peak in the mainlobe, also at 17.58 kHz.

As one additional exercise for the reader, what if we combine zero-padding (`Nfft = 4*N`) with the longer observation time `N` = 256? Where is the peak now? (Try it!)

The use of the DFT/FFT for spectral analysis is a very common method in radar signal processing, and has multiple applications. As illustrated by the Doppler processing example presented in Section 8.4 (which includes Table 8.2 and Figures 8.3 and 8.4), it can be used to extract a target velocity estimate from a received CW signal return. In Chapter 7, we also saw the FFT being used for estimating the target range using an LFMCW and stretch processing.

The utility of the FFT is not limited to CW waveforms. Section 8.5 examines the frequency-domain implementation of matched filtering of LFM pulses, which is used to improve the estimation of target range. This involves fast-time sampling of the received radar data.

Last, for a pulse-Doppler radar system, the FFT is used to measure target velocity via slow-time sampling. Pulse-Doppler processing is discussed further in Section 8.7.

8.5 Pulse-Compression Waveforms

As stated in Chapter 7, pulsed waveforms are inherently suitable for range estimation based on measurement of the round-trip time delay. This can be accomplished in the absence of any modulation, as opposed to the case of CW where modulation is required for measuring range. When there are multiple targets present, however,

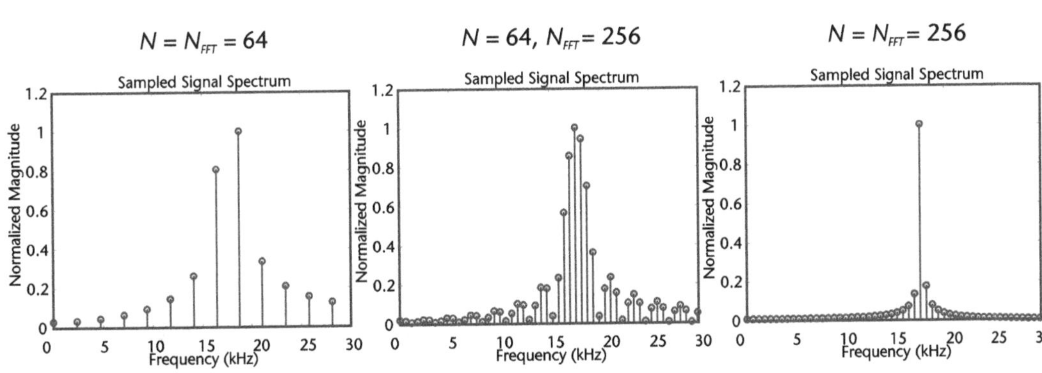

Figure 8.4 Excursions on estimation of spectrum peak for Doppler/velocity measurement example.

one additional condition is necessary for accurate range measurement—namely, the target returns must not overlap. Figure 8.5 illustrates the case of two targets closely spaced; on the left, the range separation between them is such that the difference in time delay is greater than a pulse width; thus, the return pulses do not overlap in time and the radar is able to make range measurements on each one (assuming that they are both detected).

On the right of Figure 8.5, this condition is not met; therefore, the returns overlap and the radar will not be able to tell the targets apart, all else being equal. This minimum separation δ_R is known as the radar *resolution* in the range dimension (it can also be defined in any dimension measured by the radar such as angle or Doppler). Recall that the range resolution is related to the pulse width (τ); it is, in fact, defined as

$$\delta_R = \frac{c\tau}{2} \tag{8.23}$$

A narrower pulse width therefore improves the range resolution; however, narrowing the pulse (assuming that all other waveform parameters remain unchanged) lowers the SNR and therefore degrades detection performance. The radar thus has seemingly competing and opposite objectives: increase the pulse width to increase SNR or decrease it to improve resolution.

Pulse compression is an approach to decouple these objectives by enabling the transmission of a long pulse, yet through modulation and processing on receive, short-pulse resolution can be achieved. Common pulse compression modulation options are FM and PM. Either option increases the pulse bandwidth B, which ultimately is the reason for improved range resolution. We note here that the range separation defined in (8.23) is the absolute minimum distance required to resolve targets; practical considerations often require greater separations for consistent performance. Other factors that affect the ability to resolve targets include the relative phase between them, as well as any potential windowing applied in the range dimension, as it is often done for LFM waveforms.

The amount of resolution improvement is the time-bandwidth product (TBP), defined as

$$TBP = \tau B \tag{8.24}$$

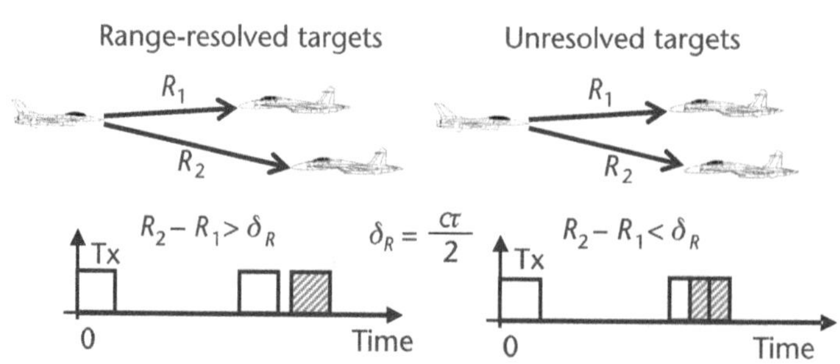

Figure 8.5 Impact of pulse width in measurement of closely spaced targets.

Coincidentally, the TBP is also the signal processing gain (SNR gain) achieved by the pulse-compression process (neglecting any windowing losses). Recall from Chapter 2 that for an unmodulated pulse, the waveform bandwidth is approximately $B = 1/\tau$; therefore the *TBP* equals 1 for this case. PM waveforms can typically achieve TBPs in the order of tens to hundreds, while FM waveforms can be orders of magnitude higher.

8.5.1 PM Waveforms

PM consists of applying a sequence of discrete phase states across the pulse. A minimum of two states is needed, giving rise to biphase sequences. This type of modulation is also known as binary phase shift keying (BPSK). The states are typically denoted as 0° and 180°, although any pair of values will work as long as they are 180° apart. Many other options with multiple states (polyphase sequences) exist [2], but we only consider biphase codes herein.

The pulse is divided into K subpulses or chips, each of width τ_c, where K is the length of the phase sequence that will be applied. One bit of the phase sequence is then applied to each chip. One benefit of the 0°/180° choice is that only the 180° (or π) phase state needs to be applied, and that just means inverting the sign of the amplitude, since $\cos(\omega t + \pi) = -\cos(\omega t)$ by trigonometric identities.

The top of Figure 8.6 shows a pulse divided in $K = 7$ chips. The sequence of '+' and '–' signs at the top denote the 0° and 180° applied phase states, respectively. Each chip can itself be thought of as an unmodulated pulse, with a bandwidth approximately equal to the inverse of its width. Thus, the application of this PM sequence increases the pulse bandwidth by a factor of K, and its *TBP* is

$$TBP = \tau B = \left(K\tau_c\right)\left(\frac{1}{\tau_c}\right) = K \tag{8.25}$$

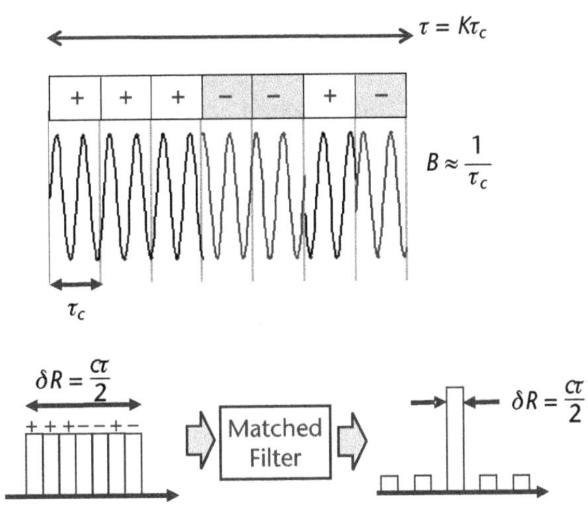

Figure 8.6 Binary phase codes for pulse compression: waveform (top) and signal processing (bottom).

The range (or fast-time) processing of this pulse is achieved via matched filtering (MF), which is illustrated at the bottom of Figure 8.6. Matched filtering may be implemented in either the analog or digital domain. A digital MF is assumed here, since it is most commonly found in modern systems. The MF belongs to the class of linear, shift-invariant (LSI) systems that are uniquely described by their impulse sequence function $h(n)$, where n is the sample time index. By definition, the output sequence $y(n)$ of the filter given an input sequence $x(n)$ is calculated as the convolution of x and h, or

$$y(n) = x(n) * h(n) = \sum_{m=-\infty}^{\infty} x(m)h(n-m) = \sum_{m=-\infty}^{\infty} x(n-m)h(m) \tag{8.26}$$

Even though the limits of the summation go to infinity, in practice both $x(n)$ and $h(n)$ are finite in length.

Equation (8.26) holds for any LSI filter; what makes a filter matched is when the impulse sequence is defined as a time-reversed replica of the input sequence. It has been shown [3] that the matched filter maximizes the output SNR.

Assume that N is the length of $x(n)$; the impulse sequence for the matched filter is then

$$h(n) = \alpha x^*(N-n) \tag{8.27}$$

where $*$ denotes complex conjugate, and α is a scaling factor that we can assume to be equal to 1. Replacing (8.27) in (8.26) results in

$$y(n) = \sum_{m=-\infty}^{\infty} x(m)x^*(m+N-n) \tag{8.28}$$

which the reader may recognize as the autocorrelation of $x(n)$ at a sample lag of $N - n$. Equation (8.28) shows that convolution with a MF is mathematically equivalent to a correlation operation, which leads this range-processing step sometimes referred to as *range correlation*. The operation may be more accurately described as a cross-correlation, since the matched impulse sequence is never an exact replica of the input—at the very least, the received pulse sample is corrupted by the presence of receiver noise.

The output of the matched filter at the bottom of Figure 8.6 shows an output pulse which has been narrowed (*pulse-compressed!*) by a factor of K, plus additional, smaller outputs on either side. We call the entire output sequence the *range profile,* and its shape is fully determined by the PM code. It comprises the actual compressed pulse as the range mainlobe, plus the additional outputs referred to as range sidelobes. Each of these output samples occupies a *range gate* or *range bin.* If the input pulse has been critically sampled, meaning one sample per chip, or $F_s = B = 1/\tau_c$, the range bins have a width equal to the range resolution

$$\delta_R = \frac{c\tau}{2K} = \frac{c\tau_c}{2} = \frac{c}{2B} \tag{8.29}$$

Ideally, range sidelobes should be as low as possible, since they are wasted energy and do not contribute to the output pulse (mainlobe). They also make it more difficult to detect other nearby small targets. One measure of performance of pulse-compression waveforms is therefore the mainlobe-to-peak sidelobe ratio; for a binary code sequence of length K, the maximum ratio is K itself. A subset of biphase codes that meets this condition is the family of so-called Barker codes, named after their inventor Ronald H. Barker [4]. Only Barker codes up to length 13 are known to exist; these are listed in Table 8.3. The reader will recognize the 7-bit Barker code as the example illustrated in Figure 8.6.

Modeling of the matched filtering operation is simply achieved by implementing (8.26) and (8.27), or equivalently, (8.28). Assuming row vectors, with $p(n)$ being the ideal pulse sequence, the respective commands in MATLAB are

```
>> y = conv(x,fliplr(conj(p)));      % Convolution approach
```
(8.30)

```
>> y = xcorr(x,p);                   % Correlation approach
```
(8.31)

where the time reversal of the impulse response is obtained by the flip left-to-right function `fliplr`. For column vectors, the flip up-to-down function `flipud` should be used instead.

Equations (8.30) and (8.31) produce identical results when both input vectors `x` and `p` have equal lengths. Otherwise, the `xcorr` function zero-pads the shorter vector to match the longer one, resulting in a few additional zeroes in the output compared to the output of `conv`.

In Simulink, this operation is easily implemented using the XCORR Correlation block from the DSP System Toolbox/Statistics sublibrary, which works similarly to the MATLAB function of the same name—except it does not zero-pad.

Figure 8.7 shows the output plots resulting from the MATLAB script *bicode_examp*. The code uses the 7-bit Barker code as an example, and follows both the autocorrelation and the convolution approach, resulting in identical outputs (albeit plotted using different functions `bar` and `stem`). The model assumes critical sampling.

Table 8.3 Known Barker Codes

Length	*Codes*	
2	+ −	+ +
3	+ + −	
4	+ + − +	+ + + −
5	+ + + − +	
7	+ + + − − + −	
11	+ + + − − − + − − + −	
13	+ + + + + − − + + − + − +	

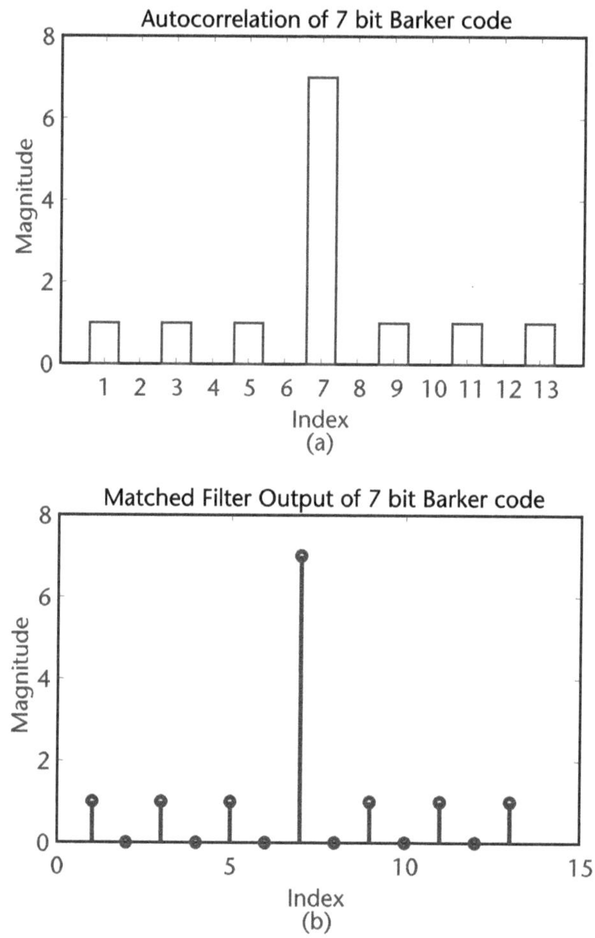

Figure 8.7 Pulse compression of 7-bit Barker code: autocorrelation output (a); matched filter convolution output (b).

A Simulink implementation is shown in Figure 8.8. The *PulseCompress* model utilizes the XCORR block for the cross-correlation of the input sequence. The impulse sequence of the matched filter is labeled "Biphase Code." The Ideal Signal block represents a noiseless received pulse, and the full range profile of length 13 is shown in the Display at the top right, after envelope detection (i.e., after taking the absolute value).

The bottom of Figure 8.8 shows a Noisy Signal of an arbitrary length greater than just the pulse samples, plus Gaussian-distributed receiver noise; this noisy input is also matched-filtered, but only the center five output range gates are displayed at the bottom right. The following lists a few observations that can be made of this section of the model; the first two are just Simulink bookkeeping.

- All signals have been created as frames (groups of time samples). The frame rates are determined by their respective lengths [e.g., one frame per 7 seconds for the top two constant blocks (Ideal Signal and Biphase Code) and one frame per 11 seconds for the bottom one (Noisy Signal)].

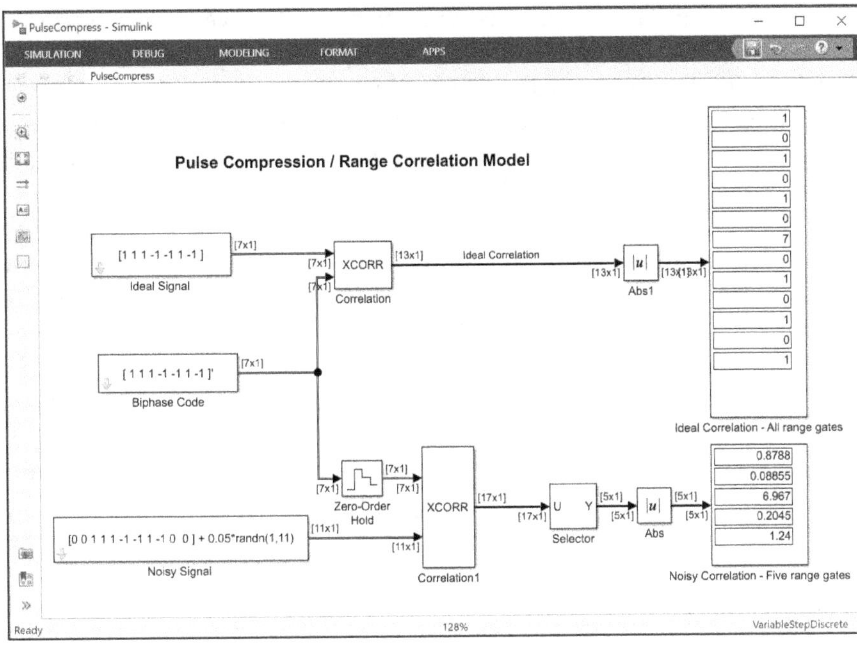

Figure 8.8 Simulink model of range correlator/matched filter for 7-bit Barker code.

- Because of the difference in frame rates, the Zero-Order Hold block has been inserted at the bottom to match the two rates.
- The length of the correlator output (17 in this case) is always a function of the lengths of the inputs (output length is $N + M - 1$ for input lengths N and M); however, the first six output samples are partial correlations as the input sequence enters the correlator one sample each time step. This initial transition time can be thought of as the filter charge-up time, which does not produce a valid output. Similarly, the last six output samples are also the result of partial correlations as the input signal samples exit the correlator. Typically, the radar processor only wants to deal with range gates that are the product of full-correlation calculations; otherwise, there would be no full pulse-compression gain applied to the target. The processor therefore discards those samples at the beginning and at the end. The purpose of the Selector block is to preserve only the fully correlated range gates.
- Last, notice that even with the added receiver noise, the target location (pulse-compressed output) is easily detectable to the naked eye by inspection of the Noisy Correlation Display block. This was to be expected, as the SNR at the input to the filter was already a very healthy 26 dB (20 * log10(1/0.05)), and the filter added an additional 10 * log10(7) ~ 8.5 dB of SNR gain.

8.5.2 Linear Frequency Modulation Waveforms

Linear frequency modulation (LFM) is another popular modulation choice for pulse compression waveforms, and can achieve much higher time-bandwidth products than PM waveforms. The basics of LFM were discussed in Chapter 7 in the context of CW signals. For pulsed signals, the pulse width τ is analogous to the modulation interval T_m of the LFMCW.

In principle, the time-convolution operation described in Section 8.5.1 for PM pulse compression may also be used for LFM pulses. The most common approach, however, is to implement the convolution in the frequency domain. The use of FFTs results in more computationally efficient processing than brute-force correlation. This alternate method is motivated precisely because of the typically higher waveform bandwidths, which require higher sampling rates, resulting in more samples for the same pulse widths.

A well-known property of FTs is that a convolution operation in one domain becomes a multiplication operation in the other domain. The frequency-domain matched filter takes advantage of this property by computing the FFT of both the received signal and the MF impulse sequence, then taking the Hadamard product (element-by-element multiplication) of the two sequences, followed by an inverse FFT to recover the time-domain MF output. This process is illustrated in Figure 8.9.

The input sequence to the matched filter in general will consist of signal-plus-noise samples,

$$x_m = s_m + n_m \tag{8.32}$$

where m is now chosen as the time index to avoid confusion with the noise sample n. Ideally, the transform S_k^* of the signal sequence—actually, its complex conjugate per (8.27)—is already precomputed and stored in memory, and premultiplied by a weighting sequence W_k (more on this shortly). This product $W_k S_k^*$ is the impulse sequence of the matched filter, even though technically the weighting introduces a mismatch relative to the input. After the FFT of the input sequence is calculated, it is multiplied (element-by-element) with the impulse sequence, and inverse transformed back to the time-domain.

Ignoring momentarily the weighting sequence (or equivalently, assuming uniform weights $W_k = 1$ for all k), Figure 8.10 shows the time and frequency-domain sequences highlighted in Figure 8.9 for a single, noiseless ($x_m = s_m$), unit-amplitude LFM pulse with the parameters specified in Table 8.4. These plots are obtained by running the MATLAB script *test_LFM_MF_FFT.*

These values result in a TBP of 100. Notice also that there is an oversampling factor (osf) of 10. While this is insightful for M&S applications, real systems typically do not oversample by that much, as it entails processing more data than necessary.

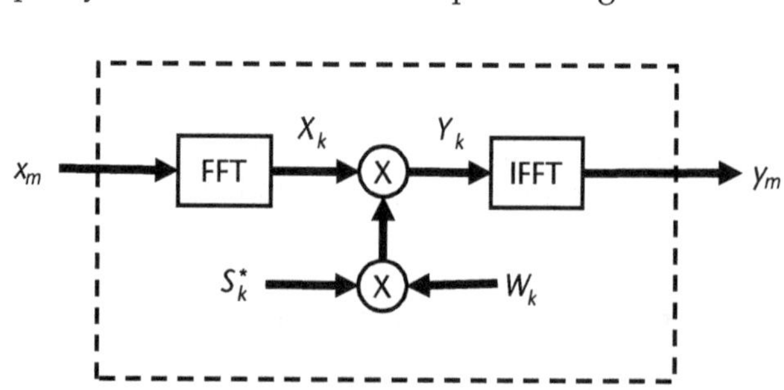

Figure 8.9 FFT-based matched filter implementation.

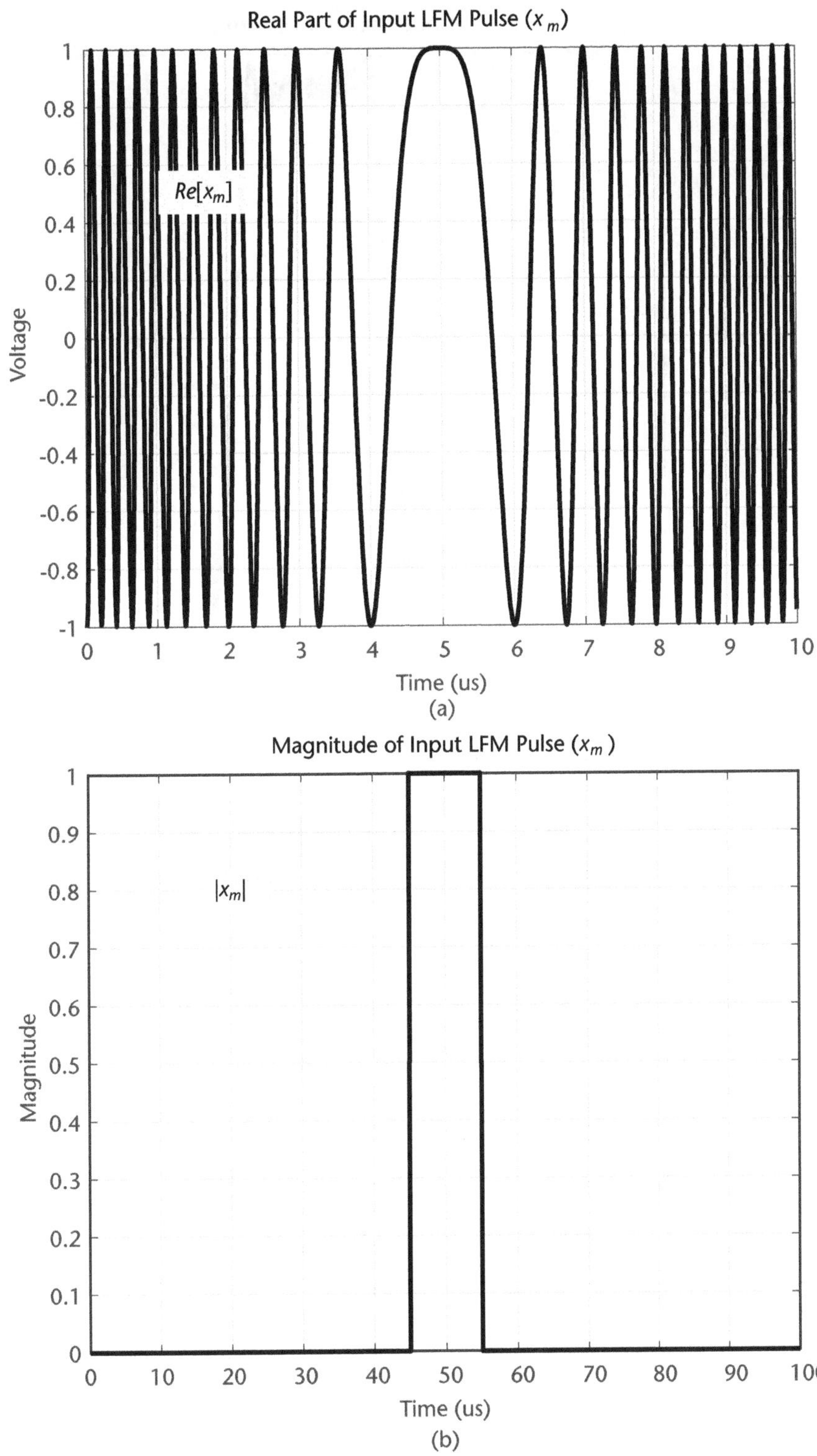

Figure 8.10 Time- and frequency-domain plots of LFM matched filter example using uniform weights: (a) Real part of LFM pulse, (b) magnitude of input LFM pulse, (c) spectrum of input pulse, (d) normalized weighted LFM spectrum, (e) normalized MF output spectrum, and (f) MF output pulse.

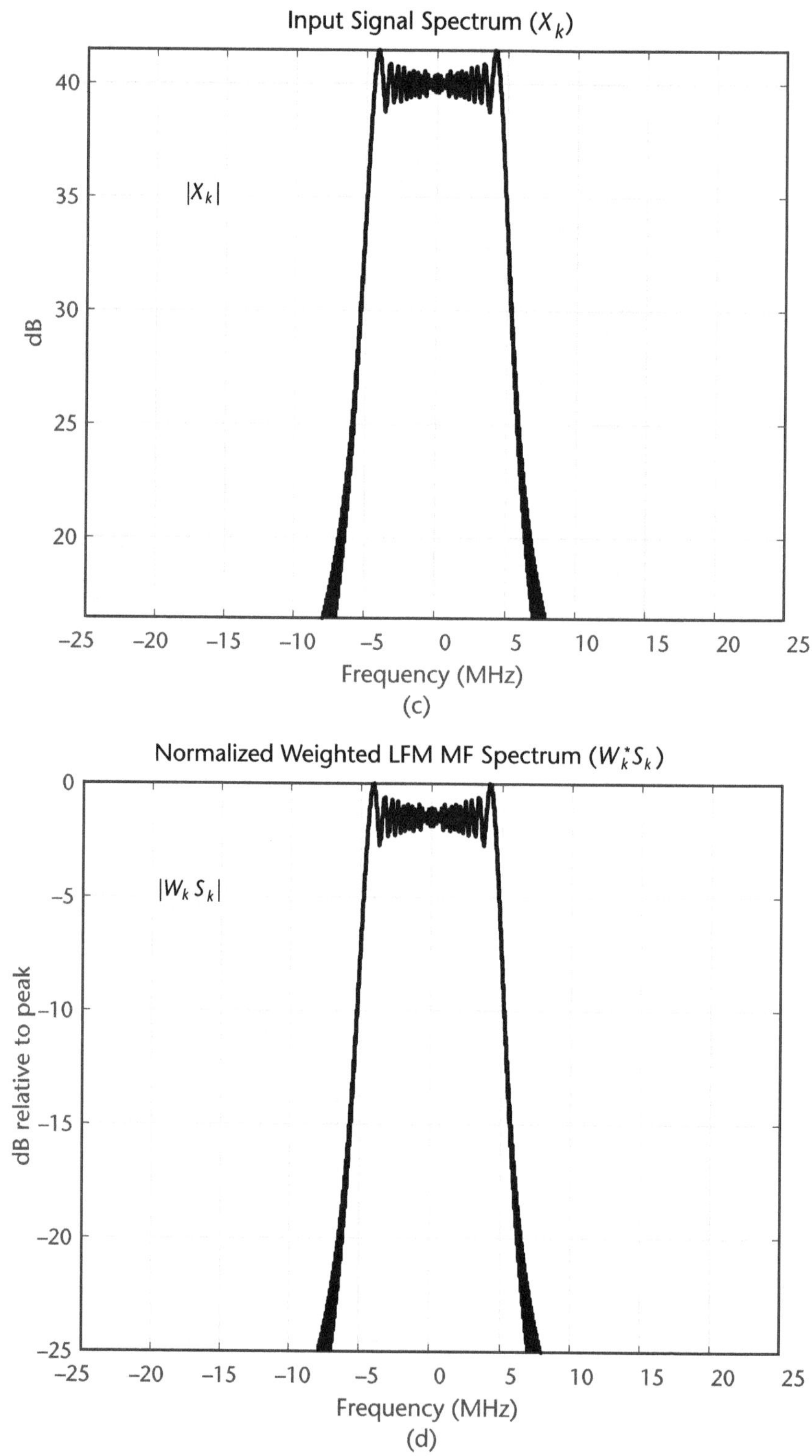

Figure 8.10 *(Continued)*

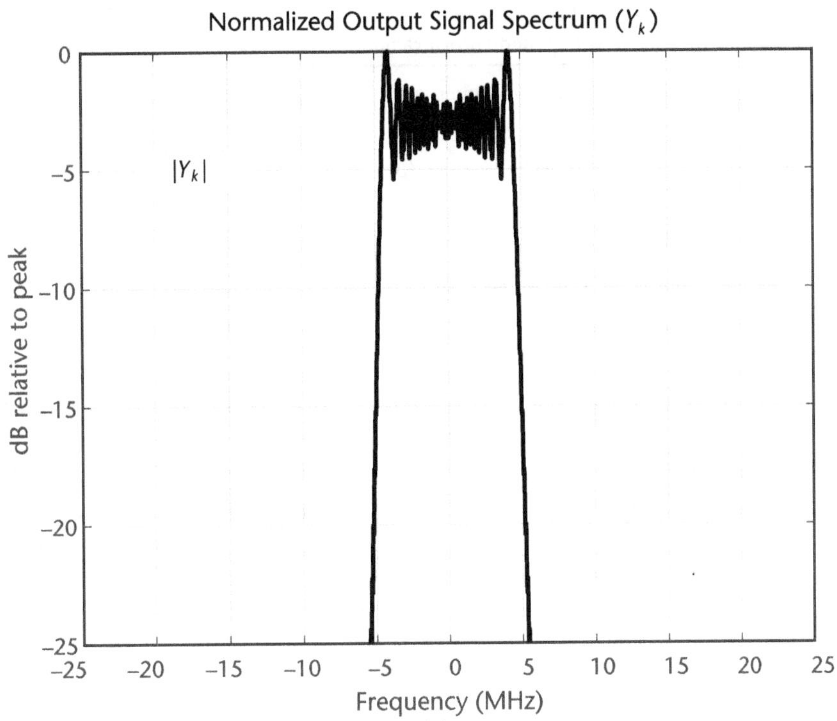

(e)

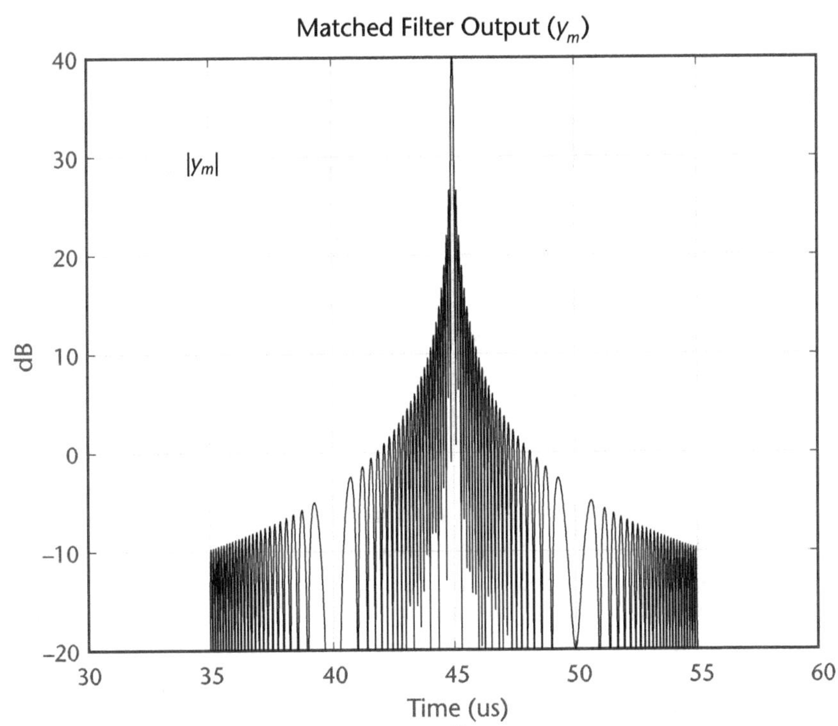

(f)

Figure 8.10 *(Continued)*

Table 8.4 Simulation Parameters for FFT-Based Matched Filter Example

Example Parameters	*Value (units)*
Radar pulse width (τ)	10 μs
LFM bandwidth (β)	10 MHz
Sampling rate (F_s)	100 MHz
Sample collection time	100 μs

Figure 8.10(a) shows the real part of the 10-MHz LFM pulse, with frequency variation from −5 to +5 MHz over the pulse width. The envelope of the pulse is shown in Figure 8.10(b) with the time axis covering the entire PRI. The leading edge of the pulse is at 45 μs from the start of the PRI; this corresponds to the round-trip time delay and therefore the target location. All remaining plots are in units of decibels, with the frequency plots all normalized to a 0-dB peak. After the FFT, the magnitude of the pulse spectrum is shown in Figure 8.10(c); notice that its shape is somewhat rectangular with some ripples at the top and that the majority of the energy is within +/−5 MHz (as expected). The magnitude spectrum of the matched filter is in Figure 8.10(d); for uniform weights, it is identical to that of the input pulse. Their Hadamard product is in Figure 8.10(e); the ripples from each individual spectrum have reinforced each other; the roll-off of the product is noticeably steeper. The inverse FFT brings the output spectrum back to the time domain, shown in Figure 8.10(f). Notice the pulse compression with a sharp peak at 45 μs indicating the correct target location, along with a range sidelobe structure on both sides of the peak. Furthermore, the peak value is at 40 dB, which corresponds to an amplitude gain of 100, the TBP for this waveform.

Figure 8.11(a) shows a close-up of this output pulse, normalized by its peak value. This range profile approximates a sinc function, which is not too surprising since its FT approximates a rectangular spectrum—recall row 2 of Table 8.1, which shows the rectangle function and the sinc function as FT pairs (the domains are inverted in this case). Herein lies one major issue with the matched filtering as described in this example; just like the sinc function, the response exhibits high-range sidelobes, with those nearest to the peak only about 13 dB down. This is a problem when the radar is attempting to detect a small signal return in the vicinity of a much larger signal. One solution is the application of a window function to lower the range sidelobes, in a manner analogous to the Doppler sidelobes discussion in Section 8.4. To reduce sidelobes in the range (or time) domain, however, the weights are now applied in the frequency domain. For a Hamming window function, the resulting range profile is shown in Figure 8.11(b), with peak sidelobes reduced by about 24 dB relative to the no-weighting case. In the example script *test_LFM_MF_FFT*, the window is applied by setting the flag `Weight _ on` to 1 in line 19.

Figure 8.12 shows the Hamming-weighted MF spectra and resulting output pulse, which can be compared to Figures 8.10(d–f).

The peak pulse output is now at 34.65 dB, which corresponds to an amplitude gain of 54. This is a 5.35-dB reduction compared to the 40-dB peak obtained with

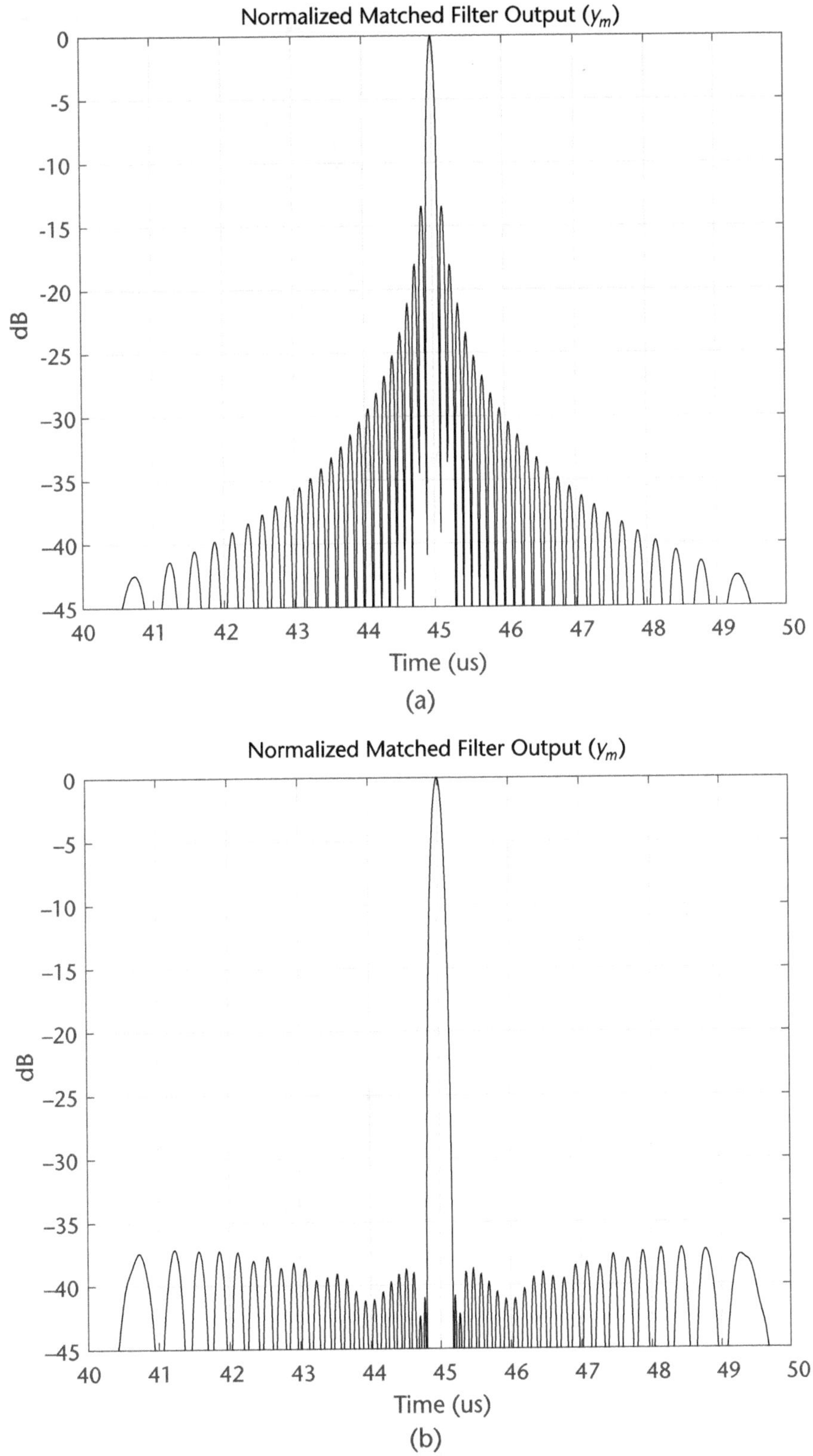

Figure 8.11 Matched filter responses: (a) uniform window and (b) Hamming window.

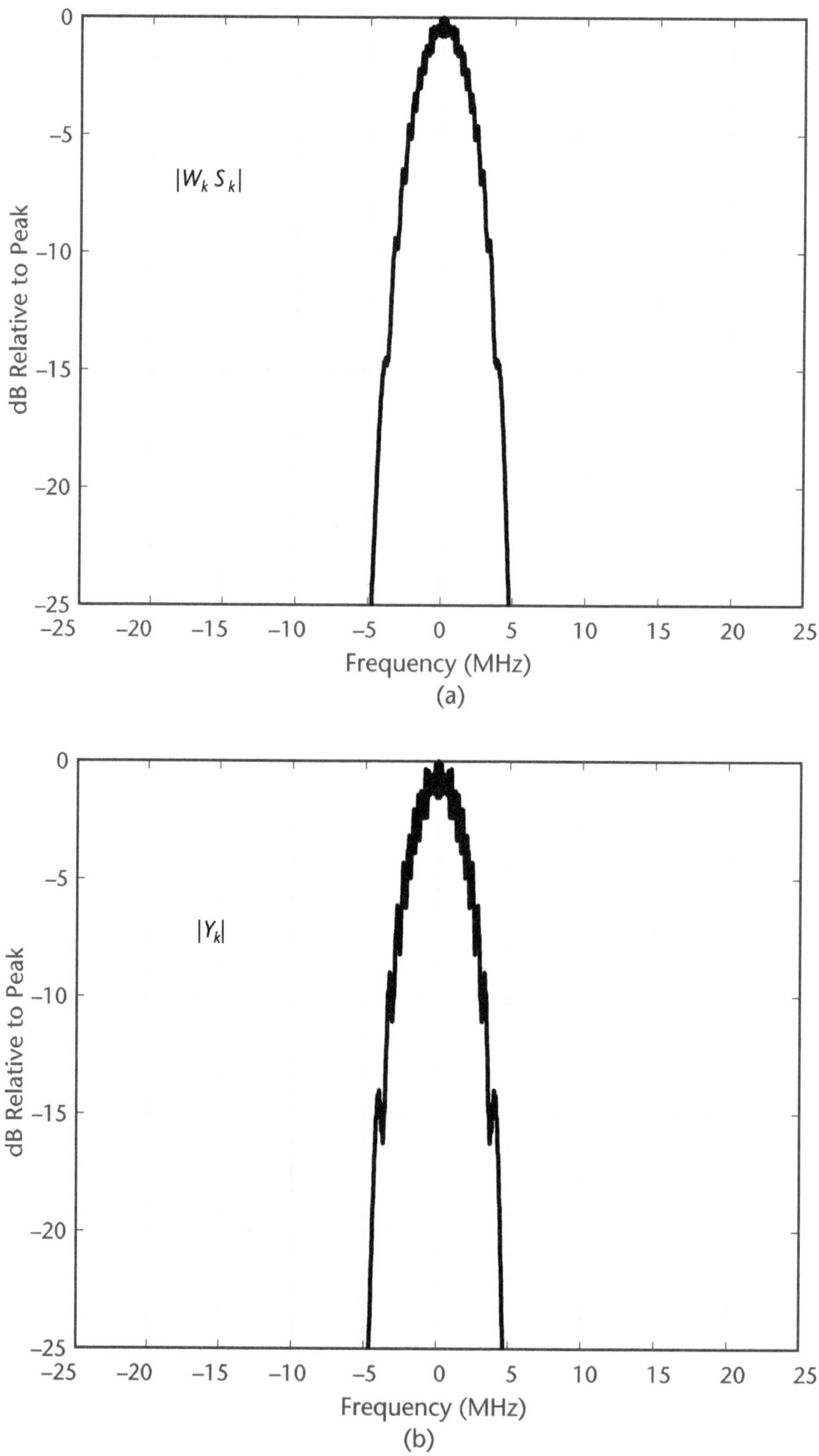

Figure 8.12 Time- and frequency-domain plots of LFM matched filter example using Hamming weights: (a) Normalized weighted LFM MF spectrum (W_k*S_k), (b) normalized output signal spectrum (y_k), and (c) matched filter output (y_m).

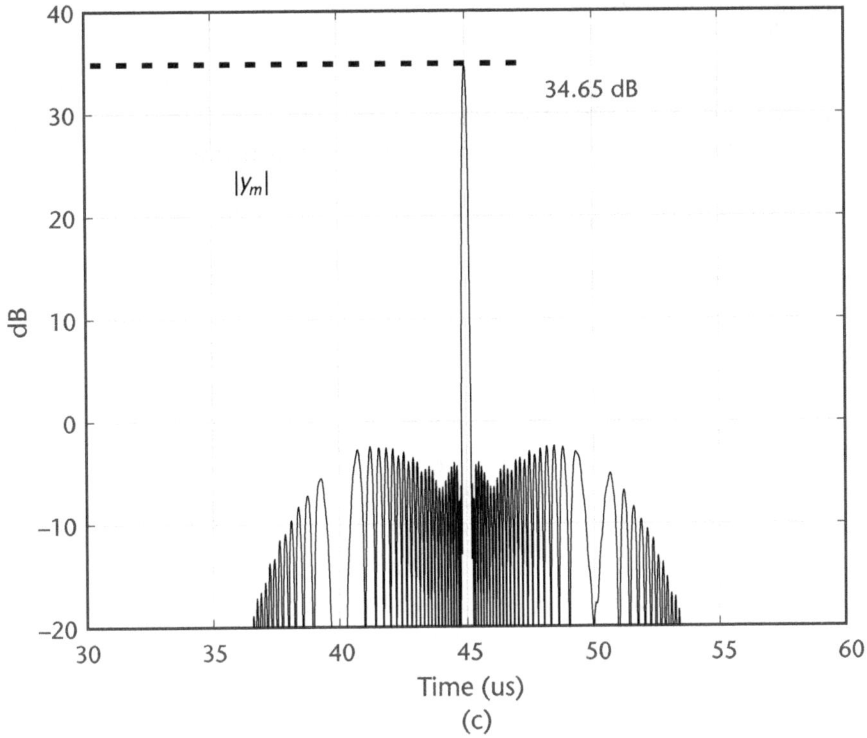

Figure 8.12 *(Continued)*

uniform weights. This reduction over the full TBP due to the window function is calculated as

$$G_{MF} = (TBP)\left(\frac{1}{N_w}\sum_{k=0}^{N_w-1} w_k\right) = (100)(0.54) = 54 \tag{8.33}$$

where N_w is the number of samples in the window function. One key point is that this is not the number of points in the FFT. The number of points in the FFT cover the entire sampling frequency interval (100 MHz), but the LFM bandwidth is only 10% of this interval. The number of points in the window function should match this fraction and be properly positioned so that they fully overlap the LFM spectrum.

Another key parameter is the number of points on the FFT itself: It should be at least as long as the sum of the lengths of both the input sequence x_m and the MF impulse response $h_m = s_m^*$; often, the next higher power of 2 is chosen, for efficient computation of the FFT. Lasty, the oversampling factor osf is accounted for by dividing the MF output by osf—otherwise, it would yield an artificially higher signal gain. These and other implementation details may be gleaned by examining the function *LFM_MF_FFT* called by the example script.

Sections 8.6 and 8.7 bring all these concepts together and summarize range-Doppler processing from a systems perspective.

8.6 Range Processing

The timing diagram shown in Figure 8.13 describes how the range gates are processed and located in space. We assume a PM waveform for the purposes of this discussion. A set of N contiguous range gates form a range window of length $N\delta_R$. This occurs within a single PRI, underscoring the fact that this processing is in fast time.

A pulse is transmitted (and the receiver blanked) at time 0, then a PRI later the next pulse is transmitted. Between those two events (trailing edge of pulse 1 and leading edge of pulse 2), the system is in listening mode, meaning that it is searching for targets. At some point in time within that interval, the ADC will start taking fast-time samples; this start time defines the location of the range window. For a PM code length of K chips, the first $K - 1$ outputs of the matched filter/range correlator are due to partial correlations (filter charge-up time as discussed earlier), which may be discarded. At the Kth sample time, the first full, valid output is generated, which corresponds to range gate 1 (RG1 in the figure). Each additional input sample produces an additional range gate output, such that for a range window of N gates, a total of $K + N - 1$ fast-time samples must be collected during the PRI. As an example, to form 10 range gates using our 7-bit Barker code, the ADC must take 16 samples each PRI. Ideally, the processor will form as many range gates it needs to cover the entire unambiguous range interval R_u with a single window. This could mean thousands of range gates, however, so if the computational resources are not sufficient for this, the search region may need to be partitioned in multiple range windows that will sweep the interval over multiple dwells.

8.7 Doppler Processing

The second stage of pulse-Doppler processing is performed over slow-time samples, or samples at an interval of a PRI. A good way to visualize this process is for the radar to populate the 2-D array shown in Figure 8.14. Section 8.6 describes how fast-time samples are collected within a single PRI, then matched-filtered to

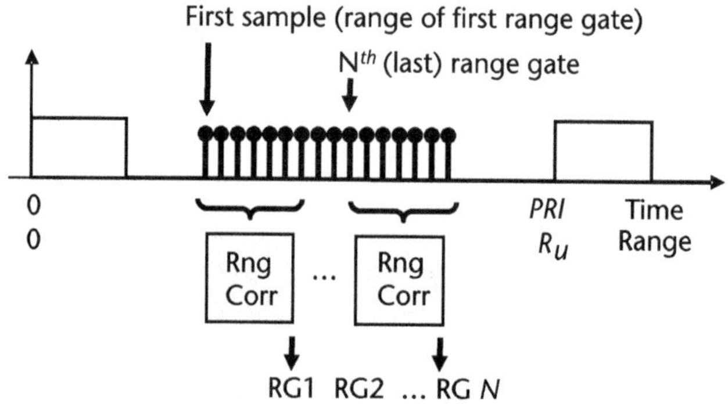

Figure 8.13 Timing diagram of fast-time/range processing.

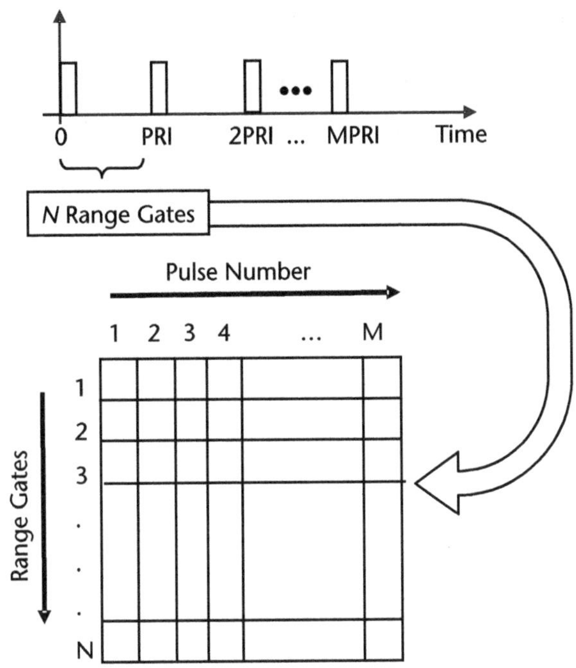

Figure 8.14 Data collection and arrangement of fast-time/slow-time samples.

produce N range gates. One instance of this calculation corresponds to populating the first column of the array. The process must then be repeated on successive PRIs, each providing data for an additional column. After M PRIs, we have a full N-by-M array with range gates in the vertical (row) dimension, and PRIs (pulse numbers) in the horizontal (column) dimension. Doppler processing is then performed per range gate (on each row) by calculating an FFT over the M slow-time samples. Assuming no zero-padding, an M-point FFT results in M samples in the Doppler frequency domain.

After execution of the FFTs, the array is now a matrix of N range bins by M Doppler bins (or range gates and Doppler filters, consistent with the concept of the FFT output as a filter bank). This array of complex voltages is known as the range-Doppler matrix or range-Doppler map (RDM). The RDM is how pulse-Doppler radars "see" the world. This is the main data product out of the signal processor, which is then sent to the detector and data processor for subsequent manipulation.

8.8 Concluding Remarks

This chapter highlighted the individual steps in processing radar returns in the range and Doppler dimension, with a focus on the fundamental concepts of matched filtering for range processing and the Fourier transform for Doppler processing. Several stand-alone examples were presented to provide insight on the modeling of isolated features. These topics are revisited in Chapter 12, where models of full end-to-end range-Doppler processing chains for two different waveform cases are presented and examined.

References

[1] Harris, F. J., "On the Use of Windows for Harmonic Analysis with the Discrete Fourier Transform," *Proceedings of the IEEE*, Vol 66, No. 1, Jan. 1978.

[2] Keel, B. M., "Polyphase Codes," in *Principles of Modern Radar, Volume I: Basic Principles*, M. A. Richards, J. A. Scheer, and W. A. Holm (eds.), Raleigh, NC: Scitech Publishing, Inc., 2010, pp. 824–29.

[3] Whalen, A. D., *Detection of Signals in Noise*, San Diego, CA: Academic Press, Inc., 1971.

[4] Barker, R. H., "Group Synchronizing of Binary Digital Systems," in *Communication Theory*, W. Jackson (ed.), London: Butterworth Scientific Publications, 1953, pp. 273–287.

CHAPTER 9

Monopulse Processing

9.1 Introduction

This chapter expands on the monopulse method for angle measurement first introduced in Chapter 3. Monopulse may be classified into two main categories: The amplitude-comparison technique employs simultaneous squinted beams, where the angle of arrival (AoA) of the target return is extracted from the relative amplitudes of the signal in each beam/receiver channel. In contrast, the phase-comparison version of monopulse processing is an interferometric approach to angle estimation. The relative phase between nonsquinted antenna elements is shown to contain information on the AoA. For both types of monopulse systems, sum and difference signals are formed by combining the multiple antenna elements through a monopulse feed network. Furthermore, the monopulse processor normalizes the difference signal by the sum signal, forming the so-called monopulse ratio. The angle is estimated based on this ratio, rather than from a direct measurement of relative amplitudes or relative phases. Thus, the distinction between the two techniques is often hard to make by only looking at the processor. There are many implementation variations depending on the type of antenna utilized (e.g., reflectors versus horns versus planar arrays), the feed network design, and the exact formulation of the receiver/processor; however, both amplitude- and phase-comparison monopulse provide comparable performance [1, 2]. As a first-order model, we specifically consider the phase-comparison version of monopulse processing.

Section 9.2 starts by describing a mathematical model of a target return signal arriving at an antenna array consisting of two elements, and a basic angle estimator is defined. Section 9.3 extends the derivation to an arbitrary number of elements in a linear array. For additional insight, a nonmathematical description of monopulse is presented in Section 9.4 using a phasor representation of the signals of interest. Last, Section 9.5 features a Simulink implementation of a one-dimensional monopulse processor model; example results are shown and compared to the performance bound predicted by the measurement uncertainty typically used in tracking radars; this is also known as the Cramer-Rao lower bound (CRLB). The chapter concludes in Section 9.6 with some final remarks and suggestions for extending the basic model.

9.2 Monopulse Processing of a Two-Element Array

Consider two identical, isotropic antenna array elements separated by a distance d, as illustrated in Figure 9.1. The antenna plane is on the x-axis, such that its boresight

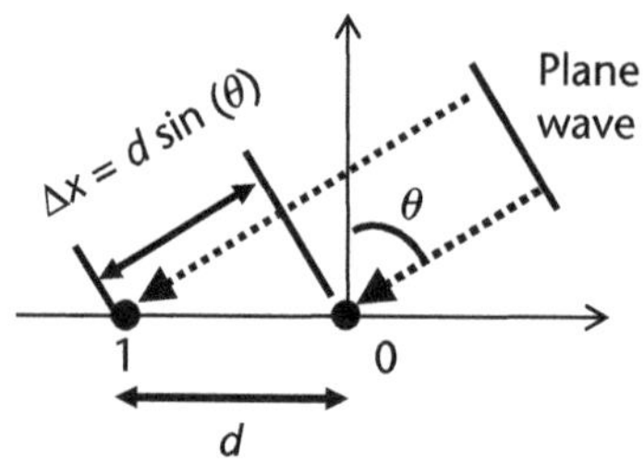

Figure 9.1 Geometry for a two-element monopulse antenna.

(orthogonal direction) coincides with the positive y-axis. An incident plane wave arrives at an angle θ relative to boresight.

The plane wave arrives first at element 0, which is taken as our reference. We define the signal at that point as

$$S_0 = Ae^{j\omega t} \tag{9.1}$$

From that reference point, the plane wave must travel an additional distance Δx to arrive at element 1. From the geometry in Figure 9.1,

$$\Delta x = d\sin(\theta) \tag{9.2}$$

The signal at element 1 is therefore

$$S_1 = Ae^{j(\omega t+\Delta\varphi)}, \text{ where } \Delta\varphi = \left(\frac{2\pi}{\lambda}\right)\Delta x \tag{9.3}$$

Notice from (9.2) and (9.3) that information on the AoA θ is explicitly embedded in the relative phase $\Delta\varphi$ of the signals at each element.

We now define the sum or sigma signal as

$$\Sigma = S_1 + S_0 = Ae^{j(\omega t+\Delta\varphi)} + Ae^{j\omega t} = Ae^{j\omega t}\left(e^{j\Delta\varphi} + 1\right) \tag{9.4}$$

Likewise, we define the difference or delta signal as

$$\Delta = -j\left(S_1 - S_0\right) = -j\left(Ae^{j(\omega t+\Delta\varphi)} - Ae^{j\omega t}\right) = -jAe^{j\omega t}\left(e^{j\Delta\varphi} - 1\right) \tag{9.5}$$

The reason for adding the $-\pi/2$ phase shift in the form of $(-j)$ to the definition is discussed shortly. We now define the monopulse ratio, or monopulse error slope as the real part of the delta-to-sum ratio:

$$\rho = \text{Re}\left[\frac{\Delta}{\Sigma}\right] \tag{9.6}$$

The real part is taken because in general, delta and sum are complex quantities. We now replace (9.4) and (9.5) into (9.6):

$$\rho = \text{Re}\left[\frac{-jAe^{j\omega t}\left(e^{j\Delta\varphi} - 1\right)}{Ae^{j\omega t}\left(e^{j\Delta\varphi} + 1\right)}\right] = \text{Re}\left[\frac{-j\left(e^{j\Delta\varphi} - 1\right)}{\left(e^{j\Delta\varphi} + 1\right)}\right] \tag{9.7}$$

Next, we factor out the square root of the complex exponential on both sides of the ratio:

$$\rho = \text{Re}\left[\frac{-je^{j\Delta\varphi/2}\left(e^{j\Delta\varphi/2} - e^{-j\Delta\varphi/2}\right)}{e^{j\Delta\varphi/2}\left(e^{j\Delta\varphi/2} + e^{-j\Delta\varphi/2}\right)}\right] = \text{Re}\left[\frac{-j\left(2j\sin\left(\Delta\varphi/2\right)\right)}{\left(2\cos\left(\Delta\varphi/2\right)\right)}\right] \tag{9.8}$$

where the last equality stems from the definitions of sine and cosine. Last, the reader will readily recognize the final result to be a tangent function:

$$\rho = \text{Re}\left[\frac{\sin\left(\Delta\varphi/2\right)}{\cos\left(\Delta\varphi/2\right)}\right] = \text{Re}\left[\tan\left(\Delta\varphi/2\right)\right] \tag{9.9}$$

We pause briefly to note that (9.8) shows the need for defining the delta signal with the additional factor of $-j$. This quantity cancels the additional j in the numerator of (9.8), forcing the result to be a real function and survive the Re[ ] operator; otherwise, the result would be zero. This factor is further explored in Section 9.4.

Turning our attention back to (9.9), we further recall that the phase difference $\Delta\varphi$ between the two elements was given by (9.3) as proportional to the path length difference Δx, defined in (9.2). Let's now assume the separation between elements d to be $\lambda/2$, the customary choice for critical spatial sampling. We may then rewrite $\Delta\varphi$ as

$$\Delta\varphi = \left(\frac{2\pi}{\lambda}\right)\Delta x = \left(\frac{2\pi}{\lambda}\right)\left(\frac{\lambda}{2}\right)\sin(\theta) = \pi\sin(\theta) \approx \pi\theta \tag{9.10}$$

where we have used the small angle approximation $\sin(x) \sim x$ on the RHS of (9.10). Replacing (9.10) back in (9.9) results in

$$\rho = \text{Re}\left[\tan\left(\frac{\pi\sin(\theta)}{2}\right)\right] \approx \text{Re}\left[\tan\left(\frac{\pi\theta}{2}\right)\right] \tag{9.11}$$

Once again, the small angle approximation has been used in (9.11). This tangent relationship is illustrated in Figure 9.2(a). To verify the validity of the small-angle approximation, both the exact form (solid line) and the approximation (dashed line) are plotted. The normalized gain pattern of the two-element transmit array (the sum beam) is displayed in Figure 9.2(b). This is simply a cosine function, as shown by the denominator of (9.8). Due to the small size of the array, this is a fairly broad beam, with a (two-sided) HPBW of +/–30 deg, as indicated in Figure 9.2. Within this HPBW, the small angle approximation is shown to match the exact expression fairly well, with some noticeable divergence toward the edges of the beam.

The reader may also notice that within the HPBW, and in particular around 0 degrees, the monopulse ratio is approximately linear. The slope of this linear approximation can be calculated by taking the derivative of (9.11) with respect to the angle θ, then taking the limit as θ tends to zero. Alternatively, one may simply apply yet another small-angle approximation tan(x) ~ x to (9.11). Either way, the resulting slope is $\pi/2 \approx 1.57$. This is often referred to as the monopulse error slope. Whether we use the exact form, the small-angle approximation, or the linear approximation, we are dealing with a monotonic function, meaning there is a one-to-one correspondence between the measured monopulse ratio and the true AoA. Solving the approximate form of (9.11) for the AoA provides the monopulse angle estimate, calculated given the measured ratio as defined in (9.6):

$$\hat{\theta} = \left(\frac{2}{\pi}\right)\tan^{-1}(\rho) \tag{9.12}$$

9.3 Extension to an *N*-Element Array

Now consider a ULA of N elements with an interelement separation of d, as depicted in Figure 9.3. The incident signal at element 0 is once again taken as the reference, still described by (9.1). The signal at the nth element is

$$S_n = Ae^{j(\omega t + \Delta\varphi_n)}, \text{ where } \Delta\varphi_n = n\pi\sin(\theta) \approx n\pi\theta \tag{9.13}$$

Once again, small-angle approximation and $d = \lambda/2$ are assumed in the RHS of (9.13). The sum signal is now

$$\Sigma = \sum_{n=0}^{N-1} S_n = Ae^{j\omega t}\sum_{n=0}^{N-1} e^{j\Delta\varphi_n} = Ae^{j\omega t}\sum_{n=0}^{N-1}\left(e^{j\pi\theta}\right)^n \tag{9.14}$$

The terms in the summation in (9.14) form a geometric series. For $\theta = 0$, the complex exponential equals 1, and the summation is simply N. For any other value such that the exponential is not 1, we can use the generalized formula for the sum of a geometric series:

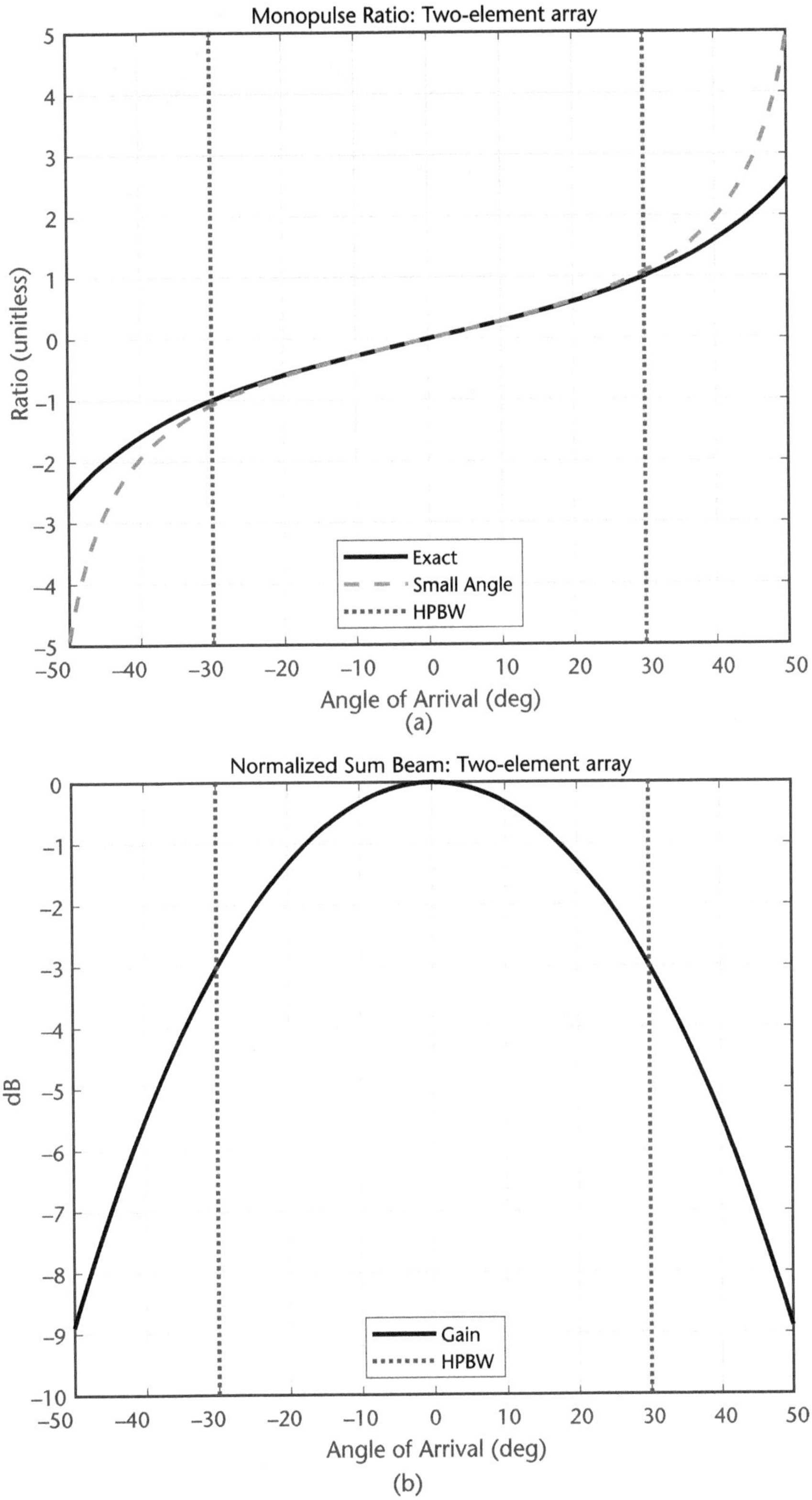

Figure 9.2 Monopulse ratio of a two-element array as a function of (a) AoA, and (b) sum beam pattern for the same array.

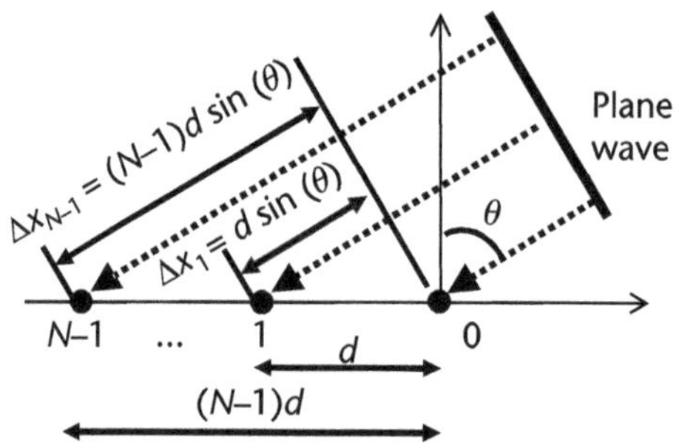

Figure 9.3 Geometry for an N-element ULA.

$$\sum_{n=a}^{b} r^n = \frac{r^a - r^{b+1}}{1-r} \qquad (r \neq 1) \tag{9.15}$$

Applying this definition to the sum signal from (9.14) results in

$$\Sigma = Ae^{j\omega t}\sum_{n=0}^{N-1}\left(e^{j\pi\theta}\right)^n = Ae^{j\omega t}\left(\frac{1-e^{j\pi\theta N}}{1-e^{j\pi\theta}}\right) \tag{9.16}$$

Likewise, the delta signal is defined as one half of the array subtracted from the other half:

$$\Delta = -j\left(\sum_{n=N/2}^{N-1} S_n - \sum_{n=0}^{N/2-1} S_n\right) = -jAe^{j\omega t}\left(\sum_{n=N/2}^{N-1}\left(e^{j\pi\theta}\right)^n - \sum_{n=0}^{N/2-1}\left(e^{j\pi\theta}\right)^n\right) \tag{9.17}$$

Applying the same definition from (9.15), we obtain

$$\Delta = -jAe^{j\omega t}\left(\frac{e^{j\pi\theta(N/2)} - e^{j\pi\theta N}}{1-e^{j\pi\theta}} - \frac{1-e^{j\pi\theta(N/2)}}{1-e^{j\pi\theta}}\right) \tag{9.18}$$

which simplifies to

$$\Delta = -jAe^{j\omega t}\left(\frac{2e^{j\pi\theta(N/2)} - e^{j\pi\theta N} - 1}{1-e^{j\pi\theta}}\right) = -jAe^{j\omega t}\left(\frac{-\left(1-e^{j\pi\theta(N/2)}\right)^2}{1-e^{j\pi\theta}}\right) \tag{9.19}$$

by noting that the numerator is a quadratic expression in $\exp(j\pi\theta N)$. We may now form the monopulse ratio from (9.19) and (9.16), resulting in

$$\frac{\Delta}{\Sigma} = \frac{j\left(1-e^{j\pi\theta(N/2)}\right)^2}{1-e^{j\pi\theta N}} \tag{9.20}$$

After some factoring and simplification, (9.20) becomes

$$\frac{\Delta}{\Sigma} = \frac{2\left(\sin\left(\pi\theta(N/4)\right)\right)^2}{\sin\left(\pi\theta(N/2)\right)} \tag{9.21}$$

Notice that the argument of the sine in the denominator is twice that of the numerator. This hints to the formula for the sine of the double angle:

$$\sin(2\alpha) = 2\sin(\alpha)\cos(\alpha) \tag{9.22}$$

Let $\alpha = \pi\theta$ (N/4) in (9.22). We may then rewrite (9.21) as

$$\frac{\Delta}{\Sigma} = \frac{2\left(\sin\left(\pi\theta(N/4)\right)\right)^2}{2\sin\left(\pi\theta(N/4)\right)\cos\left(\pi\theta(N/4)\right)} = \frac{\sin\left(\pi\theta(N/4)\right)}{\cos\left(\pi\theta(N/4)\right)} = \tan\left(\pi\theta\left(\frac{N}{4}\right)\right) \tag{9.23}$$

Thus, we arrive at the unsurprising result that the monopulse ratio for an N-element ULA is also a tangent function! The corresponding angle estimate is

$$\hat{\theta} = \left(\frac{4}{N\pi}\right)\tan^{-1}(\rho) \tag{9.24}$$

Notice that (9.24) reduces to (9.12) for the case $N = 2$, as it should. Figure 9.4 shows the corresponding monopulse curve and sum beam pattern for the case $N = 20$. Due to the much narrower beam (HPBW ~ +/−2.54 deg), the small-angle approximation is an almost perfect match to the exact form, even beyond the HPBW. In the linear region around 0 degrees, the monopulse slope m is

$$m = \frac{N\pi}{4} \approx 15.71 \tag{9.25}$$

9.4 A Nonmathematical Description of Monopulse

We now attempt to describe the monopulse ratio in nonmathematical terms, by using the concept of phasors. Generally speaking, a phasor is a representation of a complex sinusoidal, much like (9.1). Here we are concerned with the graphical representation of a phasor as a rotating vector in the complex plane, as illustrated in Figure 9.5. The phase angle θ changes over time at an angular rate of ω; where $\omega = 2\pi f$; therefore $\theta = \omega t = 2\pi ft$. For a positive frequency, the phasor rotates in a counterclockwise direction. At any one time, the projections of the phasor on the real and imaginary axes (the signals x and y in the figure) are cosines and sines of θ, respectively. The phasor is fully defined by two quantities: either by an amplitude A and phase θ or equivalently by the real and imaginary parts x and y. The relationships between these quantities are shown in Figure 9.5.

For the purposes of this discussion, let's consider the case of $N = 2$ elements as discussed in Section 9.2. We will also assume unit amplitude ($A = 1$) for simplicity,

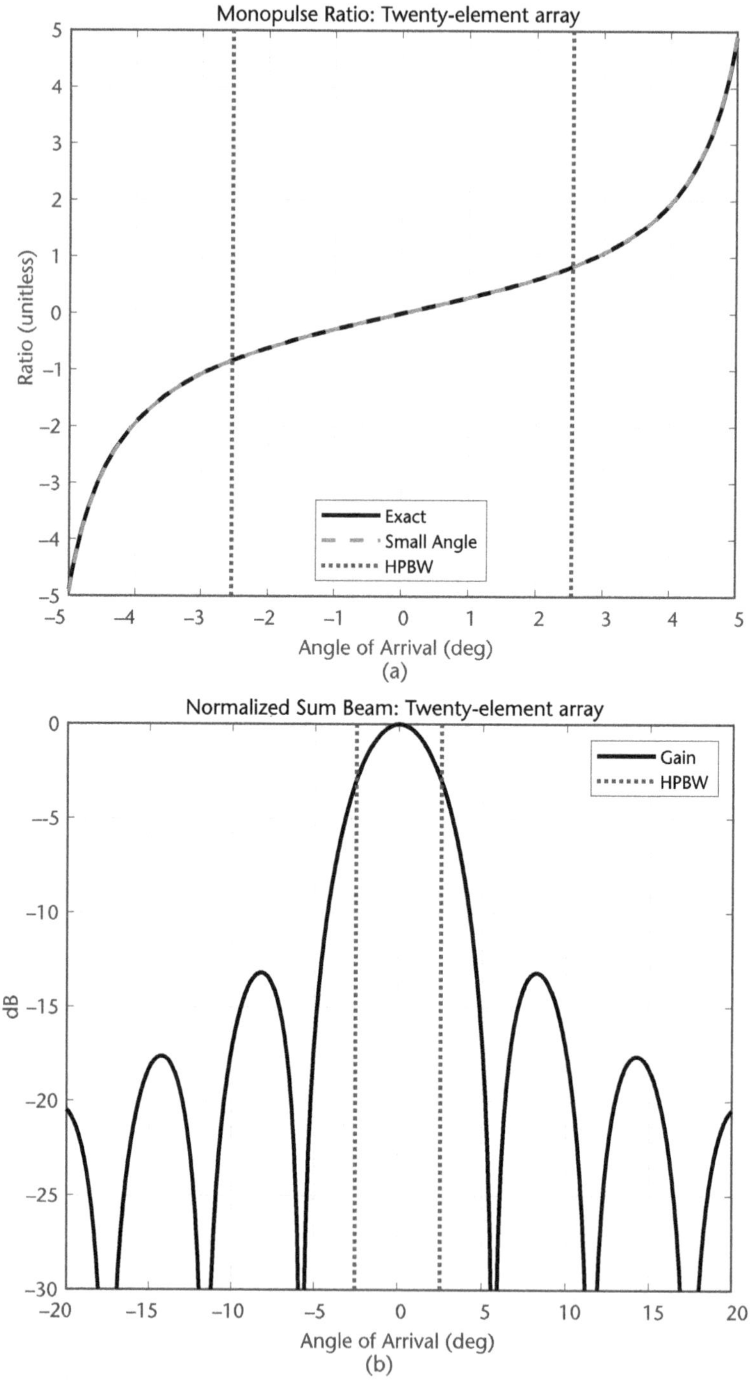

Figure 9.4 (a) Monopulse ratio of a 20-element array as a function of AoA, and (b) sum beam pattern for the same array.

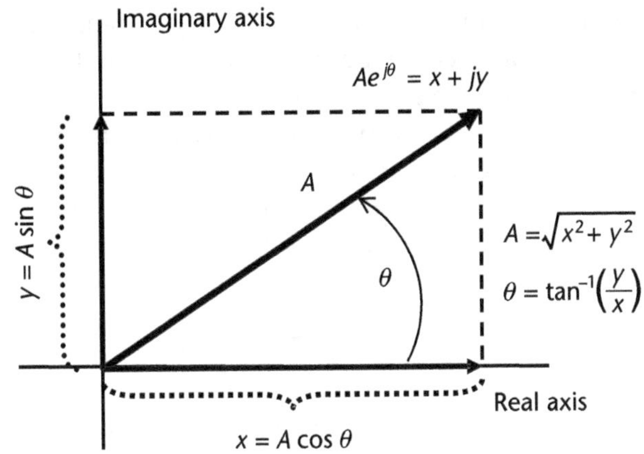

Figure 9.5 Phasor representation of a complex sinusoid.

although it is not necessary. If we depict the reference signal S_0 as a phasor of zero phase ($\theta = 0$), then the S_1 phasor will be advanced by the phase increment ($\theta = \Delta\varphi$) as shown on the upper panel of Figure 9.6.

The next step is to graphically form the sum and delta phasors by standard vector addition, as shown in the middle panel. By first defining the delta phasor without the phase shift, the resultant phasors are shown to be orthogonal to each other. The bottom panel shows the effect of the $-\pi/2$ phase shift—the negative sign meaning that the shift is in the clockwise direction. The two phasors are now

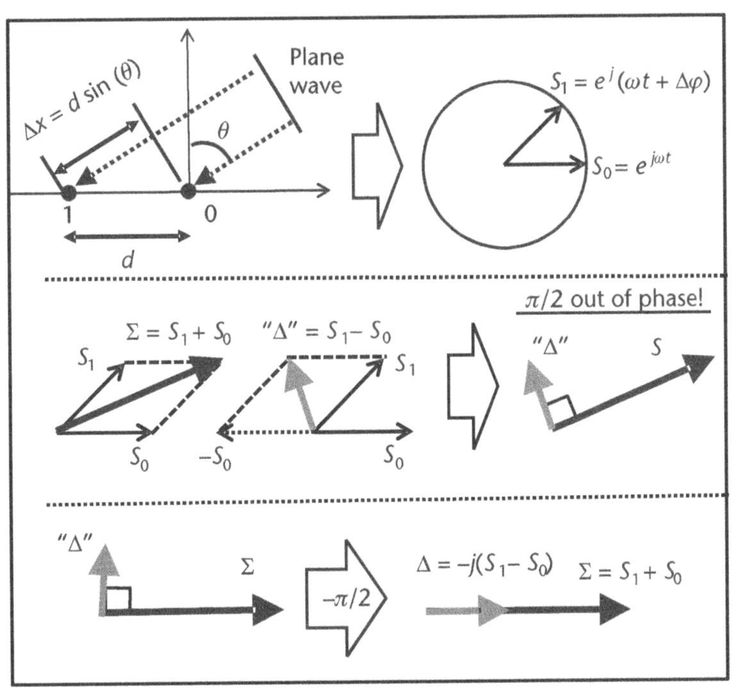

Figure 9.6 Phasor representation of received signals at a two-element array (top); sum and delta phasor shown to be orthogonal in the absence of a $-\pi/2$ phase shift (center); and sum and delta phasors aligned after the phase shift (bottom).

aligned (collinear), and furthermore they are codirectional (they point in the same direction) for the case where the signal arrives from the right. Assuming that the AoA is in the azimuth plane, the monopulse ratio obtained from codirectional phasors will be positive, yielding a positive azimuth angle. This is consistent with the typical convention that an azimuth angle is positive right.

So why is it important that the sum and delta phasors be aligned? Let's think of the monopulse ratio as the dot product of two vectors. We may then write

$$\rho = (\Delta)\bullet\left(\Sigma^{-1}\right) = |\Delta|\left|\Sigma^{-1}\right|\cos(\beta) \tag{9.26}$$

where Σ^{-1} is a phasor in the same direction as Σ, but whose magnitude is the inverse of it. If the angle β between them is +/−90 degrees, we get 0 for the monopulse ratio regardless of the vector magnitudes, which is obviously undesirable. Ideally, β should be either 0° or 180°, such that its cosine is +/−1, thereby providing a sense of direction. Once again, the $-j$ rotation in the definition of Δ accomplishes this.

If the AoA is positive (signal coming from the right in the case of azimuth, or coming from above if it is elevation), we want β and therefore ρ to be positive. The opposite is true for a negative AoA. Figure 9.7 summarizes these results for both of these cases.

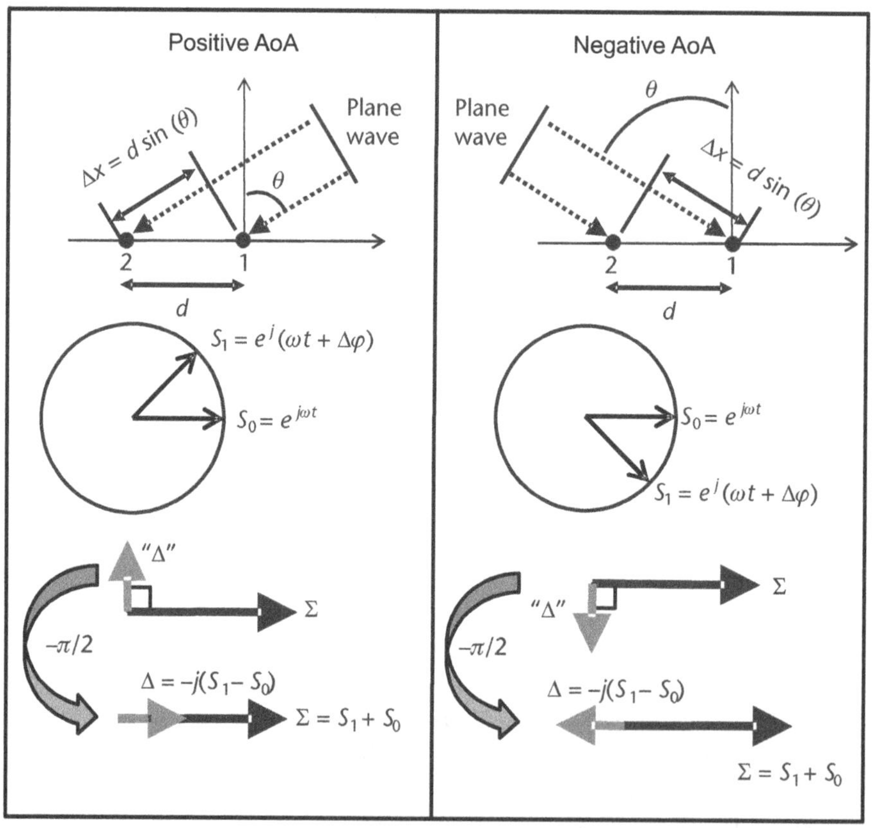

Figure 9.7 Summary of phasor representation of monopulse processing for a positive AoA (left) and for a negative AoA (right).

9.5 Simulink Model of Monopulse Processor

We now examine a Simulink implementation of a 1-D monopulse processor. The model is based on the 20-element ULA described in Section 9.3. Sum and delta antenna gain patterns are generated in a script m-file set up as a Pre-load Callback, as they only need to be loaded once. These patterns are then set up as LUTs. Figure 9.8 shows the voltage gain patterns for each. Again, isotropic elements are assumed, so there is no element pattern associated with this array. Notice that the positive lobe in the delta pattern to the right of 0° is consistent with the convention of positive right for azimuth angle.

The Simulink model file is *Monopulse_v1*; the top level is shown in Figure 9.9. The three Constant blocks contain the model inputs. The signal amplitude is provided by a constant A, which can be interpreted as the amplitude of a single pulse.

The true angle to the target "Angle" in degrees provides the beam-shape losses from the sum and delta antenna pattern LUTs, which are normalized to the sum peak transmit gain, Gt. Gr is the peak receive gain, set equal to Gt. The last input to the Monopulse Subsystem is a binary flag to add receiver noise when set to '1'; the monopulse receiver is noiseless when set to '0'. The output of the subsystem is the estimated AoA based on a single observation, which for the noiseless example shown results in a perfect estimate (2 degrees).

The second and third layers of the Simulink diagram are displayed in Figure 9.10. The top of Figure 9.10 shows the contents of the Monopulse Subsystem block, where beam-shape losses on receive are added to form the sum and delta

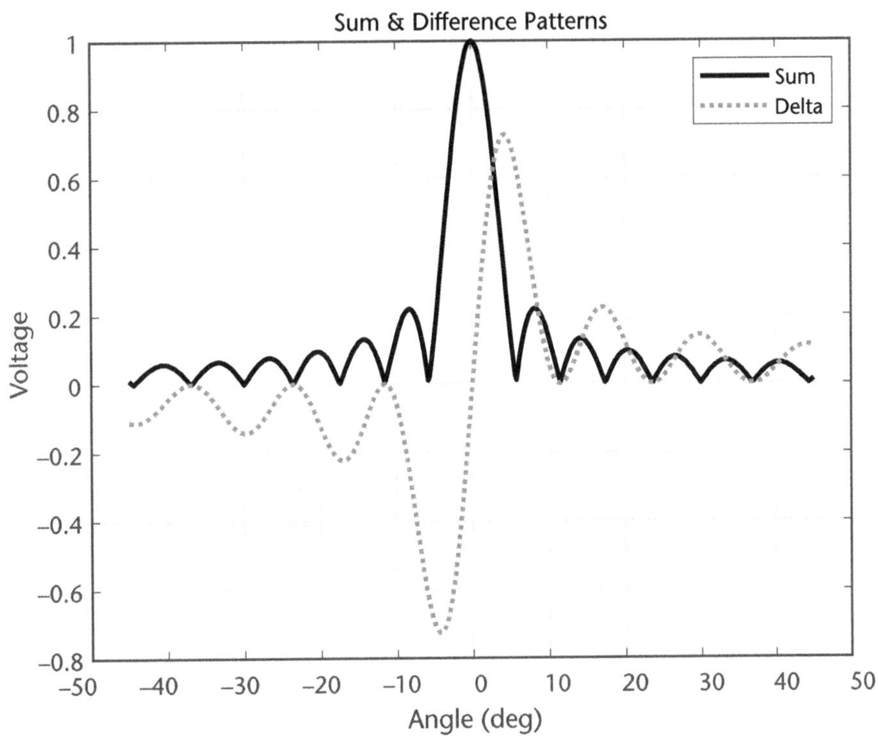

Figure 9.8 Sum and delta antenna patterns for 20-element ULA (voltage gains).

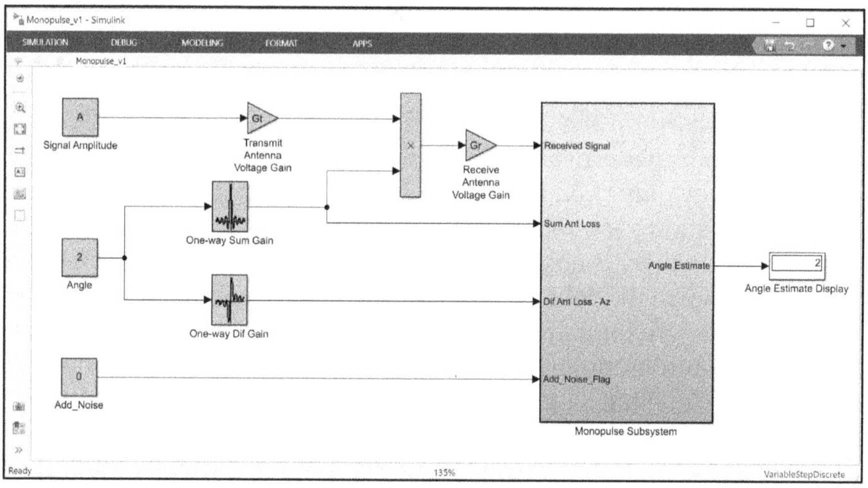

Figure 9.9 Top-level diagram of monopulse model in Simulink.

receiver channels. Independent, identically distributed (i.i.d) Gaussian receiver noise is added if the Add_Noise flag is on (set to 1). The actual AoA estimation is performed in the Az Angle Estimate subsystem, shown at the bottom of Figure 9.10. The $(6/\pi)^2$ factor includes the scaling term from (9.24) for $N = 20$, plus the conversion factor $(180/\pi)$ from radians to degrees.

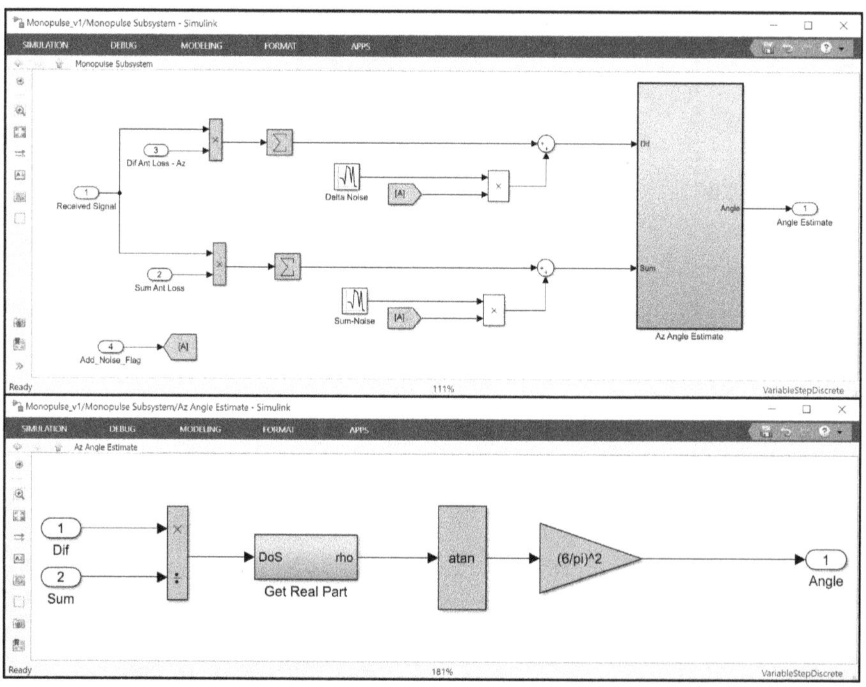

Figure 9.10 Monopulse subsystem diagram (top) and monopulse angle calculation (bottom).

Setting the Add_Noise flag to 1 and rerunning the model will introduce an estimation error that depends on SNR, among other factors. To gain additional insight on the statistical performance of this monopulse estimator, an upgraded model is shown in Figure 9.11.

This model, *Monopulse_v2_MC*, wraps the previous monopulse model in a "for" loop construct that enables Monte Carlo (MC) analysis of the model. The top of Figure 9.11 shows the DSP System Toolbox version of the random number generator block (Random Source) used to generate the Gaussian-distributed noise samples for the MC trials. This version was chosen for the convenience of creating a frame (vector) of noise samples directly from the source at time $t = 0$. Alternatively, a regular Gaussian generator and a Buffer block could be used instead—although in that case the model would need to be run long enough to fill up the buffer. Another convenience afforded by the Random Source block is the creation of complex noise samples, thereby creating two random realizations (the real and imaginary parts of each random draw) for the two receiver channels. The number of trials (set to 1,000 for this example) is also input into the MC subsystem. The output is a corresponding vector of estimated AoAs (in units of degrees) for each of the independent trials. Mean and standard deviation are then calculated for this vector. The mean value is an unbiased estimate of the true AoA. The standard deviation represents the measurement uncertainty of the monopulse estimate. Last, the CRLB for these test conditions is also calculated and displayed for comparison with the experimental measurement uncertainty. The example results in Figure 9.11 show good agreement between the mean value of the estimates (2.007°) and the

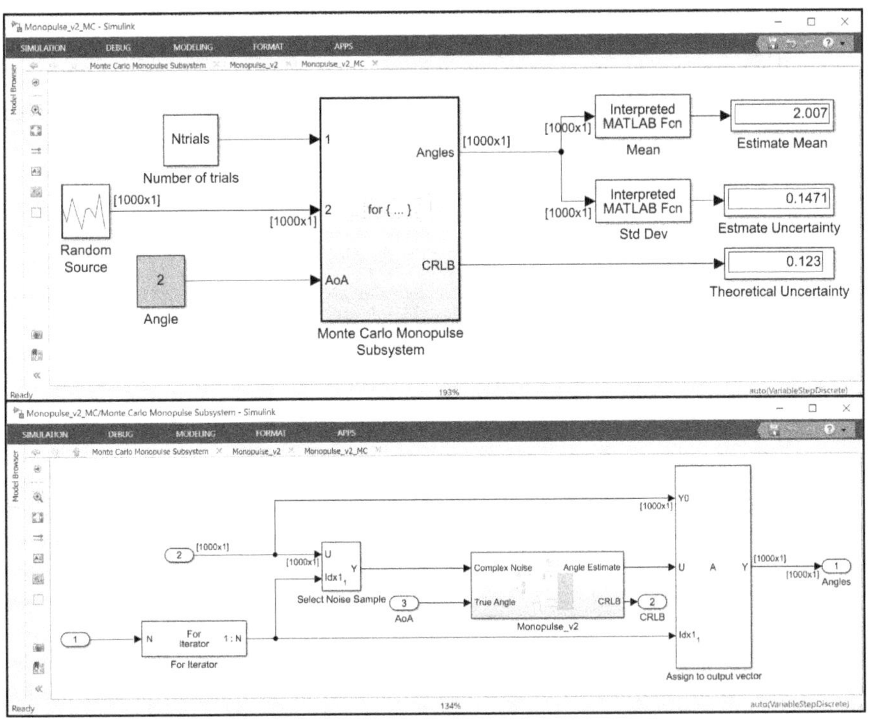

Figure 9.11 Monopulse model for Monte Carlo trials: first-level diagram (top) and second-level diagram (bottom).

true angle, and a measurement uncertainty within three-hundredths of a degree from the theoretical lower bound (0.1471° – 0.123° = 0.0241°).

The Monopulse_v2 subsystem diagram is seen in Figure 9.12 as a modified version of Monopulse_v1 from Figure 9.9. The main modification is to bring the noise samples as an input, so different realizations may be inserted within the "for" loop. The true AoA is also pulled out to the top level of the model, but only as a matter of convenience. Last, the subsystem calculates the CRLB for comparison to the MC results.

The CRLB is calculated as

$$\sigma_{\text{AoA}} = \frac{\theta_3}{k_m \sqrt{2SNR}} \tag{9.27}$$

which is a common form for the receiver noise-limited angular accuracy of a monopulse system. Several variations of this expression are widely available in the literature; this particular form is taken from Barton [3]. The numerator θ_3 is the 3-dB beamwidth or HPBW of the antenna, and k_m is a monopulse factor, typically given as a value between 1.4 and 1.6. It is in fact a normalized monopulse error slope; Barton [3] defines it as the derivative of the monopulse ratio as a function of AoA in units of beam width (i.e., angle normalized by the HPBW), or

$$k_m = \frac{\partial \rho}{\partial \tilde{\theta}} \quad \left(\tilde{\theta} = \frac{\theta}{\theta_3}\right) \tag{9.28}$$

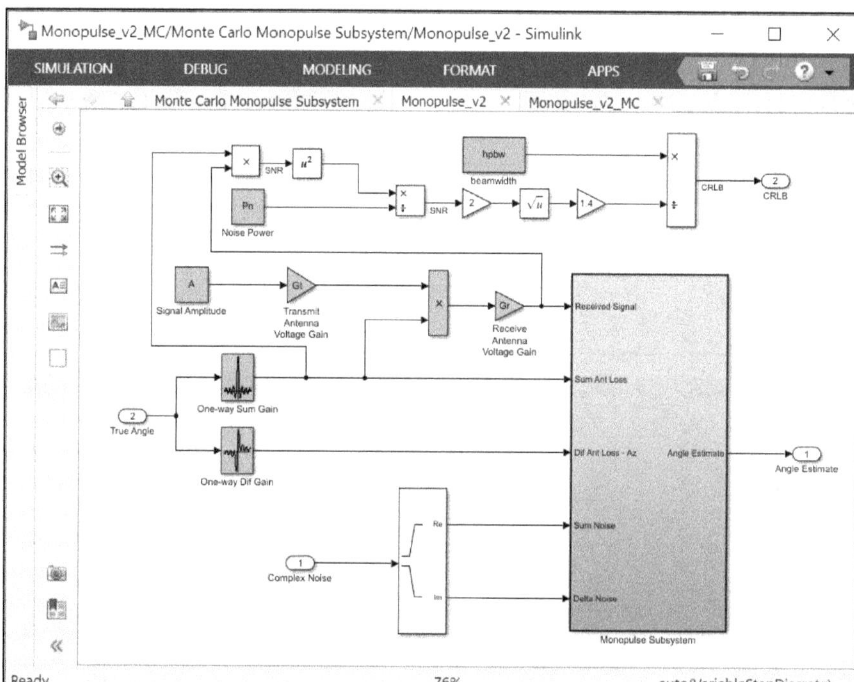

Figure 9.12 *Monopulse_v2* block diagram showing CRLB calculation.

Applying the chain rule to (9.28), we have

$$k_m = \left(\frac{\partial \rho}{\partial \theta}\right)\left(\frac{\partial \theta}{\partial \tilde{\theta}}\right) = (m)(\theta_3) \tag{9.29}$$

where m is the unnormalized monopulse slope defined by (9.25). For our N-element ULA example, the HPBW is approximately 0.886 times the first null beamwidth, which is $2/N$. We therefore calculate the monopulse factor as

$$k_m = m\theta_3 = \left(\frac{N\pi}{4}\right)\left(\frac{2}{N}\right)0.886 = \frac{0.886\pi}{2} \approx 1.4 \tag{9.30}$$

We thus use 1.4 in our calculation. Notice that this quantity is independent of the number of elements in the array because of the normalization; however, the expression for HPBW used in (9.30) is only accurate for long arrays, meaning for $Nd >> \lambda$.

9.6 Concluding Remarks

The first-order monopulse model introduced in this chapter will be integrated into a larger engagement model presented in Chapter 15. Other than a two-channel implementation (azimuth and elevation), the larger model is identical to that presented here. Several upgrades to this simple model could be added to improve its fidelity, however. Two such upgrades are described as follows:

- In the array definition used to define our monopulse model, isotropic elements were assumed; thus, the transmit antenna pattern shown in Figure 9.8 is what antenna engineers refer to as the array factor (AF). One simple upgrade that would be relatively straightforward to implement would be to add an antenna element pattern corresponding to a real radiator. Common options for element voltage patterns $E(\theta)$ fall within the family of cosine functions raised to some power, which have the form

$$E(\theta) = G_0 \cos^x\left(h(\theta)\right) \tag{9.31}$$

  for some values of G_0 and x, and where argument $h(\theta)$ is some function of the angle off-boresight. As an example, if we let $G_0 = 2$, $x = 1$, and $h(\theta) = (\pi/2)\sin(\theta)$, we obtain the two-element pattern from Figure 9.2. A good example of a real radiator is obtained by letting $x = 0.5$ and $G_0 \approx 1.28$. This is a good approximation to a half-wave dipole pattern, arguably one of the simplest and most widely used class of antennas.
- Another potential upgrade is the application of window functions to the antenna aperture. Uniform illumination is assumed in the ULA model;

however, this results in relatively large antenna sidelobes. As discussed in Chapter 3, amplitude tapers are often applied across the aperture to minimize sidelobe level at the expense of some widening of the antenna beam width and a moderate SNR loss. Common choices are Taylor weights for the sum beam and Bayliss weights for the delta beam [4]; the application of these windows affects the monopulse slope characteristic and must therefore be accounted for in the formulation of the angle estimate.

Additional upgrades to make the model more realistic may include error sources, such as amplitude and phase errors in the array elements as discussed in Chapter 3. This translates into gain and phase mismatches in the receiver channels, although other sources of error in the channels themselves could also be incorporated. We look at selected examples of H/W errors in Chapter 10; one such example is gain and phase imbalance between I/Q channels. Similar models could be applied to monopulse channels. Bickel's Sandia Report [5] provides a good discussion of these errors and approximations.

References

[1] Sherman, S. M., *Monopulse Principles and Techniques,* Norwood, MA: Artech House, 1984.

[2] Howard, D. D., "Tracking Radar," Chapter 18 in *Radar Handbook,* Second Edition, M. I. Skolnik (ed.), Boston: McGraw-Hill, Inc., 1990, pp. 18.8–18.21.

[3] Barton, D. K., *Modern Radar System Analysis,* Norwood, MA: Artech House, 1988, pp. 401–402.

[4] Elliott, R. S., *Antenna Theory & Design,* Wiley-IEEE Press, 2003, pp. 157–189.

[5] Bickel, D. L., *Precomparator and Postcomparator Errors in Monopulse,* Sandia Report SAND2013-1287, February 2013.

CHAPTER 10

Transmitter and Receiver Components

10.1 Introduction

Many waveform-level computer simulations operate at baseband frequencies, if for no other reason than to keep sampling requirements at reasonable rates. Thus, the RF transmitter and receiver chains are seldom modeled in a full-up system simulation—or at least, they are modeled less often and not as completely as the digital receiver section.

This chapter partially addresses this gap, by introducing models for selected analog components. These are not circuit-level models, however. There are many other tools such as SPICE and its many variants that are better suited for this than MATLAB, although the two can be combined [1]. Here we confine ourselves to functional models that represent the output of the component given some input.

The block diagram in Figure 10.1 provides a good roadmap for selecting analog components to model. It shows a coherent radar system where the LOs in the exciter [termed stable local oscillator (STALO) and coherent oscillator (COHO)] are phase-locked to a master oscillator, typically a crystal oscillator with low phase noise (more on phase noise later). These LO signals drive mixers in both the transmitter and receiver for signal upconversion and downconversion, respectively. The example assumes a 3-GHz COHO mixing with a 7-GHZ STALO to generate a 10-GHz transmit RF.

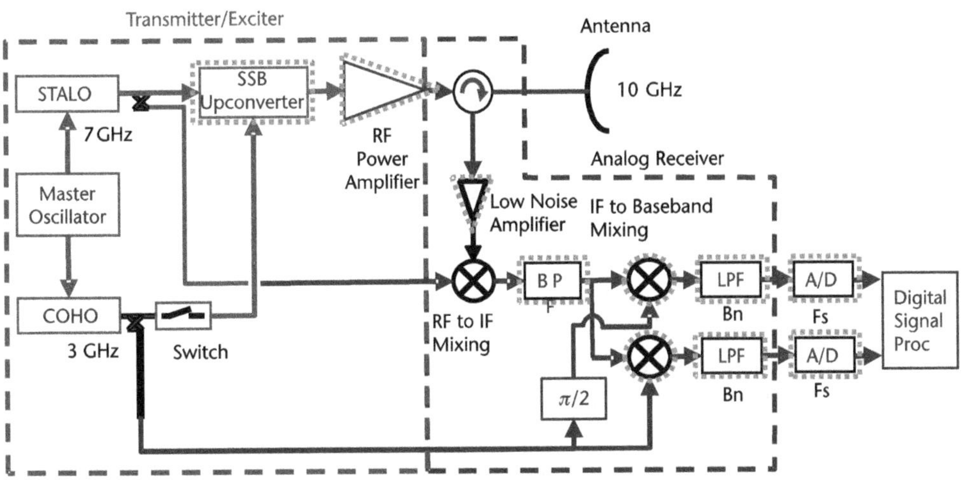

Figure 10.1 Generic transmitter/receiver block diagram for coherent radar.

Amplifiers are used throughout to increase the strength of weak signals, and filters are used to reject unwanted signals. In the last down-conversion stage, I and Q channels are formed to extract phase information later on. Finally, A/D converters digitize these analog signals for further processing by the digital processor.

Sections 10.2 to 10.7 present modeling examples of the aforementioned analog components, which are highlighted with dotted boxes in Figure 10.1. The specific topics of Sections 10.2 to 10.7 are listed as follows:

- Section 10.2: Single-sideband (SSB) upconverter (complex mixer);
- Section 10.3: Amplifiers;
- Section 10.4: Oscillator phase noise;
- Section 10.5: I/Q with channel mismatch (gain/phase imbalance);
- Section 10.6: Filters;
- Section 10.7: ADC.

10.2 SSB Upconverter

The first component considered is the upconverter highlighted in Figure 10.1. In a conventional mixing operation, a real input signal (assumed here to be a CW with fixed frequency f_{in}) multiplies the LO signal, which is also a CW but at a different frequency f_{LO}. The mixer output s_{out} is the sum of two sinusoids, one whose frequency is the sum of the two input frequencies and a second whose frequency is their difference; thus,

$$\begin{aligned} s_{out}(t) &= \cos\left(2\pi f_{in} t\right)\cos\left(2\pi f_{LO} t\right) \\ &= 0.5\cos\left(2\pi\left(f_{in} + f_{LO}\right)t\right) + 0.5\cos\left(2\pi\left(f_{in} - f_{LO}\right)t\right) \end{aligned} \tag{10.1}$$

This is a fundamental result stemming from trigonometric identities. Typically, only one of these terms is desired—the sum term for an upconversion operation, or the difference term for downconversion. It is also worth noting that these terms are often referred to as the upper and lower sidebands, respectively. Another name for this traditional mixer is a double-sideband (DSB) mixer for this reason. In either case, the unwanted term is discarded by filtering. Figure 10.2 shows a Simulink implementation of the DSB mixer implemented as an upconverter.

The *Simple_Mixer* model takes a 7-GHz CW and upconverts it to 10 GHz using a 3-GHz LO. The lower sideband at 4 GHz is rejected using an Analog Filter Design block from the DSP System Toolbox, configured as a Chebyshev Type I filter with a passband between 8 and 12 GHz. Section 10.6 addresses filter modeling in greater detail.

Often, the filter will have stringent passband/stopband requirements (i.e., a sharp cutoff), which can make it costly. An alternate approach often used for upconversion of the exciter signal to the output RF is to employ a SSB upconverter, which manipulates the signal phases to produce the desired output without an

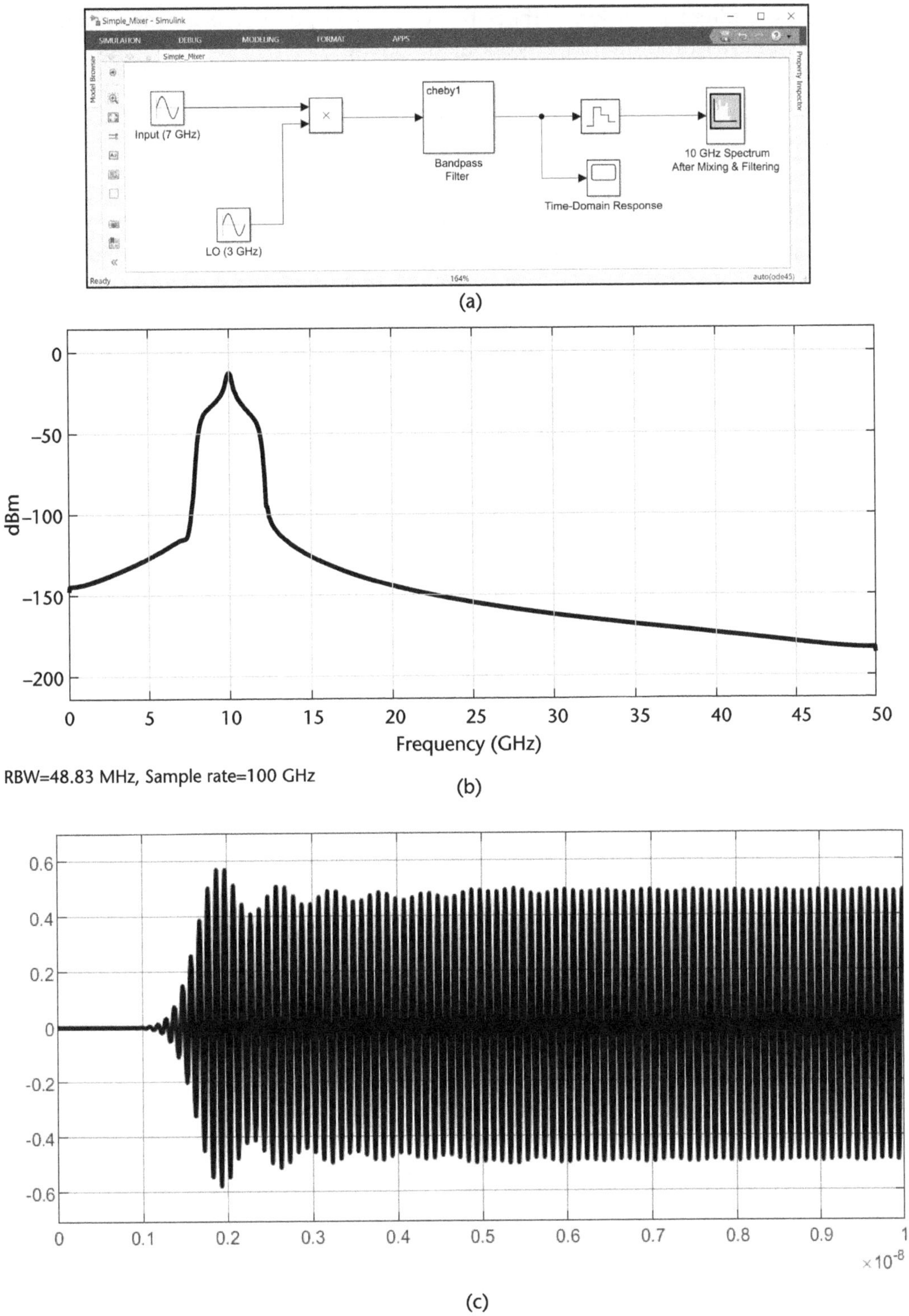

Figure 10.2 (a) Simple Simulink model of real (DSB) upconverter: block diagram, (b) output signal in frequency domain, (c) and time domain.

expensive filter [2]. The key to the SSB upconverter operation is the implementation of yet another fundamental trigonometric identity:

$$\sin(A+B) = \sin(A)\cos(B) + \cos(A)\sin(B) \tag{10.2}$$

Given the input and LO sinusoids—sine functions in (10.2)—at frequencies labeled A and B, their corresponding quadrature counterparts (cosine functions) are obtained from RF devices called quadrature hybrid couplers, or quad hybrids for short. These are a type of power-splitter that imparts a 90-degree phase shift between the two output ports.

Figure 10.3 shows the block diagram of a Simulink implementation of an SSB upconverter; it is also known as a single sideband mixer (SSM) or a quad mixer. The signal frequencies are the same as in the simple mixer example. Frequency

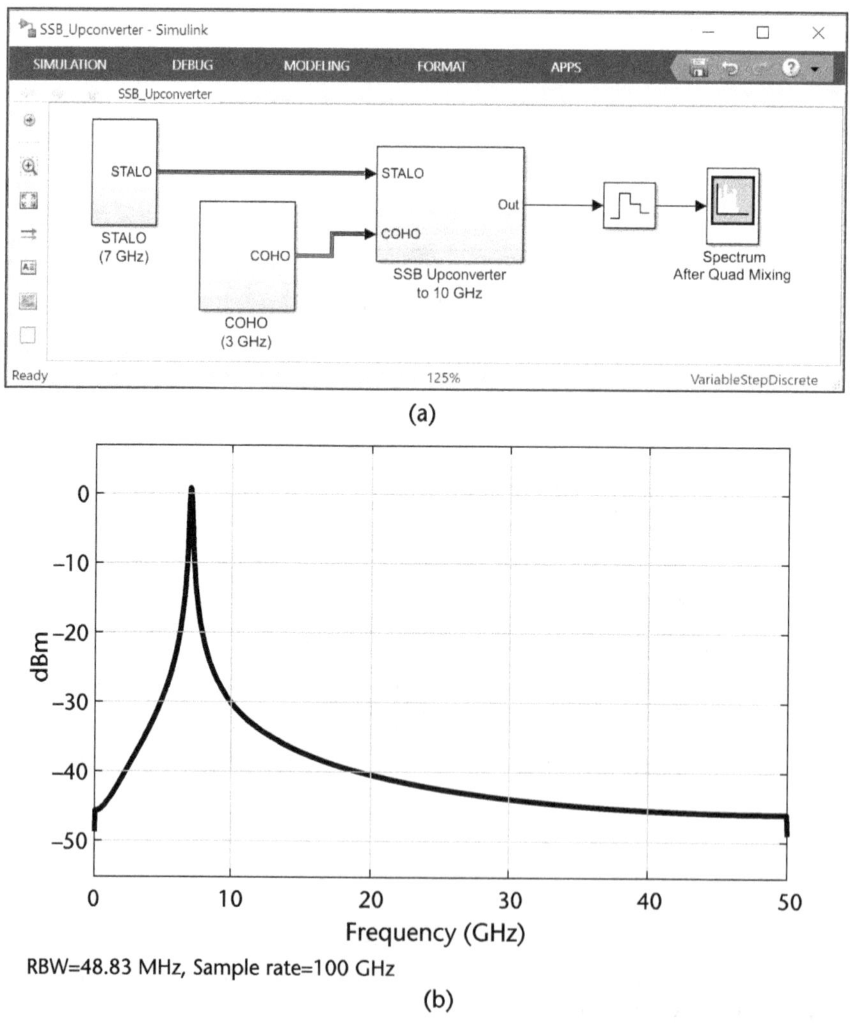

Figure 10.3 Simple Simulink model of SSB upconverter (a) block diagram, signal spectra: STALO at 7 GHz (b), COHO at 3 GHz (c), and output RF at 10 GHz (d).

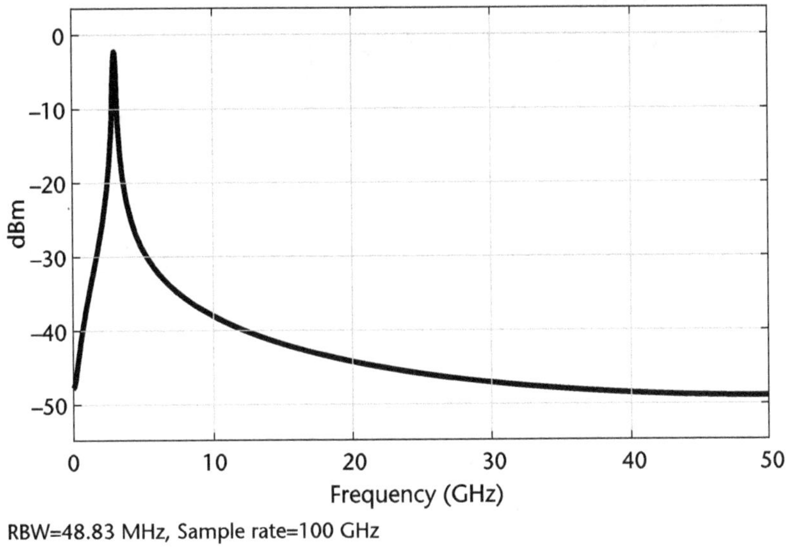

(c)

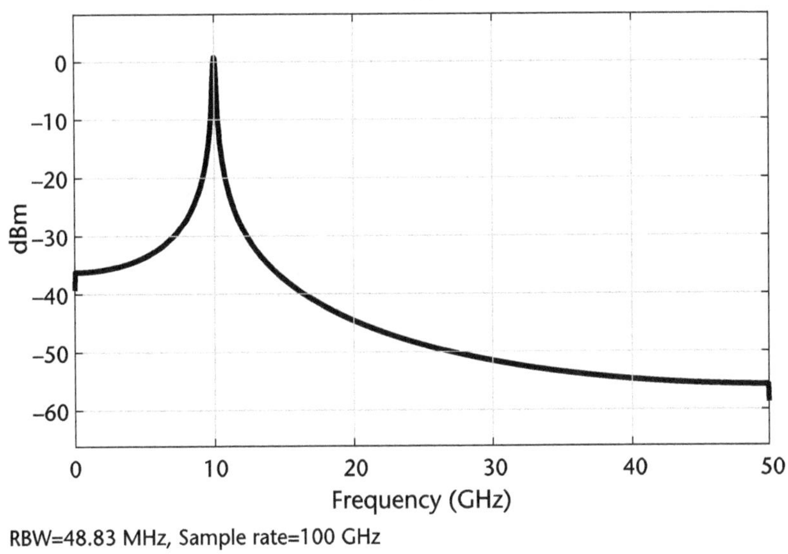

(d)

Figure 10.3 *(Continued)*

spectra for the STALO and COHO inputs and the RF output are calculated and displayed by the Spectrum Analyzer blocks.

Figure 10.4 shows the contents of the SSM subsystem as well as the contents of the STALO subsystem. The COHO subsystem is identical except for the frequency value. For convenience, the two outputs of the quad hybrid are generated at the source via two identical Sine Wave blocks. In the bottom source block (STALO-Q), a 90-degree initial phase has been added to the block parameters. Buses are then used to carry the signals to the mixer.

The mixer subsystem shows the implementation of (10.2), where the I/Q signals from the buses are the sines and cosines of (10.2), respectively. We also note that

(a) 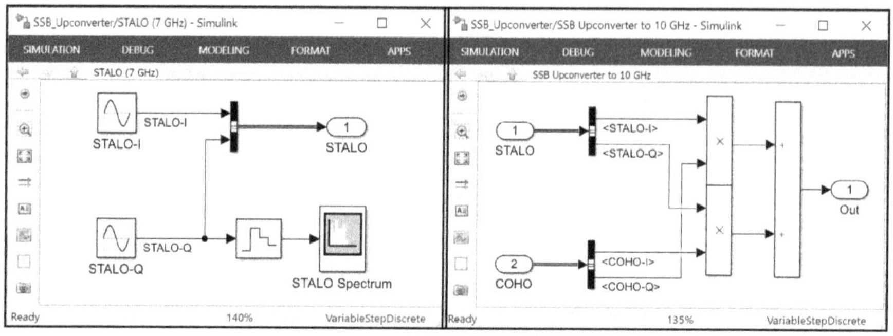 (b)

Figure 10.4 (a) STALO signal subsystem, and (b) SSB mixer subsystem.

to turn this model into a SSB downconverter, we can simply change the operation in the bottom port of the Add block (right before the subsystem output port) from an addition to a subtraction. This will result in the lower sideband being generated. (Try it!)

We now digress momentarily and use this SSB upconverter example to illustrate why in general it is undesirable to simulate radar signals at RF. In this example, we are simulating a 10-GHz signal for a very short simulation time (10 ns). Using the MATLAB stopwatch utility functions `tic` and `toc`, we timed the simulation runtime at about 0.68 seconds on a standard Windows laptop. Suppose we wanted to simulate a full radar CPI; for example, assume 128 pulses at a 10-kHz PRF. The CPI would be 12.8-ms-long. Applying the same timescale (0.68s equals 10 ns), we find that it would take 10.07 days to generate RF data for a single CPI! Admittedly, the signals are highly oversampled (by a factor of 10), and the Spectrum Analyzer blocks add significant overhead; nevertheless, the point remains that unless there is a compelling reason to model signals at RF, it is better to do it at baseband instead.

10.3 Amplifiers

The purpose of an amplifier is to scale the input signal amplitude by some fixed amount. An ideal amplifier will perform this function regardless of the input power, and without adding any distortion to the signal. Departures from this ideal behavior require additional effort in the development of the model, which goes back to the fundamental question of when is a model good enough. Figure 10.5 shows the block diagram and results from running the Simulink example amplifiers. It implements three different amplifier models of increasing fidelity at the expense of increasing complexity. The input to each model is a linear ramp to exercise the behavior of the amplifiers as the input signal level changes.

Due to the linear ramp, the time-axis on the scope window may be interpreted as the input-signal level, thereby turning Figure 10.5(b) plots into input-output characteristics for each amplifier model. In Figure 10.5(a), the ideal amplifier (top subsystem) output is displayed as the black line in the scope window, showing linear scaling of the input regardless of input signal level. In reality, however, amplifiers do saturate at some maximum input power. To capture this effect, the amp model with hard saturation (middle subsystem in Figure 10.5(a)) adds a limiter;

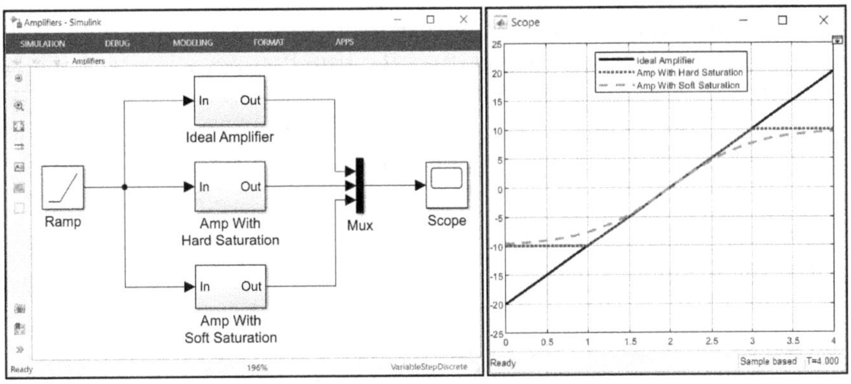

(a) (b)

Figure 10.5 Comparison of amplifier models: (a) Simulink block diagram, and (b) scope output.

this is the dotted curve, where the output is kept at some fixed value when the magnitude of the input exceeds a certain value. Although this is a better model, it is still not too realistic, as amplifiers don't exhibit the discontinuities at the hard saturation points; rather, they tend to gracefully approximate the saturation levels. The last amp model with a soft saturation (bottom subsystem in Figure 10.5(a)) shows this behavior; in this case, an Interpreted MATLAB Function block was used to implement the sigmoidal function displayed as a dashed curve, which is more representative of a real amplifier behavior. All three subsystems are shown in Figure 10.6. The specific sigmoidal function implemented as the soft saturation is displayed in the subsystem annotation.

This is a somewhat simplistic illustration of the concept of incremental model upgrades, but it bears repeating. The ideal amp model is trivial to implement and may be sufficient for most cases of interest. The amp with hard sat just requires one additional block, so it's a relatively simple upgrade that captures an important amplifier behavior at a first order. A more sophisticated and more realistic model such as the sigmoidal does require a bit more work, not only on selecting the form of the function but also on the tuning of the function parameters to obtain the right input/output behavior. Yet another level of increased complexity and fidelity

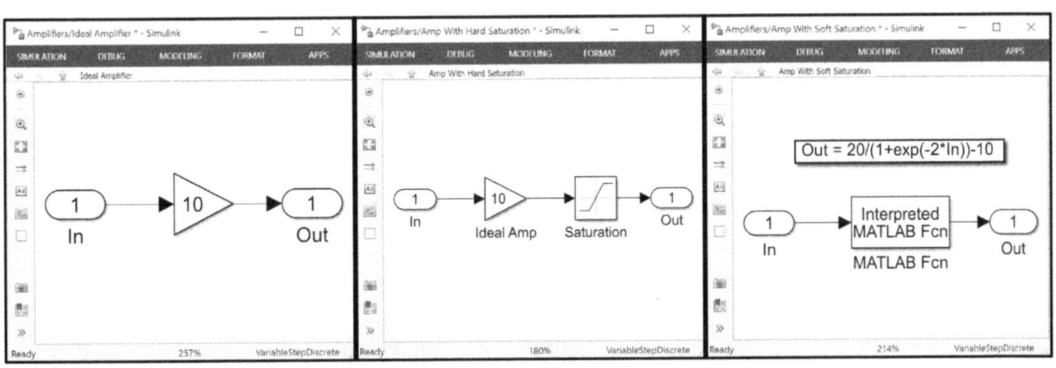

Figure 10.6 Amplifier models: ideal (left); hard saturation (middle); and soft saturation (right).

(not covered here) would be the use of measured data to obtain a suitable curve fit. MATLAB, for example, features the function `polyfit`, which takes a series of points and calculates the coefficients of a polynomial of a user-defined order that approximates the data in the least-squares sense. The user could then use `polyval` to evaluate the polynomial for an arbitrary input.

10.4 Oscillator Phase Noise

An ideal LO signal in the frequency domain would look like an impulse or delta function (i.e., all signal content would be at the LO frequency f_{LO}). Real devices, however, exhibit noise sidebands due to multiple contributors, such as random walk noise, flicker noise, and white phase noise [3]. The level of these sidebands is defined as a power density, in units of decibels relative to the carrier or decibels relative to carrier power or dBc (the carrier being the signal at f_{LO}) per unit frequency (i.e., dBc/Hz). This level typically decreases as the frequency offset from the carrier increases until it reaches some noise floor level, as depicted in Figure 10.7.

Phase-noise sidebands can be a limiting factor in radar-detection performance, in particular when the signal of interest is close in frequency to a much larger signal, such as clutter or jamming. The reason is that these noise sidebands will be replicated on each of the signals at the mixer output and will scale with their amplitudes. This effect is illustrated in Figure 10.8, where the small signal at f_2 is buried under the phase-noise sidebands of the large signal.

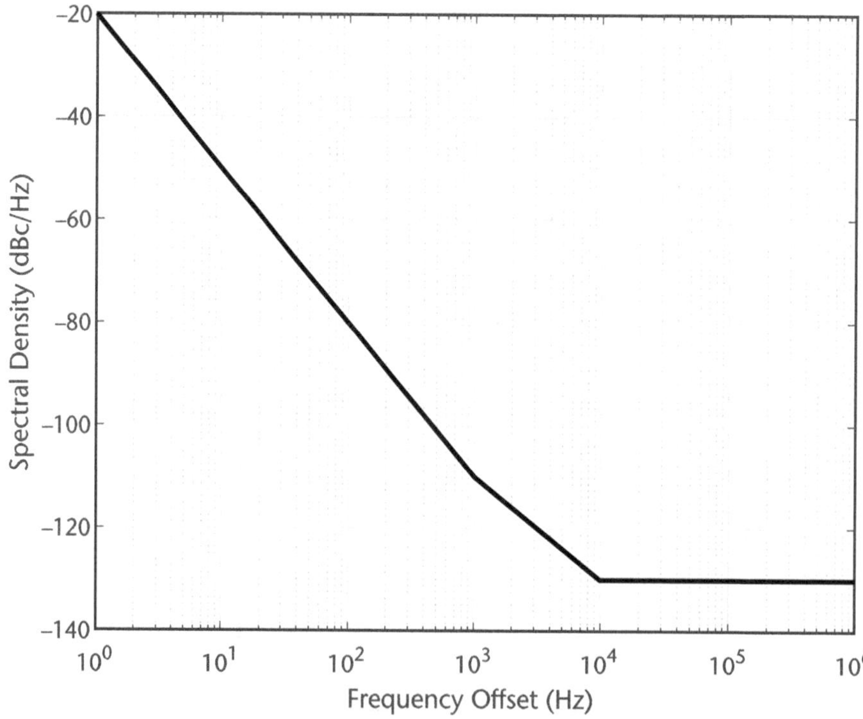

Figure 10.7 Typical phase-noise power spectrum.

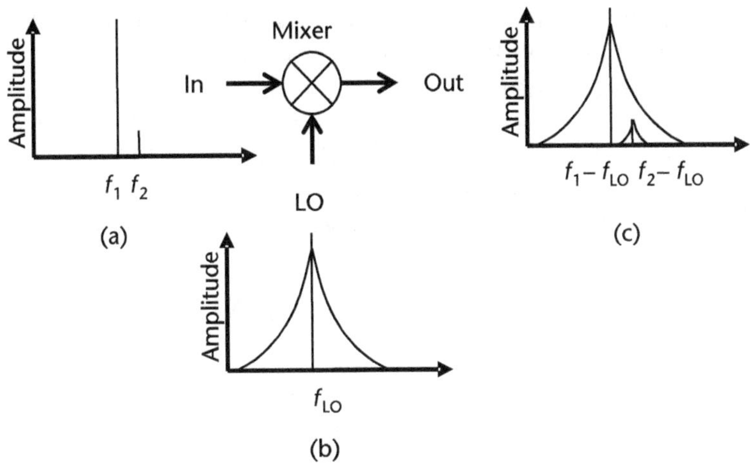

Figure 10.8 Phase-noise impact on target detection: (a) signals at mixer input; (b) local oscillator with phase noise; (c) LO phase noise on large signal masking the smaller amplitude signal. (After: http://www.circuitsage.com/pll/plldynamics.pdf.)

Vendor specifications will often come in the form of a table of single-sideband power-density values (in dBc/Hz) for several positive offset frequencies. A method for simulating phase noise that meets those specs is described as follows (and depicted in Figure 10.9).

1. Convert the power-density specs to an amplitude curve (piecewise linear in log-space). Extend the level of the maximum frequency offset (the phase noise floor) out to one-half of the assumed sampling frequency, $F_s/2$. This will be a vector of M samples, so the frequency increment between samples will be $\Delta F = F_s/(2M)$. If we denote the power spectrum vector in units of dBc/Hz as $PdB(f)$, the amplitude spectrum $A(f)$ is then obtained as

$$A(f) = \sqrt{(\Delta F)P(f)}, \text{ where } P(f) = 10^{(PdB(f)/10)} \tag{10.3}$$

   Equation (10.3) once again makes the common modeling assumption of unit impedance, such that voltage is just the square root of power.
2. Flip the one-sided spectrum and apply it to the interval $[F_s/2, F_s]$. This creates the lower sideband of the phase-noise amplitude spectrum, which

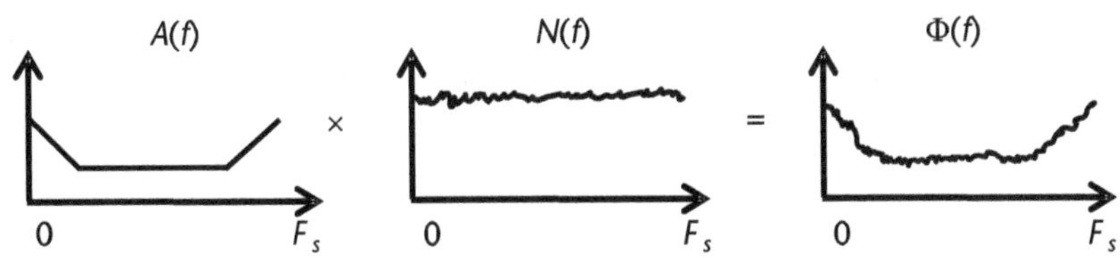

Figure 10.9 Phase-noise modeling: desired amplitude spectrum (left); white noise spectrum realization (middle); and phase noise spectrum realization from their product (right).

will be symmetric at DC. The full spectrum should look similar to the left side of Figure 10.9. The total vector length is $K = 2M$.

3. Create a realization of unit-power (unit-variance) complex white Gaussian noise vector $N(f)$ of the same length K as the phase-noise amplitude spectrum. This is shown in the middle of Figure 10.9.
4. Compute the sample-by-sample multiplication of the two vectors, resulting in the phase-noise spectrum vector $\Phi(f)$ as shown on the right of Figure 10.9.

$$\Phi(f) = A(f)N(f) \tag{10.4}$$

The last two steps are not depicted in Figure 10.9.

5. Perform an inverse K-point FFT to obtain the phase-noise process in the time domain:

$$\phi(t) = K * IFFT(\Phi(f)) \tag{10.5}$$

The scaling by K in (10.5) is intended to cancel the $1/K$ scaling inherent in the inverse FFT operation. This will preserve the desired phase noise level relative to the carrier power in the Fourier domain.

6. Last, the mixer-output function in the time domain including phase noise is given as

$$x(t) = s(t)e^{j\phi(t)} \tag{10.6}$$

where $s(t)$ is the ideal (noiseless) mixer output.

A MATLAB® model that implements these steps authored by Alex Bar-Guy [4] is available at the MATLAB Central File Exchange website. All models submitted to this website are freely available for download; it is also provided with this book's software, along with the MATLAB script *test_phase_noise*, which sets up the inputs and calls Bar-Guy's function *add_phase_noise*. Table 10.1 lists the phase-noise specifications.

The results of the script are displayed in Figure 10.10: When the input flag `VALIDATION_ON` is set to 1, the Bar-Guy function creates the plot in Figure 10.10(a), showing the nominal phase noise levels. When the flag is set to 0, the function

Table 10.1 Phase-Noise Specifications for the Model Example

Frequency Offset	*Phase Noise Power Density*
1.0 kHz	–84 dBc/Hz
10 kHz	–85 dBc/Hz
100 kHz	–88 dBc/Hz
1.0 MHz	–99 dBc/Hz
10 MHz	–112 dBc/Hz

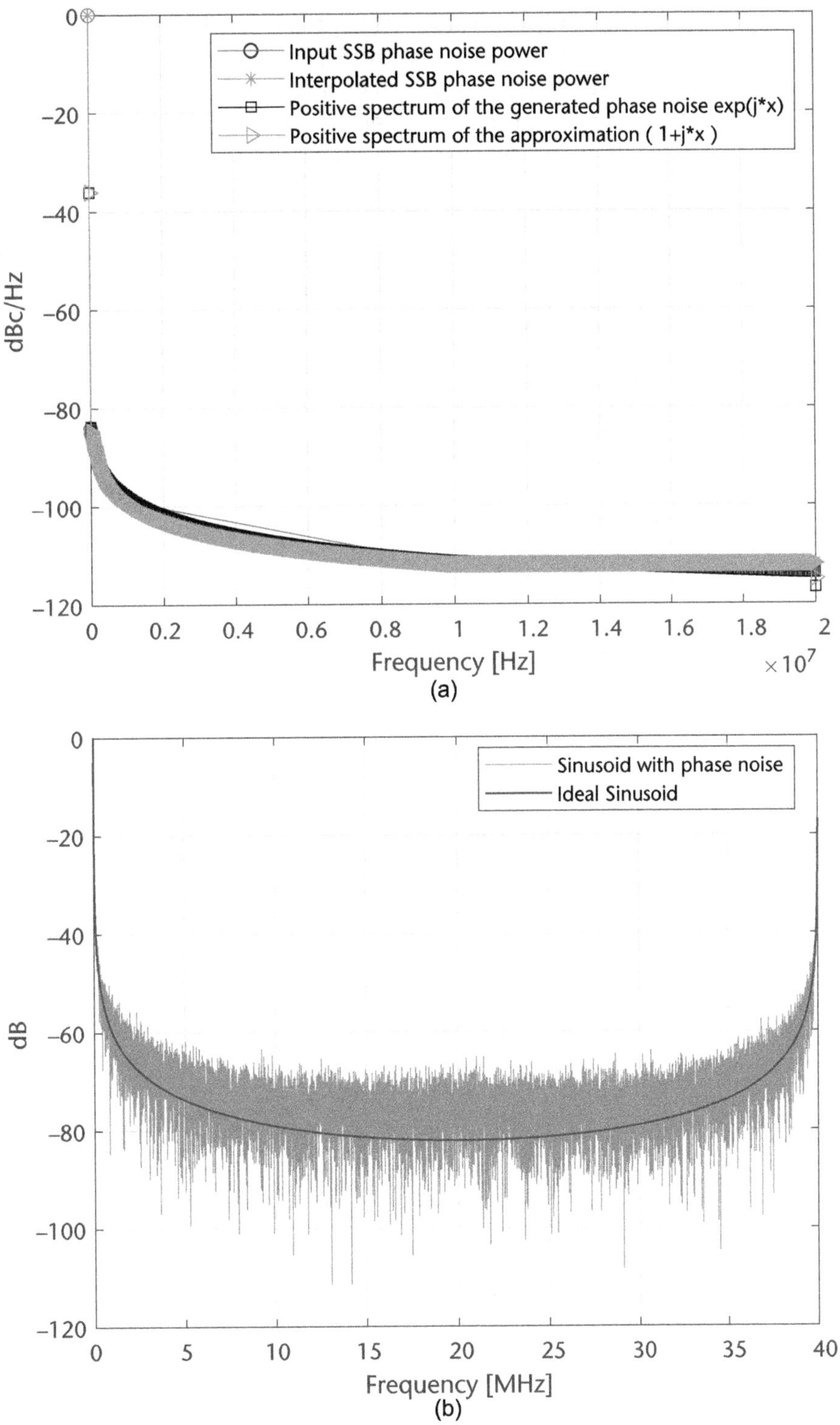

Figure 10.10 (a) MATLAB phase-noise model: specified power density levels, and (b) single phase-noise realization. (*Source:* [4].)

outputs the mixer output sinusoid corrupted with phase noise, and the script creates the plot in Figure 10.10(b), comparing it to the ideal sinusoid.

To the naked eye, it may seem that the phase-noise level in Figure 10.10(b) is much higher than the predicted levels in Figure 10.10(a). A couple observations are warranted; they are described as follows:

- The predicted levels are a power density (dBc/Hz), while the actual levels are in units of power (dBc). In the flatter portion of Figure 10.10(b) (between 10 and 30 MHz), the average phase noise power is –75 dBc. The spectrum is based on 10,000 samples, so the spacing between samples (i.e., the FFT filter bandwidth) is 4 kHz, or 36 dBHz. The spectral density is therefore –75 – 36 = –111 dBc/Hz, which compares favorably with the noise floor value of –112 dBc/Hz specified in Table 10.1.
- Near the carrier ($f < 0.5$ MHz), the phase-noise power is indeed higher than the specified curve. The reason is that this spectrum was generated with an FFT using uniform weights. The phase noise is therefore riding on top of the high sidelobes of the sinc function. If we apply a taper function such as a Chebyshev window, the phase noise will drop to the expected levels. This can be observed in Figure 10.11, where an 80-dB Chebyshev window was applied. At 0.1 MHz, for example, the phase noise spec calls for –88 dBc/Hz, which amounts to about –50 dBc for the 4 kHz FFT binwidth (including the spectral broadening due to the Chebyshev window).

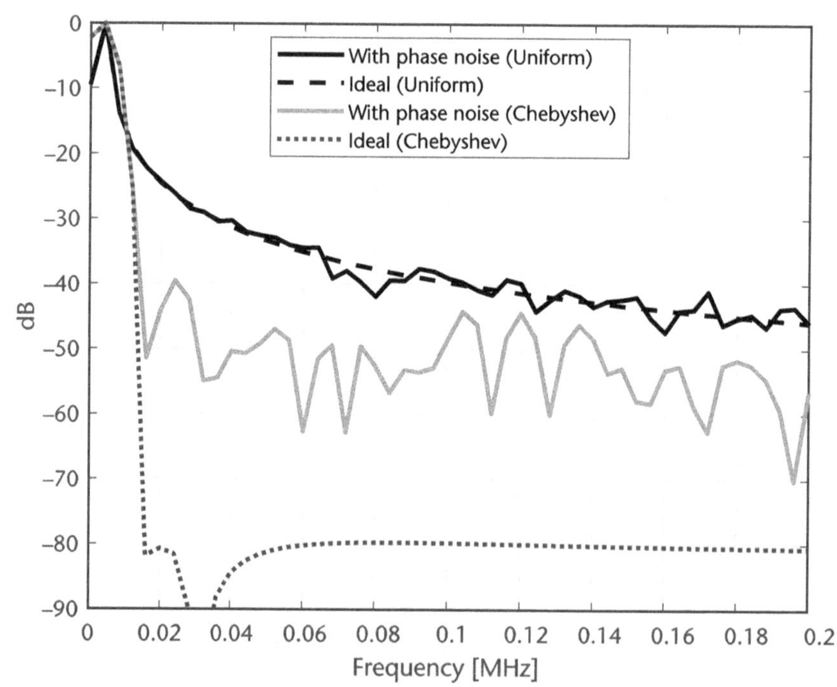

Figure 10.11 Effect of carrier sidelobe level in observed near-in phase noise power.

To conclude the topic, we acknowledge the existence of several empirical models for oscillator phase noise in the literature; the best known being Leeson's equation [5]. It models the oscillator as a linear time-invariant (LTI) system, which has some significant shortcomings. Alternate models have been developed to account for nonlinearities as well as for time-variant behavior; a good discussion and comparison of these models is found in [6].

10.5 I/Q Channel Mismatch

This section considers the formation of I/Q receiver channels, which are common in coherent pulsed systems. As seen in Figure 10.1, this is the result of splitting the IF signal and applying two mixers, with one of the LO signals having an additional $\pi/2$ phase shift relative to the other. This is equivalent to a complex mixing operation. For an input IF signal $s(t)$, the downconverted mixer output (baseband signal) $b(t)$ may be written as

$$b(t) = s(t)e^{-j\omega_{LO}t} = I(t) + jQ(t) \tag{10.7}$$

Thus, the I/Q signals are

$$I(t) = s(t)\cos(\omega_{LO}t) \tag{10.8a}$$

$$Q(t) = -s(t)\sin(\omega_{LO}t) \tag{10.8b}$$

where ω_{LO} is the LO frequency in radians per second. For the specific case where the IF signal is a sinusoid at an intermediate frequency ω_{IF}, (10.8a) and (10.8b) may be rewritten using trigonometric identities as

$$I(t) = \cos(\omega_{IF}t)\cos(\omega_{LO}t) \xrightarrow{LPF} 0.5\cos((\omega_{IF} - \omega_{LO})t) = 0.5\cos(\omega_{BB}t) \tag{10.9a}$$

$$Q(t) = -\cos(\omega_{IF}t)\sin(\omega_{LO}t) \xrightarrow{LPF} 0.5\sin((\omega_{IF} - \omega_{LO})t) = 0.5\sin(\omega_{BB}t) \tag{10.9b}$$

where the arrow implies low-pass filtering to reject the sum term, and $\omega_{BB} = \omega_{IF} - \omega_{LO}$ is the baseband frequency in radians per second. Notice that (10.9a) and (10.9b) have equal amplitudes and that the signals differ in phase by exactly $\pi/2$. Spectral analysis of these signals (taken as the real and imaginary parts of a complex baseband signal) will result in a single peak located at f_{BB}. These are still analog channels, however. They are fabricated to have equal gains and phases within the manufacturing tolerances, which means they will typically have some level of mismatch in one or both parameters. I/Q channel mismatch will manifest as a lower-level image signal located at the mirror frequency ($F_s - f_{BB}$). A general model that captures these imperfections is

$$I(t) = \alpha_I A\cos\left(\omega_{BB}t + \theta_I\right) + \beta_I \tag{10.10a}$$

$$Q(t) = \alpha_Q A\sin\left(\omega_{BB}t + \theta_Q\right) + \beta_Q \tag{10.10b}$$

where A is the nominal peak-signal amplitude and may include factors such as amplifier gains. The mismatch parameters are described in Table 10.2.

The DC bias voltages are undesirable offsets that may leak into the channel due to imbalances in the electronic components.

This model requires independent parameters to be set for each channel, and therefore has more knobs to set. A simplified model can be used by setting the I-channel as the reference and defining the gain and phase mismatches in the Q-channel relative to it. The revised model is

$$I(t) = A\cos\left(\omega_{BB}t\right) + \beta_I \tag{10.11a}$$

$$Q(t) = A(1+\varepsilon)\sin\left(\omega_{BB}t + \Delta\theta\right) + \beta_Q \tag{10.11b}$$

A MATLAB model of this channel mismatch is implemented in the script *test_gain_phase_mismatch*, which implements the independent parameters model represented by (10.10a) and (10.10b). An example result is shown in Figure 10.12 for a baseband signal at 10 MHz with a 1.0° phase mismatch, resulting in an image approximately 40-dB down from the desired signal. A second version of the script *test_gain_phase_mismatch2* implements the simplified model from (10.11a) and (10.11b), which provides identical results. These models illustrate the impact of channel imbalance in the dynamic range of a system. If the image signal is large enough to raise above the receiver noise floor (which has not been included in the model), it could be detectable and falsely identified as a target.

Table 10.2 I/Q Channel Mismatch Parameters

Parameters	*Description (units)*	*Ideal Values*
$(\alpha_I \quad \alpha_Q)$	Gain scaling factors (unitless)	(1.0, 1.0)
$(\theta_I \quad \theta_Q)$	Phase offsets (radians)	(0.0, 0.0)
$(\beta_I \quad \beta_Q)$	DC bias voltages (volts)	(0.0, 0.0)
$\varepsilon = \dfrac{\alpha_Q}{\alpha_I} - 1$	Relative gain mismatch (unitless)	0.0
$\Delta\theta = \theta - \theta_I$	Relative phase mismatch (radians)	0.0

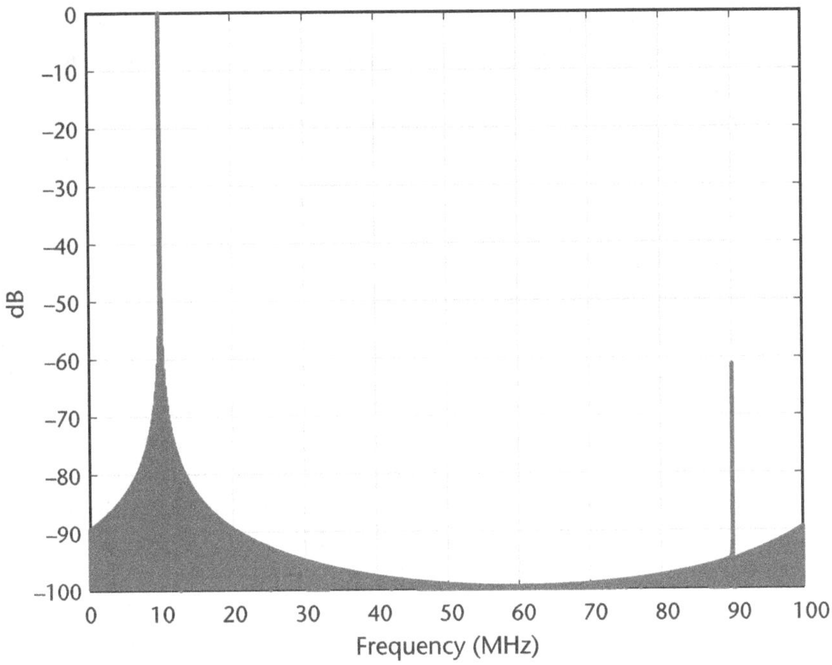

Figure 10.12 MATLAB I/Q channel mismatch model: example result showing a −41 dB image due to a 1° phase mismatch.

10.6 Filtering

Filters perform frequency selection, by accepting (passing) or rejecting (stopping) signals based on their spectral content. Stopping really means applying a very low gain, such that the output can be considered negligible relative to the signals of interest. We categorize filters based on their frequency response using terms such as low-pass, high-pass, bandpass, and band-stop. Any of these may be implemented in the analog or the digital domain [7]. A full treatise of filter theory and design is outside the scope of this book, but we will mention the transfer function $H(\cdot)$ as the fundamental mathematical expression that defines the filter behavior. It relates the filter output to its input and is commonly expressed via the Laplace transform $H(s)$ in the complex frequency domain for analog filters, or via the Z-transform $H(z)$ for digital filters. The transfer function under either case can be expressed in the Fourier domain: $H(f)$ for continuous time or $H(\omega)$ for discrete ($\omega = 2\pi f/F_s$ in this context), in which case it is known as the frequency response.

Modeling filters pretty much consists of implementing the transfer function. This is arguably easier to perform in the frequency domain; a common approach is to obtain the frequency spectra of both the input signal and the filter (i.e., the frequency response), multiply them together to obtain the output spectrum, and then convert it back to the time domain. The filter may also be implemented directly in the time domain by calculating its impulse response and convolving with the input signal, or (equivalently) by calculating the filter coefficients and implementing the filter difference equations.

Both MATLAB and Simulink offer a wide variety of functions and blocks to perform most common types of filters: Infinite impulse response (IIR) filters such as Butterworth, Chebyshev, and elliptical, as well as finite impulse response (FIR) digital filters. One key point in performing filtering operations in computer simulations is that even if the model is intended as an analog filter, the simulation deals with sampled data. Thus, it is important to pay attention to the relationship between the simulation sampling frequency and the filter frequency response, to avoid aliasing and other undesirable sampling phenomena. A couple of selected filtering examples are now presented.

The MATLAB script *butter_lpf_examp* illustrates some principles as well as the logistics of setting up the model. Figure 10.13 shows the basic structure of the model, which takes two sinusoids through a Butterworth LPF. Named after their creator, British engineer Stephen Butterworth, Butterworth filters have a maximally flat passband. The LPF is implemented using the function `butter`, which might be used to model either an analog or a digital filter. The script shows both for the purposes of illustrating the syntax differences of one versus the other, although only the digital version is used for the filtering exercise. The script also shows the use of functions `freqs` and `freqz` to generate the complex frequency response of the analog and digital filters, respectively. These are useful for visualization and verification, although they are not required for the filter implementation. The actual filtering is performed using the `filter` function which takes the coefficients calculated by `butter`, as well as the input signal vector.

The LPF is designed as a 20th-order filter with a cutoff frequency of 2 MHz and a sampling frequency of 25 MHz. The input to the filter is the addition of one sinusoid [Figure 10.13(a)] at a frequency of 1 MHz (within the filter's passband) and a second one [Figure 10.13(b)] at 3.5 MHz (outside the passband). The filter block [Figure 10.13(c)] shows the filter response, as well as the input and output spectra—the input with two peaks at 1.0 and 3.5 MHz, and the output with a single peak at 1.0 MHz. The 1.0 MHz output [Figure 10.13(d)] shows a delay in the output signal envelope. This is the so-called group delay τ_{grp}, which is defined as the negative of the slope of the filter phase function (i.e., the derivative of the filter phase with respect to the angular frequency [8]); thus,

$$\tau_{grp} = -\frac{d\Phi(\omega)}{d\omega} \tag{10.12}$$

For a FIR filter with linear phase, the group delay is a constant. This digital Butterworth filter, however, is of the IIR type, and there is some nonlinearity in the phase function (which is plotted by the script but not shown here). The script does calculate the group delay for the 1.0-MHz sinusoid, which is about 1.1 μs, or 28 samples. From the plot in Figure 10.13, 28 samples is where the output sinusoid reaches it first peak, so this is a good sanity check to verify correct operation.

A corresponding Simulink model *LPF_Example* is shown in Figure 10.14. The signal parameters are the same as the MATLAB version, as is the filter cutoff frequency. The Lowpass Filter block from the DSP System Toolbox was selected

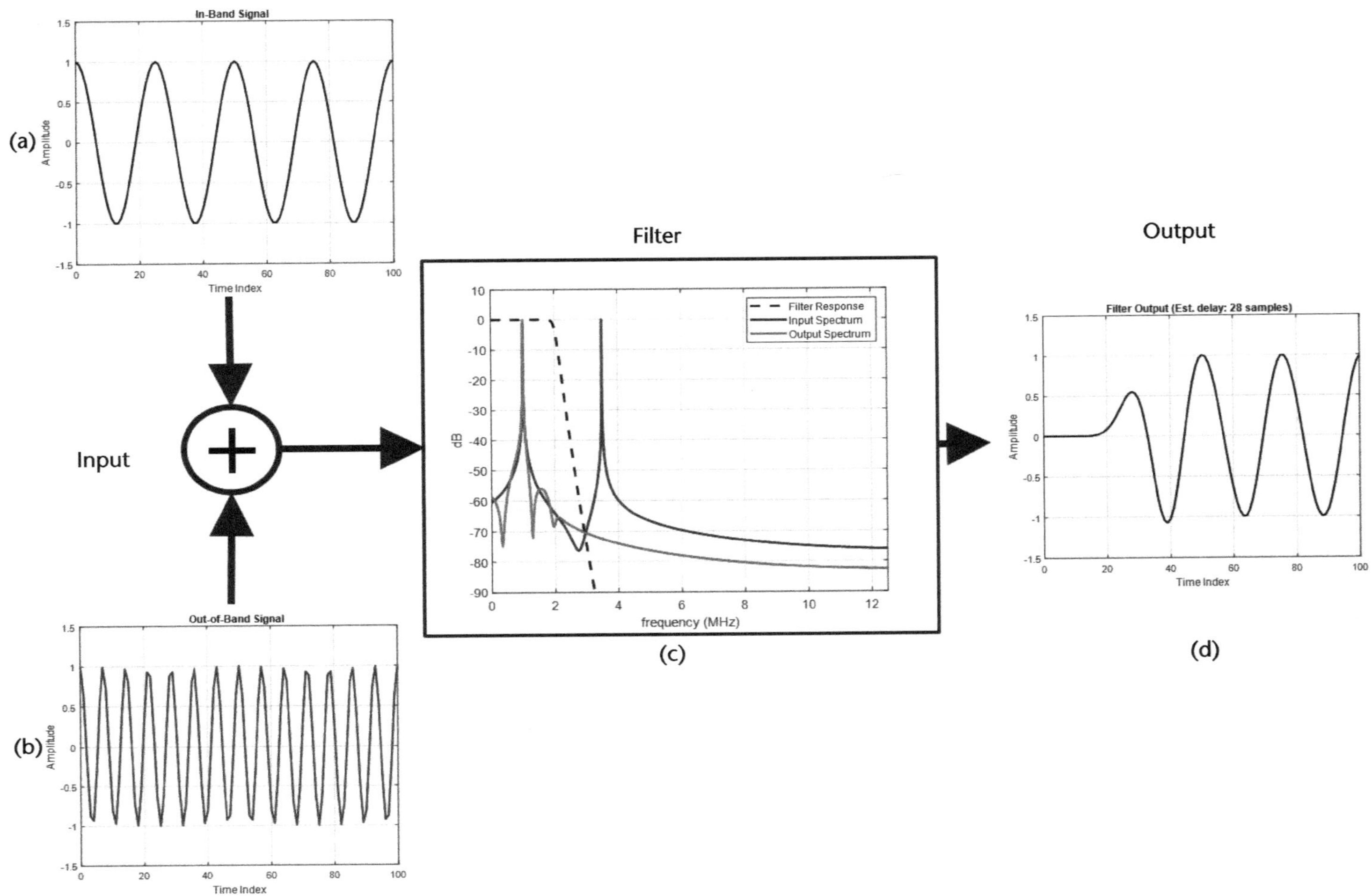

Figure 10.13 MATLAB LPF model: (a) in-band signal and (b) out-of-band signal are added and input to the (c) LPF with the frequency response shown (dashed line). The (d) filter output is the in-band signal with group delay.

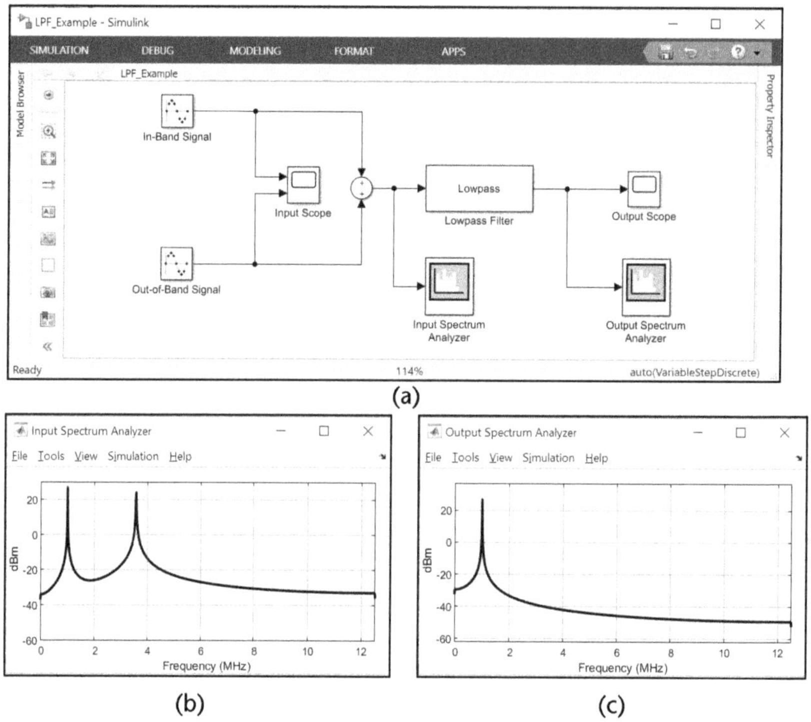

Figure 10.14 Simulink LPF model: (a) block diagram, (b) input signal spectrum, and (c) output signal spectrum.

for this implementation, among many possible choices, and set as a FIR filter (IIR was also an option).

As the spectral plots show, the filter effectively rejects the out-of-band signal. In reality, it is still present at the output but reduced by the stopband attenuation, which is another block parameter (set to 100 dB in this case). It is not noticeable in the output plot because of the high sidelobes of the 1.0-MHz signal—the spectrum analyzer did not use a window function. If we change the window option (by navigating to View → Spectrum Settings in the analyzer window toolbar) from Rectangular to Blackman-Harris, for example, we will be able to see it. (Try it!)

Last, there is always the option to develop a brute force filter model without resorting to library functions or blocks (or if a programming tool other than MATLAB is being used). Again, the key is to first characterize the frequency response of the desired filter. Many good textbooks on signal processing [7, 9] provide the necessary formulas for the desired filter types.

10.7 ADC

An ADC takes a continuous-time signal with continuous amplitude (typically a voltage), and discretizes it in both the time and amplitude dimensions. A perfect sample and hold is assumed here, where the step size of the hold is the sampling interval, or the inverse of the ADC sampling frequency F_S. The step size for the amplitude quantization is defined by two key ADC parameters: the peak-to-peak voltage V_{pp} and the number of bits Nbits. V_{pp} is the full-scale input range (FSR) of the ADC. Here we assume a symmetric FSR around 0, such that $V_p = V_{pp}/2$ is the magnitude of the peak input voltage; any voltage with a magnitude greater than V_p saturates the ADC, and the output magnitude is clipped to V_p. The total number of quantized voltages (ADC levels) is 2^{Nbits}. The quantization step size (also known as the voltage value of the least significant bit or vlsb) is

$$q = \frac{V_p}{2^{\text{Nbits}-1}} \tag{10.13}$$

The smallest (most negative) quantized output is usually mapped to the smallest input voltage $-V_p$, while the largest quantized output is mapped to $(V_p - q)$. As an example, for a 3-bit ADC with a peak voltage of 5V, there will be eight output levels with a vlsb of $q = 1.25$ V/count. The lowest output level is –5V, and the largest is +3.75V.

In addition, it is also desirable to have a 0V input voltage map to an output level of 0V. There are a couple of ways, however, in which this may be accomplished, depending on where the level transitions occur. The two common conventions are known as *mid-tread* and *mid-riser* [10]. As the name implies, in the mid-tread convention, the 0V input (the midpoint of the FSR), occurs on the tread of the ADC staircase function. In the second convention, it occurs on a staircase riser (i.e., at a transition point).

Table 10.3 shows the input and output voltages for the seven transitions corresponding to the 3-bit case with $V_p = 5$V. For the mid-tread ADC, the 0V input occurs in the tread between transitions 4 and 5, while for the mid-riser it occurs at the transition 4 riser. For the mid-riser ADC, the output voltages match the inputs at the transition points.

Figure 10.15 shows the input-output staircase function for the 3-bit ADCs under both conventions. They were created by running the MATLAB script *one_line_ADC_model*. As the name implies, the quantization of an input data vector is accomplished in one line of code, given by

```
output = min(max(input - mod(input,vlsb),-Vp ), Vp-vlsb );        (10.14)
```

As written, (10.14) implements the mid-riser convention without any modifications. For the mid-tread convention, the input vector in the equation needs to be shifted by adding one-half the vlsb to each of its values.

Table 10.3 Mid-Tread and Mid-Riser Input-Output for 3-Bit ADC

Transition index	1	2	3	4	5	6	7
Input (mid-tread)	–4.375	–3.125	–1.875	–0.625	0.625	1.875	3.125
Input (mid-riser)	–3.75	–2.50	–1.25	0	1.25	2.50	3.75
Output	–3.75	–2.50	–1.25	0	1.25	2.50	3.75

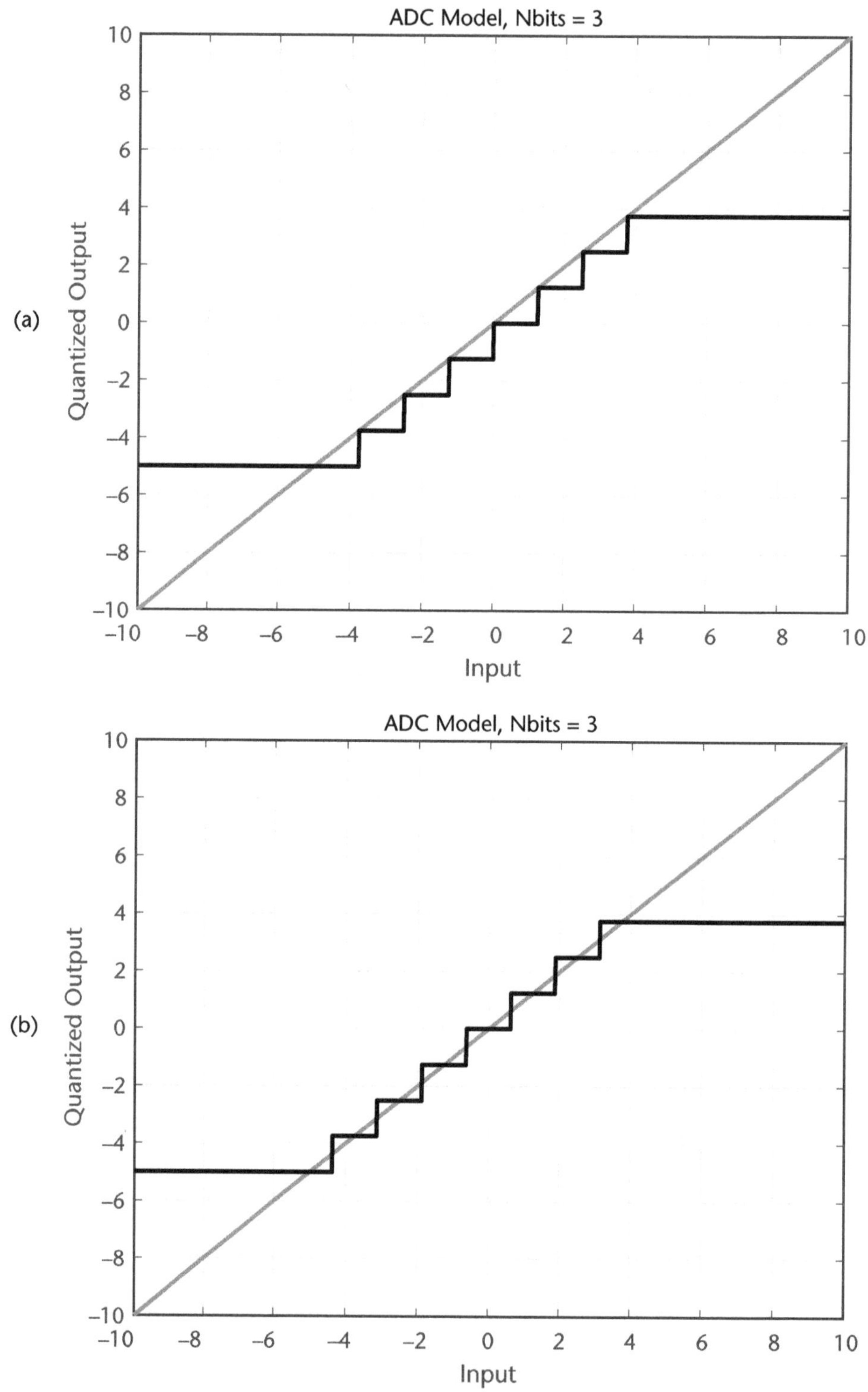

Figure 10.15 3-bit, ADC implementations in MATLAB: (a) mid-riser convention, and (b) mid-tread convention, with $V_p = 5$v in both examples.

There are several ways in which this operation may be implemented in Simulink, and there are existing quantizer blocks available in the DSP Systems Toolbox. Three options are compared in the Simulink model *ADC_Models*, which is displayed in Figure 10.16. In Figure 10.16, the top panel shows the overall block diagram, where a linear ramp is used to stimulate the quantizer subsystem, and the bottom panel shows the contents of the subsystem, with the three quantizer options.

The Uniform Encoder block implements a mid-riser ADC. The associated block parameters are V_p and Nbits; also, the block takes in voltage inputs and outputs actual counts, so if quantized volts are desired, an additional scaling is needed at the output. There is an output type parameter that must be set to signed integer to output negative values. The remaining options implement mid-tread ADCs. In addition to Nbits, the Idealized ADC Quantizer block takes separate parameters for minimum and maximum voltages (i.e., $-V_p$ and $+V_p$), so the FSR need not be symmetric. A box must be checked to output negative values. Although it is not obvious, the output is also in counts, so additional scaling is needed just like for the uniform encoder. Lastly, the Quantizer block has the fewest parameters to set, as it only requires a quantization interval q (i.e., the vlsb). The main drawback of this block is that it does not limit the output—the staircase goes on forever on both directions. To correct this, a Saturation block was added with limits set to $[-V_p, V_p - q]$. With these settings, both mid-tread options work identically.

Figure 10.17 shows two example results. The 3-bit ADC is shown on the left, which matches the MATLAB model results, and to appreciate the improvement

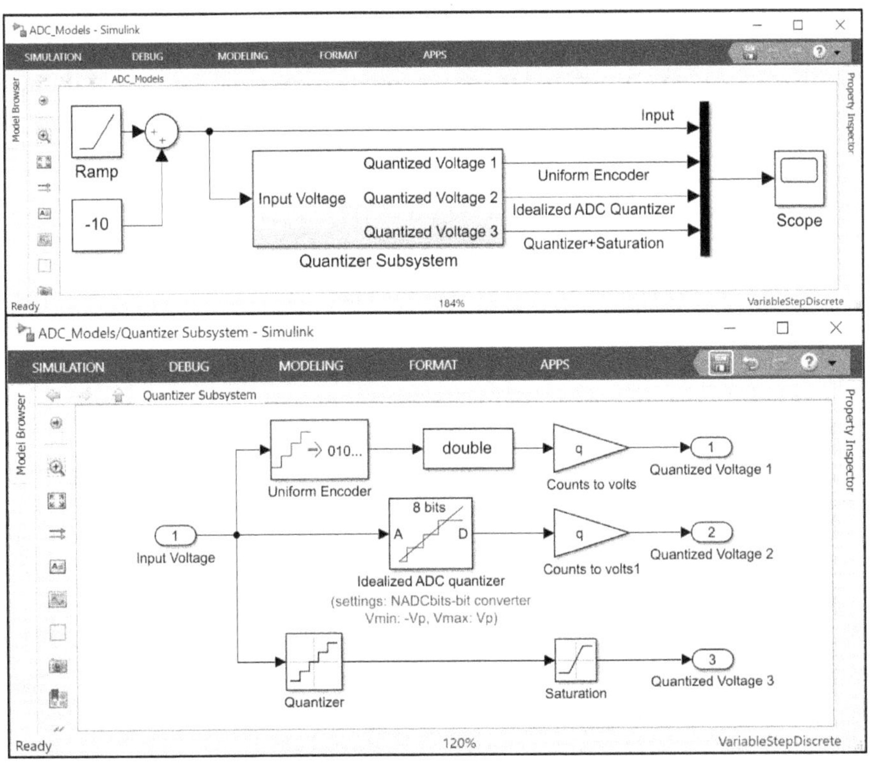

Figure 10.16 ADC implementations in Simulink: top-level block diagram (top) and quantizer subsystem options (bottom).

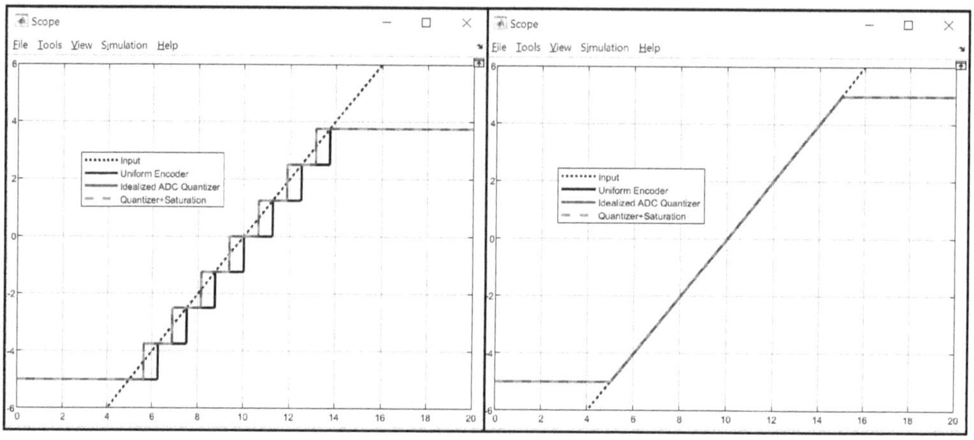

(a) (b)

Figure 10.17 Example Simulink model results: (a) 3-bit ADCs, and (b) 8-bit ADCs.

obtained from having more bits available, the right side shows the results when using 8 bits. More ADC bits mean less quantization noise and therefore a higher signal-to–quantization noise ratio (SQNR). The quantization noise power (variance) is [11]

$$\sigma_Q^2 = \frac{q^2}{12} \tag{10.15}$$

where the vlsb q is defined in (10.12). For $V_p = 5$V, this results in 0.1302 and 1.2716e-4 volts squared for 3 and 8 bits, respectively. The theoretical root-mean-square (rms) SQNR (rms-signal to rms-noise) in decibels is

$$SQNR = 6.02N_{\text{bits}} + 1.76 \tag{10.16}$$

which is a modest 19.82 dB for a 3-bit ADC, but a much more significant 49.92 dB for an 8-bit ADC. As seen in (10.16), each additional bit provides an extra 6 dB of SQNR. Many modern radars typically have 12 or 16 bits.

Related to ADC operation, it is common in radar systems to include some form of gain control to restrain the dynamic range (i.e., the size of input signals, so they may pass through the ADC within the FSR and avoid saturation). Examples of these gain control circuits include sensitivity time control (STC) and automatic gain control (AGC). Although we do not present any specific models of these circuits in this chapter, we do include a simplified AGC model as part of the engagement simulation in Chapter 15.

10.8 Concluding Remarks

Radar and EW simulations, especially at the engagement level, tend to focus on the baseband representation and processing of signals; there is therefore reduced emphasis in modeling analog components in the RF transmit and receive chain. There are also practical considerations and limitations regarding sampling frequencies, data record size, and simulation runtimes. In principle, however, any analog component can be modeled at the desired level of fidelity. Selected examples were shown to illustrate this point. In the engagement model showcased in Chapter 15, none of these effects have been incorporated, with the exception of ADC quantization. Adding most of these, however, would take relatively little work if the simulation objectives require their inclusion.

References

[1] Attia, J. O., *PSPICE and MATLAB for Electronics: An Integrated Approach*, Boca Raton, FL: CRC Press, 2002.

[2] Jorgesen, D., "IQ, Image Reject & Single Sideband Mixer Primer," online tutorial, Marki Microwave, Inc., 2018, https://www.markimicrowave.com/assets/appnotes/IQ_IR_SSB_Mixer_Primer.pdf.

[3] Stephens, D. R., *Phase-Locked Loops for Wireless Communications: Digital, Analog and Optical Implementations*, Second Edition, New York: Kluwer Academic Publishers, 2002, pp. 397–400.

[4] Bar-Guy, A., "Phase Noise," version 1.0.0.0, MATLAB Central File Exchange, https://www.mathworks.com/matlabcentral/fileexchange/8844-phase-noise.

[5] Leeson, D. B., "A Simple Model of Feedback Oscillator Noise Spectrum," *Proceedings of the IEEE*, Vol. 54, No. 2, 1966, pp. 329–332.

[6] Poddar, A. K., U. L. Rohde, and A. M. Apte, "How Low Can They Go?: Oscillator Phase Noise Model, Theoretical, Experimental Validation, and Phase Noise Measurements," *IEEE Microwave Magazine*, Vol. 14, No. 6, Sept.–Oct. 2013, pp. 50–72.

[7] Proakis, J. G., and D. G. Manolakis, *Introduction to Digital Signal Processing*, New York: Macmillan Publishing, 1988.

[8] Keel, B. M., "Fundamentals of Pulse Compression Waveforms," in *Principles of Modern Radar, Vol I: Basic Principles*, M. A. Richards, J. A. Scheer, and W. A. Holm (eds.), Raleigh, NC: Scitech Publishing, Inc., 2010, pp. 794–795.

[9] Oppenheim, A. V., and R. W. Schafer, *Discrete-Time Signal Processing*, Upper Saddle River, NJ: Prentice-Hall, 1989.

[10] "IEEE Standard for Terminology and Test Methods for Analog-to-Digital Converters," IEEE Std. 1241™-2010, IEEE Instrumentation & Measurement Society, January 14, 2011.

[11] Scheer, J. A., "A/D Conversion Effects," in *Coherent Radar Performance Estimation*, J. A. Scheer and J. L. Kurtz (eds.), Norwood, MA: Artech House, 1993, pp. 73–83.

CHAPTER 11

Target Detection

11.1 Introduction

Radar detection is the process of identifying the presence of a target return in the radar receiver data, which will also include unwanted signals such as receiver thermal noise, clutter returns, and potentially jamming (EA) and other EMI sources. This conglomerate of unwanted signals obstructing target detection is referred to as interference, although this chapter focuses on receiver noise. All of these can be considered stochastic processes—even the target returns, as the RCS may fluctuate randomly over time. Thus, a proper representation of their statistical behavior in the form of PDFs is key to assessing the radar performance in detecting signals of interest (targets) while avoiding falsely identifying interference signals as targets (false alarms).

Hypothesis testing is the branch of statistics that deals with determining which of a number of assumptions on the data may be true. In radar, there are two main assumptions or hypotheses: Either a target is absent (null hypothesis or H_0) or a target is present (alternate hypothesis or H_1). The assumption of either case being true leads to defining a PDF for the data under examination. Furthermore, the type of processing applied to the raw radar data samples up to the detection decision point also affects the statistics of the signals involved. Ultimately, the detection decision consists of comparing the signal level to some threshold value. If the signal exceeds the threshold, a detection is declared, which means the target present or H_1 hypothesis is considered to be true.

11.2 Data Processing and Detector Types

The threshold comparison operation can be performed in either the analog or the digital domain, although modern radar systems perform this function almost exclusively in the digital domain (i.e., by examining data samples of the receiver channel signals coming out of the ADC). Detection is hardly ever performed over a single ADC sample, however; as discussed in Chapter 2, target detectability improves with higher SNR, and integration of multiple samples increases the SNR when a target is present. Thus, multiple samples are collected over the radar observation time (dwell time), and these samples are combined (integrated) in a number of ways such that SNR and therefore target detection are maximized.

Figure 11.1 shows three block diagrams corresponding to common receiver/detector configurations. In all three, the signals received through the antenna travel

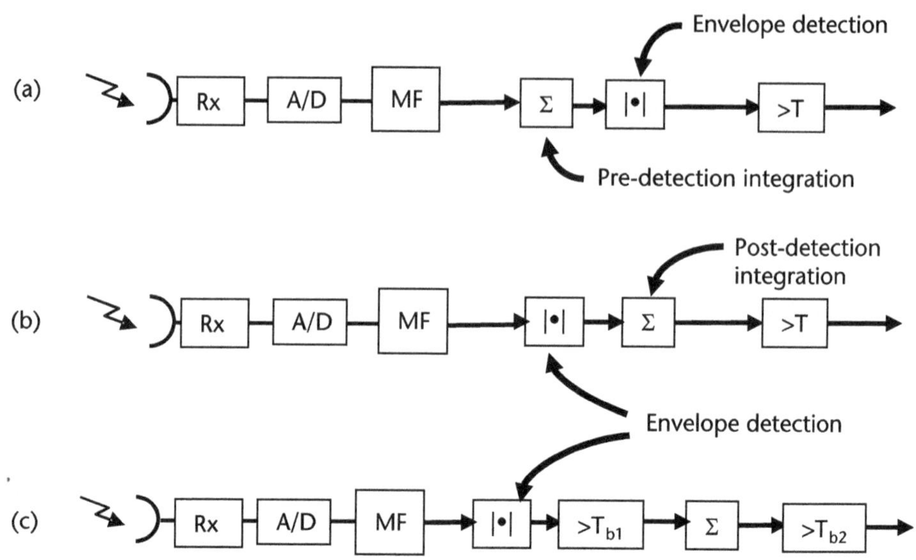

Figure 11.1 Common detector configurations: (a) coherent, (b) noncoherent, and (c) *M*-of-*N*.

through the analog receiver chain—depicted as the 'Rx' block in the diagram—get digitized through the A/D block and typically are fast-time processed through a matched filter (the 'MF' block).

In a coherent detector configuration, the integration block follows the matched filter, so the data samples are added in their complex form, thereby preserving their phase information. This coherent integration operation precedes the envelope-detection block, where the operator $|\cdot|$ denotes an absolute value operation. This operation extracts the instantaneous peak amplitude or magnitude (envelope) of the input signal while discarding its phase; if the input is $A\cos(\omega t + \theta)$, for example, the output will be A. This envelope or magnitude is then compared to the detection threshold for a target-present or -absent decision. The noncoherent detector, by contrast, performs envelope detection first before integrating. This is therefore a noncoherent, or postdetection integration (PDI). Last, the binary detector implements a two-level threshold comparison. The envelope-detected signal is tested against a threshold T_{b1}, which results in a binary outcome based on whether the threshold was exceeded (logical '1') or not (logical '0'). This is repeated multiple times, say N times, where each one of these first-level comparisons can be considered a trial. Integrating all the binary outcomes equates to counting all the first-level crossings or successes (i.e., counting all the 1s) and the summation is compared to a second threshold T_{b2}, which represents the minimum number of crossings, say M, required to declare a detection. The rule of requiring M successes out of N trials gives this detection approach its name.

These different detection approaches may all be combined in different ways. For example, the Doppler-processing FFT is an example of coherent integration over slow-time samples in a range bin, which generates a RDM as a result of a CPI. It is not uncommon to have multiple CPIs envelope-detected, then noncoherently integrated (PDI) on a cell-by-cell basis. Furthermore, the PDI'd RDM could then be fed to an M-of-N detector.

Finally, note that the envelope detection described herein provides a signal magnitude, strictly speaking. Practical implementations may often calculate the square of the magnitude instead. The latter is known as a "square-law" detector, as opposed to a linear (or linear-law) detector for the former. Both are detailed in the following sections.

11.3 Noise and Target Statistics

11.3.1 Noise Distributions

Recall that assessment of radar detection performance depends on the statistical behavior of the received signals, which is captured by their PDF. We start by considering thermal noise, which is always present in any real receiver. The standard assumption is that noise is $N(0,\sigma_G^2)$, meaning that it is Gaussian- or normal-distributed, with zero mean and variance σ_G^2. The subscript G is used here to distinguish the Gaussian variance from that of other distributions, introduced shortly. Assuming an I/Q receiver with matched channels, the noise processes x and y on each of the in-phase and quadrature channels are assumed to be i.i.d., meaning that they are each $N(0,\sigma_G^2)$. We can therefore define a *complex* noise process $n = x + jy$, which is complex-Gaussian, zero-mean with total variance equal to $2\sigma_G^2$.

Under these conditions, the type of envelope detector will determine the PDF of the noise going into the threshold comparison. If a square-law detector is used, the magnitude squared $z = |n|^2$ is exponential-distributed, while for a linear detector, the magnitude $r = |n|$ is Rayleigh-distributed instead. Furthermore, the pertinent parameters (mean and variance, or equivalently, standard deviation) for these distributions can all be defined in terms of the variance of the underlying Gaussian distribution. Figure 11.2 shows the PDFs and example plots for all three cases (assuming $\sigma_G = \sigma_E = \sigma_R = 1$). Notice that the exponential and Rayleigh PDFs are only defined for $x > 0$. Table 11.1 summarizes their relationships.

It is worth noting that the parameters σ_G and σ_E in the PDF definitions of Figure 11.2 do correspond to the standard deviations of the Gaussian and exponential PDFs, respectively. For the exponential PDF, the standard deviation is also

$$f(x) = \frac{1}{\sigma_G\sqrt{2\pi}} e^{-\frac{(x-\mu)^2}{2\sigma_G^2}} \qquad f(x) = \frac{1}{\sigma_E} e^{-\frac{x}{\sigma_E}} \qquad f(x) = \frac{x}{\sigma_R^2} e^{-\frac{x^2}{2\sigma_R^2}}$$

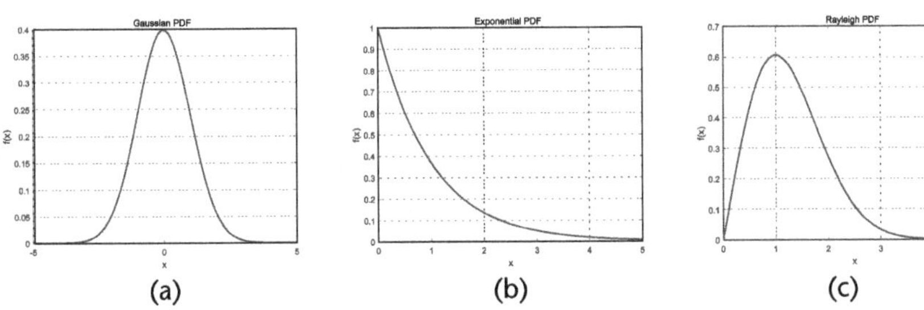

Figure 11.2 Common PDFs associated with radar receiver noise: Gaussian (a), exponential (b), and Rayleigh (c).

Table 11.1 Relationships Between Common Noise PDFs

Variable	*Physical Meaning*	*PDF*	*Mean*	*Standard Deviation*
$n = x + jy$	Complex receiver noise	Gaussian	$0 + j0$	$\sqrt{2}\sigma_G$
$z = \lvert n \rvert^2 = x^2 + y^2$	Instantaneous noise power (magnitude squared); output of square-law detector	Exponential	$2\sigma_G^2$	$2\sigma_G^2$
$r = \sqrt{z} = \lvert n \rvert = \sqrt{x^2 + y^2}$	Instantaneous noise envelope (magnitude); output of linear detector	Rayleigh	$\sqrt{\frac{\pi}{2}}\sigma_G$	$\sqrt{2 - \frac{\pi}{2}}\sigma_G$

the mean value (it is a one-parameter PDF), and in the context of receiver noise, it is numerically equivalent to the total power (total variance) of the underlying complex Gaussian process. For the Rayleigh distribution, on the other hand, σ_R is defined as a scaling parameter. It is not the standard deviation of the Rayleigh PDF, but rather a scaled version of it. It is the case, however, that this scaling parameter is the standard deviation of the underlying Gaussian PDF. We could therefore have replaced σ_R with σ_G in Figure 11.2, and in fact we have done that in the definitions of Rayleigh mean and standard deviation in Table 11.1. Thus, knowledge of the Gaussian variance is all the information needed to simulate all three noise processes.

11.3.2 Target Distributions

We now turn our attention to the statistics of the target return. The instantaneous amplitude of the target return A is (again, assuming unit impedance)

$$A = \sqrt{P_r} = \sqrt{\frac{P_t G_t G_r L_t L_r L_{\text{atmos}}^2 \lambda^2 \sigma G_{sp}}{(4\pi)^3 R^4}} \tag{11.1}$$

where P_r is the received power originally defined in (2.5) and repeated here for convenience. The only random term in (11.1) is the RCS value σ, so the statistical behavior of A will be dictated by that of σ (or rather, of $\sqrt{\sigma}$). The Swerling statistical models for RCSs are detailed in Chapter 5. Rather than a target-only PDF, it is more relevant to discuss the PDF of target plus noise (or target plus interference), as there is always interference present along with the target. As an illustration, let's consider a constant RCS target (Swerling 0) in thermal noise with a linear detector. The detector output will consist of Rayleigh-distributed noise plus a constant-magnitude value v due to the nonfluctuating target. The PDF for this type of signal corresponds to the Rician PDF, defined as

$$f(r) = \frac{r}{\sigma_R^2}\exp\left(\frac{-\left(r^2 + v^2\right)}{2\sigma_R^2}\right) I_0\left(\frac{rv}{\sigma_R^2}\right) \tag{11.2}$$

where $I_0(\cdot)$ is the zeroth-order modified Bessel function of the first kind. The Rician PDF is a generalized version of the Rayleigh PDF; setting $v = 0$ in (11.2) reverts back to the Rayleigh PDF from Figure 11.2. Thus, the scaling parameter σ_R in (11.2) is the same as the one for the Rayleigh PDF, which, again, is simply the standard deviation of the underlying Gaussian noise σ_G. The SNR is the ratio $v^2/(2\sigma_G^2)$.

Simulating a Rician-distributed signal is pretty intuitive if one considers the signal origin in the context of the I/Q receiver channels. Each channel would have zero-mean, equal-variance (i.i.d.) Gaussian interference, plus constant amplitude signals in quadrature with each other (i.e., the in-phase and quadrature target signals can be defined as

$$I_s = v\cos(\theta), \quad Q_s = v\sin(\theta) \tag{11.3}$$

for some arbitrary phase angle θ, which could be assumed to be uniformly distributed between 0 and 2π). If we denote the target-plus-interference signals as x_I and x_Q, then these are Gaussian-distributed random variables of equal-variance σ_G^2 and means I_s and Q_s, respectively. Notice that x_I and x_Q are not themselves identically distributed since their means are different. Under these conditions, the quantity

$$r = \sqrt{x_I^2 + x_Q^2} \tag{11.4}$$

is Rician-distributed as defined by (11.2). A three-step recipe for simulating Rician-distributed signal samples in MATLAB may be formulated as follows:

1. Generate independent $N(0,\sigma_G^2)$ samples y_I and y_Q using `randn`.
2. Let $x_I = y_I + v$, and $x_Q = y_Q$. This is a special case of (11.3) with $\theta = 0$.
3. Apply (11.4) to obtain Rician samples r.

Notice that the process described here is *not* the same as generating Rayleigh-distributed samples and adding the constant amplitude v to them, as one might be tempted to do. This would be incorrect, as it would only shift the mean value of the Rayleigh PDF but not affect its spread, which depends on the standard deviation. For both Rayleigh and Rician PDFs, the mean and standard deviation are related, and for the case of the Rician, they both depend on v.

A freely distributed MATLAB toolkit that follows the steps above was developed by Ridgway [1] and is available in the MATLAB Central File Exchange website. Example plots generated by this software package are shown in Figure 11.3.

In summary, for the case of a *linear detector* and a *nonfluctuating target* in Gaussian noise, the PDF under the H_0 hypothesis (target absent) is Rayleigh, while that of the H_1 hypothesis (target present) is Rician. Section 11.4 addresses how these PDFs are manipulated to define expressions for optimal detection, as well as performance-assessment equations in the form of probabilities of detection (P_d) and of false alarm (P_{fa}).

(a)

(b)

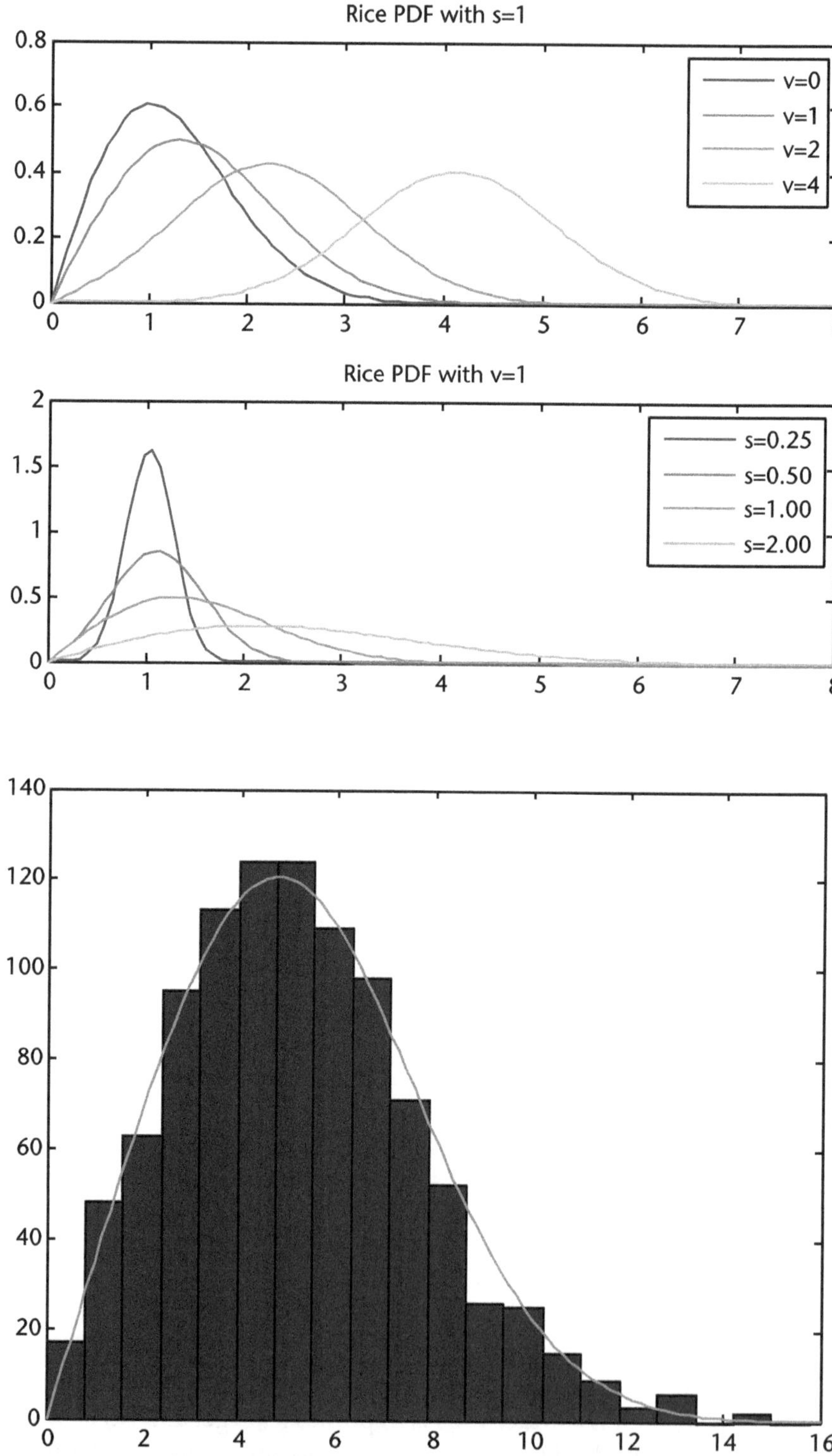

Figure 11.3 (a) Example Rician PDF plots for varying v and σ, and (b) histogram of a Rician noise realization compared to theoretical PDF. (Results obtained from [1].)

11.4 Detection Figures of Merit (P_d and P_{fa}) and the Likelihood Ratio

Figure 11.4 shows notional PDFs under the H_0 and H_1 hypotheses. For a linear detector, the interference-only PDF $f_i(x)$ is the Rayleigh PDF from the example in Section 11.3, while the target-plus-interference PDF $f_{t+i}(x)$ corresponds to the Rician PDF for nonfluctuating RCS.

A detection threshold T is set to some value x_0, and signal samples that exceed it (i.e., that lie to the right of the threshold line in Figure 11.4) are declared as detected targets. We can now formally define the main figures of merit in the detection process as follows:

1. The probability of detection (P_d) is the probability that the signal sample exceeds the detection threshold, given that a target is present. This may be written as $P(\{x > x_0\})|H_1)$. The condition of target present (H_1 is true) implies the sample x has been taken from the target-plus-interference population, whose PDF is $f_{t+i}(x)$. The subset of this population with values that exceed the threshold corresponds to the area under the curve of $f_{t+i}(x)$ to the right of x_0. This is the dark shaded area in Figure 11.4 and is calculated as

$$P_d = \int_{x_0}^{\infty} f_{t+i}(x)\,dx \tag{11.5}$$

2. The probability of false alarm (P_{fa}) is the probability that the signal sample exceeds the detection threshold, given that a target is not present, or $P(\{x > x_0\})|H_0)$. By a similar argument as before, we calculate this quantity as

$$P_{fa} = \int_{x_0}^{\infty} f_i(x)\,dx \tag{11.6}$$

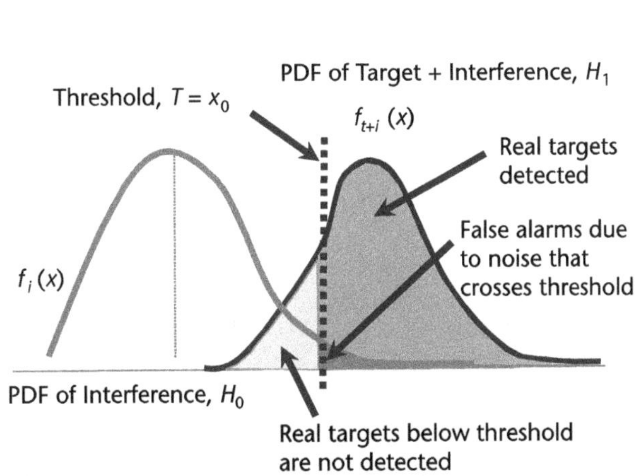

Figure 11.4 Notional H_0 and H_1 PDFs and detection threshold T.

3. For completeness, the probability of a miss (P_m) is the probability that the signal sample does not exceed the detection threshold, given that a target is present, or $P(\{x < x_0\})|\mathrm{H}_1)$. Its corresponding integral is

$$P_m = \int_{-\infty}^{x_0} f_{t+i}(x)\,dx = 1 - P_d \tag{11.7}$$

From Figure 11.4 and the associated discussion, it should be clear that a detection threshold cannot be selected to simultaneously maximize P_d while minimizing P_{fa}. Sliding the threshold to the right in Figure 11.4 decreases both, while sliding it to the left increases both. The selection of threshold value is based on some prescribed optimality criterion. Several criteria are available in detection theory [2], which involves different ways of selecting the most likely hypothesis based on the observed signal sample. This leads to the formulation of the *likelihood function* or *likelihood ratio,* defined as

$$L(x) \triangleq \frac{f_{t+i}(x)}{f_i(x)} \tag{11.8}$$

This is the ratio of the target-plus-interference PDF over the interference PDF. More general terminology would use H_1 and H_0 instead. In its most general form, the threshold comparison test becomes

$$L(x) \underset{H_0}{\overset{H_1}{\gtrless}} T_L \tag{11.9}$$

where H_1 (target present) is declared true if the likelihood function evaluated at the sample value is greater than the threshold; H_0 is declared true otherwise. Equation (11.9) is called the likelihood ratio test (LRT). The observed sample x is also often referred to as the *detection* statistic.

Revisiting the previous example of a nonfluctuating target in Rayleigh-distributed noise, it is pretty straightforward to show that the LRT has the form

$$\log\left(I_0\left(\frac{v|x|}{\sigma_G^2}\right)\right) \underset{H_0}{\overset{H_1}{\gtrless}} \tilde{T} \tag{11.10}$$

where log(·) is the natural logarithm function. The following couple of observations are worth noting about (11.10):

- The tilde notation on the threshold value in (11.10) is intended to indicate that it is not the same as that of (11.9), as it has gone through a number of transformations.

- As a reminder that the observed sample x was defined as the output of a linear detector, the magnitude operator $|\cdot|$ has been added explicitly to (11.10).

Since both the natural log and modified Bessel function are monotonically increasing functions, removing them [by applying the inverse functions on both sides of (11.10)] does not change the result of the LRT. Furthermore, the scaling term (v/σ_G^2) applied to $|x|$ may also be moved to the opposite side of the inequalities, such that a final, and much simpler form for the LRT is achieved:

$$|x| \underset{H_0}{\overset{H_1}{\gtrless}} T \tag{11.11}$$

This is a pleasing result, as it matches the coherent detector diagram at the top of Figure 11.1. The missing piece is the actual calculation of the threshold T. Once again, an optimality criterion must be chosen. In radar systems, the most common option is arguably the Neyman-Pearson (NP) criterion, developed by the Polish mathematician Jerzy Neyman and the British statistician Egon Pearson. The objective of the NP criterion is to maximize P_d for a specified P_{fa}. The detection threshold $T = x_0$ is therefore the value that satisfies (11.6) once P_{fa} is defined.

Going back to our linear detector example, where the interference PDF is Rayleigh, the integral in (11.6) is evaluated as

$$P_{fa} = \int_T^{\infty} \frac{x}{\sigma_G^2} \exp\left(\frac{-x^2}{2\sigma_G^2}\right) dx = \exp\left(\frac{-T^2}{2\sigma_G^2}\right) \tag{11.12}$$

The exponential expression on the RHS of (11.12) may now be solved for T, yielding

$$T = \sqrt{-2\sigma_G^2 \log\left(P_{fa}\right)} \tag{11.13}$$

As a second example, consider a square-law detector instead, where the interference is exponentially distributed as in Figure 11.2. The P_{fa} integral now becomes

$$P_{fa} = \int_{T_E}^{\infty} \frac{1}{\sigma_E} \exp\left(\frac{-x}{\sigma_E}\right) dx = \exp\left(\frac{-T_E}{\sigma_E}\right) \tag{11.14}$$

which results in the threshold

$$T_E = -\sigma_E \log\left(P_{fa}\right) = -2\sigma_G^2 \log\left(P_{fa}\right) \tag{11.15}$$

Notice that the threshold T_E for the quadratic detector in (11.15) is the square of the threshold T for the linear detector in (11.13). This is only true for the single observation case.

Back to the linear detector, given the threshold from (11.13), the probability of detection is obtained from (11.5), using the Rician PDF:

$$P_d = \int_T^{\infty} \frac{x}{\sigma_G^2} \exp\left(\frac{-\left(x^2 + v^2\right)}{2\sigma_G^2} \right) I_0\left(\frac{vx}{\sigma_G^2} \right) dx \tag{11.16}$$

This integral does not have a closed-form solution; however it can be written in terms of Marcum's Q function, defined as [3]

$$Q_M(\alpha, \beta) = \int_{\beta}^{\infty} t \exp\left(\frac{-\left(t^2 + \alpha^2\right)}{2} \right) I_0(\alpha t)\, dt \tag{11.17}$$

By making the substitutions $\alpha = v/\sigma_G$ and $t = x/\sigma_G$, we can rewrite (11.16) as

$$P_d = Q_M\left(\frac{v}{\sigma_G}, \frac{T}{\sigma_G} \right) \tag{11.18}$$

Last, from (11.15) and the definition for SNR as $v^2/(2\sigma_G^2)$, we can define P_d in terms of P_{fa} and SNR as

$$P_d = Q_M\left(\sqrt{2SNR}, \sqrt{-2\log\left(P_{fa}\right)} \right) \tag{11.19}$$

Marcum's Q-function is available in the MATLAB Signal Processing Toolbox. The function call equivalent to (11.19) in MATLAB syntax is

```
Pd = marcumq(sqrt(2*SNR),sqrt(-2*log(Pfa)));
```
(11.20)

In this context, the SNR term is understood as the SNR of the detection statistic (i.e., the single observation that is subject to the threshold test). It could correspond to a single signal sample but more commonly is the result of coherent integration of multiple samples, such as an FFT integrating over multiple pulses. In the case of M pulses integrated coherently, the SNR input to the function is $M * SNR_1$, where SNR_1 is the *single pulse* SNR.

11.5 Receiver-Operating Characteristic Curves

Formulas such as (11.19) allow for the creation of receiver-operating characteristic (ROC) curves, which show system performance by relating P_d to SNR and

P_{fa}. Several references [4–6] feature pages and pages of ROC plots created under different conditions. As an illustration, Figure 11.5 shows the ROC curves resulting from implementing (11.20) for four different P_{fa} values over a range of SNR values. As a good sanity check, we confirm that the SNR value for 90% Pd in the 10^{-6} curve is 13.2 dB.

Since SNR corresponds to a single observation in this formulation, we can refer to this as the $N = 1$ case.

11.6 Noncoherent Integration

We now consider performance when $N > 1$ observations are integrated noncoherently, as per the noncoherent detector configuration from Figure 11.1. The signal model is the same as in the $N = 1$ case, still assuming a nonfluctuating target—so the unknown signal amplitude v remains constant during the N observations. Likewise, the noise statistics (σ_G^2) are assumed to remain constant throughout.

The natural extension of (11.10) is therefore

$$\sum_{n=1}^{N} \log\left(I_0\left(\frac{v z_n}{\sigma_G^2} \right) \right) \underset{H_0}{\overset{H_1}{\gtrless}} T_N \tag{11.21}$$

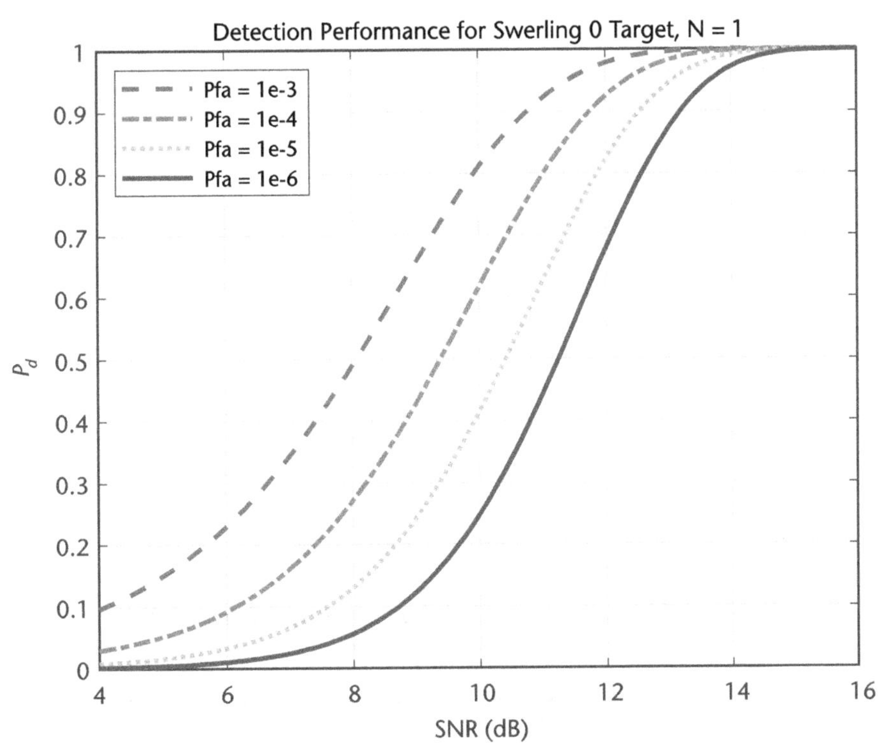

Figure 11.5 Example ROC curves for a nonfluctuating target (Swerling 0), single observation ($N = 1$).

where $z_n = |x_n|$ is the envelope of the nth sample. Equation (11.21) is hard to simplify because the scaling terms and nonlinearities are applied to each term in the summation and not to the summation itself. The approach instead is to use approximations to the nonlinearities found in this optimal detector form. These approximations give rise to the linear and square-law detectors.

For small arguments ($x << 1$), $\log(I_0(x)) \approx x^2/4$. Within a scaling constant, this is the square-law detector. On the other hand, for large arguments ($x >> 1$), $\log(I_0(x)) \approx x$, which is the linear detector. The two approximate implementations are illustrated in Figure 11.6.

Since signal levels are usually not known a priori, it begs the question of which one to use. Fortunately, their performance is nearly identical; DiFranco and Rubin [4] show that the SNR required to achieve a given P_d for one detector relative to the other is in the order of a couple tenths of a decibel, and it is in fact identical for $N = 1$. Thus, the choice is often a matter of convenience.

To illustrate the equivalency of the two detectors for $N = 1$ (at least from a false-alarm standpoint), the MATLAB script m-file `test_pfa_ray.m` creates a large (~100M) set of complex noise samples, calculates their magnitude and magnitude-squared, and compares them to their corresponding optimal thresholds (11.13) and (11.15). Table 11.2 shows the observed P_{fa} calculated from the simulation, which matches well with the desired P_{fa} for each case. It also shows that both detectors have identical performance. Since the initial random seed is not being captured, multiple runs of the script will yield different values, but will all be close to the desired result. (Try it!)

Analysis of the square-law detector is often more straightforward and more mathematically tractable, as the exponential PDF is typically easier to manipulate than the Rayleigh PDF. From a modeling standpoint, they are just as easy to implement. In real system hardware, however, square law might be easier to implement. In an I/Q receiver, each channel is magnitude-squared separately and then added together, which naturally results in the square-law detector. To implement a linear detector, an additional square root operation would need to be implemented. These options are illustrated in Figure 11.7.

In this I/Q configuration, which is pretty standard in coherent radar systems, the square-law detector is simpler to implement, as it avoids the square-root

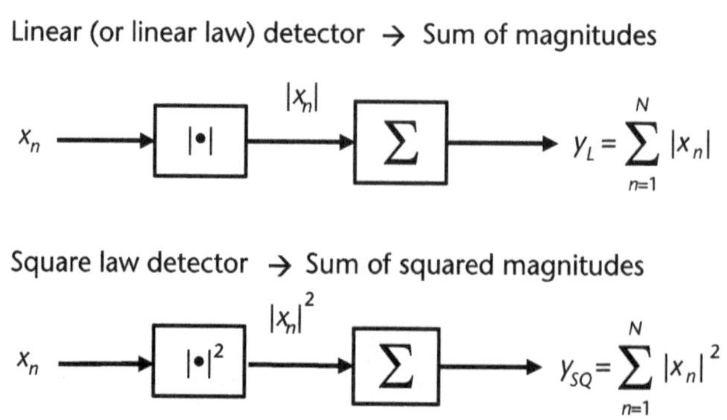

Figure 11.6 Approximations to optimum detector for noncoherent integration: linear detector (top) and square-law detector (bottom).

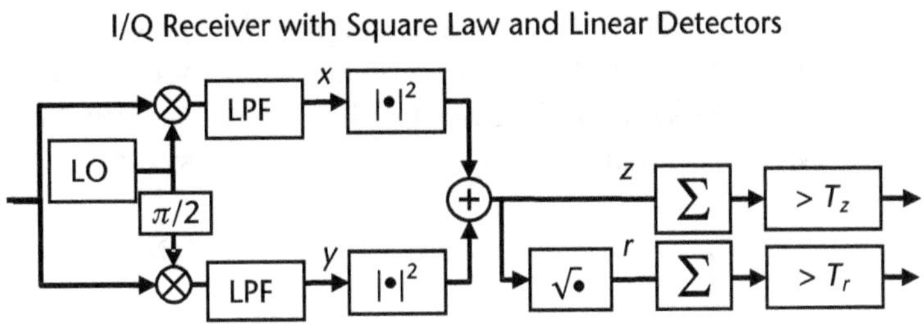

Figure 11.7 Square-law and linear detector implementations in an I/Q receiver. In the absence of a target return, the output sample z is exponentially distributed, while r is Rayleigh distributed.

operation for the linear detector. On the other hand, the square-law detector outputs tend to require more bits to represent than their linear counterparts. In other words, power values require more dynamic range than amplitudes.

The following development of performance calculations (P_d and P_{fa}) for the noncoherent detector assumes a square-law detector. In the interference-only case, the PDF of each sample is exponential-distributed. The sum of N such i.i.d. exponential random variables has an Erlang distribution,

$$f(x) = \frac{x^{N-1} \exp(-x/\sigma)}{\sigma^N (N-1)!} \tag{11.22}$$

The so-called rate parameter σ in (11.22) is the standard deviation (also the mean) of the underlying exponential random variables. From Table 11.1, recall that this is equal to $2\sigma_G^2$.

P_{fa} is found by integrating (11.22) as in (11.6), which results in

$$P_{fa} = 1 - G_L(T,N) \tag{11.23}$$

where $G_L(\cdot)$ is the lower incomplete gamma function, defined as

$$G_L(T,N) = \frac{1}{(N-1)!} \int_0^T t^{N-1} \exp(-t)\, dt \tag{11.24}$$

Table 11.2 Simulated P_{fa} Results for Linear and Square-Law Detectors Using `test_pfa_ray`

	Observed P_{fa}	
Desired P_{fa}	*Square-Law Detector*	*Linear Detector*
1e-3	1.0e-3	1.0e-3
1e-4	1.0410e-04	1.0410e-04
1e-5	1.0280e-05	1.0280e-05
1e-6	1.0700e-06	1.0700e-06

Comparing (11.24) to (11.22), we note the change of variable $t = x/\sigma$, which indicates that the threshold T is also normalized by σ. This form of $G_L(\cdot)$ is consistent with the MATLAB definition of its `gammainc` function, which enables a straightforward application of (11.23). In MATLAB syntax:

```
Pfa = 1-gammainc(t,N);   or   Pfa = gammainc(T,N,'upper');      (11.25)
```

Although there is no closed-form solution for T in terms of P_{fa} as was the case for $N = 1$, the formulation in (11.25) lends itself to a recursive solution for finding T given desired values for P_{fa} and N. One such solution is provided in the MATLAB function m-file `get_gammaincT.m`, based on code provided by Bill Melvin of GTRI. The function starts with an initial estimate and implements Newton's method (also called the Newton-Raphson method) to recursively arrive at a solution within a desired accuracy. As a sanity check, the case $N = 1$ should result in

$$T = -\log\left(P_{fa}\right) \tag{11.26}$$

which is the normalized version of (11.15).

Once T is found, we proceed to calculate P_d by following (11.5) and integrate the PDF of the sum of N i.i.d. samples under the H_1 hypothesis, which are Rician-distributed. The derivation of this result is outside the scope of this book, but several references such as DiFranco and Rubin [4] do an excellent job of walking through the derivation. As might be expected, the result also involves the Marcum-Q function, is presented here in MATLAB syntax for convenience:

```
Pd = marcumq(sqrt(2*N*SNR),sqrt(2*T),N);      (11.27)
```

Again, SNR is that of a single observation (considered constant through all N), and T is the solution to (11.25). This is a generalized form of (11.20), which is also valid for $N = 1$.

Figure 11.8 shows example ROC curves for several values of N. This plot was created by the script m-file `plot_swerling0_Pd_MultiObs_NCI.m`, which uses the aforementioned function `get_gammaincT` to find the threshold for each value of N.

The horizontal displacement from one curve to another for a given P_d represents the apparent SNR improvement obtained by noncoherent integration (NCI). This SNR improvement is often included in the RRE as a factor N^k, where $0.5 < k < 1$ ($k = 1$ for *coherent* integration). For example, at 90% P_d, the required SNR for $N = 8$ is 6 dB, which is 7.2 dB lower than the 13.2 dB required for $N = 1$. This results in $k = 0.7973$ for this example, since $10*\log_{10}(N^{0.7973}) = 7.2$ dB for $N = 8$.

Again, families of ROC curves such as those in Figure 11.8 have been computed for different combinations of N, P_{fa}, and target fluctuation models and are available from many sources [4–6]. Before the days of MATLAB, radar engineers and analysts would typically flip through these books and read off P_d values from the curves of interest, since the numerical solutions of these integrals were

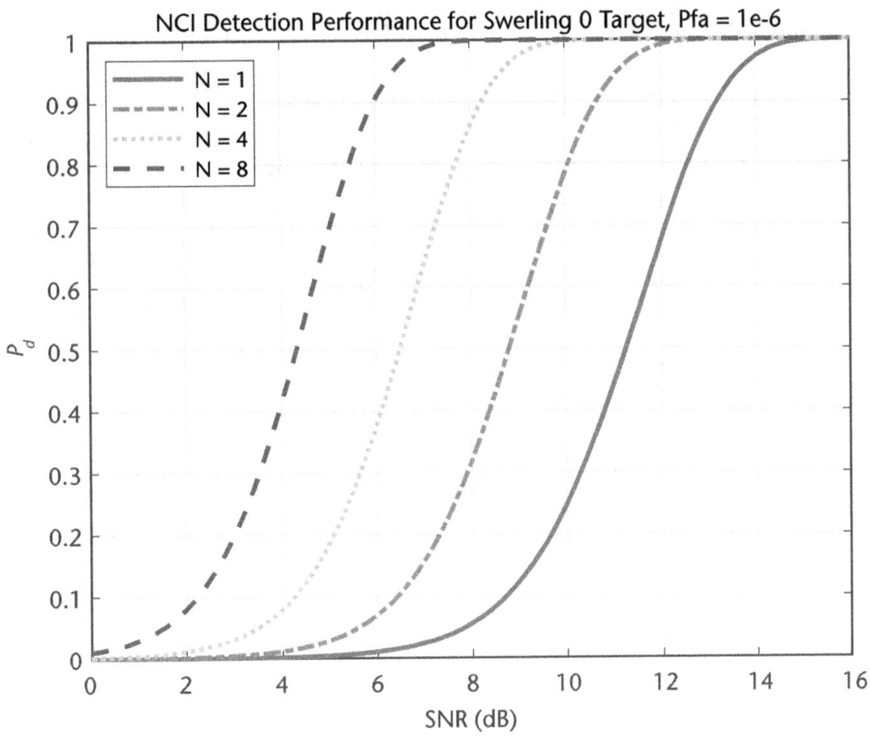

Figure 11.8 Example ROC curves for a nonfluctuating target (Swerling 0), multiple observations noncoherently integrated (NCI).

computationally costly. Over the years, several approximations to these calculations were developed, enabling quick generation of results that were close enough. One well-known example is Albersheim's equation [7], developed by the German-American engineer Walter Albersheim and published in 1981. It calculates the required SNR to meet desired values of P_d, P_{fa}, and N, assuming a Swerling 0 target RCS and a linear detector. It may be expressed as follows:

$$\left.\begin{aligned} A &= \ln\left(0.62/P_{fa}\right) \\ B &= \ln\left(P_d/\left(1-P_d\right)\right) \\ C &= 6.2 + \frac{4.54}{\sqrt{N+0.44}} \\ D &= A + 0.12 * A * B + 1.7 * B \\ snrdB &= -5 * \ln(N) + C * \ln(D) \end{aligned}\right\} \tag{11.28}$$

This formula was coded in the MATLAB function m-file `albersheim.m`. As an example, the calculated value for 90% P_d and 10^{-6} P_{fa} is 13.1145 dB, which is within 0.1 dB of the true value of 13.2 dB. Albersheim's equation may also be modified to solve for Pd instead, given SNR, P_{fa}, and N.

11.7 Detection Performance for Fluctuating Targets

We now consider detection of targets with fluctuating RCS. As discussed in Chapter 5, the RCS for Swerling 1 and Swerling 2 targets is modeled as an exponential-distributed random variable. While the Swerling 1 RCS remains constant over multiple observations within a scan (slow fluctuation rate), the Swerling 2 RCS varies from observation to observation (fast fluctuation rate). The P_d derivation for these cases is rather involved and also outside the scope of this book. Readers may refer to [4] for the detailed derivation. We will simply show the results, which are

$$P_d = 1 - G_L(T, N-1) + \left(1 + \frac{1}{N * SNR}\right)^{N-1} \exp\left(\frac{-T}{(1 + N * SNR)}\right) G_L\left(\frac{T}{\left(1 + \frac{1}{N * SNR}\right)}, N-1\right) \tag{11.29}$$

for the Swerling 1 case, and

$$P_d = 1 - G_L\left(\frac{T}{1 + SNR}, N\right) \tag{11.30}$$

for the Swerling 2 case. The SNR in these cases is understood to be the average SNR per observation, since in general it fluctuates. These equations are valid for any $N \geq 1$. In fact, for the Swerling 1 case when $N = 1$, (11.29) simplifies to

$$P_d = \exp\left(\frac{-T}{1 + SNR}\right) \tag{11.31}$$

since $G_L(X, N-1) = 1$ for $N = 1$. Replacing (11.26) in (11.31) results in

$$P_d = \left(P_{fa}\right)^{\frac{1}{1+SNR}} \tag{11.32}$$

which is a compact and convenient form relating all three quantities of interest and lending itself to quick ROC calculations.

The corresponding equations for Swerling models 3 and 4 are considerably more complicated, as they stem from evaluating a more complex PDF (chi-squared with 4 DOFs). The results are

$$P_d \cong \left(1 + \frac{1}{N * SNR/2}\right)^{N-2} \left(1 + \frac{T}{(1 + N * SNR/2)} - \frac{N-2}{N * SNR/2}\right) \exp\left(\frac{-T}{(1 + N * SNR/2)}\right) \tag{11.33}$$

for Swerling 3, and

$$P_d = 1 - \frac{N!}{(1+SNR/2)^N} \sum_{k=0}^{N} \frac{(SNR/2)^k}{k!(N-k)!} G_L \left(\frac{T}{(1+SNR/2)\sqrt{N+k}}, N+k-1 \right) \tag{11.34}$$

for Swerling 4. Notice that (11.33) is given as an approximation for $N > 2$, but exact for $N = 1$ and $N = 2$.

Table 11.3 summarizes the P_d calculations for all five Swerling types but provides the equations in MATLAB syntax for easy modeling and generation of results.

We also point the reader to a GUI-driven MATLAB suite that generates ROC curves for all Swerling models. This software, developed by Douglas Dougherty [8], is based on the Meyer and Mayer derivations of P_d formulas, which were verified to match the DiFranco and Rubin versions from Table 11.3. It is included in this book's software and is also available for download at the MATLAB File Exchange web site. The GUI command is `meyer`, and upon execution it calls the main executive function `meyerfun`. The script m-file `plot_Pd` (not part of the original suite) was developed to create ROC curves using `meyerfun` directly without going through the GUI. Example results are shown in Figure 11.9.

Table 11.3 MATLAB Formulas for P_d and P_{fa} of Swerling Targets (Based on [4])

Swerling Model	*Probability of Detection/Probability of False Alarm*	*Remarks*
0	`Pd = marcumq(sqrt(2*N*SNR),sqrt(2*T),N);`	Requires Signal Processing Toolbox

1	`temp = 1 + 1./(N*SNR);`
	`Pd = 1 - gammainc(T,N-1)+ temp.^(N-1) ...`
	`.*exp(-T./(1+N*SNR)).*gammainc(T./`
	`temp,N-1);`
		–SNR = average SNR per observation
		–T = normalized detection threshold
		–N = number of noncoherently integrated samples

2	`Pd = 1 - gammainc(T./(1+SNR),N);`	

3	`temp2 = N*SNR/2;`
	`Pd = (1 + 1/temp2)^(N-2)*(1 + T/`
	`(1+temp2)...`
	`- (N-2)/temp2)*exp(-T/(1+temp2));`	

4	`sumI = 0;`
	`for k = 0:N`
	`sumI = sumI + ...`
	`((SNR/2)^k)/`
	`(factorial(k)*factorial(N-k))...`
	`*gammainc(T/(1+SNR/2),N+k);`
	`end`
	`Pd = 1 - sumI*factorial(N)/(1+SNR/2)^N;`	

All	`Pfa = 1 - gammainc(T,N);`	

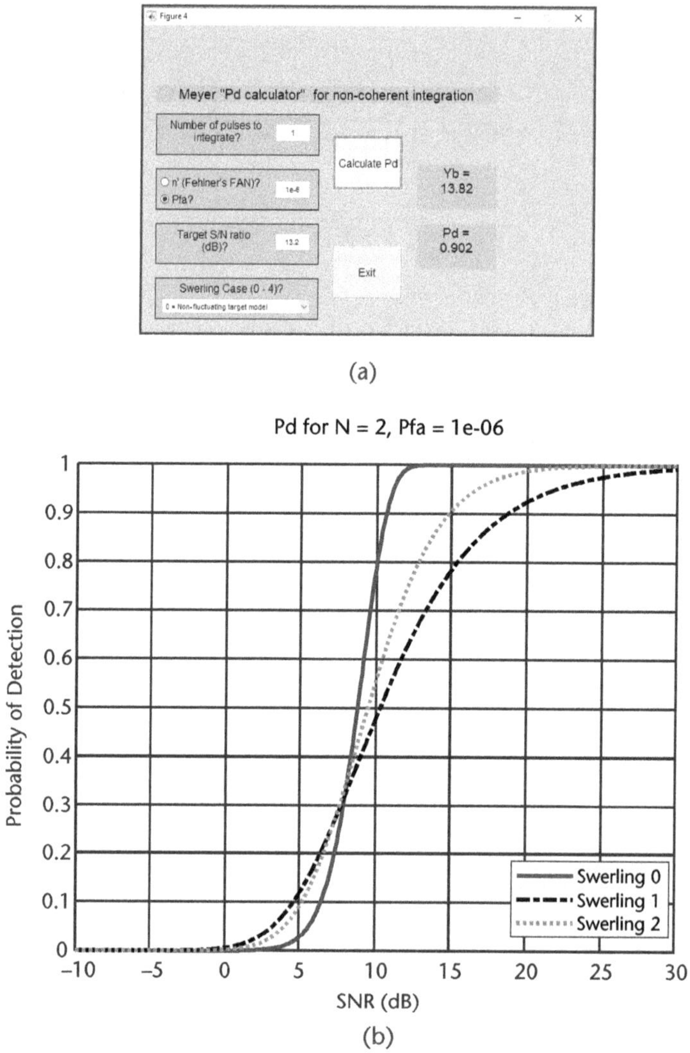

Figure 11.9 MATLAB freeware developed by Dougherty [8] for P_d calculations: (a) `meyer` GUI, and (b) example ROC curves using `plot _ Pd`, which calls `meyerfun`.

Last, similar to Albersheim's equation for Swerling 0 targets, there are empirical equations to approximate performance under fluctuating target conditions. The best known is arguably Shnidman's equation, developed by David Shnidman [9]. Just as Albersheim's, it calculates the SNR needed to meet required values of P_d, P_{fa}, and N but supports all four Swerling models in addition to Swerling 0. This equation was coded in the MATLAB function m-file `shnidman.m`. Figure 11.10 shows an example result along with the corresponding result for Albersheim's version.

11.8 Constant False-Alarm Rate Detectors

Implementation and analysis of the optimum NP detector and the corresponding linear and square-law approximations all depend on perfect knowledge of the

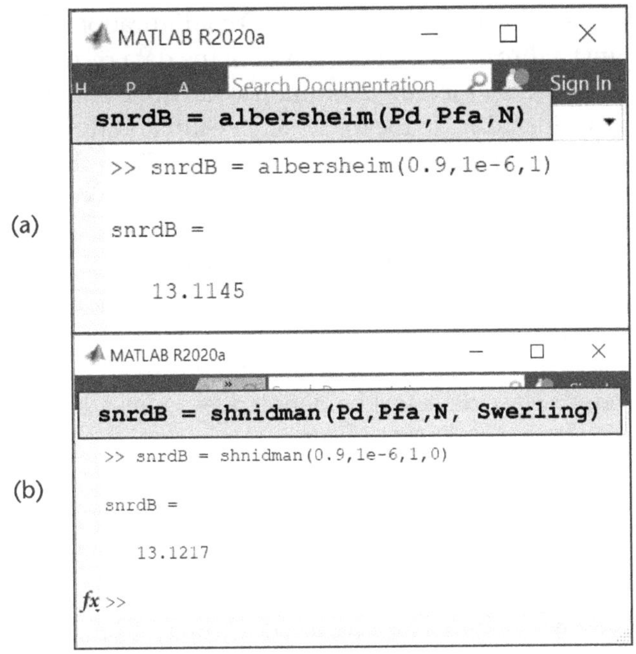

Figure 11.10 MATLAB implementation of Shnidman's equation: (a) Example Albersheim's calculation for Swerling 0, and (b) corresponding Schnidman's result for the same case.

interference statistics (such as the variance σ_G^2 for thermal noise and additional parameters for clutter and jamming) for the calculation of the detection threshold. In practice, these are never perfectly known, so the sensible approach is to estimate the interference level from the data itself. Once an estimate is obtained, the detection threshold is calculated as a function of it, in the spirit of the NP criterion. And just like the NP detector, the goal is to maintain a constant false-alarm rate (CFAR). This gives rise to the family of CFAR detection algorithms.

The most common variation of CFAR algorithms is the cell averaging-CFAR (CA-CFAR). As the name implies, the algorithm computes an interference estimate by averaging a number of cells or signal samples. These signal samples may be processed over a single dimension such as range or Doppler, or over a 2-D space such as the RDM obtained from pulse-Doppler signal processing. Figure 11.11

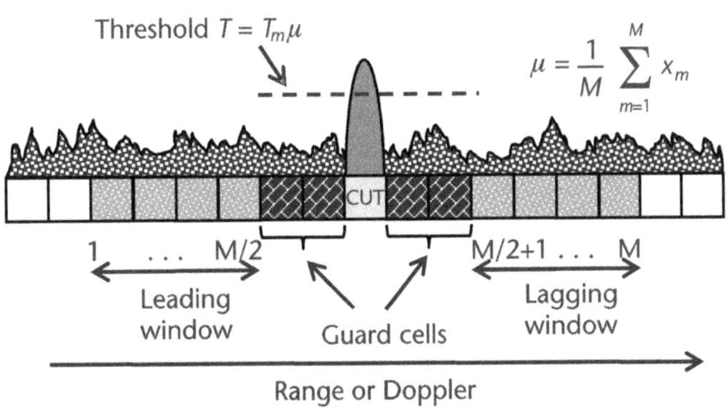

Figure 11.11 Depiction of the 1-D CA-CFAR algorithm.

illustrates the CA-CFAR method in a 1-D configuration. The signal cell to be examined against a threshold is at the center of the data record and is referred to as the cell-under-test (CUT). It is also common to have one or more guard cells around the CUT to avoid including in the interference estimate any possible target energy from the CUT that may spill into neighboring cells. Thus, the guard cells are excluded from the estimate. Finally, groups of cells beyond the guard cells on either side of the CUT are reference windows used to calculate an average interference level.

The window on the side that precedes the CUT is often referred to as the leading window, while the other is the lagging window. (Figure 11.11 assumes a right-to-left data flow.)

Let the signal values in the reference cells (which are outputs of either a linear or a square-law detector) be denoted as x_m, $m = 1$ to M. The CFAR detection threshold is then

$$T = T_m \mu, \quad \mu = \frac{1}{M}\sum_{m=1}^{M} x_m \tag{11.35}$$

where μ is the interference estimate resulting from the cell averaging, and T_m is a threshold multiplier (also called the CFAR constant) intended to meet the desired P_{fa}. For a square-law detector and $N = 1$, we readily see that T_m is (11.26), since multiplication by the estimated mean noise level would approximate the ideal NP threshold defined by (11.15). This threshold is used to test the CUT. At the next time step, all signal values shift one position, and the process repeats again for the new signal sample in the CUT. The reference windows along with the guard cells and the CUT thus may be seen as a sliding window across the received data.

If the interference estimate from the CA-CFAR is close to the real interference average, CFAR performance will approximate that of the NP detector. A second version of the `test_pfa_ray` script m-file can be used to illustrate this. This second version `test_pfa_ray_ca` does not assume knowledge of the noise statistics but estimates it by cell-averaging the samples. Because the averaging is done over a very large number of samples, the performance of this CA-CFAR example is near identical to the NP optimal detector. In practice, only a limited number of cells is used to form the reference window. This makes the CFAR detector less efficient than the NP detector. The detection threshold ends up being slightly higher than the corresponding NP threshold that meets the same desired P_{fa}; therefore the CFAR SNR (SNR_{CA}) required to achieve the same P_d is also slightly higher than the NP SNR (SNR_{NP}). This SNR increase is interpreted as a CFAR loss, and is typically included as one of the receive loss terms in the RRE loss budget. It is formally defined as [10]

$$L_{\text{CA-CFAR}} \triangleq \frac{SNR_{\text{CA}}}{SNR_{\text{NP}}} \tag{11.36}$$

Assuming a Swerling 1 or 2 target and a single observation ($N = 1$), the average P_d obtained with a square-law detector, followed by a CA-CFAR detector with a reference window size of M is [10]

$$P_d = \left(1 + \frac{T_m/M}{1 + SNR}\right)^{-M} \tag{11.37}$$

where T_m is the CFAR constant (not the full threshold). Setting SNR = 0 in (11.37) provides the equation for P_{fa}:

$$P_{fa} = \left(1 + \frac{T_m}{M}\right)^{-M} \tag{11.38}$$

which may now be solved for the threshold multiplier T_m, resulting in

$$T_M = M\left(P_{fa}^{-1/M} - 1\right) \tag{11.39}$$

The reader may contrast (11.39) with (11.26). A fair question to ask is how large must M be for the CA-CFAR threshold to be close enough to the NP threshold. Figure 11.12 compares the NP threshold multiplier from (11.26) to the CFAR constant as M increases.

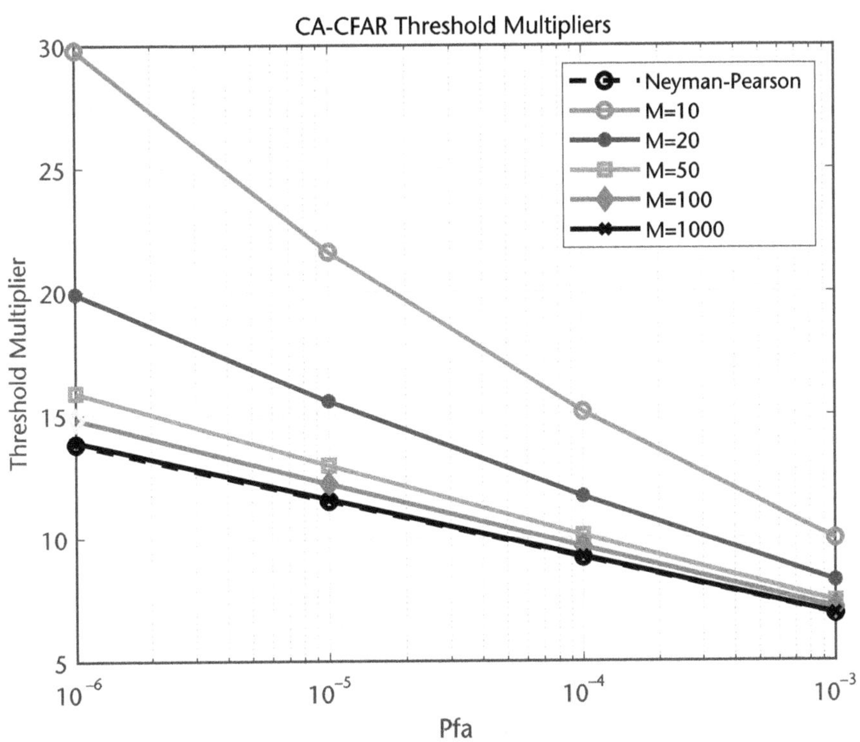

Figure 11.12 Comparison of CA-CFAR constants (threshold multipliers) against the ideal NP multiplier.

For $M \approx 1{,}000$, the difference is negligible, although that may still be too many samples for most practical cases. The case $M = 50$ seems reasonable, as it implies 25 cells on each side of the CUT. This is all assuming homogeneous interference.

Figure 11.13 shows a Simulink implementation of the 1-D CFAR algorithm illustrated in Figure 11.11.

The input vector consists of 25 samples of zero-mean Gaussian noise with standard deviation $\sigma = 0.1$, plus a single unit-amplitude target sample. The samples are envelope-detected, then compared to a threshold T. The CA-CFAR processing is modeled as a binary (0/1) CFAR Mask sliding window, where the locations with a 1 are included in the averaging, while the locations with a 0 (the CUT and one guard cell on each side) are excluded. There are five 1s on each leading and lagging window, therefore $M = 10$. The CFAR mask is correlated with the input vector, so each correlator output corresponds to a different average. The Selector blocks that discard a few samples at the beginning and the end of each vector are intended to discard partial correlations where there is not enough data to form interference estimates with the full 10 samples.

The CFAR constant is 9.95, which corresponds to a P_{fa} of 10^{-3} per (11.38). The multielement Display blocks in the lower right of the block diagram (Figure 11.13) show the calculated interference estimates at each sample time, along with the corresponding detection thresholds, CUTs, and the Boolean results (0 or 1) from the threshold comparisons. It is observed that the target (in the center of the data vector) is detected, while there are no false alarms. In addition to the numerical displays, the model features an Array Plot block in the upper right that displays the data and the CFAR threshold values. The resulting plots are shown in Figure 11.14.

The behavior of the threshold as it passes over the target is typical of the CA-CFAR algorithm. As the algorithm examines cells in the vicinity of the target, the target cell itself is included in the reference windows, which biases the threshold to a higher value than it would have otherwise. Once the target cell is itself the CUT, only noise contributes to the interference estimate, so the threshold drops considerably.

There are several variations to the standard CA-CFAR detector that provide robust performance against heterogeneous interference environments, such as

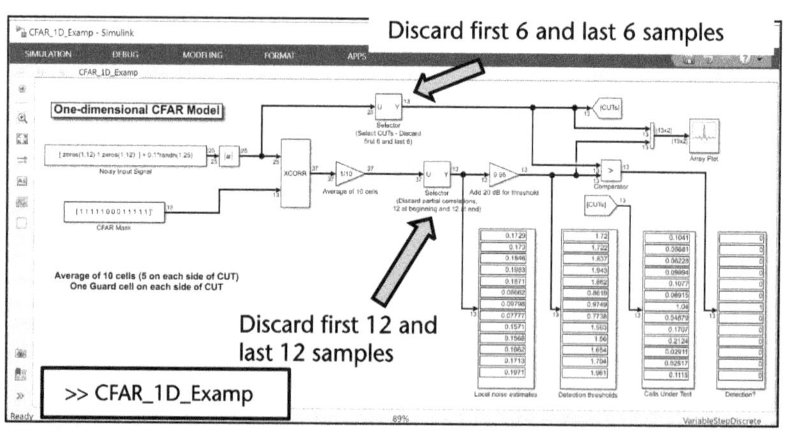

Figure 11.13 Simulink model of a 1-D CFAR detector.

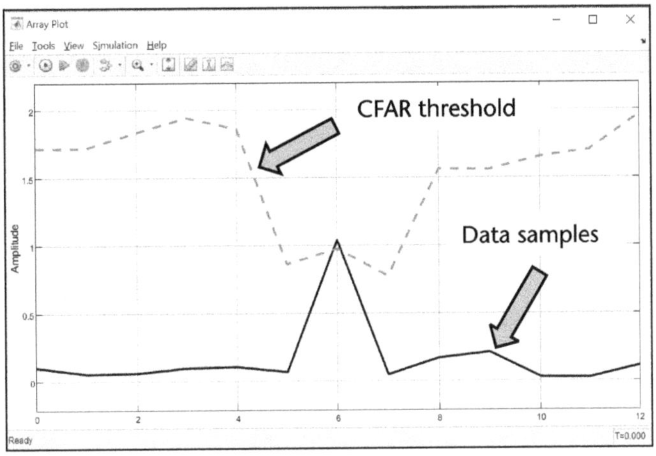

Figure 11.14 CFAR threshold and data samples from Simulink 1-D CFAR model.

greatest-of and smallest-of CA-CFAR (GOCA-CFAR) and (SOCA-CFAR), as well as order statistics CFAR (OS-CFAR) and related variants. Keel [10] provides a comprehensive survey and discussion on these variants.

11.9 Binary (M-of-N) Detection

Having discussed in detail the different P_d and P_{fa} calculations for the different detectors, we may now revisit the performance of the M-of-N detector. If the N trials/observations are independent, and the SNR is the same for all, then the single-observation (also called single-look) P_d—now termed $P_{d(1)}$—will be the same for each trial. The overall probability of detection then follows the binomial distribution, and is calculated as

$$P_D = \sum_{k=M}^{N} \binom{N}{k} P_{d(1)}^k \left(1 - P_{d(1)}\right)^{N-k} \tag{11.40}$$

where the binomial coefficient

$$\binom{N}{k} = \frac{N!}{k!(N-k)!} \tag{11.41}$$

denotes the number of ways in which you can get exactly k successes out of N independent trials. In MATLAB, it is calculated by the function `nchoosek(N,k)`.

Thus, the summation in (11.40) is the probability of succeeding (i.e., detecting) at least M times out of N. Likewise, the overall probability of false alarm is

$$P_{FA} = \sum_{k=M}^{N} \left(\frac{N!}{k!(N-k)!}\right) P_{fa(1)}^k \left(1 - P_{fa(1)}\right)^{N-k} \tag{11.42}$$

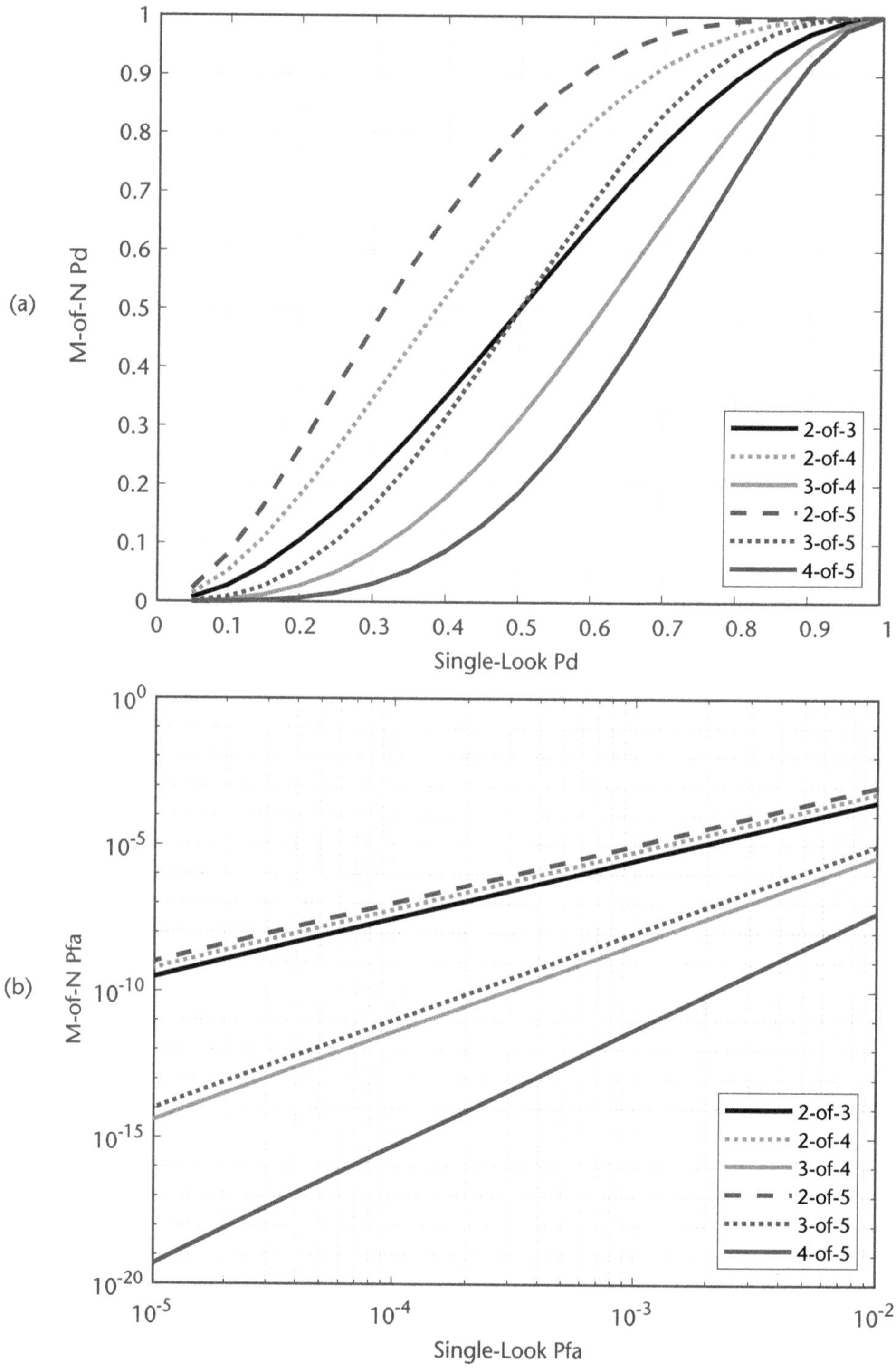

Figure 11.15 Binary detector performance examples for different combinations of M and N: (a) probability of detection, and (b) probability of false alarm.

where $P_{fa(1)}$ is that of a single look. Since typically $P_{fa(1)} << 1$, the first term in the summation gives a good approximation to (11.42),

$$P_{\mathrm{FA}} \approx \binom{N}{M} P_{fa(1)}^{M} \tag{11.43}$$

Figure 11.15 shows M-of-N P_d and P_{fa} as a function of their single-look counterparts, for different combinations of M and N. These were calculated using the function `get _ binomial _ Pd`, which works for both probabilities.

References

[1] Ridgway, G., "Rice/Rician Distribution," version 1.2.0.0, MATLAB Central File Exchange, https://www.mathworks.com/matlabcentral/fileexchange/14237-rice-rician-distribution.

[2] Whalen, A. D., *Detection of Signals in Noise*, San Diego, CA: Academic Press, Inc., 1971.

[3] https://mathworld.wolfram.com/MarcumQ-Function.html.

[4] DiFranco, J. V., and W. L. Rubin, *Radar Detection*, Dedham, MA: Artech House, 1980.

[5] Meyer, D. P., and H. A. Mayer, *Radar Target Detection, Handbook of Theory and Practice*, New York: Academic Press, 1973.

[6] Blake, L. V., *Radar Range-Performance Analysis*, Norwood, MA: Artech House, 1986.

[7] Tufts, D. W., and A. J. Cann, "On Albersheim's Detection Equation," *IEEE Trans. Aerospace and Electronic Systems*, Vol. AES-19, No. 4, July 1983, pp. 643–646.

[8] Dougherty, D., "Meyer & Mayer Probability of Detection Calculator," version 1.0.0.0, MATLAB Central File Exchange, https://www.mathworks.com/matlabcentral/fileexchange/1569-meyer-mayer-probability-of-detection-calculator.

[9] Shnidman, D. A., "Determination of Required SNR Values," *IEEE Trans. Aerospace & Electronic Systems*, Vol. AES-38, No. 3, July 2002, pp. 1059–1064.

[10] Keel, B. M., "Constant False Alarm Rate Detectors," in *Principles of Modern Radar, Vol I: Basic Principles* (M. A. Richards, J. A. Scheer, and W. A. Holm, eds.), Raleigh, NC: Scitech Publishing, Inc., 2010, pp. 589–620.

CHAPTER 12

Pulse-Doppler and FMCW Signal Processors

12.1 Introduction

This chapter builds on concepts and examples from Chapters 2, 5, and 7 to 11 to string together a few models of some of the better-known waveform processing chains. In Chapter 7, LFM was introduced as a common choice for CW, as well as the method of stretch processing for range estimation. Chapter 7 also discussed pulse waveforms and the main parameters associated with them. Chapter 8 examined the processing of pulse waveforms in the range domain (via matched filtering) and in the Doppler domain via DFT or its fast version, the FFT. We now pull these functions together to generate a range-Doppler Map (RDM), which is then processed through a CFAR detector to look for targets, as described in Chapter 11.

Section 12.2 revisits the FMCW waveform and corresponding range-Doppler processing, culminating with a Simulink model that takes digitized I/Q samples, forms an RDM, and ends with detection reports out of the CFAR detector. Two different versions are presented.

Section 12.3 follows a similar format, this time with a pulse-Doppler waveform employing pulse compression via binary phase modulation (BPM).

12.2 FMCW Processing

Figure 12.1 shows a generic block diagram for an FMCW radar system. Contrast this block diagram to that of the fixed-frequency CW radar in Figure 7.2.

While the single-frequency CW system employs a single antenna and a circulator to achieve transmit/receive isolation, the FMCW system example employs two separate antennas for transmit and receive. These two isolation approaches are not specific to the waveform employed; in practice, either method is used for either system. In terms of similarities, they are both homodyne systems that use a copy of the transmit waveform to downconvert the received signal to baseband, and they both implement I/Q demodulation [1]. The LPFs that follow the mixers reject the high-frequency outputs (the sum components); the filter-noise bandwidth Bn is assumed to match the waveform bandwidth. The major difference really is in the digital-signal processing section; while the fixed-frequency system only requires processing in one dimension for measuring Doppler, the FMCW system is intended to measure target range and Doppler. Thus, 2-D processing (fast-time and slow-time) is needed, as described in Chapter 8.

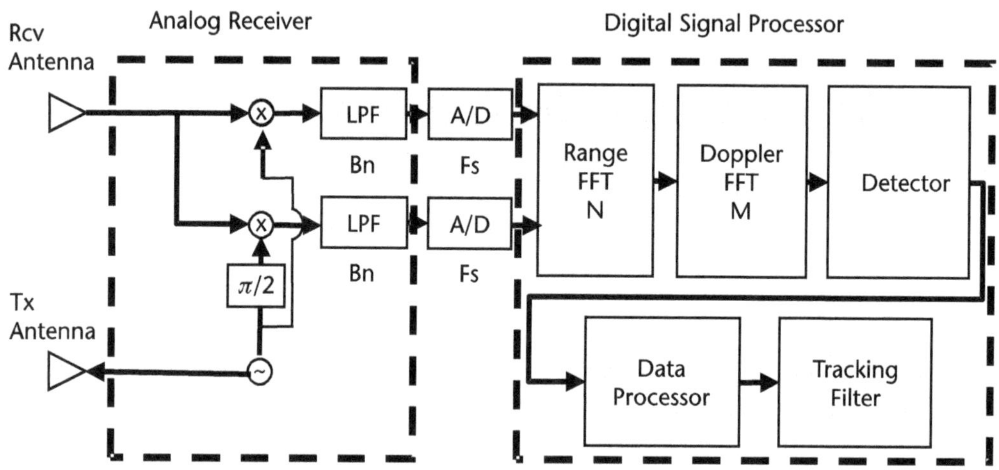

Figure 12.1 Block diagram of a typical FMCW radar.

As Figure 12.1 indicates, the range-Doppler processing to be implemented for this notional FMCW radar consists of FFTs on both dimensions. The resulting data product is a complex RDM, a 2-D matrix of N range gates and M Doppler filters that is sent to the data processor. Functions performed in the data processor include envelope detection, CFAR detection, and measurement extraction. These measurements are eventually sent to the tracker for further processing, as described in Chapter 13.

Modeling of the system depicted in Figure 12.1 focuses on the digital (baseband) portion of the diagram (i.e., to the right of the A/D blocks). It is therefore instructive to understand the mathematical model of the baseband LFM waveform, starting from the RF transmit and receive form.

12.2.1 LFM Waveform Model

Chapter 7 presented a complex formulation for a general radar waveform model in Table 7.1 and furthermore introduced (7.1) as the complex baseband representation of such a waveform; it is repeated here for convenience:

$$s_c(t) = z(t)\exp\left(j2\pi \mathbf{f} t\right) \tag{12.1}$$

where $z(t)$ is the complex envelope including any waveform modulation, and $\mathbf{f}$ is the frequency function, which is a constant for a fixed-frequency CW.

Assuming an upchirp or upsweep LFM waveform (i.e., a positive linear slope as depicted in Figure 7.5), we can write the transmit RF waveform as

$$x_T(t) = \hat{A}\exp\left(j2\pi\left(f_0 t + \frac{1}{2}m\tilde{t}^2\right)\right),\quad m = \frac{\Delta F}{T_m},\quad \tilde{t} = \mathrm{mod}\left(t, T_m\right) \tag{12.2}$$

where f_0 is the carrier frequency, m is the LFM slope, and ΔF and T_m are the frequency deviation and modulation/sweep interval (respectively), as defined in

Chapter 7. The transmit signal amplitude $\hat{A}$ is essentially the squared root of the ERP, $\sqrt{P_t G_t L_t}$, as defined in Chapter 2.

The *instantaneous frequency* of $x_T(t)$ is by definition

$$f_{\text{inst}}(t) \triangleq \frac{1}{2\pi}\frac{d\Theta}{dt} = f_0 + m\tilde{t} \tag{12.3}$$

where Θ is the phase function of $x(t)$ (i.e., the argument of the complex exponential). Notice from (12.3) that within a modulation interval T_m, the instantaneous frequency varies linearly from f_0 to $f_0 + \Delta F$. This is the behavior depicted in Figure 7.5 using 0 as a reference instead of f_0.

An alternate, often used form is to add a $-\Delta F/2$ offset to the argument of (12.2), such that the instantaneous frequency varies from $f_0 - \Delta F/2$ to $f_0 + \Delta F/2$. Under that convention, f_0 truly becomes the center frequency of the waveform, which is often preferred. For simplicity, however, we will follow the definition in (12.2).

We may rewrite (12.2) as

$$x_T(t) = \hat{A}\exp\left(j\pi m\tilde{t}^2\right)\exp\left(j2\pi f_0 t\right) = z_T(t)\exp\left(j2\pi f_0 t\right) \tag{12.4}$$

which is now in the form of (12.1), thereby defining the transmit complex envelope $z_T(t)$.

The received signal reflected from a target at a range R is simply a time-delayed, scaled version of either (12.2) or (12.4), which we will write as

$$\begin{aligned} x_R(t) &= \tilde{A}\exp\left(j\pi m\left(\tilde{t} - t_d\right)^2\right)\exp\left(j2\pi f_0\left(t - t_d\right)\right) \\ &= \tilde{A}\exp\left(j2\pi\left(0.5m\left(\tilde{t} - t_d\right)^2 - f_0 t_d\right)\right)\exp\left(j2\pi f_0 t\right) \\ &= z_R(t)\exp\left(j2\pi f_0 t\right) \end{aligned} \tag{12.5}$$

with the round-trip time delay $t_d = 2R/c$ as previously defined in (7.4). The received signal amplitude $\tilde{A}$, in contrast to $\hat{A}$ from (12.2), includes the additional terms from the RRE as discussed in Chapter 2.

In the homodyne FMCW system, the LO signal used to downconvert the received signal depicted by (12.5) is the transmit waveform itself, namely (12.4). Due to their complex representation, however, one of them must be conjugated such that their product results in a difference signal (i.e., their frequencies subtract). Given our assumption of an upchirp LFM, we choose to conjugate the receive signal in our mathematical model, such that the resulting beat frequency is positive when we define it as $(f_T - f_R)$. This is depicted in the top panel of Figure 12.2. The complex baseband waveform at the output of the mixer is

$$x_{\text{BB}}(t) = x_T(t)x_R^*(t) = z_T(t)z_R^*(t) \tag{12.6}$$

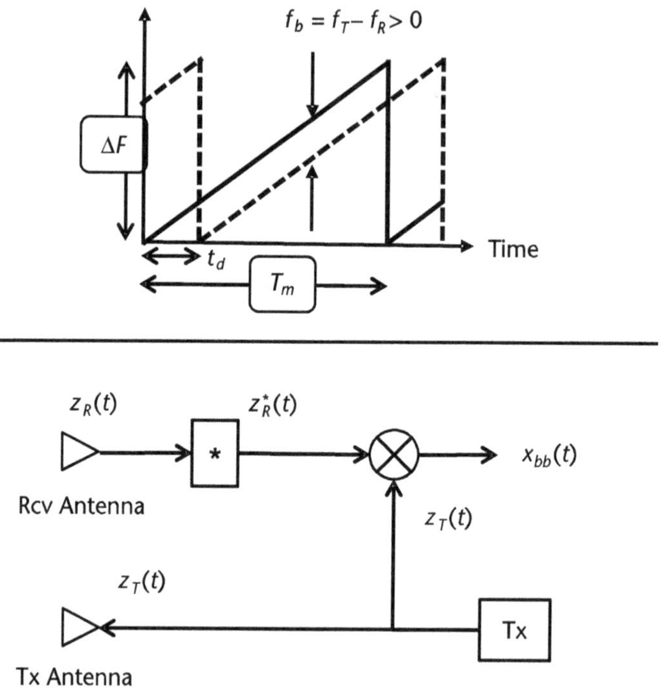

Figure 12.2 Mathematical model of homodyne downconversion using complex baseband representation: beat frequency definition (top), block diagram of downconversion (bottom). The '*' operator denotes complex conjugation.

where the last equality falls directly from (12.4) and (12.5). A graphical representation of (12.6) using only complex baseband notation is depicted in the bottom panel of Figure 12.2.

Writing the baseband waveform explicitly,

$$\begin{aligned} x_{\mathrm{BB}}(t) &= \hat{A}\tilde{A}\exp\left(j\pi m\tilde{t}^2\right)\exp\left(-j2\pi\left(0.5m\left(\tilde{t}-t_d\right)^2 - f_0 t_d\right)\right) \\ &= A\exp\left(j2\pi\left(-0.5m\left(\tilde{t}^2 - 2t_d\tilde{t} + t_d^2\right) + 0.5m\tilde{t}^2 + f_0 t_d\right)\right) \\ &= A\exp\left(j2\pi\left(mt_d\tilde{t} + f_0 t_d - 0.5mt_d^2\right)\right) \end{aligned} \tag{12.7}$$

For a stationary radar and a stationary target at range R, t_d is a constant, such that the last two terms in the argument of (12.7) become a fixed range-induced phase φ. The last row of (12.7) may thus be rewritten as

$$x_{\mathrm{BB}}(t) = A\exp\left(j2\pi\left(f_b\tilde{t} + \varphi\right)\right),\; f_b = mt_d = \left(\frac{2m}{c}\right)R \tag{12.8}$$

We now assume motion between the radar and the target in a linear, constant-velocity trajectory such that range may be written as

$$R(t) = R_0 + \dot{R}t \tag{12.9}$$

where R_0 is the target range at $t = 0$, and $\dot{R}$ is the range rate, which is negative for an approaching target. Replacing this expression in the RHS of (12.8) results in a redefinition of the beat frequency as proportional to R_0, plus an additional term dependent on the range rate:

$$\left(\frac{2m}{c}\right)R = \left(\frac{2m}{c}\right)\left(R_0 + \dot{R}t\right) = \left(\frac{2mR_0}{c}\right) + \left(\frac{2m\dot{R}}{c}\right)t = f_b + \left(\frac{2m\dot{R}}{c}\right)t \tag{12.10}$$

Reexamining the phase term, writing it explicitly in terms of R, and replacing it with (12.9) results in

$$\begin{aligned}\varphi &= f_0 t_d - 0.5mt_d^2 = f_0\left(\frac{2R}{c}\right) - 0.5m\left(\frac{2R}{c}\right)^2 = \left(\frac{2R}{\lambda}\right) - \left(\frac{2m}{c^2}\right)R^2 \\ &= \left(\frac{2R_0}{\lambda}\right) + \left(\frac{2\dot{R}}{\lambda}\right)t - \left(\frac{2m}{c^2}\right)\left(R_0 + \dot{R}t\right)^2 \\ &= -f_d t + \left(\frac{2R_0}{\lambda}\right) - \left(\frac{2m}{c^2}\right)\left(R_0^2 + 2R_0\dot{R}t + \dot{R}^2t^2\right),\ f_d = \frac{-2\dot{R}}{\lambda}\end{aligned} \tag{12.11}$$

where the Doppler frequency shift is defined in terms of range rate as in (8.7). We may now rewrite (12.8) for a moving target using (12.10) and (12.11):

$$\begin{aligned}x_{\mathrm{BB}}(t) &= A\exp\left(j2\pi\left(f_b\tilde{t} - f_d t + m\left(\frac{2\dot{R}}{c}\right)t + \left(\frac{2R_0}{\lambda}\right) - \left(\frac{2m}{c^2}\right)\left(R_0^2 + 2R_0\dot{R}t + \dot{R}^2t^2\right)\right)\right) \\ &= A\exp\left(j2\pi\left(f_b\tilde{t} - f_d t + \left[\left(\frac{2R_0}{\lambda}\right) - \left(\frac{2mR_0^2}{c^2}\right)\right] - \left[\frac{4mR_0\dot{R}}{c^2}\right]t + \left[\left(\frac{2m\dot{R}}{c}\right)\left(1 - \frac{\dot{R}}{c}\right)\right]t^2\right)\right) \\ &= A\exp\left(j2\pi\left(f_b\tilde{t} - f_d t + \gamma(t)\right)\right)\end{aligned} \tag{12.12}$$

Following a similar derivation, the complex baseband representation for a downchirp LFM is

$$x_{\mathrm{BB}}(t) = A\exp\left(j2\pi\left(f_b\tilde{t} + f_d t + \gamma(t)\right)\right) \tag{12.13}$$

where now a positive Doppler shift adds to the beat frequency instead of subtracting from it. This will actually be the model followed in the Simulink implementations presented in Section 12.2.3.

Regardless of the chirp slope adopted, the end result from either the third row in (12.12) or from (12.13) consists of a complex exponential at a total frequency formed by the combination of the beat frequency and the Doppler frequency, plus a phase function $\gamma(t)$. This function is the collection of terms in square brackets in the second row of (12.12). These terms are either constant, linear in t, or quadratic in t, thereby forming a quadratic equation. Left unchecked, this motion-induced quadratic phase function can potentially smear the target energy across neighboring range-Doppler cells, degrading detection performance. Motion compensation algorithms are often applied to the data to cancel this term as best possible. The effect of this term is illustrated in the Simulink model results in Section 12.2.3.

12.2.2 LFM Waveform Processing

Given the baseband waveform model from the last row in (12.12), its corresponding processing may now be seen as separable operations in fast time (within a sweep) and slow time (from sweep to sweep). Within a sweep, the Doppler frequency term changes slowly relative to the beat frequency. (Under many operational conditions, the beat frequency can be orders of magnitude higher than the Doppler frequency.) Thus, within a single sweep the beat-frequency term dominates and is the one estimated by an FFT over fast-time samples. This is the stretch or deramping processing described in Section 7.3 and illustrated in Figure 7.6 [2]. From sweep to sweep, however, the Doppler frequency induces a phase rotation that can be extracted by performing an FFT over slow time, analogous to the pulse-Doppler FFT across PRIs described in Section 8.7. Thus, a 2-D range-Doppler matrix is formed, as shown in Figure 12.3. Compare Figure 12.3 to its pulse-Doppler counterpart depicted in Figure 8.14.

12.2.3 Simulink Models of the LFMCW System

Two different versions of the LFMCW system implementations in Simulink are now presented. The first version is vector-based, meaning that at each simulation time step (fast-time sample), all corresponding slow-time samples are generated. This may be visualized as generating an entire row of the matrix depicted in Figure 12.3 at every simulation time step. This approach enables a faster runtime compared to the second version we will examine, which is sample-based (i.e., a single fast-time sample is generated at a time).

The top-level block diagram for the vector-based implementation *FMCW_Processor1* is shown in Figure 12.4.

It starts with a signal model to which noise is added, followed by the signal processor (Range-Doppler FFT), envelope detector, and CFAR detector. The CFAR outputs are sent to an *Interpreted MATLAB* Fcn block that plots the RDM and CFAR detections.

The signal subsystem shown on the lower left of Figure 12.4 creates a complex signal with two sinusoids in quadrature. The frequency parameter of the sinusoids is set to the LFM beat frequency f_b, while the phase parameter is set to the target Doppler frequency f_d times a vector of time samples separated by the modulation interval T_m. This signal model can be written as

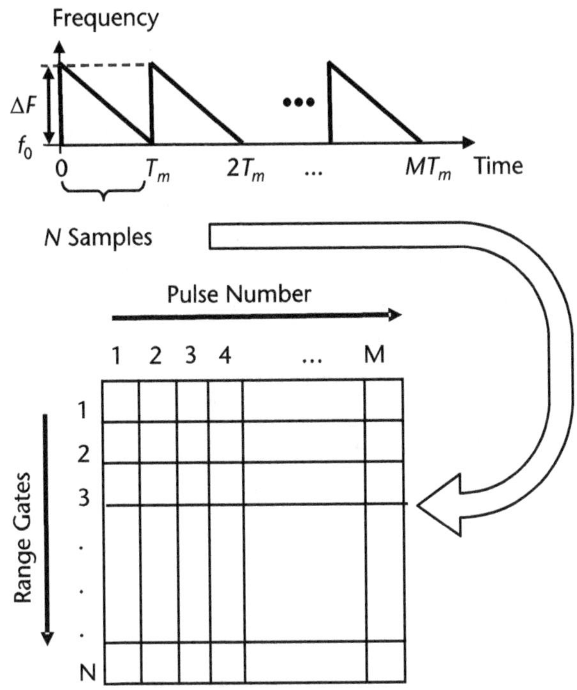

Figure 12.3 Data collection and arrangement of fast-time/slow-time samples in a LFMCW system.

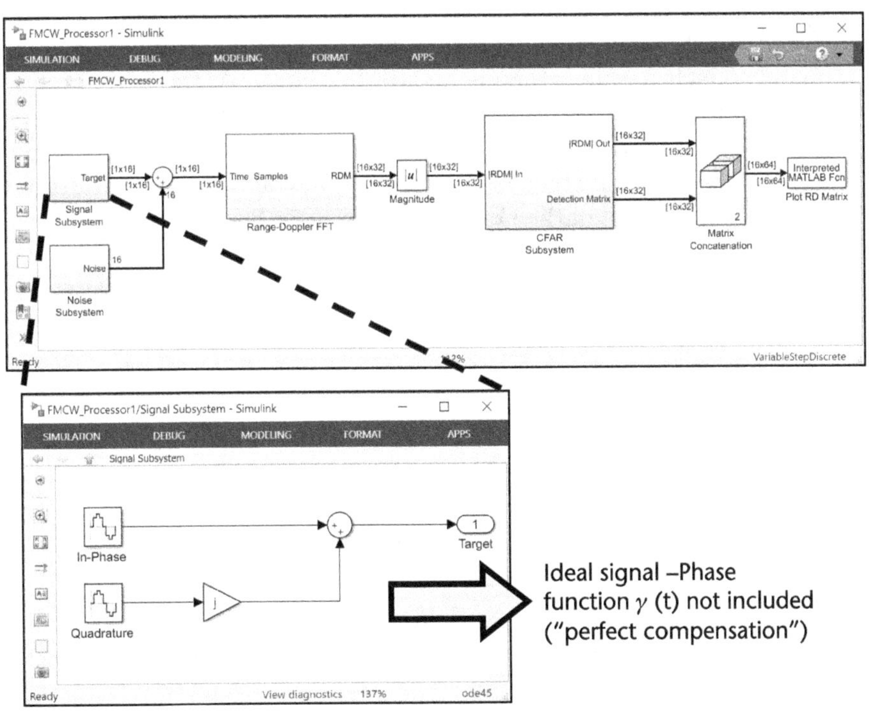

Figure 12.4 Simulink implementation of FMCW system (vector-based).

$$s_{BB}(t) = A\exp\left(j2\pi\left(f_b t + f_d\left[0{:}M-1\right]T_m\right)\right) \tag{12.14}$$

Comparing this signal model to (12.13), we make the following observations:

- At each simulation time step t, (12.14) is an M-element vector where each element corresponds to a slow-time sample for each of the M sweeps in the simulated dwell. In this particular example, $M = 16$ as can be seen on the signal dimensions displayed in Figure 12.4.
- The Doppler shift term is not changing over fast time; so, while it will yield the right answer with the slow-time FFT, this expression is an oversimplification that ignores the effect of Doppler on the beat frequency term that will be processed by the range FFT.
- The quadratic phase function $\gamma(t)$ is not represented in (12.14) at all, which implies it has been perfectly canceled by a hypothetical motion-compensation processing.

In summary, this is an idealized signal model that does not capture the effects described by the last two observations. The effect of these simplifying assumptions is observed and discussed later in this section (Section 12.2.3).

After adding complex Gaussian noise to the signal, the composite signal vector goes into the Range-Doppler FFT block, whose contents are shown in Figure 12.5.

Since each vector corresponds to a single fast-time value (i.e., a single row of the 2-D matrix in Figure 12.3), the first step is to accumulate all rows within the sweep. A total of $N = 32$ fast-time samples are collected over the sweep; thus the model collects all N vectors using a Buffer block. At the end of the N simulation time steps, the full N-by-M matrix is ready for the range and Doppler FFTs. Prior to each FFT, the option to add amplitude weights on each dimension is available via Window Function blocks. As a default, boxcar windows (uniform weights) are applied. The output of the Doppler FFT is an M-by-N (16-by-32 in this case) RDM.

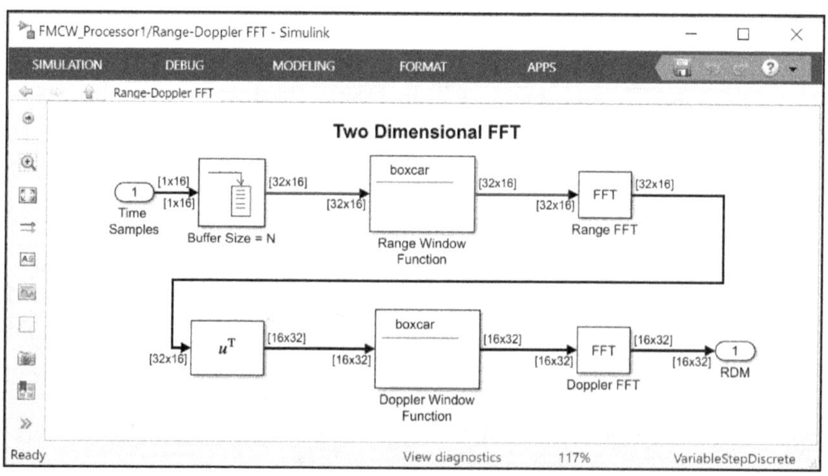

Figure 12.5 Range-Doppler signal processor for the FMCW system (vector-based).

This complex matrix is envelope-detected (linear detector) and the magnitude-only RDM goes into the CFAR subsystem, which is shown in Figure 12.6.

The classic sliding window CA-CFAR implementation described in Chapter 11, although possible to implement here, is replaced by a much simpler approach that is nevertheless suitable for this basic model. A global noise level estimate is obtained by cell-averaging over all range bins in each Doppler filter, thereby obtaining a set of $M = 16$ averages. We then pick the median of these values as representative of the noise level. Assuming that the noise environment is homogeneous across Doppler, this scheme results in a reasonable estimate. If clutter were simulated, its spectrum would typically be confined to a relatively few Doppler filters—assuming that its bandwidth is narrower than the modulation frequency Fsweep. The median filter would discard the filters containing clutter, as their noise estimate would be biased to higher values than filters with noise only. A similar argument can be made for Doppler filters containing target returns. The median filter rejects up to $M/2$ filters containing either clutter or target returns.

The threshold multiplier (CFAR constant) is based on (11.38) for $P_{fa} = 10^{-6}$ and $N = 32$ samples in the average. Furthermore, we take the square root since the model employs a linear detector, while (11.38) assumes a square-law detector.

The CFAR threshold is compared to all cells in the RDM; the resulting Detection Matrix is a binary (0/1) array where the location of each logical '1' indicates a threshold crossing.

We will present the result of this model after we introduce the sample-based version of the FMCW system, *FMCW_Processor2*, so we may compare the two. The top-level block diagram of this second model is shown in Figure 12.7.

Comparing this model to Figure 12.4, we observe that other than the signal model at the beginning, the top level diagram looks almost identical to the vector-based version. The signal model now generates a single fast-time sample at every simulation time step. The simulation time (provided by a Clock block) is fed to an *Interpreted MATLAB Fcn* block, which runs the function m-file *fmcw* displayed in the lower left of Figure 12.7. In the function, the variable `tau` is the modulation interval T_m. The function implements the expression

$$x_{\mathrm{BB}}(t) = \exp\left(j2\pi\left(mt_d\tilde{t} - f_0t_d - 0.5mt_d^2\right)\right) \tag{12.15}$$

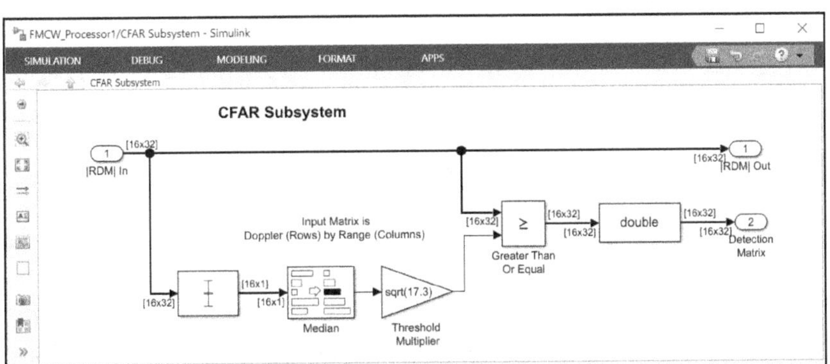

Figure 12.6 Simple CFAR detector for the FMCW system (vector-based).

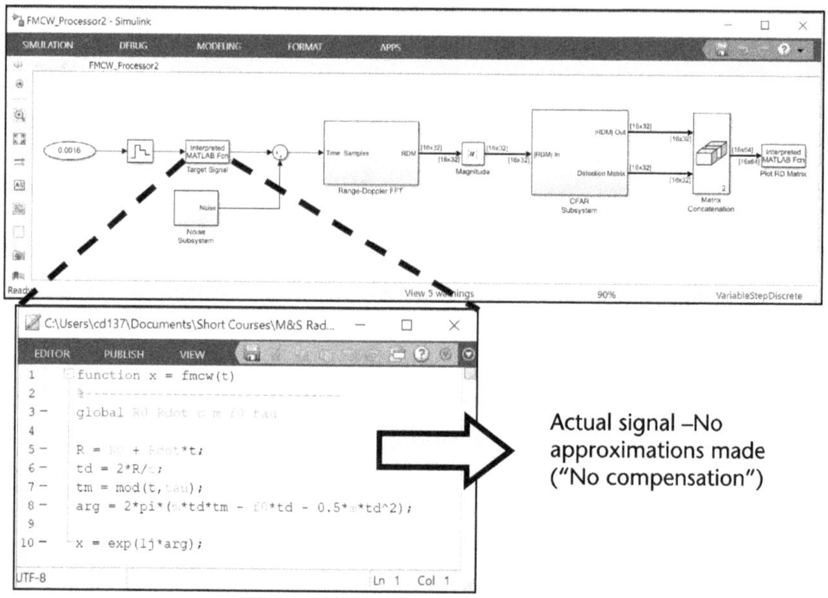

Figure 12.7 Simulink implementation of FMCW system (sample-based).

Comparing this expression to (12.7), the only difference (other than unit amplitude) is that the second term in the argument (which gives rise to the Doppler frequency term) is negative in (12.15). This stems from (12.15) being defined for a downchirp LFM, while (12.7) corresponds to an upchirp. That being said, this form of the baseband LFM signal maintains all the nuances that the vector-based signal model ignored; the Doppler term changes over fast time, and the quadratic phase terms are included. Indeed, they are embedded in the terms of (12.15). This implementation therefore represents the actual baseband LFM signal without any approximations or simplifications.

The signal processor is fundamentally the same as in version 1, except now the inputs to the subsystem are scalar samples; thus, an additional Buffer block is needed to collect all the fast-time samples for a single sweep. The block diagram is shown in Figure 12.8; contrast this version with that of Figure 12.5.

Once the N-element fast-time vector has been collected, the range FFT is performed, and the slow-time buffer collects all M sweeps for the Doppler FFT. The CFAR subsystem is identical to that shown in Figure 12.6 so it is not repeated here.

We now compare the results of both models. Each is set up to have the target show up at range bin 15, Doppler bin 4. The target range and velocity are calculated based on those locations. The resulting RDMs are shown in Figure 12.9.

Not surprisingly, the vector-based model runs faster, as intended. It is also noted that on both versions, the target is detected (threshold-crossing locations are indicated by circles overlaid on the RDM) without any false alarms. The vector-based version, shown in Figure 12.9(a), shows all target energy nicely focused within a single range-Doppler cell. This is a direct consequence of the signal model simplifications discussed earlier in this section. The sample-based version, however, shows noticeable spillover on neighboring cells, in particular in the range

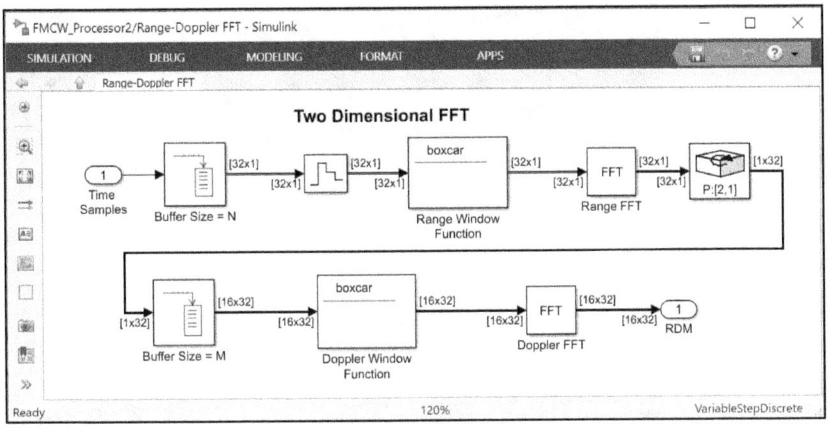

Figure 12.8 Range-Doppler signal processor for the FMCW system (sample-based).

dimension. There is enough signal strength on a neighboring range cell, in fact, that it also exceeds the threshold and is flagged as a detection. In this particular scenario, the target SNR was high enough that the spillover or smearing did not prevent detection. The contiguous detections would normally be taken care of by a clustering step that is typical of detection logic (not included in this model). A simple clustering algorithm would consolidate contiguous threshold crossings into a single detection report sent to a subsequent tracker.

Under different circumstances and simulation parameter values, however, the effects of ignoring motion compensation may not be so benign, and the target could be missed.

12.2.4 Processed Overlap of FMCW Systems

The two system models presented in Section 12.2.3 are simple examples intended to illustrate the basics of FMCW waveform and signal-processing modeling. They do overlook, however, an important aspect of FMCW processing, which is accounting for the processed overlap between transmit and receive frequency sweeps. Let's review Figure 12.2. The processed overlap is the amount of time where the receive LFM ramp coincides with the transmit ramp within the transmit modulation interval T_m. That is the portion of time where the correct beat frequency is available at the output of the mixer. On the remaining portion of the interval, there is an inversion of the frequency ramps relative to one another due to the ramp periodicity, resulting in a large frequency difference, which is rejected by the LPF that follows the mixer in Figure 12.1.

Ideally we would like to process the receive signal over the entire interval T_m. This maximum processing duration yields the best achievable range resolution, (8.29) with $B = \Delta F$. From Figure 12.2, however, the signal is only available over the interval $T_m - t_d$. We thus define an overlap factor as

$$\alpha_{\text{of}} = \frac{T_m - t_d}{T_m} = 1 - \frac{t_d}{T_m} \tag{12.16}$$

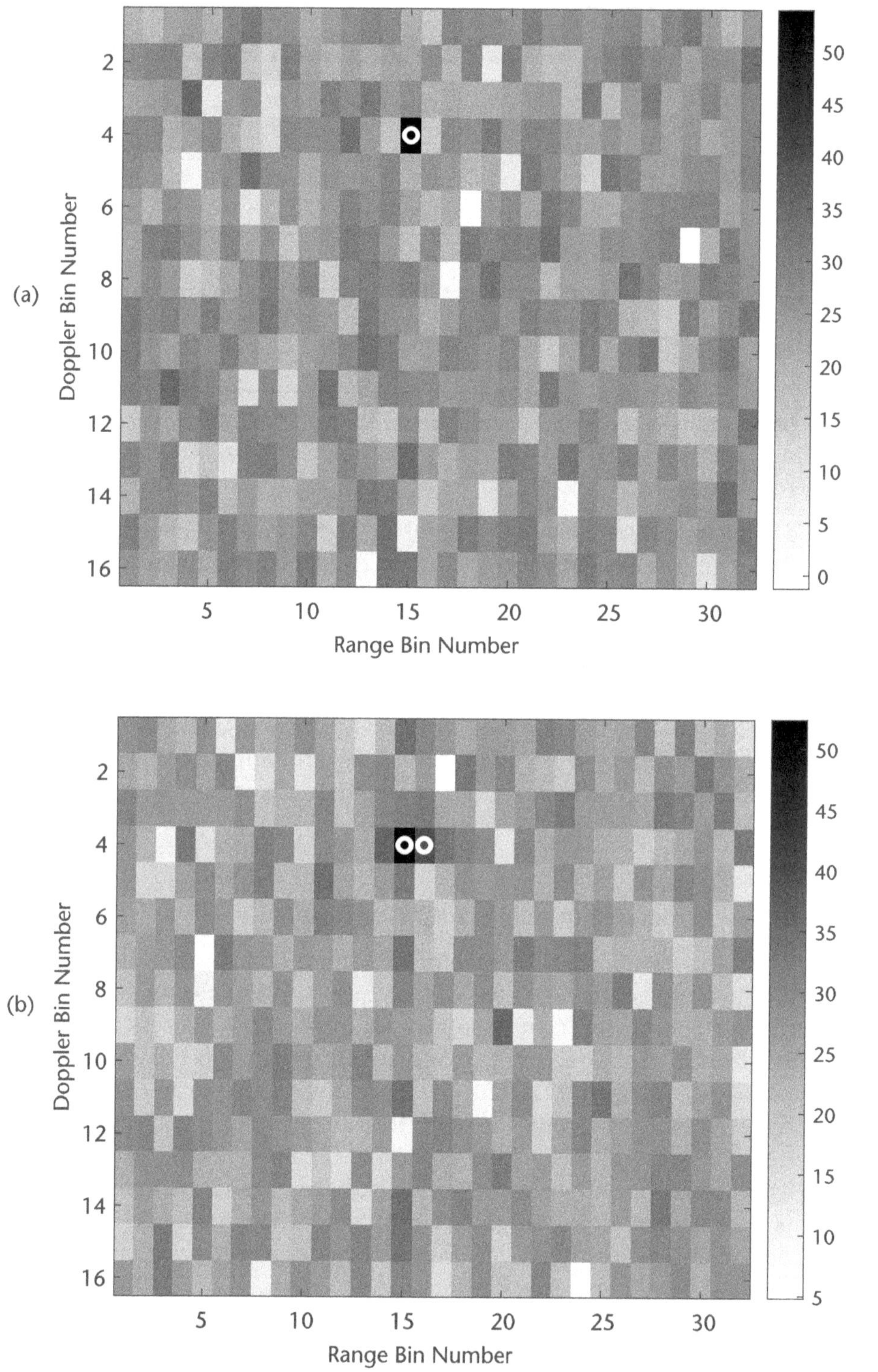

Figure 12.9 Range-Doppler Maps for FMCW system models: (a) vector-based, and (b) sample-based.

It is expressed as a fraction of the modulation interval; therefore, $0 < \alpha_{of} < 1$. For a moving target, it varies with range, as the time delay does.

This reduction in processing time causes both a reduction in SNR (fewer target samples are integrated), and a degradation in range resolution (since the actual bandwidth processed is less than ΔF) [3]. Both effects are quantified by the overlap factor as follows:

$$SNR_{actual} = \alpha_{of}^2 SNR_{ideal}, \text{ and } \delta_{R(actual)} = \frac{c}{2\Delta F \alpha_{of}} \tag{12.17}$$

The SNR loss in (12.17) assumes that while the target samples are only collected during the overlap duration, noise samples are still collected over the entire interval T_m. This is a reasonable assumption when the target range is unknown, as would be the case in a radar search mode. When in track mode, or in a cued search mode where the radar has been given an estimated location, the ADC may not start taking samples until the approximate time when the target is expected to show up. In that case, the receiver takes an equal number of target and noise samples, and the SNR would only be reduced by a factor of α_{of}.

In terms of the Simulink system model, setting up the timing of the received waveform correctly will automatically take care of both of these effects. In the previous examples presented and discussed in Section 12.2.3, a small size of fast-time samples ($N = 32$) was chosen arbitrarily for illustration purposes. The simulation sampling rate was set based on taking these few samples over the modulation interval, resulting in an unrealistically low value. In reality, the sampling rate for complex data should be at least as high as the waveform bandwidth, ΔF. Setting $F_s = \Delta F$ results in a much higher number of fast-time samples. In fact, the number becomes the time-bandwidth product (TBP), which makes sense since N corresponds to the coherent integration gain of the range FFT. Thus,

$$N = TBP = (T_m)(\Delta F) \tag{12.18}$$

A modified version of the sampled-based system model (*FMCW_Processor2b*) incorporates these two changes. Equation (12.18) is implemented in the Init function (Init Callback) of the Simulink model. For the waveform parameters chosen (ΔF = 150 MHz and T_m = 100 μs), N = 15,000 samples. Notice that this is the number of samples that will be collected over the entire modulation interval (search-mode assumption), but not all of those samples will include the target signal due to the time delay. The delay (and subsequent reduction in available signal samples) is incorporated in the target signal function m-file. An updated version *fmcw_delayed* is used in this new version, which outputs zero-valued samples for $t < t_d$. Other than calling this new function from the Interpreted MATLAB Fcn block, there are no other changes to the Simulink block diagram; therefore, it is not shown again.

Figure 12.10 shows the code for the updated signal function *fmcw_delayed* and the resulting RDM from running the sampled-based model *FMCW_Processor2b*.

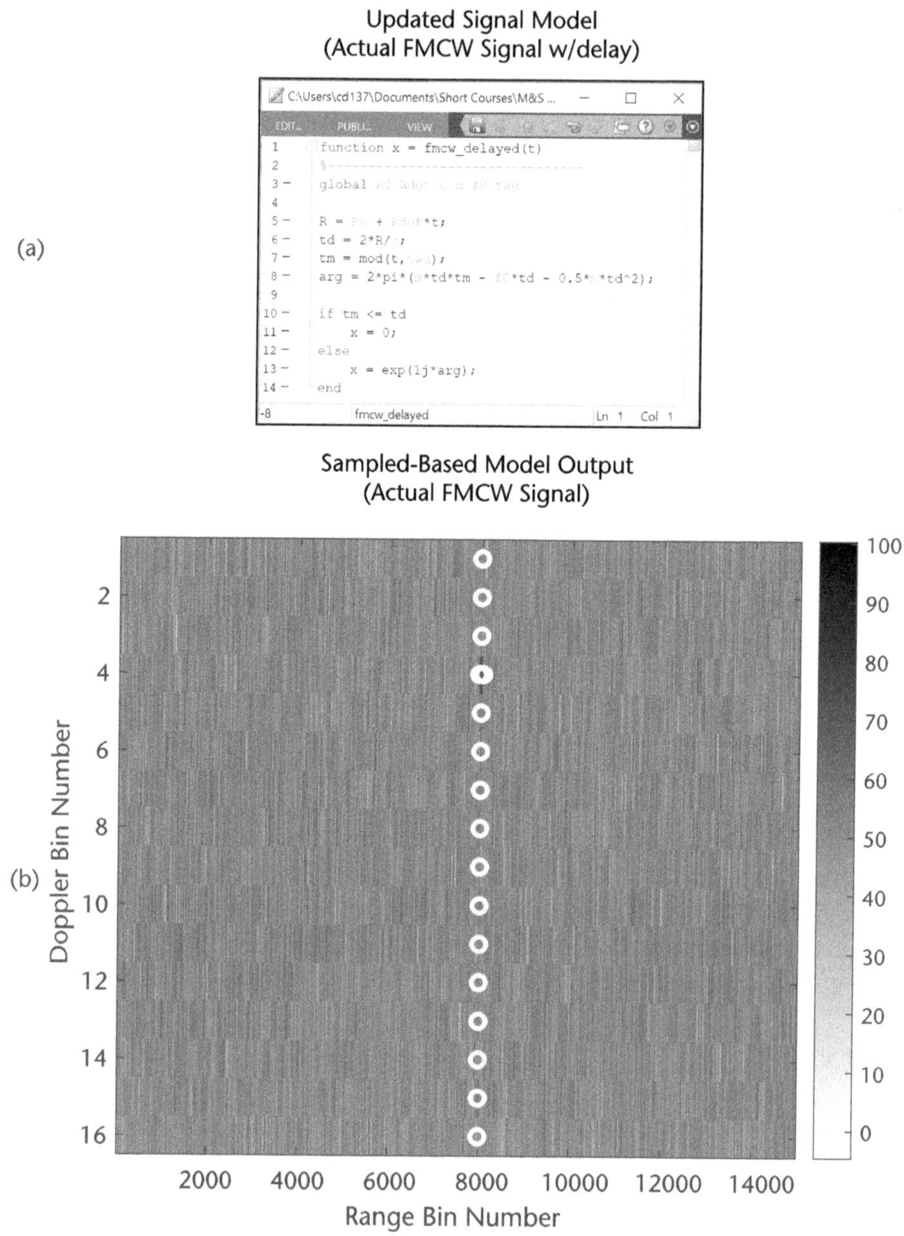

Figure 12.10 Upgraded Simulink implementation of FMCW system (sampled-based): (a) waveform model function, and (b) output rdm.

The target range is set to 8 km, so it is expected to show up at range bin 8000 (the range bin width is 1m).

There are additional threshold crossings in the target range gate, which are due to Doppler sidelobes. These are now detectable above the noise level due to the much higher coherent gain from the range FFT. Even with the SNR loss due to the reduced overlap, the gain is about 20-dB higher than in the previous simple example. On this previous example, the coherent gain from the range FFT was

$10 * \log_{10}(32) = 15$ dB, while now it is $10 * \log_{10}(\alpha_{\text{of}}^2 N) = 10 * \log_{10}((0.47)^2(15{,}000))$ = 35.1 dB. The traditional way to reduce the Doppler sidelobe level is to apply a window function to the slow-time data prior to the FFT. Figure 12.11 shows the signal processor subsystem, where the Window Function block before the Doppler FFT has now been modified to apply a Hamming window.

The resulting RDM shows that the Doppler sidelobes are no longer crossing the detection threshold. The penalty for the window function (aside from an additional SNR loss discussed in Chapter 8) is the broadening of the Doppler mainlobe, which results in detections on adjacent filters. Again, additional clustering logic

(a)

(b)

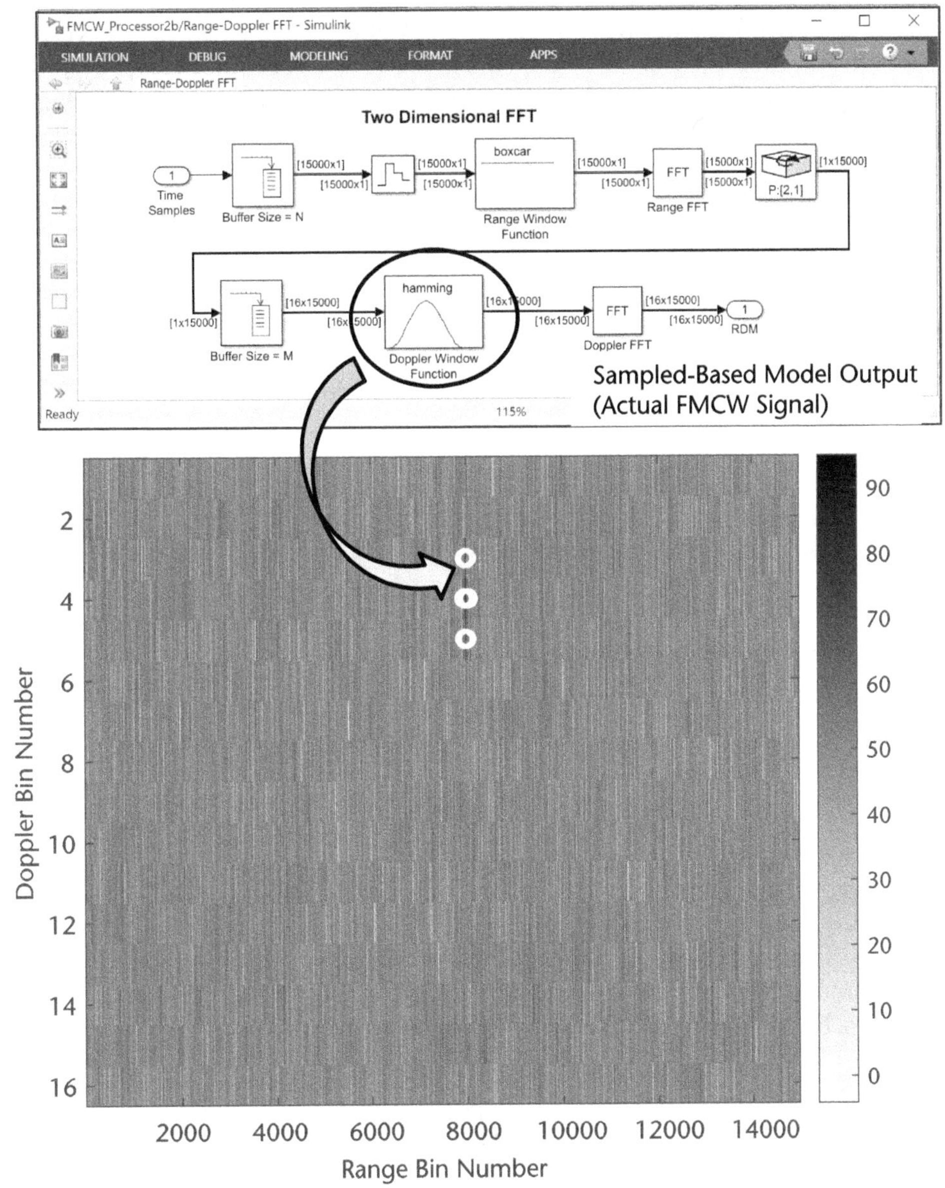

Figure 12.11 Simulink model results after applying a window function to reduce Doppler sidelobes: (a) signal processor subsystem, and (b) output rdm.

(not included in the model) would prevent redundant detections to be reported to a tracker.

12.3 Pulse-Doppler Processing

We now turn our attention to the pulse-Doppler waveform processing. The system to be modeled follows the digital signal processor diagram from Figure 1.4. Just like the FMCW system, the signals are modeled at baseband.

Even though Simulink features a convenient Pulse Generator block, it is still instructive to examine the mathematical model for a baseband coherent-pulse waveform. We start with the basic form of a square pulse $p(t)$ of width τ, which was defined by (7.2) and (7.3). This is the causal form of the square pulse, meaning that it is defined from $t = 0$ to $t = \tau$. We use this basic building block to create a train of M pulses separated by the pulse repetition interval, *PRI*. This may be written as the summation of multiple delayed copies of $p(t)$:

$$x_p(t) = \sum_{n=0}^{M-1} p(t - nPRI) \tag{12.19}$$

Equation (12.19), scaled by the transmit amplitude, is actually the transmit complex envelope $z_T(t)$. Thus,

$$z_T(t) = \hat{A}x_p(t) = \hat{A}\sum_{n=0}^{M-1} p(t - nPRI) \tag{12.20}$$

The full RF transmit waveform is still (12.1), with $\mathbf{f} = f_0$. Likewise, the received signal after a round-trip delay t_d is

$$\begin{aligned} x_R(t) &= \tilde{A}x_p(t - t_d)\exp\left(j2\pi f_0(t - t_d)\right) \\ &= \tilde{A}x_p(t - t_d)\exp\left(-j2\pi f_0 t_d\right)\exp\left(j2\pi f_0 t\right) \\ &= z_R(t)\exp\left(j2\pi f_0 t\right) \end{aligned} \tag{12.21}$$

The carrier term is removed by the heterodyning process, such that only $z_R(t)$ remains (within a scaling constant). Writing the complex exponential term in the baseband signal explicitly in terms of target range R,

$$\begin{aligned} x_{\text{BB}}(t) &= Ax_p(t - t_d)\exp\left(-j2\pi f_0 t_d\right) \\ &= Ax_p(t - t_d)\exp\left(-j2\pi f_0\left(\frac{2R}{c}\right)\right) \end{aligned} \tag{12.22}$$

Assuming a constant-velocity, linear-motion target model as in (12.9), the complex exponential in (12.22) becomes

$$\exp\left(-j2\pi f_0\left(\frac{2R}{c}\right)\right)=\exp\left(-j2\pi\left(\frac{2R_0+2\dot{R}t}{\lambda}\right)\right)=\exp\left(j2\pi f_d t+\phi_0\right) \tag{12.23}$$

where f_d is the Doppler frequency as defined in (12.11) and $\phi_0 = -2R_0/\lambda$ is a constant phase term.

The final form of the baseband signal for a moving target is

$$x_{\text{BB}}(t) = Ax_p(t-t_d)\exp\left(j2\pi f_d t+\phi_0\right) \tag{12.24}$$

The pulse train term x_p in (12.22) could have also been written in terms of (12.9), but it doesn't really provide any additional insight, other than explicitly showing that t_d changes over time. Over the course of a CPI, as long as the change in time delay (change in range) is small enough to prevent the target from moving from one range bin to another (a phenomenon called *range migration*), the effect of a time-varying t_d in fast-time is negligible. Over slow-time, however, the Doppler frequency term that stems from it is the key to obtaining a target velocity estimate.

The signal-processing details for creating the RDM are covered in Chapter 8, with the CPI data collection in the form of a fast-time/slow-time matrix illustrated in Figure 8.14, which is similar to the FMCW counterpart from Figure 12.3. In the Simulink model presented in Section 12.3.1, the matrix samples will be generated one column at a time (i.e., all fast-time samples in a single PRI will be created at every simulation time step).

12.3.1 Simulink Model of Pulse-Doppler System

The top-level block diagram of the model *PulseDop_Processor1* as well as the signal model subsystem are shown in Figure 12.12. The waveform is phase-modulated using a Barker code of length $K = 7$. The system forms $N = 10$ range gates; therefore the number of fast-time samples that must be collected at each PRI is $K + N - 1 = 16$ (the same numerical example discussed in Section 8.6). In the signal subsystem, a frame of 16 time samples is thus created every simulation time step.

Notice the DSP Sine Wave block, which enables the creation of a frame-based signal. By setting the Output complexity parameter to Complex, this block provides the complex exponential portion of (12.19). The square pulse with the Barker modulation is provided by the bottom path of the subsystem. It starts with a zero vector of length 16, followed by an Overwrite Values block that inserts the Barker-7 code as a sequence of ±1's. This code is shown in Table 8.3 and Figure 8.6. Last, the Zero-Order Hold block forces an output sampling rate equal to the PRI.

The signal processor subsystem is shown in Figure 12.13. It consists of a MF followed by a Doppler FFT.

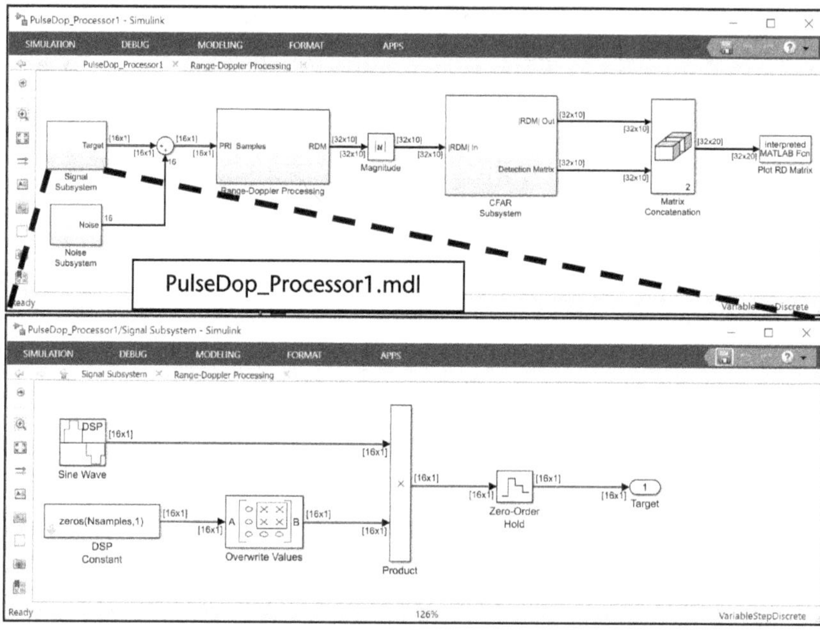

Figure 12.12 Simulink implementation of pulse-Doppler system (frame-based).

The MF is a custom block placed in a user-defined library. It is implemented using an XCORR correlation block just as in the example from Figure 8.8. The impulse-response sequence is contained in the vector RCcode provided by a DSP Constant block. At the output of the MF, a Buffer block collects all $M = 32$ PRIs and passes them to the Doppler FFT to complete the RDM. The CFAR subsystem is identical to that used in the FMCW models and is therefore not shown here. The only difference is in the CFAR constant used here, as the number of range bins used to estimate the noise floor is different than for the FMCW example. Applying (11.38) for 10 range bins and $P_{fa} = 10^{-6}$ results in $T_m = 29.81$. As in the case of the FMCW models, both the RDM and the detections matrix are sent over to a plotting function.

Figure 12.14 shows the resulting RDM and overlaid detection symbol. The target is located in range bin 6 and Doppler bin 10 and was detected by the CFAR process with no false alarms.

The SNR in this example at the RDM is about 30 dB, so target detection was certainly expected. The range sidelobes are 20 $*$ log10(7) = 16.9 dB below the pulse-compressed peak. This sidelobe level happens to be just below the threshold; otherwise, the range sidelobes would have been flagged as detections as well. In that case, some additional clustering or range sidelobe rejection logic would be needed to handle this situation.

Another interesting observation is that the range sidelobe profile is almost visible to the naked eye in the target Doppler filter. From Figure 8.7, we know that the range sidelobe pattern for the Barker-7 code alternate in magnitude 0/1 relative to a mainlobe peak of magnitude 7. On either side of the target at range bin 6, we can observe noticeably darker shades of gray (relative to the noise background) on

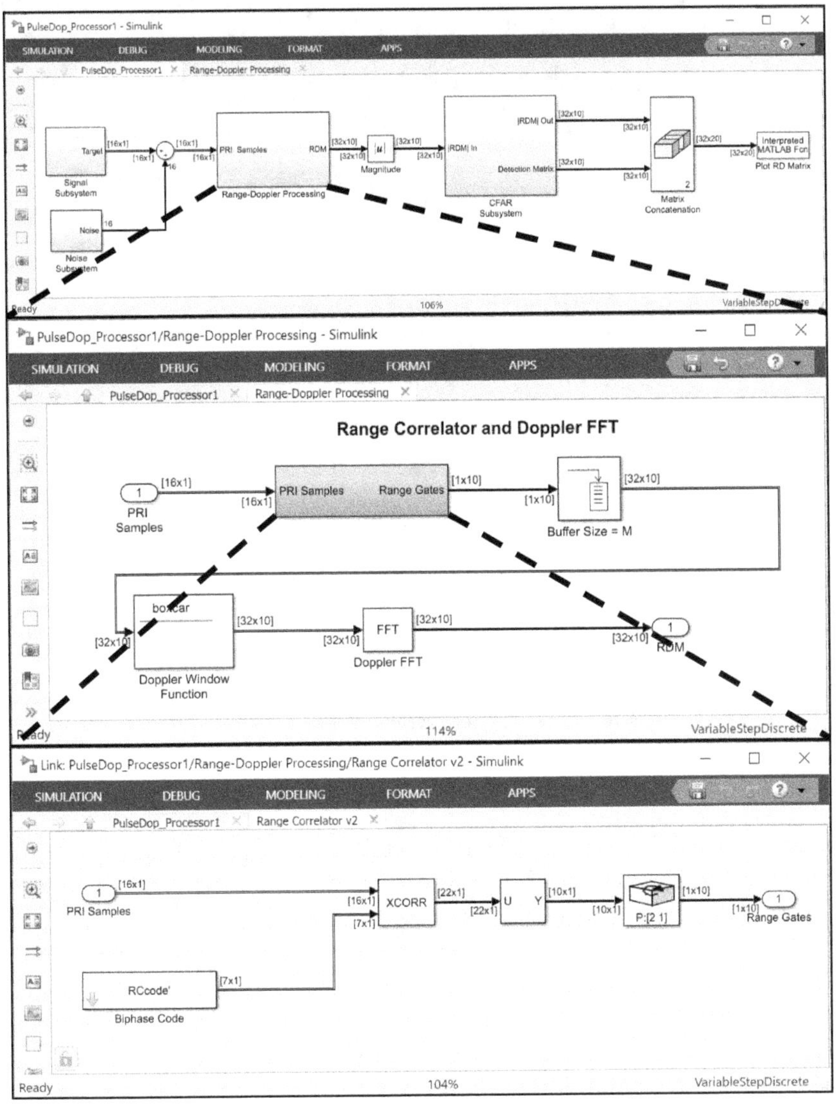

Figure 12.13 Range-Doppler signal processor for the pulse-Doppler system (frame-based).

range bins 2, 4, 8, and 10. These are the Barker-7 range sidelobes, thus providing a nice visual sanity check for this waveform.

12.4 Radar-Processing Timeline and Swerling Fluctuation Models

Collection and processing of pulse-Doppler data over a CPI results in a single RDM. As discussed in Chapter 11, many systems often perform PDI and add N CPIs noncoherently before attempting target detection. We now tie this processing to the Chapter 11 terminology of ROC curves based on N pulses noncoherently integrated and reinterpret the term "pulse" more broadly as a "sample" or

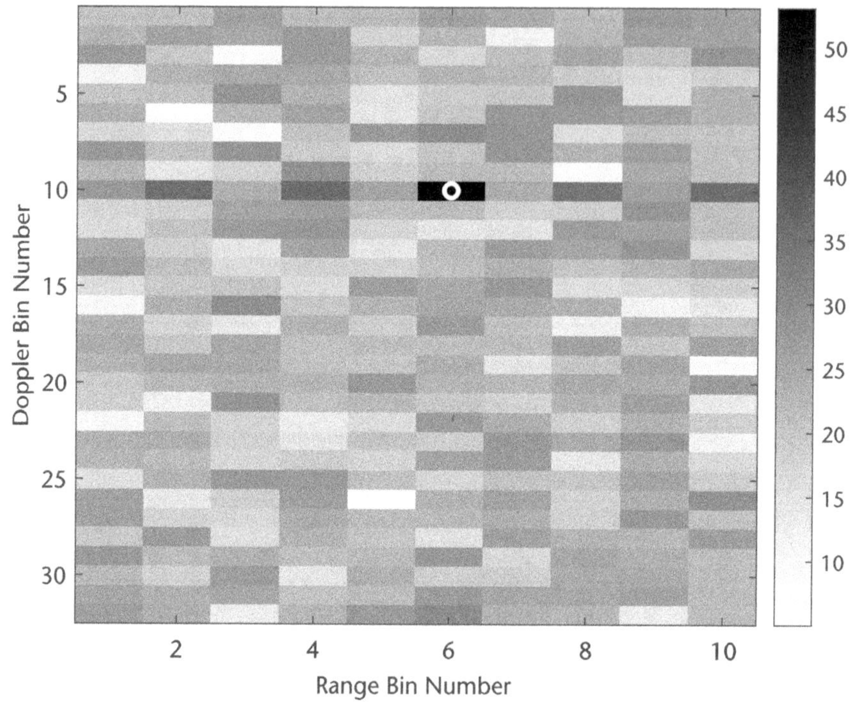

Figure 12.14 RDM for pulse-Doppler system model.

"observation." Each range-Doppler cell in the PDI'd RDM is the result of N samples noncoherently integrated. The single observation case ($N = 1$) therefore corresponds to an RDM from a single CPI (i.e., with no PDI).

With this understanding of the processing timeline for a pulse-Doppler radar, we now revisit the concept of RCS fluctuation for the Swerling models presented in Chapter 5, specifically, the meaning of scan-to-scan and pulse-to-pulse fluctuations, as in the Swerling 1 and Swerling 2 models, respectively (same argument applies to Swerling models 3 and 4).

Clearly, for a single observation ($N = 1$), it must be true that RCS remains constant throughout the CPI, as both Swerling 1 and 2 exhibit identical performance. The fast-fluctuating Swerling 2 RCS must then change no faster than from CPI to CPI. Thus, the "pulse" in the old Swerling terminology is a CPI. That is the right interpretation in a pulse-Doppler context. For the slow-fluctuating Swerling 1, a scan would then be considered the entire group of N CPIs being PDI'd.

Figure 12.15 illustrates the interpretation of the RCS fluctuation model; the top section shows a collection of n physical pulses (lowercase n to avoid confusion with the number of observations) coherently integrated to form a single observation, which is then envelope-detected and compared to a detection threshold. This is the $N = 1$ case.

Furthermore, if each pulse has a pulse-level SNR = SNR_1, then the integrated SNR will be (e.g., assuming no weighting losses) $\text{SNR}_n = n * \text{SNR}_1$. SNR_n should be the value used in the Pd formulas presented in Chapter 11 (e.g., in Table 11.3), along with $N = 1$.

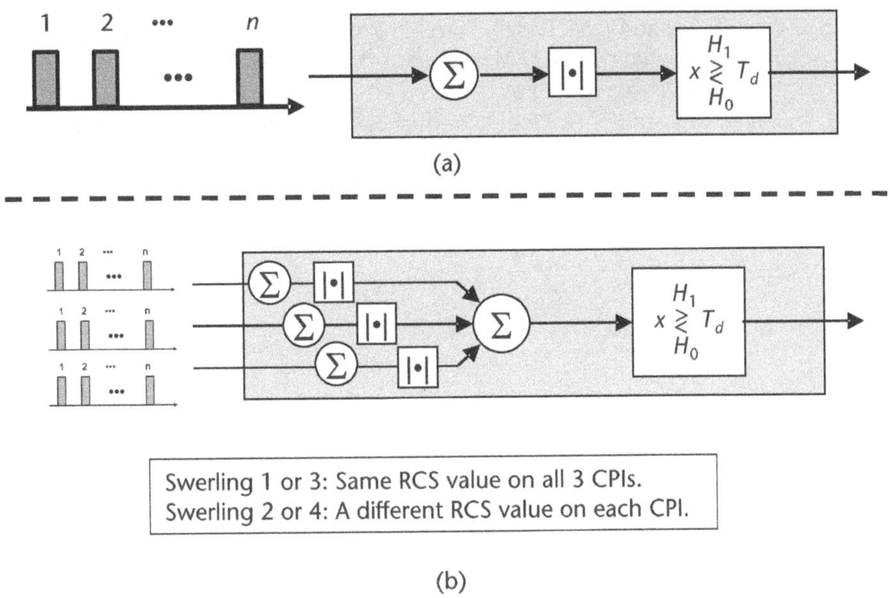

Figure 12.15 Swerling RCS fluctuation models in the context of pulse-Doppler processing: (a) single CPI = single observation (N = 1), and (b) PDI of 3 CPIs = three observations (N = 3).

Figure 12.15(b) shows the case $N = 3$, where now PDI is applied to the three individual observations before threshold comparison. A fast-fluctuation model (Swerling 2 or 4) calls for an independent RCS value applied to each of the three CPIs (fixed during the CPI). A slow-fluctuation model (Swerling 1 or 3) would use a single RCS value for all three CPIs. Notice that if the radar employs RF agility (a common way to induce fast RCS fluctuations in an otherwise slow-fluctuating target), it would also be applied from CPI to CPI (to preserve coherency within a CPI). This is consistent with the model presented here.

12.5 Concluding Remarks

The signal-processing models presented here represent a straightforward implementation of the range and Doppler processing described in Chapter 8. The pulse-Doppler system model in particular will be used later on as part of the larger end-to-end model that's been alluded to a few times. It provides an indispensable component of the process, although more work is still needed. Target detections enable measurements that provide information on the target state over time. The infrastructure to maintain and update an estimate of this target state is provided by the tracking process, which is addressed in Chapter 13.

References

[1] Baker, C. J., and S. O. Piper, "CW Radar: Configuration Types," in *Principles of Modern Radar, Vol III: Radar Applications*, W. L. Melvin and J. A. Scheer (eds.), Raleigh, NC: Scitech Publishing, Inc., 2014, pp. 23–26.

[2] Keel, B. M., and J. M. Baden, "Stretch Processing," in *Principles of Modern Radar, Vol II: Advanced Techniques*, W. L. Melvin and J. A. Scheer (eds.), Raleigh, NC: Scitech Publishing, Inc., 2013, pp. 26–40.

[3] Baker, C. J., and S. O. Piper, "Frequency Modulated CW Radar: Range Resolution," in *Principles of Modern Radar, Vol III: Radar Applications*, W. L. Melvin, J. A. L. Melvin, and J. A. Scheer (eds.), Raleigh, NC: Scitech Publishing, Inc., 2014, pp. 44–52.

CHAPTER 13

Target Tracking

13.1 Introduction and Basic Terminology

The IEEE defines tracking as "the process of following a moving object or a variable input quantity. In radar, target tracking in angle, range, or Doppler frequency is accomplished by keeping a beam or angle cursor on the target angle, a range mark or gate on the delayed echo, or a narrowband filter on the signal frequency, respectively" [1].

In a more general sense, we may think of tracking as the process of maintaining knowledge of the state of a given target over time. In this context, a target is any object of interest within the radar sensor field of view. The state typically comprises the target position and velocity, perhaps acceleration, plus any ancillary information such as SNR or RCS. Estimates of these parameters are collected as elements of a state vector $\hat{\mathbf{x}}_k$ where k is the target index. Strictly speaking, $\hat{\mathbf{x}}_k$ is an estimate of the true target state, $\mathbf{x}_k$, which is unknown. The radar makes measurements (or observations) of the true state, but these observations are corrupted by noise and other error sources. Radar measurements, such as range, angles, and sometimes range rate, are collected in a measurement vector $\mathbf{z}_m$, where m is the target-detection index. The target-state estimate $\hat{\mathbf{x}}_k$ is formed from these measurements. Although $\hat{\mathbf{x}}_k$ is an *estimated* state vector, for brevity we herein refer to it simply as the state vector.

Thus, the target-state vector is initialized in the tracker, and then updated over time with additional radar measurements. As will be described later, the tracking formalism includes a measurement covariance matrix $\mathbf{R}_m$, which captures the level of uncertainty in the raw measurements due to noise and other factors.

The tracking process may thus be considered as filtering or smoothing these noisy measurements over time, resulting in a more accurate estimate of the true target state. This filtering process may be performed in hardware (i.e., a dedicated tracking circuit, more typical of older systems), but nowadays it is more common to implement it in software, in the form of a tracking algorithm. Even with this filtering, there is some residual level of uncertainty in the state vector $\hat{\mathbf{x}}_k$. Again, the true state is never known perfectly. This uncertainty is captured by $\mathbf{P}_k$, the state-error covariance matrix. All of these vector and matrix quantities are part of the formalism of tracking algorithms, which are detailed later in this chapter.

The state information on a single target ($\hat{\mathbf{x}}_k$, $\mathbf{P}_k$) is maintained in a so-called track file, and the specialized software that manages the potentially multiple track files over their life cycle (creation, maintenance, and deletion) is sometimes referred to as the Track File Manager (TFM). Table 13.1 summarizes the discussion on these quantities, which are revisited during the modeling example discussions.

13.2 Radar Tracking Modes

There are multiple strategies for the radar to pursue the acquisition of additional measurements needed to maintain a target track. These strategies are implemented in radar modes, which dictate the radar scheduling and general behavior. Two common tracking modes are single-target track (STT) and track while scan (TWS). In STT, as the name implies, the radar points at a single target continuously and tracks only that target. The track file is continually updated—provided that the target continues to be detected—with measurements taken at the radar processing rate (radar dwell).

In TWS, the radar can track multiple targets simultaneously, or track a target while continuing to search for additional targets. In older radars, this is accomplished by pointing the beam sequentially in the desired directions (time-sharing). Modern sensors capable of synthesizing multiple beams are able to track and search simultaneously by splitting the aperture (power-sharing). Phase-array radars (PARs) with element-level or subarray-level control are the enabling technologies to accomplish this.

These radar modes may be performed separately by dedicated radars; for example, a search radar may hand off a target detection to a tracking radar, which then performs continuous tracking or STT. Systems such as airborne fire-control radar (FCR), on the other hand, are able to transition from a search mode into either STT or TWS depending on mission parameters and engagement conditions.

Table 13.1 Multidimensional Variables Associated with Radar Measurements and Target State Estimates

Symbol	*Definition*	*Remarks*
$\mathbf{z}_m$	Measurement vector from the *m*th detection	Collection of all measurements associated with a detection (threshold crossing)
$\hat{\mathbf{x}}_k$	State vector of the *k*th track file	Contains the best current estimate of the *k*th target state (typically position and velocity and potentially other parameters)
$\mathbf{R}_m$	Measurement error covariance matrix	Captures the uncertainty of the measurements made on the *m*th detection
$\mathbf{P}_k$	State error covariance matrix	Captures the uncertainty in the knowledge of the state of the *k*th target

13.3 Tracking Initiation and Management Process

Modern radars will hardly ever transition from a single raw detection directly into track. It would be inefficient to invest tracking resources into what may well be a false alarm; thus, a series of steps are first taken to ensure that only real objects are tracked. The exact number and complexity of the steps varies from system to system, but they will tend to perform the following general functions:

1. *Alert:* This is the target-detection process applied to a search dwell. It consists of the initial declaration of the presence of a potential target as indicated by a threshold crossing (e.g., as the result of a CFAR processor operating on a RDM).
2. *Confirmation (validation):* Typically, a valid target detection is not declared until it is detected a subsequent time, either by scheduling a confirmation dwell immediately after the initial alert, or on the next TWS scan. Often, binary or m-of-n logic is applied in this step to ensure some target persistence, as opposed to a spurious detection caused by a noise spike.
3. *Correlation (association):* Once the persistence of the detection has been confirmed (i.e., that it is real), the next step is to compare the measurements from this detection to existing track files. The track file association process answers the question: Is this something I have seen before?
4. *Track file initialization:* If the detection does not associate with any existing tracks, a new track file may then be started. The new measurements may be used to initialize the track file state vector, often involving some form of coordinate transformation. Other system-dependent conditions may need to be met, such as checking whether the tracker has not reached the maximum number of track files it is designed for.

Most of these steps are captured in the example logic-flow diagram depicted in Figure 13.1. This is a simplified diagram intended to convey some simple conditions for transitioning between a generic volume search mode and a TWS mode that supports multiple target tracking.

Figure 13.1 intends to show a continuous decision loop moving through multiple beam positions within a search mode, while reacting to detections within a beam/dwell and taking the appropriate action based on existing conditions. It includes the track-initiation steps listed earlier in this section, plus a few additional management actions. The Track File Management block at the beginning is intended to periodically check the status of existing track files. The TFM that executes these actions may be an operator (human-in-the-loop), or it could be a software entity running in the mission computer of a missile seeker, for example.

Track files typically have a defined lifetime, or period of time between measurement updates where the state information is still considered valid. If a track file is approaching its lifetime, the TFM may request a dedicated track dwell to update the state information with current measurements. A track file that exceeds its lifetime becomes stale, and its data is deleted, thereby freeing resources to the system. This may be the case for track files whose corresponding

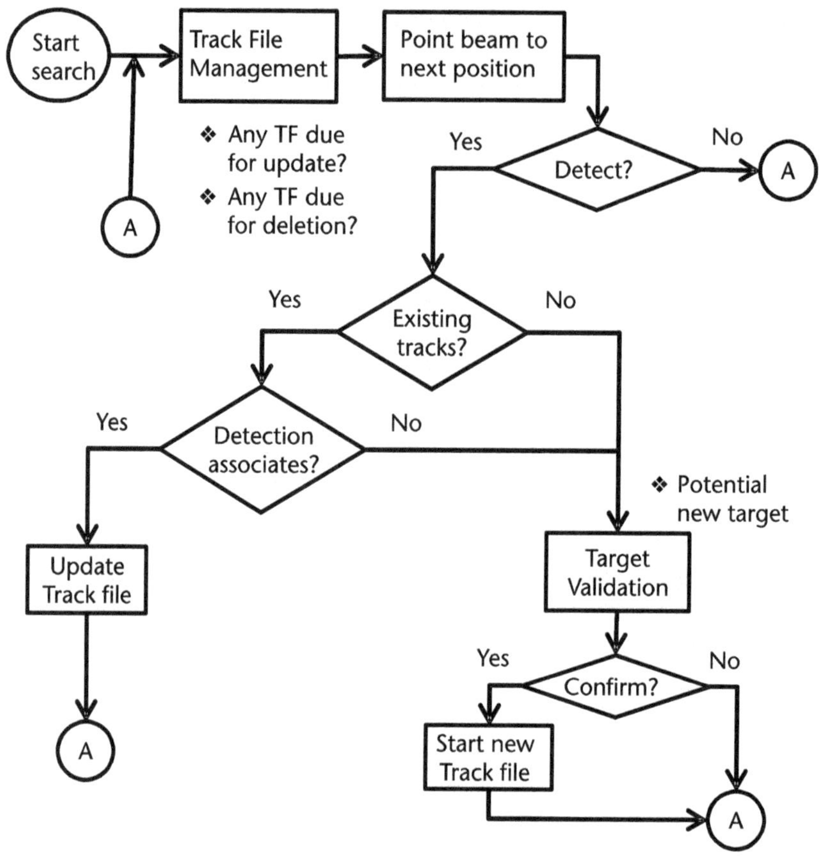

Figure 13.1 Generic flow diagram of track-file management and initiation process.

targets have escaped the sensor field-of-view, or that might have been taken out by a weapon.

The TFM actions just mentioned are not shown explicitly in Figure 13.1, as this is intended as a relatively simple example. Additional logic paths and decision points to cover all foreseeable operational conditions will make this diagram considerably more complex.

13.4 Tracking M&S Considerations

A multiple target-tracking algorithm, in general, is a pretty complex piece of software. The actual tracking filter is usually based on rigorous mathematical formalism, such as a Kalman filter. In contrast, the executive software that includes the TFM, scheduler, and other higher-level rule-based functions are often empirical in nature. In terms of tracking M&S, if the objective is to simulate a real system, the best possible fidelity will be achieved by running the actual tracking software as part of the simulation environment.

If the actual software is not available or does not yet exist (for example, if the model does not correspond to an actual system), then a surrogate would be the next best option. Surrogate in this context means a simplified and/or notional version of the desired tracking algorithm. If the tracking code is created from scratch, it will likely be incomplete at first. It may cover about 80% of the operational space but occasionally may run into unanticipated conditions, either crashing or providing nonsensical results. Over time, the model software will need to be fine-tuned and stabilized so that it can perform robustly over all operational conditions.

13.5 Modeling Examples

Our first modeling example addresses the challenge of associating multiple detections to existing track files (i.e., step #3 of the tracking management process discussed in Section 13.3). This example implements a form of the nearest-neighbor (NN) algorithm. The second example looks at the processing of a single-target tracking filter over multiple dwells, and implements the classic discrete-time Kalman filter (DTKF) algorithm.

13.5.1 Data Association Model (Nearest Neighbor)

In a multiple-target scenario, a single radar processing dwell may yield multiple detections and corresponding measurements. A set of rules is needed to intelligently assign these measurements to existing track files. Any set of measurements that does not correlate or associate to a track file becomes a candidate for initializing a new track file. There are several schemes that may be used to accomplish this, typically based on some distance metric [2].

In the NN implementation presented here, the distance metric is related to the likelihood that a measurement set belongs to a given track file, assuming the measurements belong to a multivariate Gaussian PDF. This is defined as

$$f\left(\mathbf{z}_m, \hat{\mathbf{x}}_k, \mathbf{S}_{mk}\right) = \frac{1}{\sqrt{2\pi\left|\mathbf{S}_{mk}\right|}} e^{-\frac{1}{2}\left(\mathbf{z}_m - \hat{\mathbf{x}}_k\right)^T \mathbf{S}_{mk}^{-1}\left(\mathbf{z}_m - \hat{\mathbf{x}}_k\right)} \quad (13.1)$$

where we now refer back to the matrix-vector terminology first introduced in Section 13.1. The covariance matrix $\mathbf{S}_{mk}$ is

$$\mathbf{S}_{mk} = \mathbf{R}_m + \mathbf{P}_k \quad (13.2)$$

and $|\mathbf{S}_{mk}|$ is its determinant.

As a simplifying assumption, (13.1) and (13.2) assume that the measurement and state vectors (and corresponding covariance matrices) not only have the same dimensions, but that they are are defined in the same space. Otherwise, an additional space transformation is required. The DTKF example will address the general case.

We now define the log-likelihood (or more accurately, the negative log-likelihood) of (13.1) as

$$\delta_{mk} = -\ln\left(f\left(\mathbf{z}_m, \hat{\mathbf{x}}_k, \mathbf{S}_{mk}\right)\right) = \frac{1}{2}d_{mk}^2 + \ln\left(\sqrt{2\pi\left|\mathbf{S}_{mk}\right|}\right) \tag{13.3}$$

where

$$d_{mk} = \sqrt{\left(\mathbf{z}_m - \hat{\mathbf{x}}_k\right)^T \mathbf{S}_{mk}^{-1}\left(\mathbf{z}_m - \hat{\mathbf{x}}_k\right)} \tag{13.4}$$

The quantity d_{mk} is known as the Mahalanobis distance, after the Indian scientist and statistician P. C. Mahalanobis. It is a statistical measure of the divergence of a sample from a distribution.

A measurement vector $\mathbf{z}_m$ is assigned to (or associated with) the kth track file when it maximizes the likelihood that it belongs to the PDF with mean $\hat{\mathbf{x}}_k$ and covariance $\mathbf{S}_{mk}$ [i.e., that it maximizes (13.1)]. That is equivalent to minimizing (13.3).

The NN algorithm therefore consists of calculating (13.3) for all available measurements and track files and pairing those with the smallest value. Notice that the second term in (13.3) is proportional to the covariance determinant (i.e., it is a minimum when the combined uncertainty around both the measurements and the state is the lowest among all available combinations).

The MATLAB m-file *simple_association* implements this method by creating a K-by-M matrix $\mathbf{C}$ whose elements are given by (13.3) for all combinations of K track files and M measurement vectors in a given scenario. The model is based on MATLAB code provided by George Brown from GTRI. The algorithm recursively inspects the matrix and finds the smallest element that defines a measurement-to-track association. The row and column corresponding to the selection are eliminated from further consideration, and the process is repeated until all associations are made. All unassigned measurements at the end of the process are flagged for potentially initializing new track files.

Another concept related to the measurement association is that of a tracking gate. The tracking gate is a threshold that must be met by a measurement-to-track distance before the pair is considered a candidate for association. This additional restriction can be easily included in the aforementioned implementation by inserting an additional row and column in the distance matrix $\mathbf{C}$, with all elements set to the desired threshold value.

The following example illustrates the process. Assume a scenario where two track files and three measurement sets are present, with the corresponding distance matrix calculated as

$$C = \begin{bmatrix} 1.2 & 14.7 & 0.01 \\ 3.4 & 8.1 & 4.2 \end{bmatrix} \tag{13.5}$$

First, assume that the tracking gate value is 9.0. What would be the results of the NN association?

The global minimum in (13.5) is the (1,3) element with a value of 0.01, which is below the tracking gate threshold. Thus, measurement #3 would be associated to track #1. Eliminating row 1 and column 3 from further consideration, the remaining minimum would be the (2,1) element with a value of 3.4. Therefore, measurement #1 would be associated with track #2, as their distance is also within the tracking gate. The remaining measurement #2 would be a candidate for a new track file.

Now consider a tracking gate value of 3.0. How would the results be different? The answer is that only the association with track #1 would take place. All distances for track #2 fall outside the gate; thus track #2 does not get associated, and both measurements #1 and #2 become candidates for new track files.

The MATLAB example is executed by running the script *test_association*, which sets up data for two existing tracks and four sets of measurements, then calls *simple_association*. The intended results are for the first two measurements to associate with the first two tracks and the remaining measurements to be designated for new track files.

Once the measurement-to-track associations are identified, the next step in the tracking process is to update the tracks with the new information provided by the measurements. The mechanics of doing that are explored in Section 13.5.2.

13.5.2 STT Model (Kalman Filter)

Tracking algorithms are sets of mathematical equations that provide a means to estimate the state of a process under some optimality criteria. The process of interest here of course is the behavior of a target moving in space. Many highly sophisticated algorithms have been developed over the last 60-plus years; here we will focus on the venerable Kalman filter. This is a classic algorithm that recursively calculates a state estimate by minimizing the mean-squared error. Named after Hungarian-American engineer Rudolf Kalman, it was first described in his seminal paper published in 1960 [3], and it is widely used in radar, sonar, and multiple other applications. We specifically consider here the DTKF. Countless articles and tutorials on the DTKF are available online as well as in printed form. An excellent online resource is the introductory paper by Welch and Bishop [4].

The DTKF comprises two sets of equations: time-update (or time-propagation) equations and measurement-update equations. The time updates project forward in time both the state vector and the error covariance, while the measurement updates blend in the new information provided by the measurement into a new and improved state.

In presenting these equations, we refer back to the terminology introduced in Table 13.1 for state and measurement vectors and covariance matrices. Since we only consider here a single-target case, we drop the track-and-measurement indexes k and m, respectively. Instead, we will adopt n as a time index in all of these terms. Thus, the state vector at discrete time n is denoted as $\hat{\mathbf{x}}_n$, and so forth.

13.5.2.1 Time-Update (Propagation) Equations

The state-propagation equation is

$$\hat{\mathbf{x}}_n^- = \mathbf{A}\hat{\mathbf{x}}_{n-1} \tag{13.6}$$

where **A** is a state-transition matrix, and the minus sign superscript on the LHS of (13.6) indicates it is an a priori estimate (i.e., before any measurement has been incorporated at time step n). As a simple illustration, consider 1-D motion with constant velocity. If we define the state vector as

$$\hat{\mathbf{x}}_n = \begin{bmatrix} p_n \\ v_n \end{bmatrix} \tag{13.7}$$

where the elements p_n and v_n are the estimated target position and velocity at time n, respectively, then the state-transition matrix would have the form

$$\mathbf{A} = \begin{bmatrix} 1 & \Delta t \\ 0 & 1 \end{bmatrix} \tag{13.8}$$

where Δt is the time step size from time $n - 1$ to time n. The error-covariance propagation equation is

$$\mathbf{P}_n^- = \mathbf{A}\mathbf{P}_{n-1}\mathbf{A}^H + \mathbf{Q} \tag{13.9}$$

where **Q** is known as the process noise covariance, and H is the Hermitian (complex-conjugate transpose) operator. The process noise covariance is intended to capture error sources in the propagation model, which are often unknown and unknowable. In practice, **Q** is usually treated as a tuning parameter.

Once a track file is established, the track is time-updated at some regular interval within the radar timeline, whether there is a target measurement available or not. If at any given dwell the target is detected and the measurements associate with the track, the track is first time-updated, then measurement-updated.

13.5.2.2 Measurement-Update Equations

The first step in updating the target state with a new measurement is to calculate the so-called Kalman gain of the filter:

$$\mathbf{K}_n = \mathbf{P}_n^-\mathbf{H}^H\left(\mathbf{H}\mathbf{P}_n^-\mathbf{H}^H + \mathbf{R}\right)^{-1} \tag{13.10}$$

where **H** is a transformation matrix (often called the observation matrix) that takes a vector in state space and brings it into measurement space. Notice that the term in parenthesis is a more general form of (13.2).

Another way to think about the observation matrix is that it models the sensor, in that it forms the measurement vector by observing the true state $\mathbf{x}_n$. The general model is

$$\mathbf{z}_n = \mathbf{H}\mathbf{x}_n + \mathbf{r}_n \tag{13.11}$$

where $\mathbf{r}_n$ is measurement noise corrupting the observation. Assume the same state space definition as in (13.7). Now assume that the sensor only measures position (e.g., range) and not velocity. The observation matrix would then be

$$\mathbf{H} = \begin{bmatrix} 1 & 0 \\ 0 & 0 \end{bmatrix} \tag{13.12}$$

For a sensor that measures both position and velocity (such as a pulse-Doppler radar), the observation matrix would be the identity matrix.

The Kalman gain is the optimal blending factor that weighs the new information from the measurement against the existing information from the current state. It can be seen from (13.10) that the Kalman gain varies between 0 (for $\mathbf{P}_n^- \to 0$) and $\mathbf{H}^{-1}$ (for $\mathbf{R} \to 0$).

The state update is

$$\hat{\mathbf{x}}_n = \hat{\mathbf{x}}_n^- + \mathbf{K}_n\left(\mathbf{z}_n - \mathbf{H}\hat{\mathbf{x}}_n^-\right) \tag{13.13}$$

The quantity in parentheses is termed the *residual* or the *innovation*. It is new information being presented to the filter, calculated as the difference between the measurement and the current a priori state vector (transformed to measurement space). If the track state has very low uncertainty, the filter will believe the state more than the measurement, quantified by a very low Kalman gain. In the limit, $\mathbf{K}_n$ approaches zero, which means the a posteriori (i.e., updated) state is the same as the a priori state, or $\mathbf{x}_n = \mathbf{x}_n^-$. If instead, the measurement has very low relative uncertainty, then the filter will believe the measurement more. In that case, $\mathbf{K}_n$ approaches $\mathbf{H}^{-1}$ in the limit such that (13.13) becomes

$$\mathbf{x}_n \approx \mathbf{x}_n^- + \mathbf{H}^{-1}\left(\mathbf{z}_n - \mathbf{H}\mathbf{x}_n^-\right) = \mathbf{x}_\text{n}^- + \mathbf{H}^{-1}\mathbf{z}_n - \mathbf{x}_n^- = \mathbf{H}^{-1}\mathbf{z}_n \tag{13.14}$$

Lastly, the error covariance update is

$$\mathbf{P}_n = \left(\mathbf{I} - \mathbf{K}_n\mathbf{H}\right)\mathbf{P}_n^- \tag{13.15}$$

Following the same argument, (13.15) varies between $\mathbf{P}_n = \mathbf{P}_n^-$ for the $\mathbf{K}_n = 0$ extreme (trust the state), and $\mathbf{P}_n = 0$ for the $\mathbf{K}_n = \mathbf{H}^{-1}$ extreme (trust the measurement).

The preceding expressions [(13.6) to (13.15)] are all linear equations. When the processes that govern the target state and/or the measurements are nonlinear in nature, these no longer provide a valid representation, and more complex versions of the filter are required. For example, the target position and velocity may be defined in Cartesian coordinates, while measurements of target range, azimuth, and elevation angles are akin to spherical coordinates. Thus, a nonlinear transformation is required between the two spaces. One approach is the extended Kalman filter (EKF), which linearizes the estimates by calculating Jacobian matrices (partial derivatives) of the nonlinear processes. Other methods exist in the literature.

These higher-order mathematical models are beyond the scope of this text, but the reader is referred to [4, 5] for additional information.

13.5.2.3 Measurement Noise Covariance

If the radar measurements are taken independently from one another, then the measurement covariance **R** becomes a diagonal matrix. The terms in the main diagonal correspond to the variances of each individual measurement. Traditionally, these terms are estimated based on the CRLB, which is proportional to the resolution element of the dimension being measured, and inversely proportional to SNR. The CRLB σ_{AoA} for monopulse AoA, presented in Chapter 9 and defined by (9.27), is repeated here for convenience, along with the corresponding terms σ_R and σ_f for range and Doppler measurements:

$$\sigma_R = \frac{\delta_R}{\sqrt{2SNR}} \tag{13.16}$$

$$\sigma_f = \frac{\delta_f}{\sqrt{2SNR}} \tag{13.17}$$

$$\sigma_{\text{AOA}} = \frac{\theta_3}{k_m\sqrt{2SNR}} \tag{13.18}$$

The Doppler and range-resolution elements δ_f and δ_R are defined by (8.18) and (8.23), respectively.

13.5.2.4 MATLAB Model Example

The DTKF modeling example looks at the evolution of a single track file over time, given a number of detections and corresponding measurements. Simplifying assumptions include the following:

- Only one target is present;
- The target is detected 100% of the time;
- There are no false alarms.

The DTKF equations are implemented in three different function m-files: *kalman* implements a 2-D filter for range and range rate, while *kalmanAz* and *kalmanEl* implement 1-D filters for azimuth and elevation angles, respectively. The state space is the same as the measurement space (R, Rdot, Az, and El), such that the observation matrix **H** is the identity matrix. Measurement noise variances are calculated based on (13.16) to (13.18). All three DTKF functions are called from the main executive function *TrackTarget*, which serves as the TFM. The filter states are initialized based on a notional cue as would be provided by, for example, a search radar.

(a)

(b)

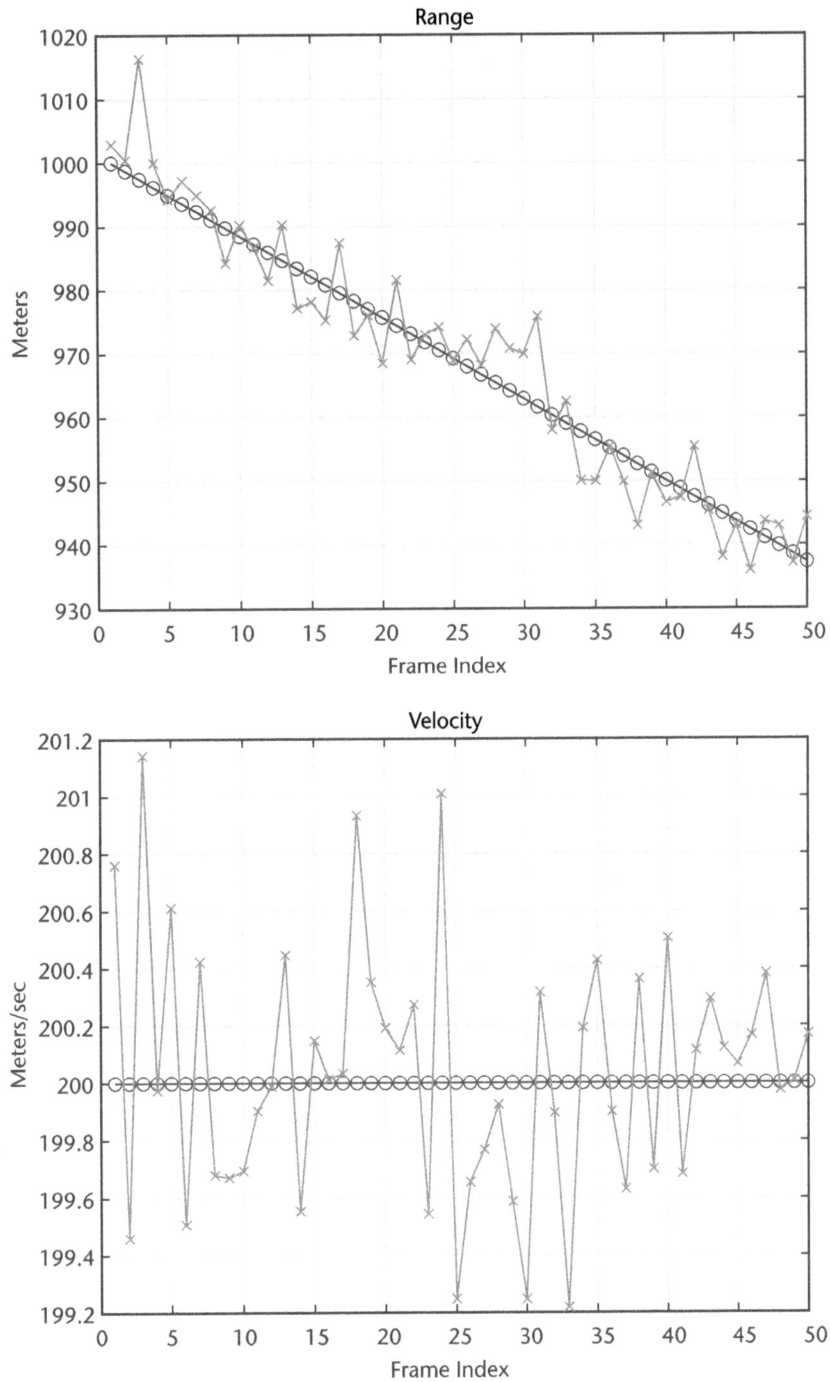

Figure 13.2 Target truth and noisy measurement inputs to DTKF model: (a) range, (b) velocity, (c) azimuth angle, and (d) elevation angle.

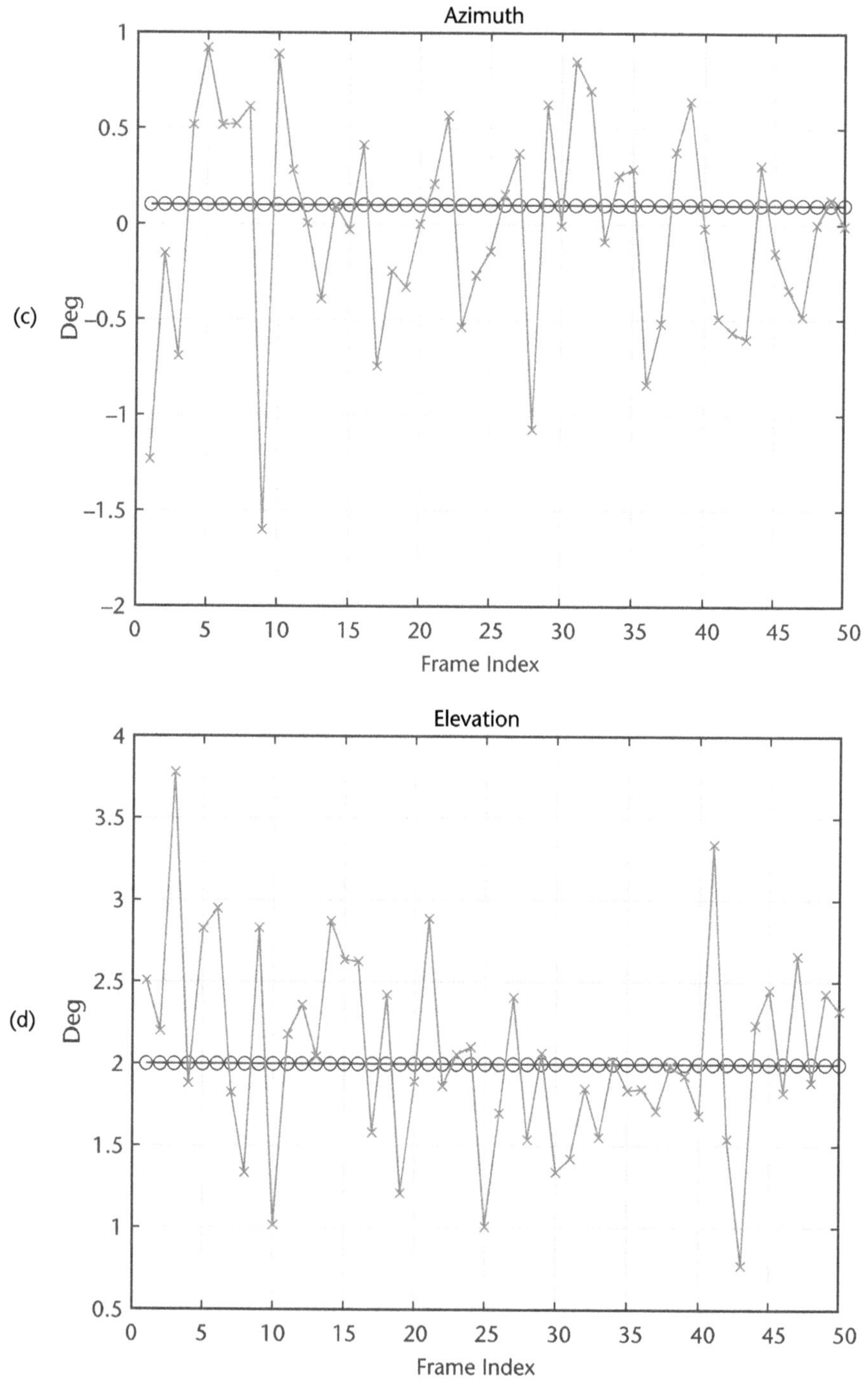

Figure 13.2 *(Continued)*

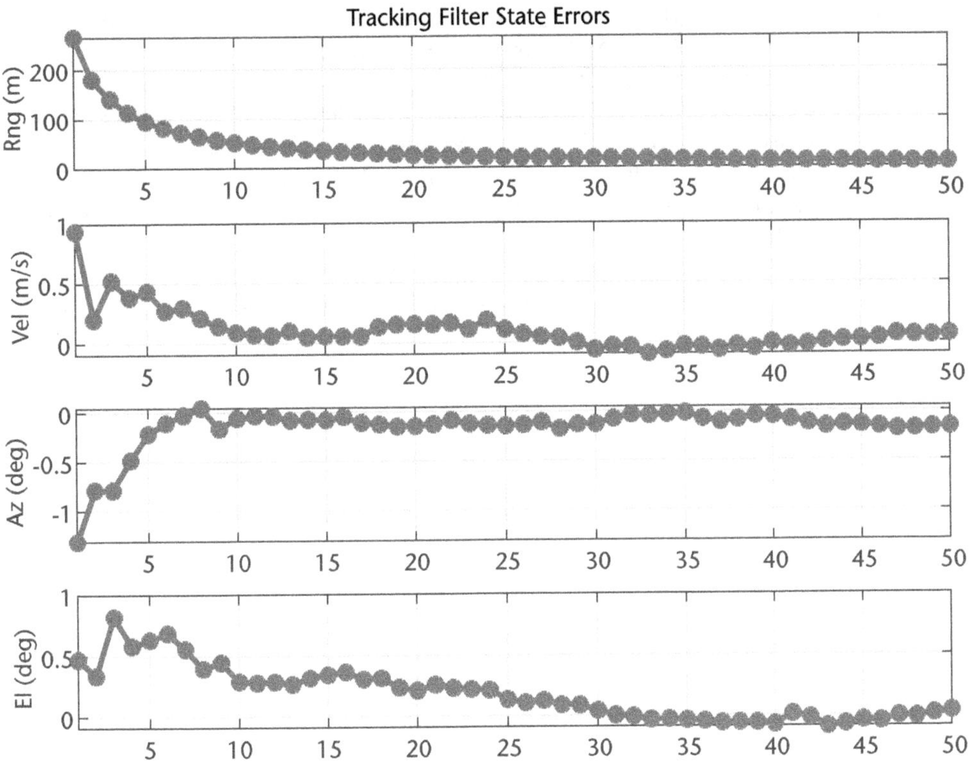

Figure 13.3 DTKF state errors: range, velocity, azimuth angle, and elevation angle (top to bottom).

The model is executed by running the script *test_tracker,* which creates the truth data based on a linear motion model with constant velocity, generates the measurements as truth-plus-noise, and calls *TrackTarget* over a total of 50 dwells. Output plots of truth and measurements for all four variables are shown in Figure 13.2. The filter-tracking errors over time (state minus truth) are shown in Figure 13.3.

13.6 Concluding Remarks

The process of target-tracking is inextricably linked to other radar functions. It includes mathematically rigorous procedures for estimating target-state information in the form of tracks, as well as higher-order functions that manage radar resources to optimally search and detect targets while maintaining said tracks. Only a couple models of basic approaches are explored here, namely the NN for data association and the DTKF for actual tracking. We revisit the DTKF model as part of the larger engagement simulation in Chapter 15.

References

[1] IEEE Aerospace and Electronic Systems Society, sponsored by the Radar Systems Panel, "IEEE Standard Radar Definitions," *IEEE Std. 686™-2008,* May 2, 2008.

[2] Blair, W. D., "Measurement-to-Track Data Association," in *Principles of Modern Radar, Vol I: Basic Principles,* M. A. Richards, J. A. Scheer, and W. A. Holm (eds.), Raleigh, NC: Scitech Publishing, Inc., 2010, pp. 760–766.

[3] Kalman, R. E., "A New Approach to Linear Filtering and Prediction Problems," *Journal of Basic Engineering,* Vol. 82, No. 1, March 1960, pp. 35–45.

[4] Welch, G., and G. Bishop, *An Introduction to the Kalman Filter,* Technical Report TR 95-041, Dept. of Computer Science, University of North Carolina, Chapel Hill, NC, Updated July 24, 2006.

[5] Blair, W. D., "Radar Track Filtering," in *Principles of Modern Radar, Vol I: Basic Principles,* M. A. Richards, J. A. Scheer, and W. A. Holm (eds.), Raleigh, NC: Scitech Publishing, Inc., 2010, pp. 757–760.

CHAPTER 14

Engagement Geometry

14.1 Introduction

One important feature of realistic engagement simulations is the ability of objects to move in space as a function of time. Simulating movement entails defining the object's position and higher derivatives (velocity, acceleration, etc.), as well as its orientation relative to some frame of reference, typically described as a coordinate system (CS). Quite often, multiple CSs are utilized in dynamic simulations. In a radar-simulation context, measurements are made relative to some CS—for example, the antenna CS—and must later be transformed to a different system.

This chapter has the following three overall objectives:

- Describe common coordinate systems and transformation methods between them;
- Define the main radar observables (range, Doppler, angles) as a function of vector components defined in the aforementioned systems;
- Present a target-generator model that will implement the calculation of these radar observables given the required input vectors.

14.2 CSs and Their Transformations

Figure 14.1 illustrates the three main CSs we focus on in this book and in our simulation development. All of these are right-handed, orthogonal coordinate systems.

The inertial CS (ICS), also referred to as a local-level (LL) CS or as a local tangent–plane (LTP) CS, is used as the principal frame of reference for all objects in

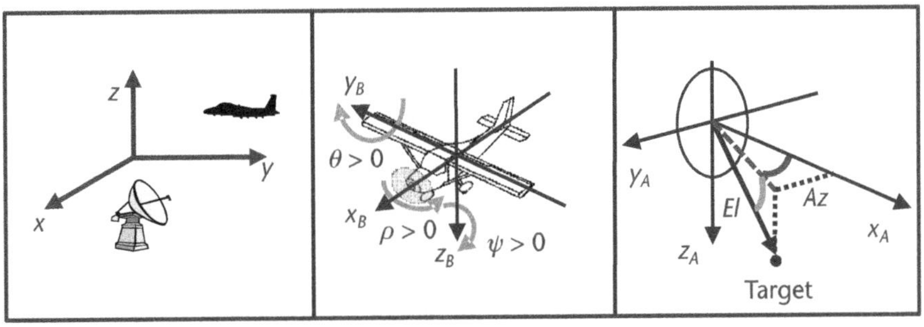

Figure 14.1 Common coordinate systems in engineering and engagement simulations: (left) inertial CS, (middle) body CS, and (right) antenna CS.

our simulation, and their motion is described relative to it. It is referred to as the "local ellipsoidal (geodetic) system" by Torge [1] and may be connected to global systems such as the Earth-centered Earth-fixed (ECEF) CS. The ICS may be envisioned as planting a stake on the ground to indicate the origin of the system, and defining object positions relative to it. The term inertial refers to a coordinate system that does not undergo any acceleration. Since the Earth is rotating, strictly speaking this CS is not an inertial frame of reference in the classical physics sense. We will treat it as such to a first approximation, however, since our focus is on engagement-level simulations, which, by definition, encompass a relatively small area with surface and/or airborne objects. If a larger geographical scenario is considered, such as involving space-based platforms, then secondary effects such as Coriolis forces, would need to be considered, as they would have a more significant impact in our calculations. The axes orientation illustrated in the inertial CS of the figure corresponds to the east-north-up (ENU) convention, where the positive x- and y-axes point due east and north, respectively. Another common convention is north-east-down (NED).

The body CS defines the orientation of every object in the simulation, whether moving or stationary, such as ground vehicles, naval vessels, and aircraft. The NED convention is shown for the example in the middle of Figure 14.1, where the north or first positive axis is parallel to the main axis (longitudinal axis) of the object (i.e., the body vector), and the origin is placed at the center of gravity (CG) of the object. The east axis or lateral axis is positive to the right of the north axis, and the down or normal axis completes the orthogonal, right-handed system [2].

Arguably the most common way of defining orientation is by the three Euler angles, named after the Swiss mathematician Leonhard Euler. They are yaw (ψ), pitch (θ), and roll (ρ); these are angles of rotation around the D, E, and N axes, respectively. The rotation conventions for positive angles are illustrated in Figure 14.1: Yaw (also called heading) and roll angles are positive to the right, while the pitch angle is positive up. We note that there are alternate definitions for Euler angles in the literature, in particular the so-called proper Euler angles; here we follow the conventions defined by Scottish physicist Peter Guthrie Tait and English mathematician George Bryan. Their version of Euler angles are often referred to as Tait-Bryan angles [3]; these angles define rotations in three distinct axes and are widely used in aerospace applications. In orthopedics and related medical fields, they are used to describe anatomical joint motion and are commonly referred to as Cardan angles, named after Italian mathematician Geronimo Cardano [4].

The antenna coordinate system (ACS) is used to define azimuth and elevation angles of objects (targets) relative to the antenna boresight. Using the NED convention for the ACS shown in Figure 14.1, the face of the antenna is placed in the east-down plane, and the positive north axis coincides with the antenna mechanical boresight. Placements of the antenna face in other planes of the ACS are common.

In principle, nothing prevents us from adopting dissimilar conventions for the three coordinate systems of interest. One may choose an ENU convention for the ICS, while adopting an NED convention for the body CS of an airborne object, for example. The price to pay is some additional bookkeeping in our CS transformations. As discussed in Section 14.2.1, coordinate transformations are primarily rotation operations; however, if dissimilar conventions are used, an additional step is required to reorder the axes. Given a vector in ENU coordinates, for example, its corresponding NED version is

$$\begin{bmatrix} N \\ E \\ D \end{bmatrix} = \begin{bmatrix} 0 & 1 & 0 \\ 1 & 0 & 0 \\ 0 & 0 & -1 \end{bmatrix} \begin{bmatrix} E \\ N \\ U \end{bmatrix} \tag{14.1}$$

It is therefore simpler and more efficient (fewer operations) to maintain a consistent convention throughout all the coordinate systems in the simulation.

Section 14.2.1 discusses the use of rotation matrices as part of the process of transforming vectors from one CS to another. We note here that an alternative to rotation matrices is to use quaternions [5], which are more efficient and numerically stable than matrices; however, they are not as intuitive and easy to understand. We therefore limit our discussion to the more familiar rotation matrices.

14.2.1 Coordinate Transformations

Transforming a vector defined in its original CS in order to reference it relative to a second CS in general requires two operations: a rotation of the CS axes and a translation of the origin. These two operations may be performed in any order, but it is often more computationally efficient to first translate, then rotate. We illustrate these operations with the following 2-D example. Consider the two coordinate systems in Figure 14.2, in which the x'-y' CS differs from the x-y CS only by a fixed offset y_0 in the y-dimension (i.e., there is no offset in the x-direction), plus an axis rotation by an angle θ. Thus, if the origin of the x-y CS is defined as [0,0], then the origin of the x'-y' CS is $[0, y_0]$. Now consider vector $\mathbf{a} = [x_0, y_0]$ defined in the x-y plane, which defines a target location in 2-D. We would like to express that position relative to the x'-y' CS.

We first consider an unrotated version of the x'-y' CS, shown using dashed lines in Figure 14.2. The unrotated CS and the original CS are identical but for a translation of the origin, which can be expressed as a translation vector $\mathbf{r} = [0, y_0]$ defined relative to the x-y CS. By vector addition and subtraction, the target vector relative to this unrotated CS becomes

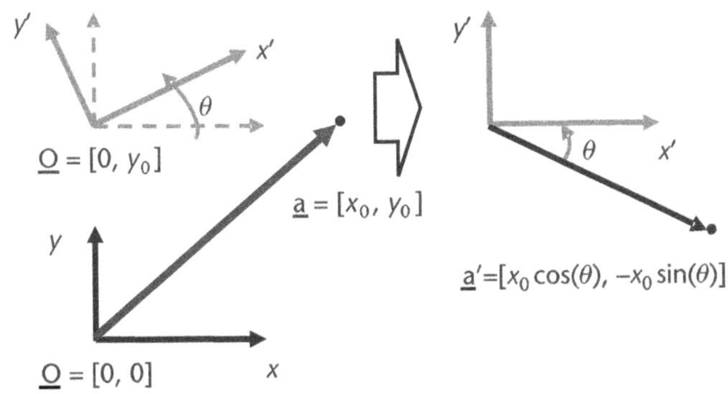

Figure 14.2 Two-dimensional example of coordinate transformation.

$$\mathbf{a}_u = \mathbf{a} - \mathbf{r} = \begin{bmatrix} x_0 \\ y_0 \end{bmatrix} - \begin{bmatrix} 0 \\ y_0 \end{bmatrix} = \begin{bmatrix} x_0 \\ 0 \end{bmatrix} \tag{14.2}$$

We now find the target vector in the rotated CS x'-y' by applying a Givens rotation matrix **T**, defined as

$$T = \begin{bmatrix} \cos(\theta) & \sin(\theta) \\ -\sin(\theta) & \cos(\theta) \end{bmatrix} \tag{14.3}$$

This matrix is named after American mathematician Wallace Givens. The final result is

$$\mathbf{a}' = \begin{bmatrix} x' \\ y' \end{bmatrix} = \mathbf{T}\mathbf{a}_u = \begin{bmatrix} \cos(\theta) & \sin(\theta) \\ -\sin(\theta) & \cos(\theta) \end{bmatrix} \begin{bmatrix} x_0 \\ 0 \end{bmatrix} = \begin{bmatrix} x_0 \cos(\theta) \\ -x_0 \sin(\theta) \end{bmatrix} \tag{14.4}$$

We note that from (14.2), (14.4) may be written as

$$\mathbf{a}' = \mathbf{T}\mathbf{a}_u = \mathbf{T}(\mathbf{a} - \mathbf{r}) = \mathbf{T}\mathbf{a} - \mathbf{T}\mathbf{r} \tag{14.5}$$

which shows that the correct result is also obtained by rotating the original target vector first, then translating the origin. The origin translation, however, must be done over the rotated version of the translation vector **r**; thus, fewer operations are required if one translates first, then rotates. For the remaining discussion, we will focus on the rotations and assume that the translations are performed as required.

Regarding rotations (2-D, in particular), there is a natural tendency to characterize them as either clockwise (cw) or counterclockwise (ccw). We discourage this terminology, however, as it is highly ambiguous. Whether a rotation appears cw or ccw is a function of a number of factors, such as the following:

- The point of view of whether the vector is rotating relative to the axes, or whether the axes are rotating relative to the vector;
- Whether the angle of rotation is positive or negative;
- Whether the rotation is right-handed or left-handed;
- The choice of CS (e.g., ENU versus NED).

We elaborate on these factors on the next few sections and offer suggestions to help readers avoid confusion.

14.2.1.1 Rotation Point of View

The example rotation depicted in Figure 14.2 is a perfect illustration of the dual point of view (POV). After the origin translation (i.e., the CS in light gray), the unrotated vector lies in the positive x-axis. After the rotation, the resulting vector is shown on the right portion of Figure 14.2, with a negative y-component. Thus, the vector appears to have been rotated clockwise relative to the x'-y' axes. The alternate POV, however, is that the axes have been rotated ccw relative to the vector. We therefore have the exact same rotation described with opposite terms. Although either POV is valid, in subsequent discussions we use the rotation-of-axis POV for consistency.

14.2.1.2 Angle of Rotation

If a positive angle, say +30 deg, results in an apparent *cw* rotation, then its negative (–30 degrees) will look like a ccw rotation, and vice versa. This ambiguity is relatively easy to solve, as we will always assume positive angles, unless specifically told otherwise.

14.2.1.3 Right-Handed Versus Left-Handed Rotations

The right-handed (RH) and left-handed (LH) rotation terminology applies to 3-D rotations, so we will consider a third axis as part of this discussion. If we add the z'-axis to the rotation illustrated in Figure 14.2, we can designate it as the axis of rotation, as the x'-y' plane rotates relative to it. We place our open right hand with our thumb pointing toward the positive z'-axis (pointing out of the page), and the remaining four fingers extended toward the positive axis of the first element in the vector (x'-axis in this example). If we then curl our fingers in the direction of the second axis (y'-axis), the curling moves in the direction of a positive rotation (rotation by a positive angle). This is therefore a RH rotation of the x'-y' axes. If a LH rotation is applied instead, the direction of the finger curling will be opposite to that of the RH rotation. Thus, a rotation that appears cw for one sense, will appear ccw for the other. In our models we deal strictly with RH rotations and RH coordinate systems.

14.2.1.4 Choice of CS (ENU Versus NED)

When considering a 2-D rotation on the first two axes, the obvious difference between the EN plane and the NE plane is that the vector elements are in reverse order. When adding the third dimension, we notice that the positive direction is reversed. These relationships are described mathematically by (14.1). They are both RH coordinate systems, however, so the rotations are consistent. For the rotation defined by (14.3), under either CS, the effect is best visualized as a RH rotation of the axes relative to the vector. When considering the specific rotation described by (14.4) in graphical terms, a major difference is that for EN coordinates, the unrotated vector lies in the horizontal axis, while for NE coordinates it lies in the vertical axis. Although the values of the elements of the rotated vector are numerically identical, the rotated axes have gone in opposite directions: ccw for EN and cw for NE. These are shown in Figure 14.3.

We can thus conclude that cw and ccw are just not good descriptors for these rotations. It is more accurate to use the RH/LH terminology.

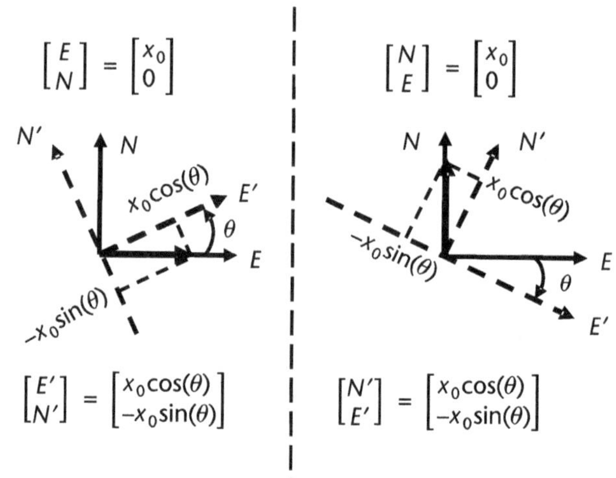

Figure 14.3 RH rotation in EN coordinates (apparent ccw rotation of axes) (left) and RH rotation in NE coordinates (apparent cw rotation of axes) (right).

14.2.1.5 Inertial-to-Body Transformation

A common challenge arises when writing down the elements of the rotation matrix; the format of cosines in the main diagonal and sines in the off-diagonal is straightforward, but the tricky part is knowing where to place the minus sign on one of the sine elements. Before presenting a rule for this, we first want to extend the 2-D definition from (14.3) to 3-D. We will do this in the context of inertial-to-body transformations, where we will define a RH rotation matrix for each of the Euler angles. Assuming an NED convention, the rotation matrices are [6]

$$\text{Yaw rotation: } \mathbf{Y}(\psi) = \begin{bmatrix} \cos(\psi) & \sin(\psi) & 0 \\ -\sin(\psi) & \cos(\psi) & 0 \\ 0 & 0 & 1 \end{bmatrix} \tag{14.6}$$

$$\text{Pitch rotation: } \mathbf{P}(\theta) = \begin{bmatrix} \cos(\theta) & 0 & -\sin(\theta) \\ 0 & 1 & 0 \\ \sin(\theta) & 0 & \cos(\theta) \end{bmatrix} \tag{14.7}$$

$$\text{Roll rotation: } \mathbf{R}(\rho) = \begin{bmatrix} 1 & 0 & 0 \\ 0 & \cos(\rho) & \sin(\rho) \\ 0 & -\sin(\rho) & \cos(\rho) \end{bmatrix} \tag{14.8}$$

In these 3-D extensions, a value of '1' is placed on the position corresponding to the axis of rotation, and zeroes on the remaining elements of the same row and column. The rule for placement of the negative sign is now presented as follows:

> "For a RH rotation and a positive angle, the negative sign goes on the row above the row with the 1 in it."
>
> —*Brian Stevens, GTRI*

In the case of the roll rotation in (14.8), the interpretation is that the rows circle back, such that the row "above" the row with 1 in it (the first row) is the third row. Notice that in the rule above, the terms "RH" and "above" are underlined and are intended to be coupled. If LH rotations are used instead, the rule above would read with "LH" and "below" replacing "RH" and "above," respectively.

These matrices are applied sequentially to a vector in inertial coordinates (after origin translation) to obtain the corresponding vector in body coordinates. The order of the rotations is important! The full transformation is

$$\mathbf{a}_{\mathbf{B}} = \mathbf{R}(\rho)\mathbf{P}(\theta)\mathbf{Y}(\psi)\mathbf{a}_{\mathbf{I}} = \mathbf{T}_{\mathbf{BI}}\mathbf{a}_{\mathbf{I}} \tag{14.9}$$

where the inertial-to-body transformation matrix $\mathbf{T}_{\mathbf{BI}}$ is the product of the three rotation matrices. The reverse body-to-inertial transformation is obtained by applying the inverse of the matrix.

14.2.1.6 Body-to-Antenna Transformation

Moving on to the body-to-antenna transformation, this very much depends on the mounting of the antenna on the platform in question. A nose-mounted antenna would have its mechanical boresight parallel with the body north axis, while for a side-mounted antenna it would be with the body east axis. A nose-mounted antenna is assumed herein. Next, the specific type of mounting must be considered when defining how the antenna is pointed. Mechanical steering requires some type of gimbal mechanism, of which there are several variations. An elevation-over-azimuth gimbal system or positioner is assumed here. In this system, as the name implies, the elevation gimbal rests on top of the azimuth gimbal. Given a pair of gimbal commands, the system will swivel in the azimuth plane first, then swivel in elevation. This option is a natural choice that leads to an intuitive understanding of the pointing behavior; azimuth and elevation gimbal commands correspond to the actual azimuth and elevation look directions of the antenna. For other gimbal systems this is not the case. An azimuth-over-elevation positioner will first swivel in elevation, then the azimuth gimbal will swivel in the slanted (no longer horizontal) azimuth plane. Thus, the gimbal commands do not match the resulting look directions [7]. An electronically steered array (ESA) works the same as an El-over-Az mechanical gimbal [i.e., the commanded look directions in Az and El (no longer gimbal commands technically) match the actual look directions].

In an NED CS, assuming a nose-mounted antenna, the rotations for azimuth and elevation gimbal angles α and η, respectively, are

$$\text{Azimuth gimbal rotation: } \mathbf{A}(\alpha) = \begin{bmatrix} \cos(\alpha) & \sin(\alpha) & 0 \\ -\sin(\alpha) & \cos(\alpha) & 0 \\ 0 & 0 & 1 \end{bmatrix} \tag{14.10}$$

$$\text{Elevation gimbal rotation: } \mathbf{E}(\eta) = \begin{bmatrix} \cos(\eta) & 0 & -\sin(\eta) \\ 0 & 1 & 0 \\ \sin(\eta) & 0 & \cos(\eta) \end{bmatrix} \tag{14.11}$$

Notice that these are also RH rotations; in fact, they are identical to the yaw and pitch rotations from (14.6) and (14.7), with the same conventions of positive angles being "right" and "up." For an El-over-Az gimbal system, the azimuth rotation must be applied first, then elevation. The full transformation is

$$\mathbf{a}_{\mathrm{A}} = \mathbf{E}(\eta)\mathbf{A}(\alpha)\mathbf{a}_{\mathrm{B}} = \mathbf{T}_{\mathrm{AB}}\mathbf{a}_{\mathrm{B}} \tag{14.12}$$

Again, the reverse transformation is performed by using the inverse of $\mathbf{T}_{\mathrm{AB}}$.

14.2.2 Additional Conventions and Assumptions

To summarize, we adopt the following conventions and make the following assumptions in our engagement simulation geometry:

- We use the NED convention for all three coordinate systems: inertial, body, and antenna.
- With regards to the body CS, we assume that the object's body vector, or main axis is aligned with the north axis, further, we assume the front of the object (e.g., the cockpit of an aircraft) looks toward the positive axis. For zero-valued Euler angles (no yaw, pit, or roll), the pilot looks straight toward the North Pole, and the aircraft is straight and level.
- With regards to the antenna CS, we assume that the antenna is mounted on the platform nose with the equivalent of an El-over-Az gimbal system. For zero-valued gimbal angles, the antenna boresight aligns with the positive North body axis.

As an additional simplifying assumption, we treat the simulated platforms as point objects, such that the origin of the antenna CS is the same as the body CS. In general, there is an origin displacement from the platform CG to the antenna location, which we ignore here. Under this assumption, the inertial-to-body and the body-to-antenna transformations are treated as a single rotation matrix, as these are linear operations. Thus,

$$\mathbf{a}_{\mathrm{A}} = \mathbf{T}_{\mathrm{AB}}\mathbf{a}_{\mathrm{B}} = \mathbf{T}_{\mathrm{AB}}\left(\mathbf{T}_{\mathrm{BI}}\mathbf{a}_{\mathrm{I}}\right) = \left(\mathbf{T}_{\mathrm{AB}}\mathbf{T}_{\mathrm{BI}}\right)\mathbf{a}_{\mathrm{I}} = \mathbf{T}_{\mathrm{AI}}\mathbf{a}_{\mathrm{I}} \tag{14.13}$$

where the inertial vector $\mathbf{a}_{\mathrm{I}}$ has already been translated to the origin of the body CS.

14.2.3 Simulink Model

The Simulink model *RotationExamp* implements a 2-D vector rotation. It is in the same vein as the example illustrated in Figure 14.2 but specifically for NED coordinates (NE in this 2-D case). The model takes an input yaw angle of 45°, builds the 2-D rotation matrix in a subsystem, then multiplies the matrix by the input vector and sends the result to a Display block. Figure 14.4 shows the top-level block diagram as well as the contents of the subsystem.

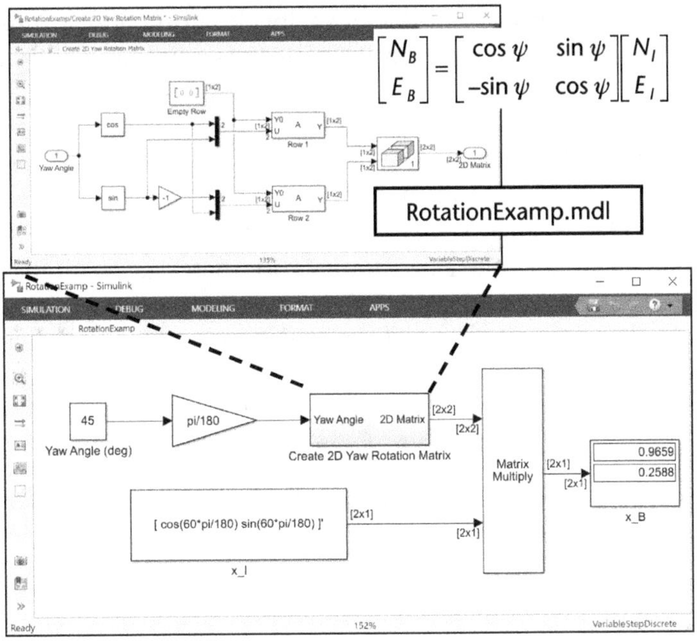

Figure 14.4 Simulink example of a 2-D coordinate transformation.

Numerical verification of this model is straightforward, as shown in Figure 14.5, both numerically as well as graphically. The original vector is rotated 60° from the inertial north axis; after a 45° rotation, it is now only 15° in the rotated (body) north axis. The graphical representation of this rotation is consistent with the discussion from Section 14.2.1.

14.3 Truth Calculation of Radar Observables

We now proceed to define the calculations for geometry-dependent target parameters observed by the radar. We assume that the following vector quantities are provided:

$$\begin{bmatrix} N_B \\ E_B \end{bmatrix} = \begin{bmatrix} \cos 45 & \sin 45 \\ -\sin 45 & \cos 45 \end{bmatrix} \begin{bmatrix} \cos 60 \\ \sin 60 \end{bmatrix}$$

$$= \begin{bmatrix} 0.7071 & 0.7071 \\ -0.7071 & 0.7071 \end{bmatrix} \begin{bmatrix} 0.5 \\ 0.866 \end{bmatrix} = \begin{bmatrix} 0.9659 \\ 0.2588 \end{bmatrix} = \begin{bmatrix} \cos 15 \\ \sin 15 \end{bmatrix}$$

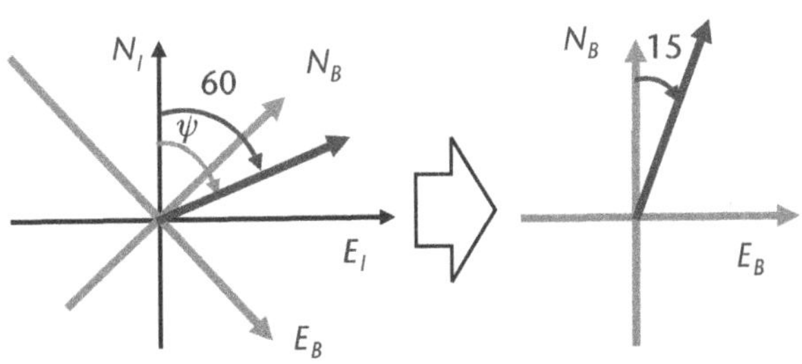

Figure 14.5 Verification of Simulink example results.

- Radar position and velocity vectors as a function of time: $\mathbf{p}_r(t)$, $\mathbf{v}_r(t)$;
- Target position and velocity vectors as a function of time: $\mathbf{p}_t(t)$, $\mathbf{v}_t(t)$.

14.3.1 (Slant) Range

Chapter 8 first introduced the LOS vector between the radar and the target, as shown in Figure 8.1, but it does not formally define it. Moreover, the LOS vector in Figure 8.1 is target-centric, meaning it originates at the target location and points in the direction of the radar. This is a natural convention used to define target Doppler. We now formally define a *radar-centric* LOS vector as

$$\mathbf{LOS}(t) = \mathbf{p}_t(t) - \mathbf{p}_r(t) \tag{14.14}$$

where now the vector points toward the target. The target range R is simply the magnitude of this vector, or

$$R(t) = |\mathbf{LOS}(t)| \tag{14.15}$$

14.3.2 Doppler Frequency

The Doppler definition presented here is consistent with (8.7) but more general, as it considers radar motion. It also uses the alternate radar-centric convention for LOS.

Figure 14.6 shows a more general version of Figure 8.1, with radar and velocity vectors pointing in arbitrary directions. Their contributions to the total radial velocity V_{LOS} are calculated using dot products as shown by (8.6), repeated here for the general case being considered:

$$V_{r_{\text{LOS}}}(t) = \mathbf{v}_r(t) \bullet \mathbf{LOS}_u(t) = |\mathbf{v}_r(t)|\cos(\theta) \tag{14.16}$$

$$V_{t_{\text{LOS}}}(t) = \mathbf{v}_t(t) \bullet \mathbf{LOS}_u(t) = |\mathbf{v}_t(t)|\cos(\phi) = -|\mathbf{v}_t(t)|\cos(\phi') \tag{14.17}$$

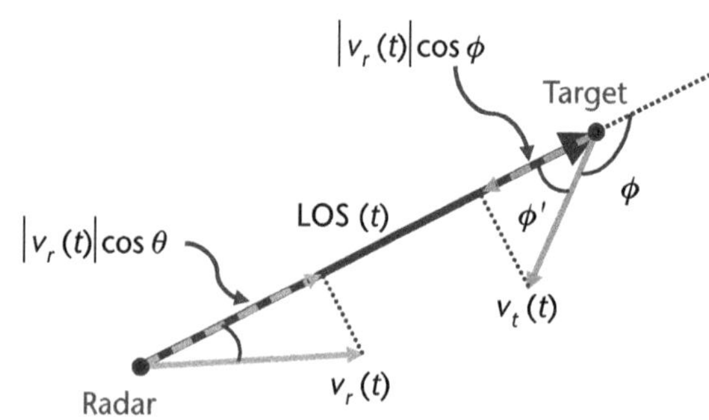

Figure 14.6 General geometry for Doppler calculation with radar and target motion.

where $\mathbf{LOS}_u$ is the normalized LOS vector, obtained by dividing (14.14) by (14.15). Last, the total radial velocity and Doppler are defined as

$$V_{\mathrm{LOS}}(t) = V_{r_{\mathrm{LOS}}}(t) - V_{t_{\mathrm{LOS}}}(t), \quad f_D(t) = \frac{2V_{\mathrm{LOS}}(t)}{\lambda} \tag{14.18}$$

14.3.3 Target Angles

The angles to the target relative to the antenna boresight may be calculated by inspection from the NED ACS depicted on the RHS of Figure 14.1. Given the LOS vector in NED antenna coordinates (axes labeled x_A-y_A-z_A in Figure 14.1), the azimuth and elevation angles are [2]

$$Az(t) = \mathrm{atan}\left(\frac{LOS_y(t)}{LOS_x(t)}\right) \tag{14.19}$$

$$El(t) = \mathrm{atan}\left(\frac{-LOS_z(t)}{\sqrt{LOS_x^2(t) + LOS_y^2(t)}}\right) \tag{14.20}$$

where the denominator in (14.20) is the magnitude of the projection of the LOS onto the NE plane (i.e., the dashed line in Figure 14.1). The negative sign in the numerator of (14.20) preserves the convention that the elevation angle is positive-up. Likewise, the azimuth angle is positive-right.

One additional angular parameter is the aspect angle used to extract an RCS value from LUT. A polar plot of RCS data is shown in Figure 5.12 and repeated in Figure 14.7 for convenience, this time with the outline of the aircraft superimposed to indicate orientation.

For this particular 1-D data set, it is the *horizontal* aspect angle that is relevant. Vector $\mathbf{b}_{xy}$ is the body vector of the aircraft, projected in the horizontal (NE) plane. Likewise, we project the radar-to-target LOS vector to the NE plane, shown as $\mathbf{LOS}_{xy}$ in Figure 14.7. In general, these are time-varying vectors, although the time

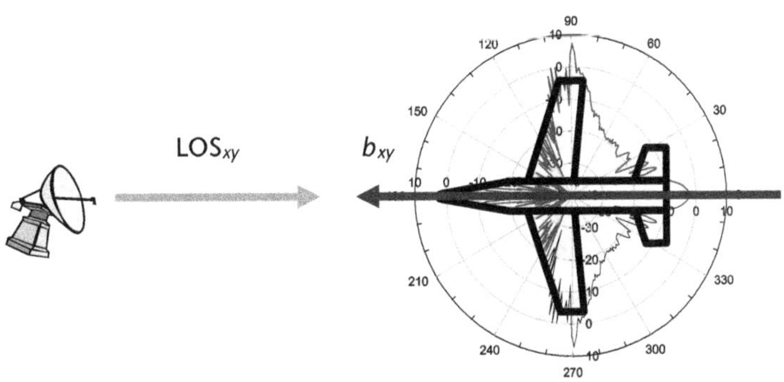

Figure 14.7 Geometry for horizontal-aspect angle calculation, used for RCS look-up.

dependency is not explicitly shown in Figure 14.7. Moreover, we use the normalized versions of these vectors (indicated by the subscript 'u' in (14.21)). The horizontal (azimuth) aspect is then calculated from the dot product of the two unit vectors, as

$$AzAspect(t) = \mathrm{a}\cos\left(\mathbf{b}_{xy_u}(t) \bullet \mathbf{LOS}_{xy_u}(t)\right) \tag{14.21}$$

It can be shown that the unit-magnitude body vector is only a function of the yaw angle ψ:

$$\mathbf{b}_{xy_u}(t) = \begin{bmatrix} \cos(\psi(t)) \\ \sin(\psi(t)) \end{bmatrix} \tag{14.22}$$

For the specific orientation depicted in Figure 14.7, the aspect angle is 180°. This is known as the nose aspect, as the aircraft is pointing its nose toward the radar. Similarly, 0° is known as tail aspect, and either 90° or 270° is the beam aspect, as the target body vector is orthogonal (broadside) to the beam of the antenna—this is typically when the RCS blooms, or reaches its peak value.

14.4 Simulink Model of Target Generator

The Simulink model *Target_Gen_v1* exercises a Target Generator subsystem intended to provide truth data on target range, Doppler, RCS, azimuth, and elevation angles over time as the radar and target move in space during the simulation execution time. Inputs to the subsystem are radar and target state with position, velocity, and orientation information in vector form. The transmit wavelength is also included for the Doppler calculations. Figure 14.8 shows the top-level block diagram for this model, as well as the contents of the Target Generator subsystem. This subsystem implements (14.14) to (14.22).

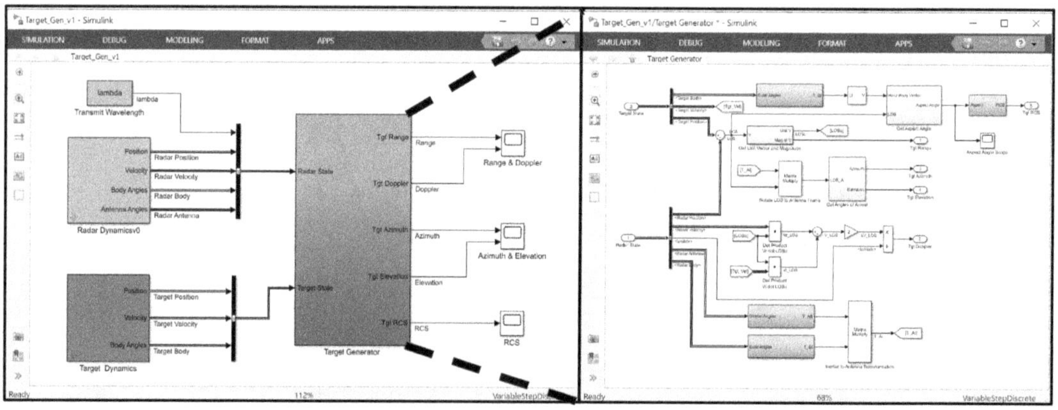

Figure 14.8 Simulink implementation of Target Generator model.

This model includes a number of features worth mentioning. The Target Generator subsystem is driven by two Dynamics blocks that provide radar and target state information over time. The Radar Dynamics block in the middle-left outputs position and velocity vectors in NED coordinates, a Radar Body vector consisting of Euler angles, and a Radar Antenna vector consisting of azimuth and elevation gimbal angles. The Target Dynamics block in the bottom left provides similar position, velocity, and orientation vectors for the target. Both of these are user-defined blocks that have been stored in a user-defined library file called *MS_Models*, discussed in the online Simulink Appendix.

Furthermore, both of these blocks are examples of masked subsystems. Also discussed in the online Simulink Appendix, a mask is a custom interface that acts and looks very similar to a Block Parameters window, where the user enters parameter values that shape the block behavior.

The parameters for these particular blocks consist of initial conditions for the radar and target state. These blocks generate simple, linear trajectories at a constant velocity. Both the velocity and Euler angle vectors remain constant throughout the simulation run. The position vector evolves over time based on the rectilinear motion model

$$\mathbf{p}_{\text{obj}}(t) = \mathbf{p}_{\text{obj}}(0) + \mathbf{v}_{\text{obj}}(0)t \tag{14.23}$$

where $\mathbf{p}_{\text{obj}}(0)$ and $\mathbf{v}_{\text{obj}}(0)$ are the initial conditions entered by the user.

While these trajectories are trivially simple, they provide a reasonable use case for testing the truth calculations contained in the Target Generator. Decoupling the Target Generator from the trajectory generator blocks facilitates the replacement of these linear trajectories with potentially more realistic and complex motion models including turns and other maneuvers. These trajectories may come from more advanced kinematic simulations, or even from real trajectory data captured and stored in a file, such as Time-Space-Position Information (TSPI) data commonly recorded in a flight test.

We also acknowledge that Simulink offers an Aerospace Blockset (toolbox) that features a six-degrees-of-freedom (6DOF) motion platform example among its blockset demos [8], in addition to Euler angle-based and quaternion-based equations of motion models and related features. The 6DOF example seems a good starting point for modeling more realistic trajectories, and we therefore refer the reader to the MathWorks website for further exploration of the toolbox capabilities.

The model is tested using a simple test geometry illustrated in Figure 14.9(a). The radar is stationary and placed on the ground at the origin [0, 0, 0] of the inertial CS, looking due east. The coordinates used in Figure 14.9 are ENU for easier visualization, although all model calculations are based on the NED convention. The target is a low-altitude aircraft flying straight and level in the due-south direction, with initial position at ENU coordinates [2,000m; 70m; 20m]. Over the simulation runtime the aircraft flies in front of the radar and continues past it.

The simulation results are various plots from Display blocks, shown in Figure 14.9(b–e). A cursory examination of the plots reveals that the calculated parameters behave in a manner consistent with the geometry. Target range is dominated by the 2000m offset in the east direction, with the north-south component decreasing at first, becoming a minimum at the crossover point, and then increasing again. Likewise, target Doppler starts as a positive value, as the projection of the velocity vector onto the LOS vector (radial velocity) points toward the radar. As the angle between the two vectors approaches 90°, Doppler gets smaller and becomes zero at the crossover point (since vectors are orthogonal). As the simulation continues beyond this point, the radial velocity points away from the radar so Doppler

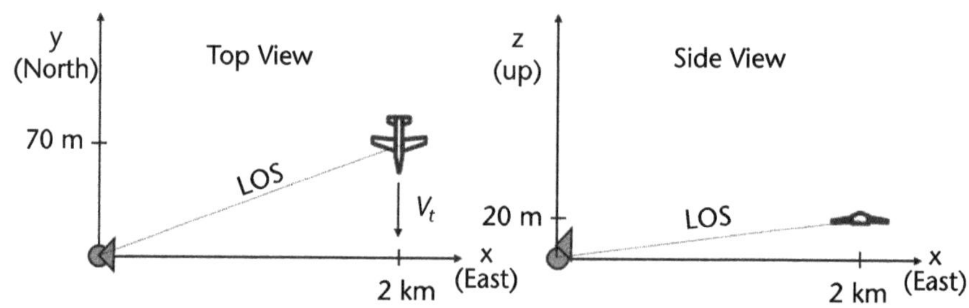

(a)

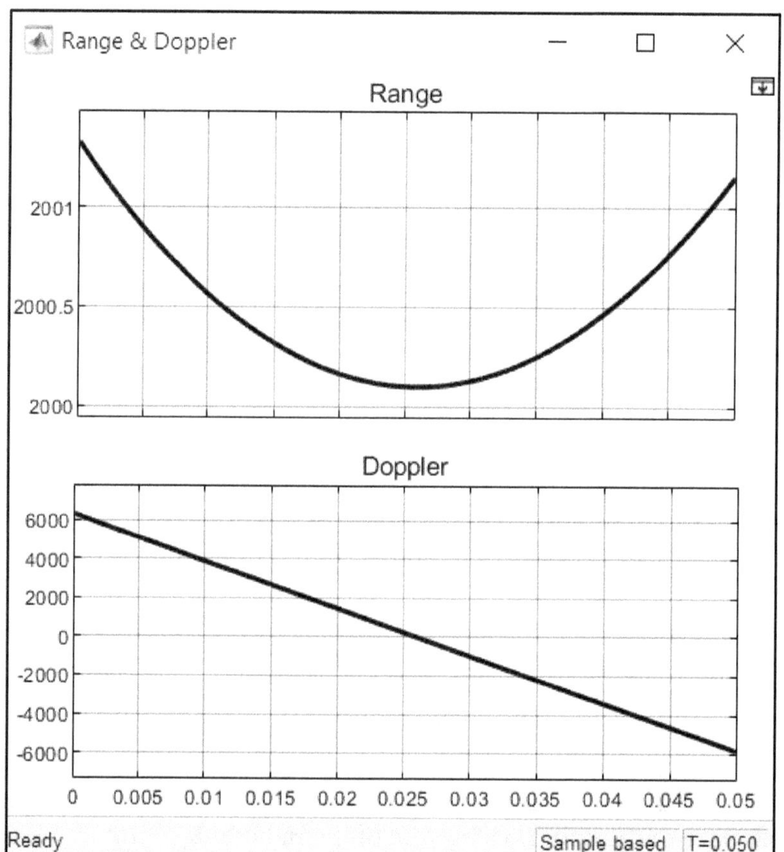

(b)

Figure 14.9 Test geometry (a) and results (b–e) for Target Generator Model.

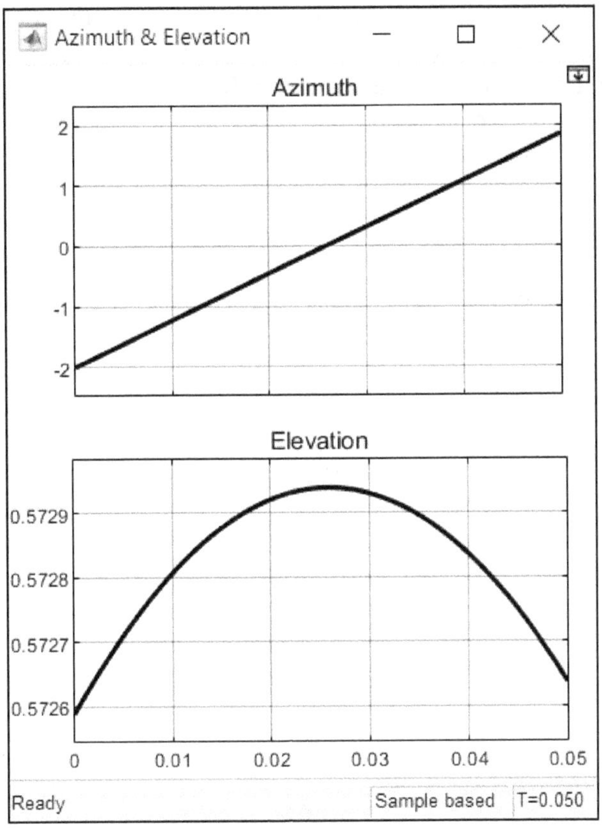

(c)

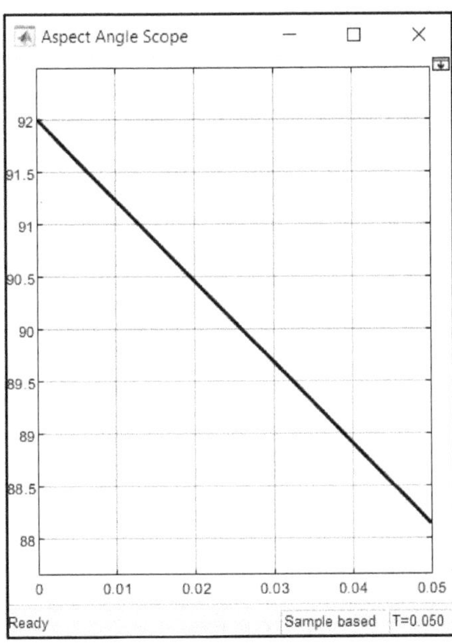

(d)

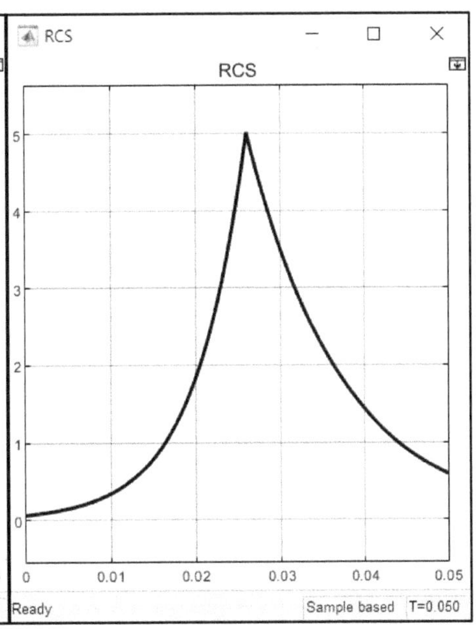

(e)

Figure 14.9 *(Continued)*

becomes negative, and its magnitude increases as the angle between the vectors moves farther away from 90°. Similar arguments can be made to explain the angle curves. Last, we notice that the target RCS reaches a peak value of 5m^2 (7 dBsm) at the crossover point, where the aspect angle reaches 90°. This too is consistent with the RCS data from the LUT. We thus feel confident that the model works as intended.

14.5 Concluding Remarks

This chapter examined a few common coordinate systems used in engagement simulations and the use of rotation matrices to transform vectors from one CS to another. It also reviewed the equations to calculate target truth information on radar observables (range, Doppler, and angles) given position and velocity information on the radar and the target. Last, the chapter presented a Simulink implementation of a Target Generator subsystem, which executes such calculations, including coordinate transformations as needed, and verifies this model against a simple 1-on-1 scenario.

In Chapter 15, this model is integrated with the larger engagement simulation that also incorporates all other subsystem models presented in earlier chapters.

References

[1] Torge, W., *Geodesy*, Third Edition, Berlin: Walter de Gruyter, 2001.

[2] American Institute of Aeronautics and Astronautics, *Recommended Practice for Atmospheric and Space Flight Vehicle Coordinate Systems*, American National Standard (ANSI/AIAA R-004-1992), Washington, D.C.: American Institute of Aeronautics and Astronautics, February 1992.

[3] https://rotations.berkeley.edu/the-euler-angle-parameterization/.

[4] Glasoe, W. M., F. A. Pena, and V. Phadke, "Cardan Angle Rotation Sequence Effects on First-Metatarsophalangeal Joint Kinematics: Implications for Measuring Hallux Valgus Deformity," *Journal of Foot and Ankle Research*, Vol. 7, No. 29, May 2014.

[5] Franklin, W. R., "Efficient Iterated Rotation of an Object," *IEEE Transactions on Computers*, Vol C-32, No. 11, November 1983.

[6] https://mathworld.wolfram.com/EulerAngles.html.

[7] Masters, G. F., and S. F. Gregson, "Coordinate System Plotting for Antenna Measurements," *Proc. Antenna Measurement Techniques Association (AMTA) Annual Symposium*, St. Louis, MO, 2007.

[8] https://www.mathworks.com/help/aeroblks/six-degree-of-freedom-motion-platform.html.

CHAPTER 15

Engagement Simulation

15.1 Introduction

The sole purpose of this chapter is to introduce the Simulink engagement model we have been mentioning throughout the book. First, Chapter 2 offered the RRE as a suitable framework for building a radar simulation and presented a simple Simulink model of an SNR calculation (*RadarEquationv1*, shown in Figure 2.6) that is, nevertheless, the foundation for a dynamic engagement simulation between a single radar and a single target. Subsequently, Chapters 3–14 presented more detailed models of radar waveforms, antennas, propagation, target RCS, signal processors, detectors, trackers, and selected H/W components. These models will replace the simple blocks from *RadarEquationv1* to make it a more interesting and useful model. These upgrades should occur in an incremental and methodical manner, for ease of testing and verification.

The remainder of this chapter is organized as follows. Section 15.2 illustrates how relatively minor upgrades can go a long way in helping simulate and visualize common radar phenomena such as range ambiguities in a pulse system. Section 15.3 extends the model to incorporate time-varying radar observables and propagation effects. This model features a prototype Target Generator subsystem as a placeholder for the Target Generator presented in Chapter 14. Next, Section 15.4 presents the full engagement model with all subsystems in place, and finally, Section 15.5 discusses example results from a notional one-on-one engagement scenario.

15.2 Extending the Radar Equation Model

The original *RadarEquationv1* model provides an instantaneous SNR calculation of a target at a fixed range. Versions 2 and 3 depicted in Figure 2.9 perform similar calculations for multiple ranges. These are all still static models, in the sense that, other than the target range, nothing else in the simulation changes. We can easily turn *RadarEquationv1* into a dynamic simulation, however, with just a few modifications. Figure 15.1 shows the upgraded model *RadarEquationv4*. This version features a Pulse Generator block that turns the static model into a true time-based simulation.

In addition to the Pulse Generator, the other major upgrade to the model is a pair of Transport Delay blocks, one on the forward path to the target and the second on the return path. These blocks add a time delay to the input signal, in this case representing the one-way propagation delay $t = R/c$. The last block in the

(a)

(b)

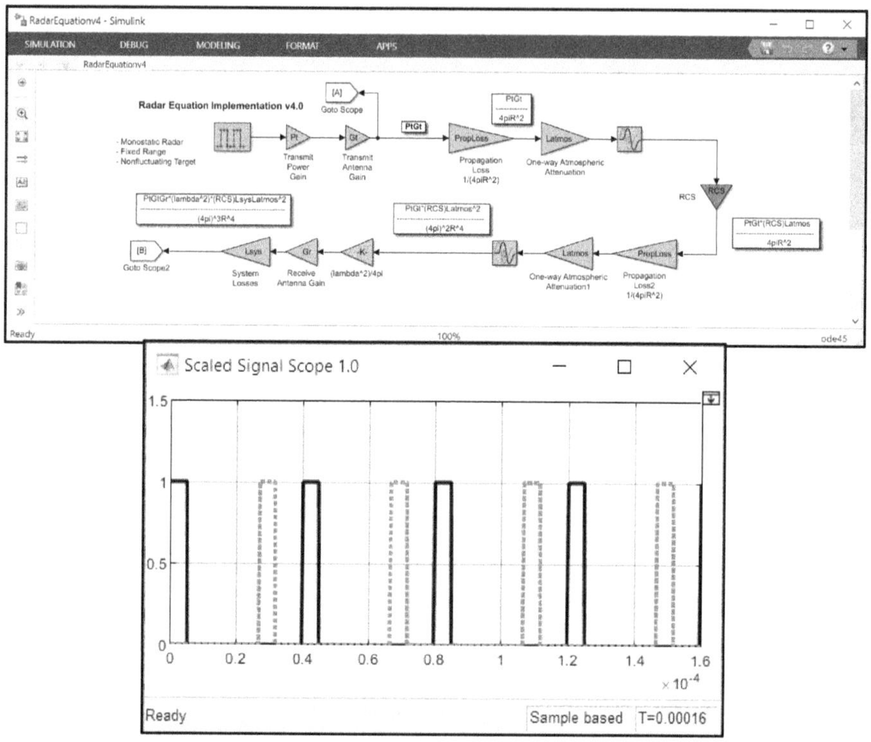

Figure 15.1 Simulink block diagram of *RadarEquationv4*, an upgraded version that simulates transmit pulses with propagation delays (a) and transmit and receive pulses from a Scope block (b).

diagram is a Go To block that sends the received pulses to a Scope block (not visible in Figure 15.1). Another Go To block branching out of the Transmit Antenna Gain block output sends the transmit pulses to the same scope. These two pulse trains are shown at the bottom of Figure 15.1; the solid black line depicts the transmit pulses. Both pulse trains have been scaled to unit magnitudes for easy visualization of the round-trip time delay; otherwise, the received pulses would be too small to be observed next to the transmit pulses due to the $1/R^4$ scaling and other losses. In conjunction with the new blocks, we updated the Init Callback m-file with the new variables needed to define the parameters for the Pulse Generator and remaining additions.

We also deleted a few blocks. Since we are no longer interested in an SNR calculation, we removed the Divide by Pn Gain block, as well as the dB Converter and Display blocks. These are all minimal changes to the model, yet the resulting simulation enables the observation of real-world radar behavior. One example is range ambiguity of target returns, as first discussed in Chapter 7 and illustrated in Figure 7.5.

In the example results shown in Figure 15.1, the target has been placed at a 4-km range from the radar. This example employs a PRF of 25 kHz, which results in a maximum unambiguous range of 6 km as defined by (7.6). Thus, the range measurement under these conditions is unambiguous. If the user now changes the target range—in line 9 of *radareqnv4_init.m*—to 8 km (try it!), the apparent range will now be 2 km; a range-ambiguous measurement. The before and after results are shown in Figure 15.2.

15.3 Initial Engagement Model

As Section 15.2 illustrated, small incremental changes to the model can result in significant improvements in capability and fidelity. These upgrades eventually transform the model into something way beyond a simple RRE calculation; we may now truly consider it an engagement simulation between a radar and a target. We therefore rename the model at this point; what might have been called *RadarEquationv7* following the previous naming convention is now *Engagement_v0*. Figure 15.3 shows the top-level block diagram.

In terms of output products, this version is not that much different from *RadarEquationv4* in the sense that its main output is the display of transmit and receive pulses. It includes, however, a number of much needed upgrades to support a fully dynamic engagement. These upgrades are listed as follows:

- A prototype Target Generator subsystem to provide truth data (range, RCS, angles) on the target;
- A 2-D antenna pattern LUT to add beam-shape loss factors to the peak antenna gains when the target is off-boresight;
- Range-dependent propagation factors (spherical spreading loss, atmospheric attenuation, and transit delays) now support time-varying range values;
- Modified scope scaling to support time-varying gains and losses.

We first consider the prototype target generator. This is actually a precursor to the fully developed Target Generator model from Chapter 14, as this engagement model predates it. The intent here is to define the interfaces to the rest of the

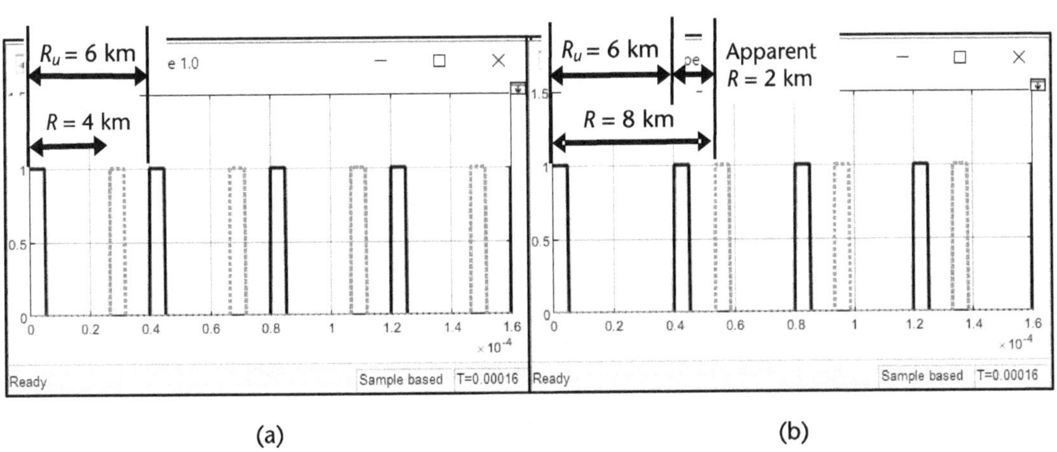

Figure 15.2 Illustration of range ambiguity using upgraded Simulink model: (a) unambiguous range, and (b) ambiguous range.

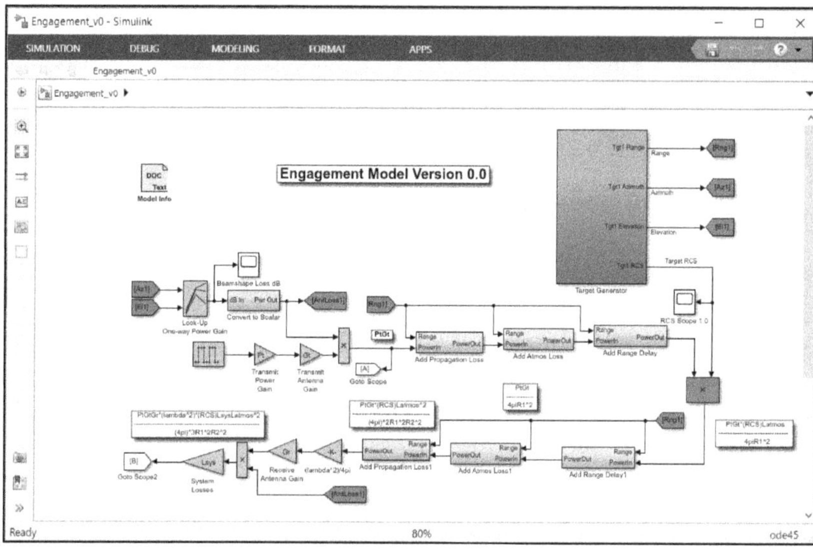

Figure 15.3 Block diagram for ***Engagement_v0***, the next step in the development of the end-to-end engagement model.

engagement model; examining this subsystem in Figure 15.4 reveals this prototype to be an empty shell, where all outputs are hardcoded except for the RCS.

The target RCS subsystem Get Target RCS consolidates three different RCS models presented in Chapter 5: a Swerling 0 (nonfluctuating RCS) model represented by a Constant block, the Swerling 1/Swerling 2 (exponential-distributed fluctuating RCS) model shown in Figure 5.10, and the LUT method from Figure 5.13. These three models feed a Multiport Switch block, where the output is selected by a control signal RCS_Type. This is a user-selected variable set in the Init Callback m-file that takes on the values 0, 1, or 2 for the Swerling 0, Swerling 1, and LUT options, respectively. The Scaled Signal and RCS Scopes show the effects from the RCS selection, as seen in Figure 15.5.

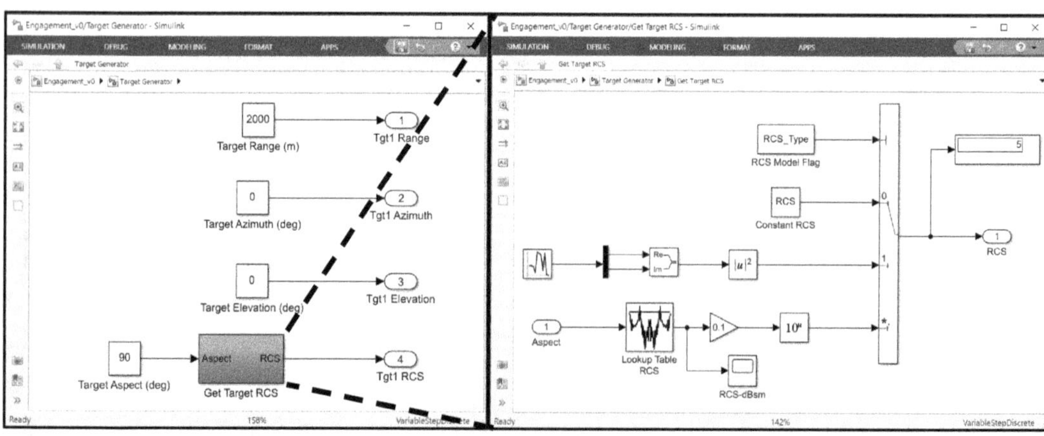

Figure 15.4 Contents of prototype Target Generator subsystem (left) and Target RCS subsystem (right).

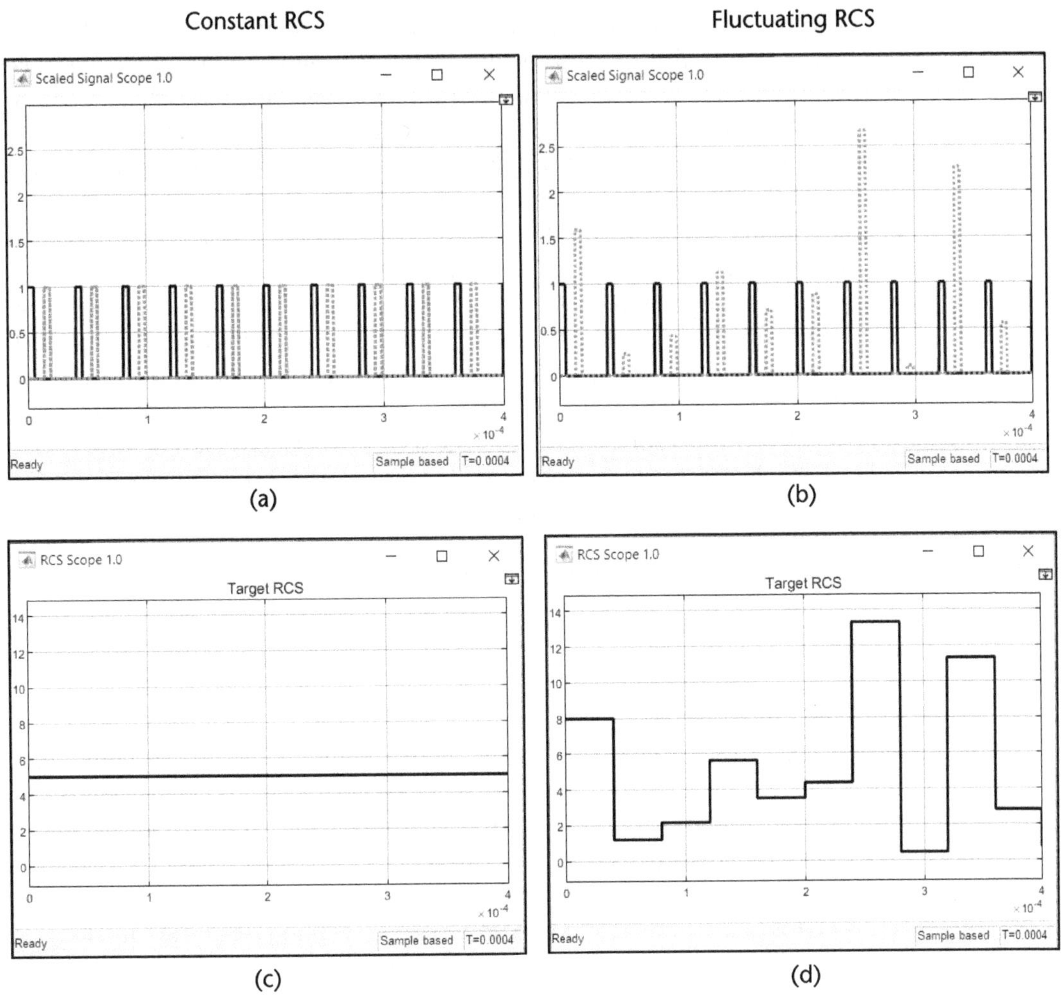

Figure 15.5 Received pulse amplitudes and RCS values obtained by selecting (a, c) the constant RCS model versus (b, d) the fluctuating RCS model.

For the fluctuating-RCS case, the fluctuation rate was set equal to the radar PRI for demonstration purposes only. Although one might be tempted to identify this fluctuation rate as pulse-to-pulse following the terminology first introduced in Chapter 5, we stress that this is not what is commonly understood as pulse-to-pulse or fast fluctuation for a pulse-Doppler system. RCS fluctuation at the PRI rate would render coherent pulse integration ineffective, as fluctuations affect both magnitude and phase. For this reason, as discussed in Chapter 12, RCS values for fast fluctuation models (Swerling 2 or 4) are assumed constant over a radar CPI. The RDM resulting from a single CPI constitutes a single observation, in the parlance of Chapter 11.

The remaining upgrades are self-explanatory. The antenna gain LUT is a 2-D extension of the RCS LUT presented in Chapter 5. The normalized gain pattern corresponds to a 2-D sinc function symmetric in azimuth and elevation, consistent with a uniformly illuminated square aperture. The values from the LUT are in decibels so they are converted to linear values; the same value is applied on receive, consistent with the assumption of identical transmit and receive antennas.

The range-dependent propagation subsystems now ingest range as a signal instead of a block parameter as was the case with *RadarEquationv1*. The same equations are implemented, however, through the signal path. The one-way time delays are now implemented using Variable Time Delay blocks that take the target range (in units of time) as a second input. Last, the scaling gains that normalize both transmit and receive pulses seen in the output Scope block have been modified to include the time-varying factors.

In summary, this *Engagement_v0* model incorporates higher levels of complexity throughout the transmitter-target-receiver signal path; however, significant work is still required to bring the model to the point where it may be considered a practical and useful analysis tool. At a minimum, it needs a receiver/signal processor model. A coherent waveform model where the signal may take complex values is another key upgrade, necessary for modeling target Doppler. All these upgrades are presented in Section 15.4.

15.4 Full Engagement Model

The final version of the engagement model is version 7, shown in Figure 15.6. *Engagement_v7* features many of the models covered in previous chapters and more.

Comparing the model in Figure 15.6 to version 0 from Figure 15.3, we certainly see some commonalities such as the Target Generator subsystem (upper right section of the diagram), the transmit antenna gain pattern LUT, and the range-dependent propagation subsystems in blue (as seen in the software). The Target Generator, however, is no longer the hardcoded prototype from version 0 but the full-fledged model presented in Chapter 14 (Figure 14.8), which takes the radar and target state inputs from the user-defined Dynamics blocks. The Radar Dynamics block in green (as seen in the software) is in fact an updated version from that shown in Figure 14.8, as this one takes in antenna-steering angles as inputs (more on this later in this section). The steering assumes an elevation-over-azimuth gimbal mounting.

In Figure 15.6, the antenna LUT is the same block as before, and the antenna pattern is the same; however, in this version 7 the values in the table are voltage gains in linear units, while in version 0 they were in decibels, later converted to a linear power gain. This points to a fundamental difference with version 7 in that we are now modeling coherent waveforms; the signal samples are now complex values.

The simple Pulse Generator source block from version 0 is now replaced by a subsystem that generates baseband pulses with biphase modulation; specifically, the seven-bit Barker code (Barker-7) introduced in Chapter 8 and depicted in Figure 8.6. The implementation of this waveform is shown in Figure 15.7; this is the block diagram inside the Transmit Pulse Train subsystem.

Each chip of the Barker-7 code comes from a Pulse Generator block with an initial delay relative to the leading edge of the pulse. This delay equals the number of chip widths that precede it. Thus, the trailing edge of one chip coincides with the leading edge of the next chip. The actual Barker-7 phase sequence (+ + + − − + −) is

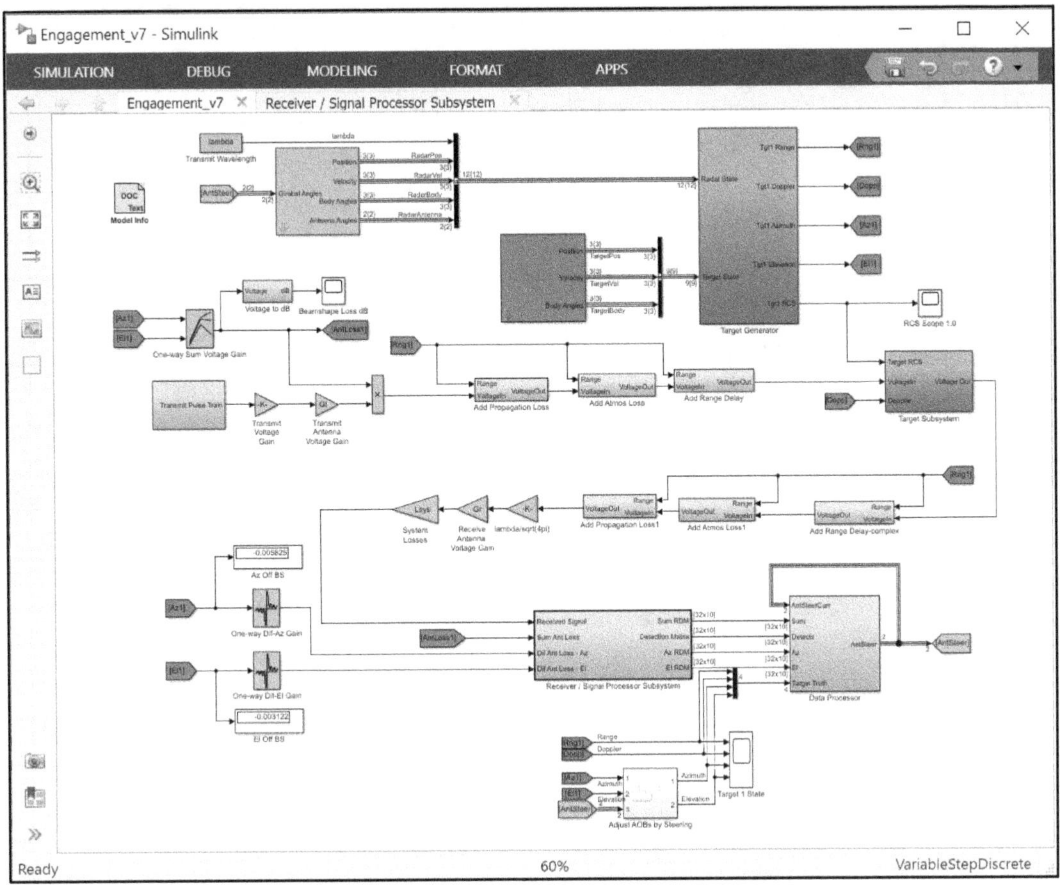

Figure 15.6 Block diagram for ***Engagement_v7***, our end-to-end engagement model.

obtained by applying the correct sequence of addition and subtraction operations top-to-bottom on the Add block that follows the chips.

Following the signal path through the one-way propagation blocks we arrived at the Target Subsystem block. This subsystem adds the instantaneous RCS and Doppler values from the Target Generator. The output of the subsystem is

$$s_{\text{out}}(t) = \left(\sqrt{\sigma}\right)s_{\text{in}}(t)e^{j2\pi f_{\text{Dop}}t} \tag{15.1}$$

where σ and f_{Dop} are the target RCS and Doppler frequency, respectively.

After propagating back to the radar, the target returns arrive at the Receiver/Signal Processor subsystem. This subsystem models a multichannel (sum, delta-Az, and delta-El) receiver to support simultaneous monopulse estimation of azimuth and elevation angles. It incorporates the monopulse model introduced in Chapter 9 and depicted at the bottom of Figure 9.10.

The transmit sum pattern LUT is also applied on receive to form the sum channel. The difference pattern from Figure 9.8 is used in the two 1-D LUTs included

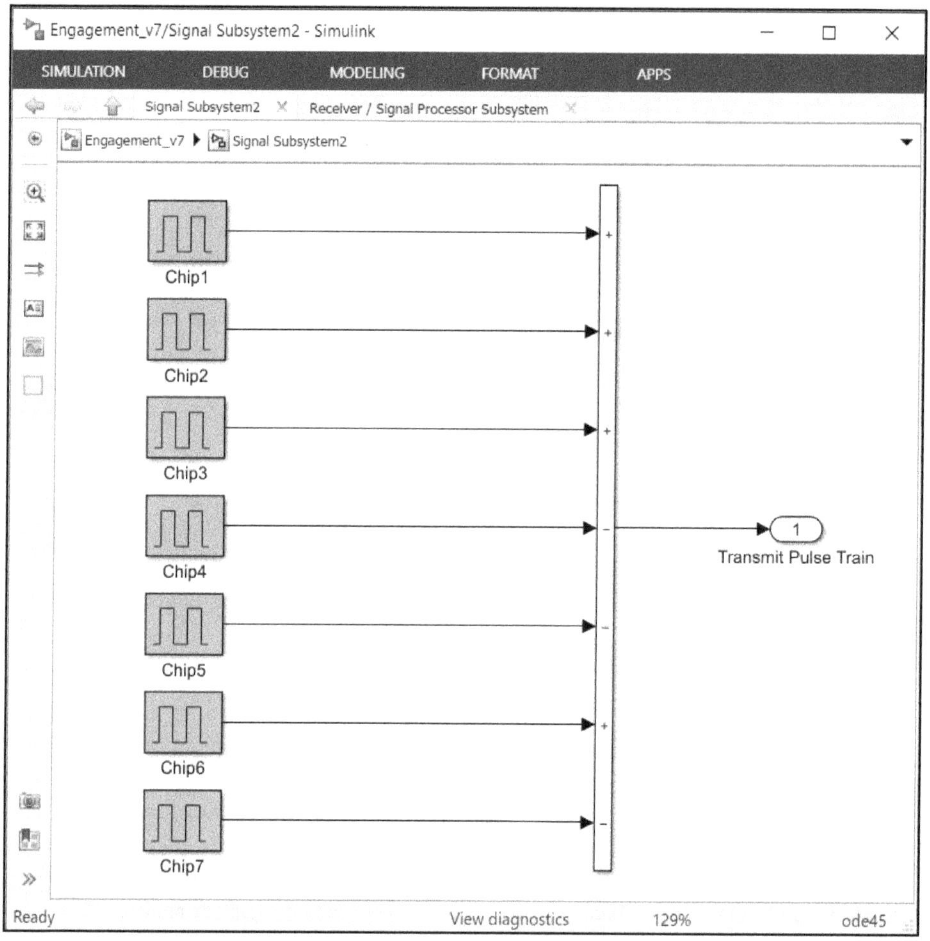

Figure 15.7 Barker-7 code implemented by superposition of delayed-pulse generators for each chip.

for the delta channels. We acknowledge there is a minor source of error introduced by using a 1-D delta pattern and a 2-D sum pattern in the monopulse calculation. A more accurate model would implement a 2-D delta pattern—this is a future growth opportunity.

Figure 15.8 shows the contents of the Receiver/Signal Processor subsystem at the top. Gains for the three receive antennas are applied to the received signal to form the inputs to the three identical receiver channel subsystems, one of which is shown at the bottom of Figure 15.8.

The continuous-time signal is sampled at the A/D rate, and then receiver-noise samples (independent for each channel) are added. Prior to the amplitude quantizer, AGC scaling is applied to avoid saturating the A/D converter. The gain-control scheme is implemented by dividing the input signal-plus-noise by the AGC scaler, meaning that the scaler is a number greater or equal to one. The quantizer model is the Uniform Encoder implementation presented in Figure 10.16; it takes a continuous voltage and outputs a quantized version based on the number of A/D bits. The quantized fast-time samples enter the Range-Doppler Processing subsystem, which generates and outputs a RDM. This subsystem is virtually identical to the

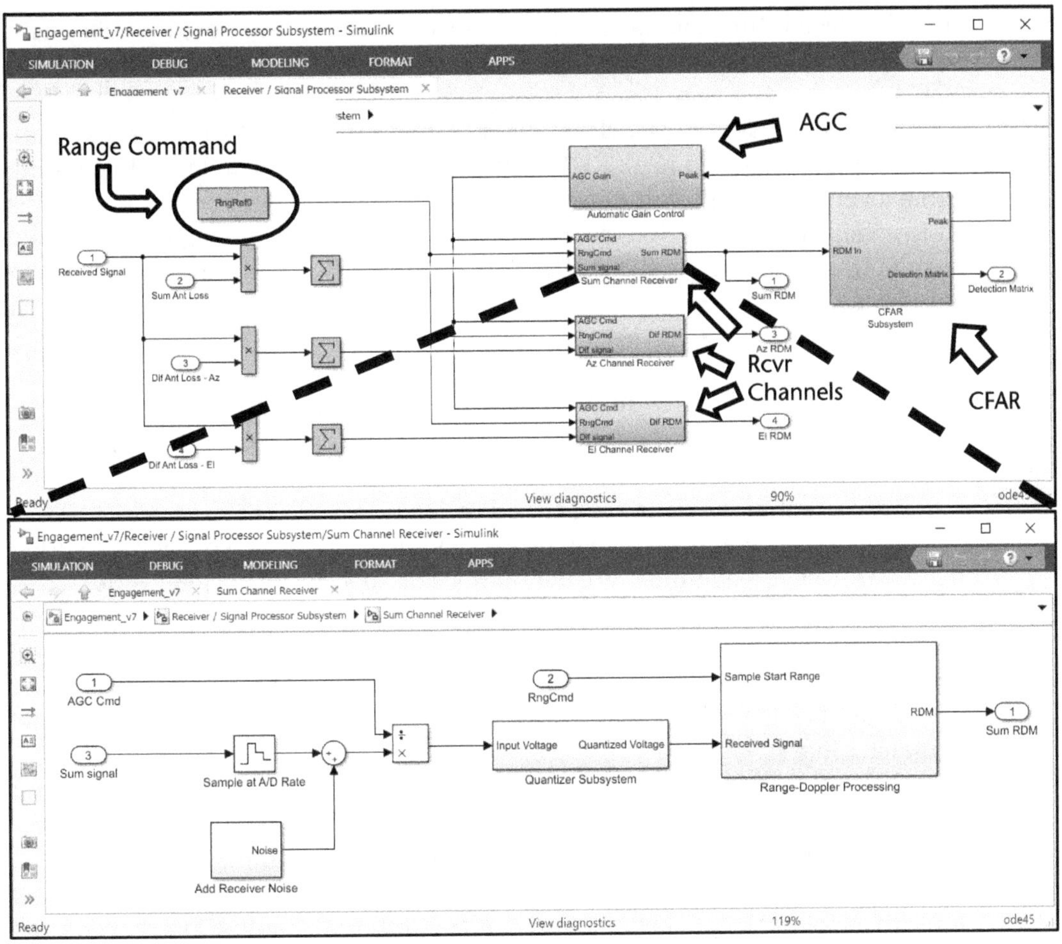

Figure 15.8 Block diagram of Receiver/Signal Processor subsystem (top) and block diagram of individual receiver channel (bottom).

model presented in Chapter 12 and shown in Figure 12.13. The only difference here is the addition of a control signal to command the timing of the samples within the PRI. The Range Command indicated at the top of Figure 15.8 dictates the start of the range window (i.e., the location of the first range gate). This range value transforms into a start time that controls the collection of fast-time samples within the desired time interval. Only those samples are sent to the matched filter. This range command is currently a constant, so the range window is fixed in space; as another growth opportunity, making this command a variable will allow the set of range gates to move within the unambiguous range interval. This can enable future schemes for range search, as well as precision range tracking (i.e., the adjustment of the range gates based on the target-estimated location).

The sum channel RDM moves into the CFAR subsystem for target detection, as well as for peak-signal monitoring, which drives the AGC subsystem. The CFAR subsystem is essentially the same as the scheme presented in Figure 12.6, with an additional modification that adds a second detection threshold to prevent threshold crossings from the range sidelobes of the pulse-compressed target. The subsystem uses this second threshold during high-SNR conditions; otherwise, the noise-based threshold is used the same as before. In addition, the subsystem captures the peak signal magnitude within the entire RDM and sends it to the AGC subsystem.

The AGC subsystem is a closed-loop system to control the input signal gain [1]. It takes the captured peak RDM value and compares it to a desired operating signal level (V_{AGC}). This value is chosen to prevent the ADC from saturating; it is set to a value equal to 6-dB below what the full-scale ADC voltage would create as an output value in the RDM. The AGC command in units of decibels can be written as

$$AGCdB_{new} = \max\left(AGCdB_{old} + 20\log_{10}\left(\frac{V_{peak}}{V_{AGC}}\right), 0\right) \quad (15.2)$$

The relevant parameters are defined in Table 15.1.

The AGC command as defined by (15.2) is always positive, so when converted to a linear scale it is a number greater than 1. The ADC input voltage is attenuated when divided by this quantity. When $V_{peak} > V_{AGC}$, additional decibels of attenuation are added to the previous command. If smaller than V_{AGC}, decibels are taken out but never to the point where the new command becomes negative. The receiver applies the AGC command equally to all channels, preserving channel gain-matching.

The Data Processor subsystem takes all three RDMs (sum, delta-Az, and delta-El) plus the binary Detection Matrix from the CFAR processor and generates measurements for the target tracker.

The target range and velocity measurements are implemented in a MATLAB m-file that looks at the detection matrix and identifies the range-Doppler cells that crossed the CFAR threshold (i.e., those cells with a value of '1' in them). The range and velocity estimates are

$$\hat{R} = R_{start} + \left(R_{bin} - 1\right)\delta_R \quad (15.3)$$

$$\hat{V} = V_{start} + \left(D_{bin} - 1\right)\left(\frac{\lambda PRF}{2M}\right) \quad (15.4)$$

where R_{bin} and D_{bin} are the indexes of the detection range-Doppler cell. R_{start} is the range corresponding to the first range bin (i.e., the range command), and δ_R is the

Table 15.1 AGC Parameters

Parameter	*Description*
$AGCdB_{new}$	New (or current) AGC command in decibels
$AGCdB_{old}$	Old (or previous) AGC command in decibels
V_{peak}	Peak signal value measured in the RDM
V_{AGC}	Desired operating point (signal level designed to prevent A/D saturation)

range bin width, typically equal or close to the range resolution. Similarly, V_{start} is the target velocity corresponding to the first Doppler bin. This value is usually zero, although in a real system it depends on the value of the LO frequency of the last mixer in the receiver chain. M is the size of the Doppler FFT, such that (PRF/M) is the Doppler bin width. Last, the factor ($\lambda/2$) converts Doppler frequency to radial velocity. Both range and velocity estimates are assumed unambiguous. Furthermore, these values correspond to the center of their respective range and Doppler bins and thus could be off by as much as half a bin width on either dimension. There are methods for refining these estimates to within a small fraction of a bin width, such as split gate tracking and centroiding [2]. None of those methods are implemented here for a single detection—yet another opportunity for growth. The only adjustment made is when there are multiple contiguous detections; in that instance, the detections are clustered into a single detection report using power centroiding for both range and Doppler.

Monopulse angle measurements are made separately for azimuth and elevation using the model presented in Figure 9.10, with a minor modification. The current antenna-steering commands (i.e., gimbal angles) are added to the measured angles—which are relative to the antenna boresight—to make the target direction relative to the radar platform body vector. The tracker works with the angles defined in this frame of reference, such that the angle state estimates may be used to correct the pointing of the antenna. Figure 15.9 illustrates this approach.

Target measurements are sent to the target tracker, which is the same DTKF introduced in Chapter 13 and defined by (13.6) to (13.15). The model only supports a single target being tracked, so a single set of measurements is reported with no data-association logic. Tracking is performed in measurement space (range, velocity, azimuth, and elevation), and the azimuth and elevation estimates become the new steering commands, which feed back into the Radar Dynamics block for repositioning the antenna. This model therefore performs closed-loop angle tracking. Recall that a future upgrade may also include closed-loop range tracking in which the range bins may adjust over time to keep the target centered on its range bin.

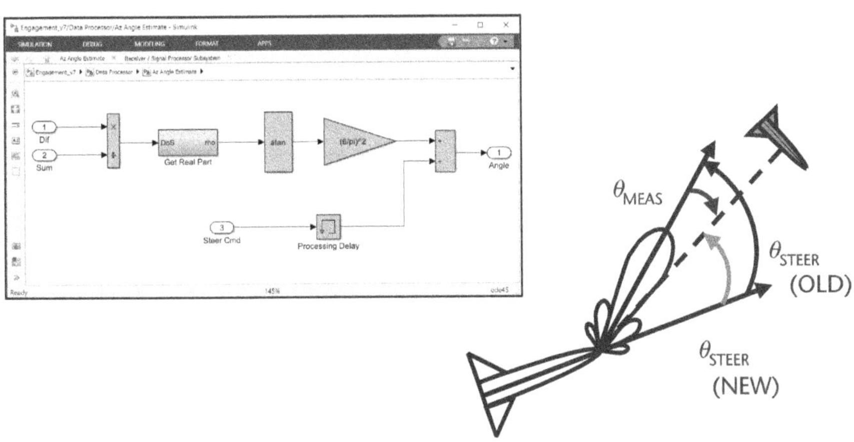

Figure 15.9 Block diagram of monopulse angle estimation model, with adjustment by steering command (left); new steering command is calculated as the previous command plus the measured angle relative to antenna boresight (right).

15.5 Example Single Radar Versus Single Target Engagement

The 1-v-1 scenario depicted in Figure 15.10 is used to test the engagement simulation; the radar platform is shown on the left, flying against a coaltitude target in a nose-on an east-west trajectory. The radar-target range is admittedly unrealistically short, but it is chosen so the measurement is unambiguous.

The simulation stop time is set long enough to collect and process 50 CPIs. Output products include RDMs for all three channels, as well as time histories for track-state errors (as shown in Figure 13.3) and raw measurement errors. Figure 15.11 shows error plots for this 1-v-1 example.

The errors on all four observables are small; however, range errors in particular are worthy of additional discussion. The range bin width for this example is 30m, so the instantaneous errors are one-half of the bin width (15m) or less. This is reasonable given that no measurement refinements are made; for a single detection (i.e., no clustering), the reported range is always the center of the range bin. Recall that the range bins are fixed in space. As the target flies toward the radar, it traverses across the bin, until it nears its edge. At that point the target energy starts straddling the adjacent range bin, and once the energy spillover is high enough, it will trigger a detection in that range bin as well. Figure 15.12 illustrates this condition, which first occurs at CPI #22. The circles in Figure 5.12 mark the threshold crossings. The centroiding of the two detections on CPI #22 causes the discontinuity observed in the range error plot. Straddling and centroiding continues until CPI #48, where the target goes back to being detected on a single bin. This causes yet another discontinuity.

In terms of angle-tracking, this was not a stressing case as the target is along the antenna boresight throughout the encounter. A better test of the antenna pointing control is to place an offset on the target, so that it is initially off boresight. (Try it!) Adding a target position offset of 80m due south (i.e., shifting the target to the right of the antenna boresight) results in an azimuth offset of about 3 degrees. Figure 15.13 shows the true target state calculated by the Target Generator and displayed in a Scope block. The angles are relative to the antenna boresight.

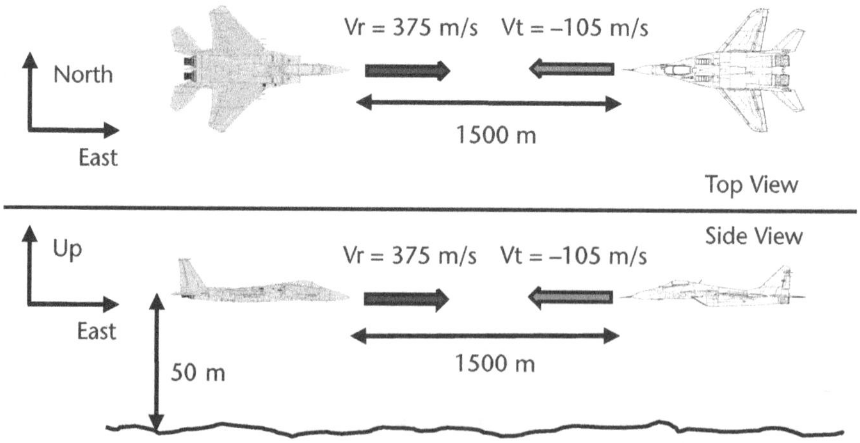

Figure 15.10 Notional radar-target engagement.

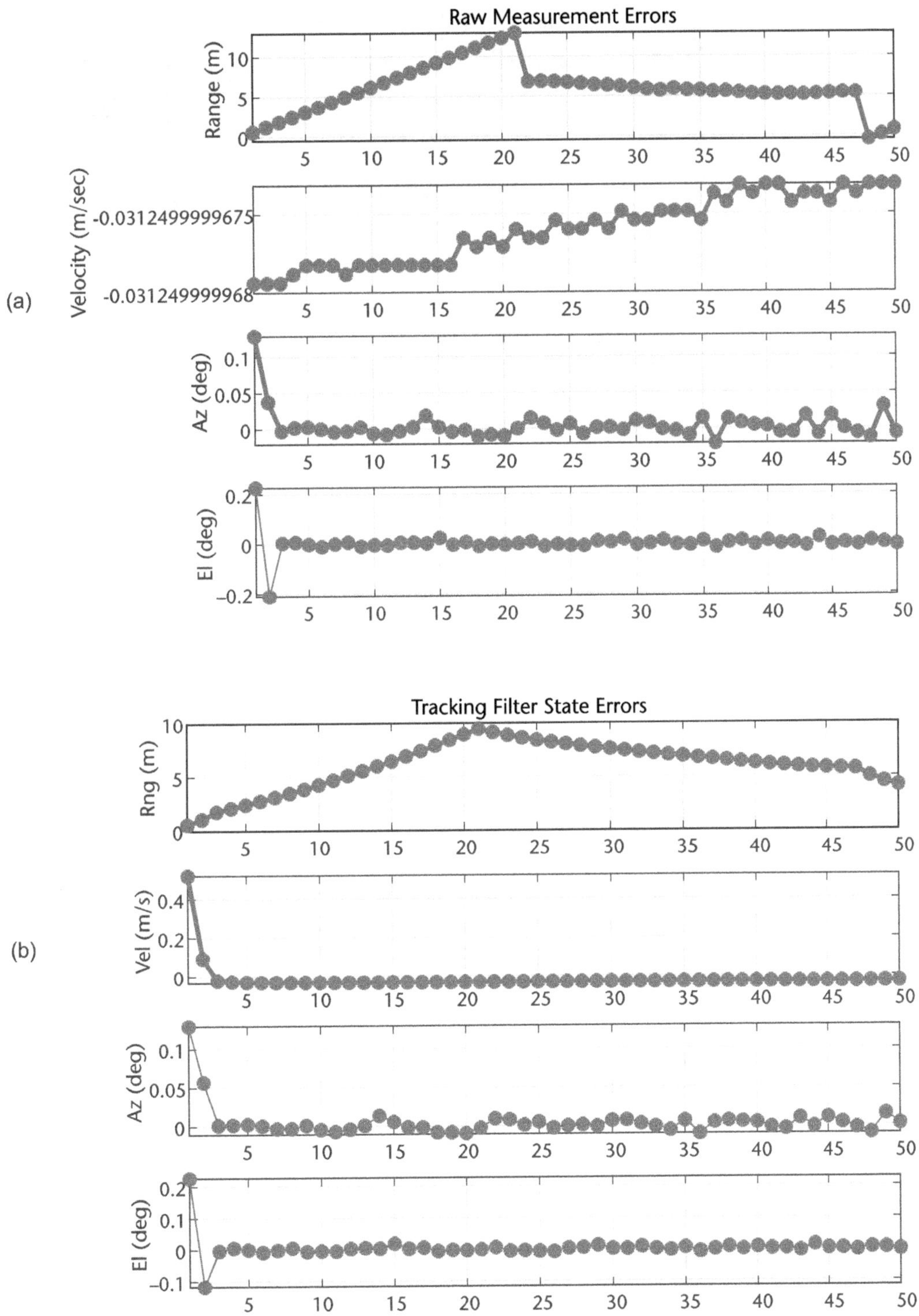

Figure 15.11 Raw measurement errors from 1-v-1 engagement (a) and tracking filter state errors (b).

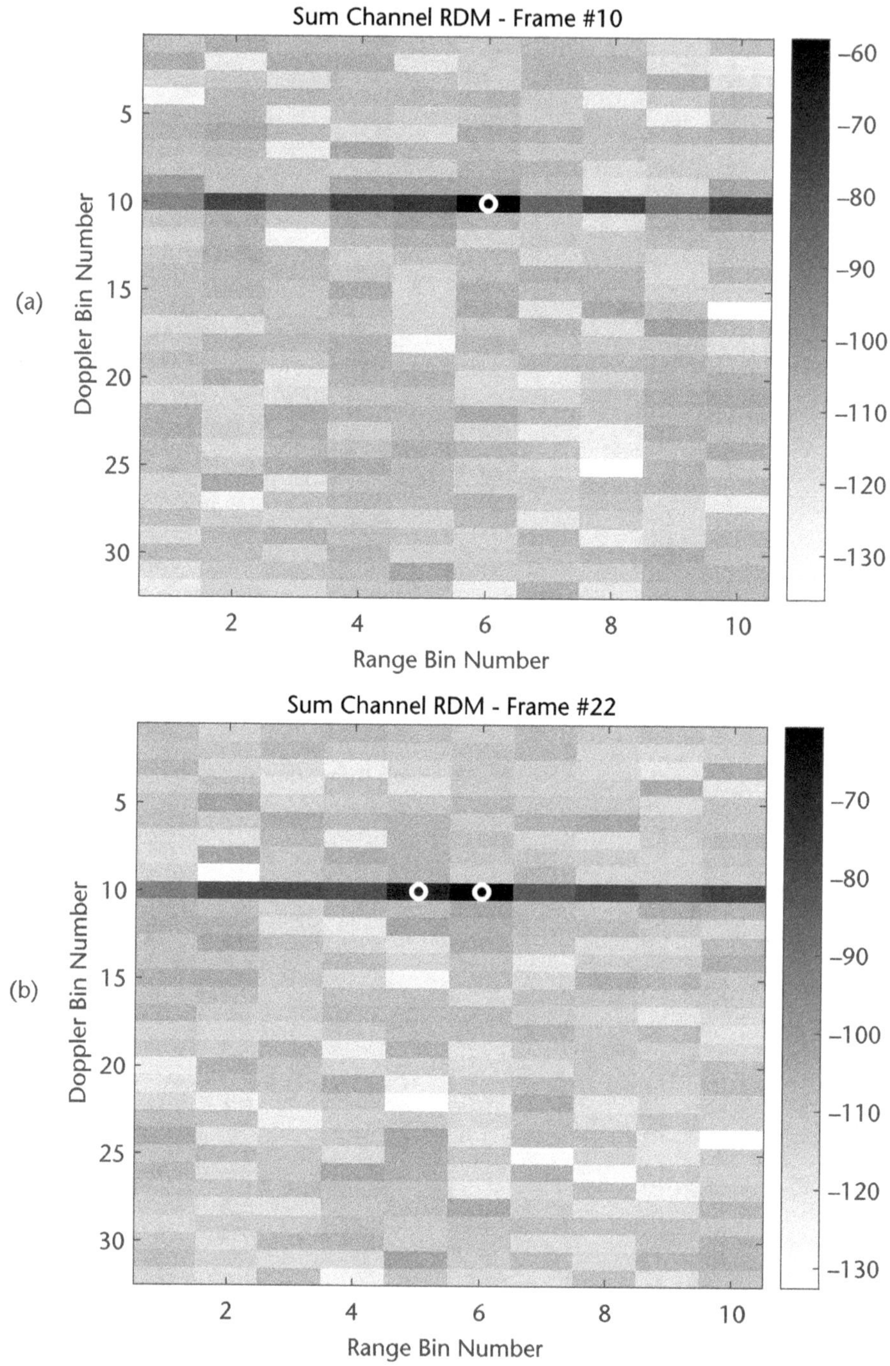

Figure 15.12 Example RDMs showing target detections: detection in a single range bin (a) and detections in adjacent bins due to range straddle (b).

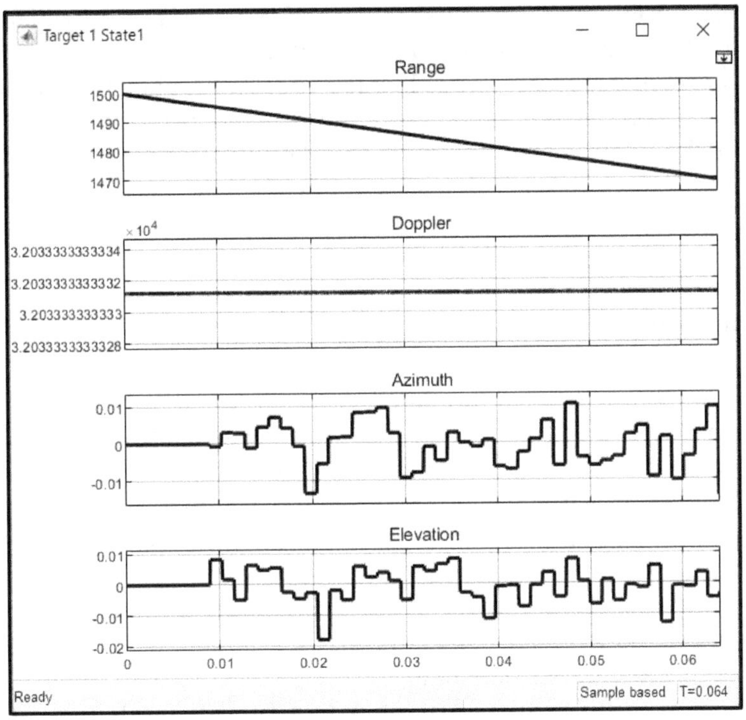

(a) Target at 0 deg Azimuth

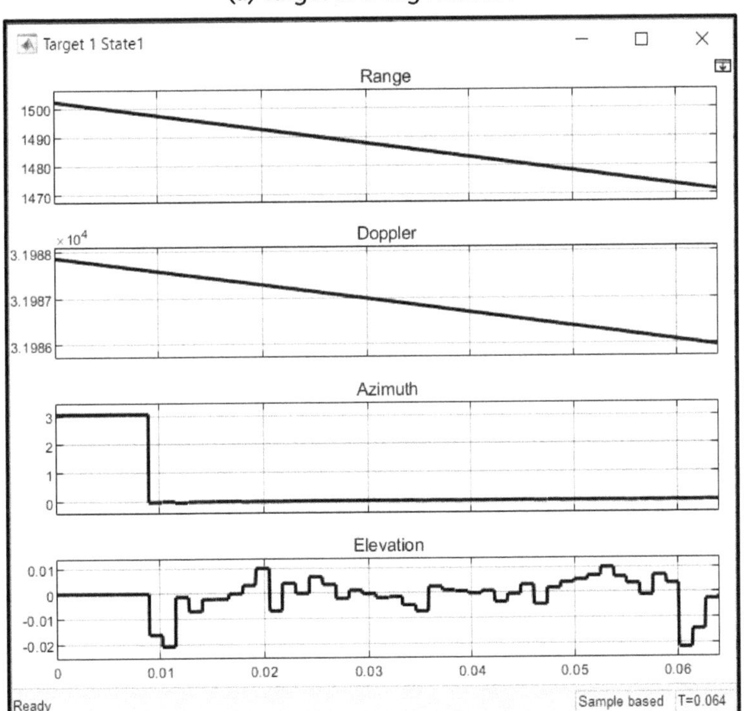

(b) Target at 3 deg Azimuth

Figure 15.13 Closed-loop angle tracking results: (a) target on boresight, and (b) target positioned 3 degrees off in azimuth.

The original collinear geometry is shown in Figure 15.13(a), where both azimuth and elevation measurements wander around 0 degrees; results from the offset geometry are in Figure 15.13(b). The gimbal angles are initially at 0 degrees, so the measured azimuth angle to the target is correctly measured at +3 degrees. Once the gimbals start adjusting based on the tracker estimates (which happens after five measurements have been processed by the DTKF), the antenna shifts 3 degrees to the right, so the target is placed (and remains) on boresight.

15.6 Concluding Remarks

This chapter presented upgraded versions of the RRE model, culminating with the end-to-end *Engagement_v7* model just described. This model simulates with some level of fidelity a dynamic engagement between a radar and a target. It took a considerable amount of effort to develop the model to its current state; however, there are still many upgrades yet to be completed to improve the model fidelity and robustness to a point where it can become a useful analysis tool. Many of those enhancements have been mentioned along the way, including, among many others, closed-loop range tracking, multiple target tracking support, and measurement-to-track association logic.

Chapters 16–18 address elements of EW modeling. This engagement model is used again to illustrate EW techniques and the impact they have on radar performance.

References

[1] Bruder, J. A., "Radar Receivers," in *Principles of Modern Radar, Vol I: Basic Principles,* M. A. Richards, J. A. Scheer, and W. A. Holm (eds.), Raleigh, NC: Scitech Publishing, Inc., 2010, pp. 408–409.

[2] Blair, W. D., M. A. Richards, and D. G. Long, "Radar Measurements," in *Principles of Modern Radar, Vol I: Basic Principles,* M. A. Richards, J. A. Scheer, and W. A. Holm (eds.), Raleigh, NC: Scitech Publishing, Inc., 2010, pp. 693–699.

CHAPTER 16

M&S of EA

16.1 Introduction

This is the first of three chapters that focus on the modeling and simulation of EW systems and techniques. EW formally stood for *electronic warfare;* many references and broad segments of the EW community still refer to it this way, and will likely do so for a while. Joint Publication 3-85, "Joint Electromagnetic Spectrum Operations" [1], formally replaced Joint Publication 3-13.1, "Electronic Warfare" [2], when the Chairman of the Joint Chiefs of Staff published it in 2020. It was at that point when the term "electromagnetic" formally replaced "electronic" in the DoD Dictionary.

In both documents [1, 2], however, the definition remained unchanged: "Military action involving the use of electromagnetic and directed energy to control the electromagnetic spectrum or to attack the enemy."

There are three categories of operations within EW:

- *Electromagnetic attack* (EA), formerly termed electronic attack, is the "Division of electromagnetic warfare involving the use of electromagnetic energy, directed energy (DE), or antiradiation weapons to attack personnel, facilities, or equipment with the intent of degrading, neutralizing, or destroying enemy combat capability and is considered a form of fires" [1]. This is what we colloquially call jamming. An older term for EA is electronic countermeasures (ECM), although ECM did not traditionally include directed energy and antiradiation weapons; these were added during the transition from ECM to EA.
- *Electromagnetic protection* (EP), formerly termed electronic protection, is the "Division of electromagnetic warfare involving actions taken to protect personnel, facilities, and equipment from any effects of friendly or enemy use of the electromagnetic spectrum that degrade, neutralize, or destroy friendly combat capability" [1]. The previous term for EP was electronic counter-countermeasures (ECCM), as a counterpart to ECM.
- *Electromagnetic support* (ES), formerly termed electronic warfare support, is the "Division of electromagnetic warfare involving actions tasked by, or under direct control of, an operational commander to search for, intercept, identify, and locate or localize sources of intentional and unintentional radiated electromagnetic energy for the purpose of immediate threat recognition, targeting, planning and conduct of future operations" [1]. The older term to go along with ECM and ECCM is electronic support measures (ESM).

This chapter focuses on EA, the first pillar of EW, and is organized as follows. Section 16.2 delves further into basic EA concepts and terminology, including types of EA systems (jammers) and example techniques. Section 16.3 focuses on the first-order modeling of repeater jammers, while Section 16.4 incorporates those stand-alone jammer models to the engagement model presented in Chapter 15 and observes their impact on radar performance. Similarly, Section 16.5 focuses on first-order modeling of noise jammers, and, subsequently, Section 16.6 examines their effects using the engagement model. Section 16.7 concludes with some additional modeling remarks.

16.2 EA Concepts

This book does not address the DE or antiradiation weapons (antiradiation missiles, for example) aspects of EA. Traditional EA is typically considered a nonkinetic type of attack, whose main targets are electronic equipment and facilities (not personnel as in the formal definition). It is in this traditional area that we will devote our efforts, thereby departing somewhat from the more comprehensive definition presented in Section 16.1.

One may categorize the different types of EA in a number of ways. One natural approach is by the type of waveforms used by jammers. The two extremes of the waveform spectrum are noise (noncoherent) and repeater (coherent) waveforms. In this context, the term coherent refers to the degree of similarity with the radar waveform. A coherent signal would seem target-like to the victim radar. Between these extremes, there is a continuum of waveforms with varying degrees of coherency (semicoherent waveforms).

Another way of looking at EA is by its intended effects. Two categories of effects are masking and deception. Masking refers to EA effects that prevent or delay detection of the protected target. Masking can also degrade the measurement and tracking accuracy of radars. In general, masking EA achieves these objectives by reducing the SNR in the radar receiver. Noise waveforms map naturally into masking techniques.

Deception EA, on the other hand, does not attempt to prevent target detection, but to present the radar with erroneous information. Deception techniques create fake signals that the radar may confuse with legitimate target returns, but whose range and/or angle and/or Doppler measurement will be incorrect. Deception EA is also known as spoofing. Deception EA is typically implemented using coherent waveforms.

A third form of effect is saturation or overloading of resources, and it is somewhat of a hybrid of the first two. Saturation of the actual RF components in the radar receiver may be achieved by high-power noise, driving the receiver into a nonlinear region of operation. Examples include driving amplifiers into signal compression or saturating the A/D forcing clipping of the input signals. Another form of saturation is accomplished by deception EA creating a very large number of false target returns, beyond the ability of the radar to sort and track all of them. The radar processor may waste valuable time and computing resources in detecting, measuring, and otherwise processing all these false signals, to the detriment of prosecuting the real target.

One last categorization of EA is by the relative geometry of the radar, the target, and the jammer. This refers to the physical location of the jammer with respect to the target it intends to protect, as well as the jammer positioning relative to the radar antenna beam. Figure 16.1 presents four cases of interest, all in which the radar is assumed to be pointing its antenna beam toward the target.

The geometry depicted in the lower right diagram of Figure 16.1 is the case where the target is carrying the jammer to protect itself; this is known as self-screening jamming (SSJ) or self-protection jamming (SPJ). A couple general observations about this case follow:

- The jammer range and Doppler measurements (in the absence of any deception in those domains) is the same as those of the target;
- The jamming signal is received through the radar antenna main beam; therefore, its angle (in the absence of any angle deception) is the same as that of the target and can be measured.

The remaining three cases are variations of off-board jamming (OBJ), meaning that the jammer is at a different location from the target. Another term used is support jamming. If the jammer is in the general vicinity of the target (e.g., flying in formation at a similar range and angle from the radar), it is called an escort jammer (EJ) or sometimes escort support jammer (ESJ). The angular separation between the target and the jammer relative to the radar is small enough that the jammer is still in the main beam of the radar antenna—its angle may still be measured. This is not the case for the remaining two geometries on the left side of Figure 16.1, where the diagrams show a relatively large angular separation, such

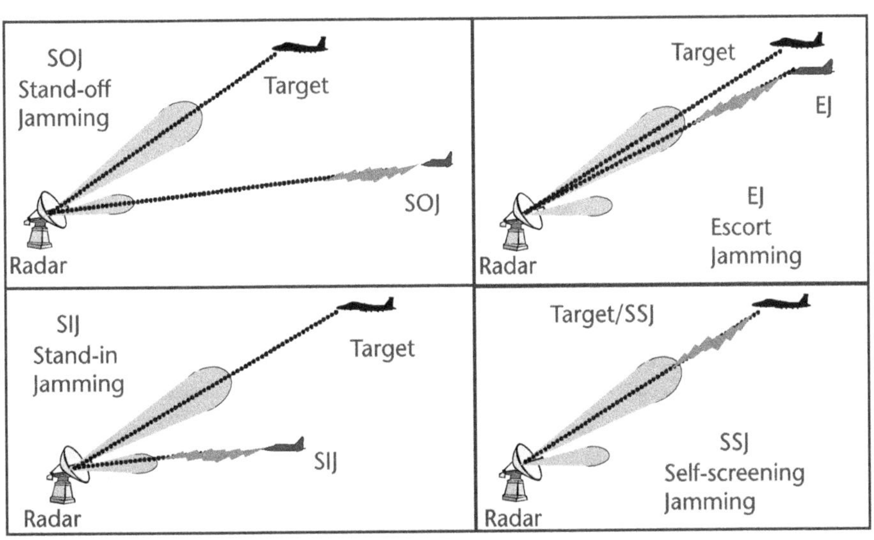

Figure 16.1 Common radar-jammer engagement geometries (counter-clockwise from the lower right): self-screening, escort, stand-off, and stand-in jamming.

that the jammer is outside the main beam, and its signal is coming through the sidelobes. These are two examples of *sidelobe jamming*; the key implication is that the radar is incapable of measuring the angle to the jammer. In the stand-off jamming (SOJ) case, the jammer is at a longer range (often a much longer range) than the target. It will typically be a high-power jammer, to overcome the $1/R^2$ one-way propagation loss. By contrast, the stand-in jammer (SIJ) in the lower left of Figure 16.1 is at a shorter range, which means that it may not need as much power to be just as effective. We will be able to model all of these geometries in our Simulink Engagement model.

We consider three general classes of jammer types; these are illustrated in Figure 16.2.

The complexity and capability of the systems shown in Figure 16.2 increase from left to right; on the left, the barrage jammer features a waveform generator whose output may not resemble the radar waveform whatsoever; it is therefore in the class of noncoherent jammers, such as noise jammers. It cannot receive, only transmit; it is therefore nonresponsive to the radar emissions. It may be always on, or it may transmit at regular intervals completely decoupled from the radar transmit schedule. In order to compensate for its lack of knowledge of the radar location and frequency of operation, this jammer must cover a large frequency extent as well as a large range extent (by radiating for a long period of time). A notional RDM processed by the radar shows this notionally wide coverage in these two dimensions.

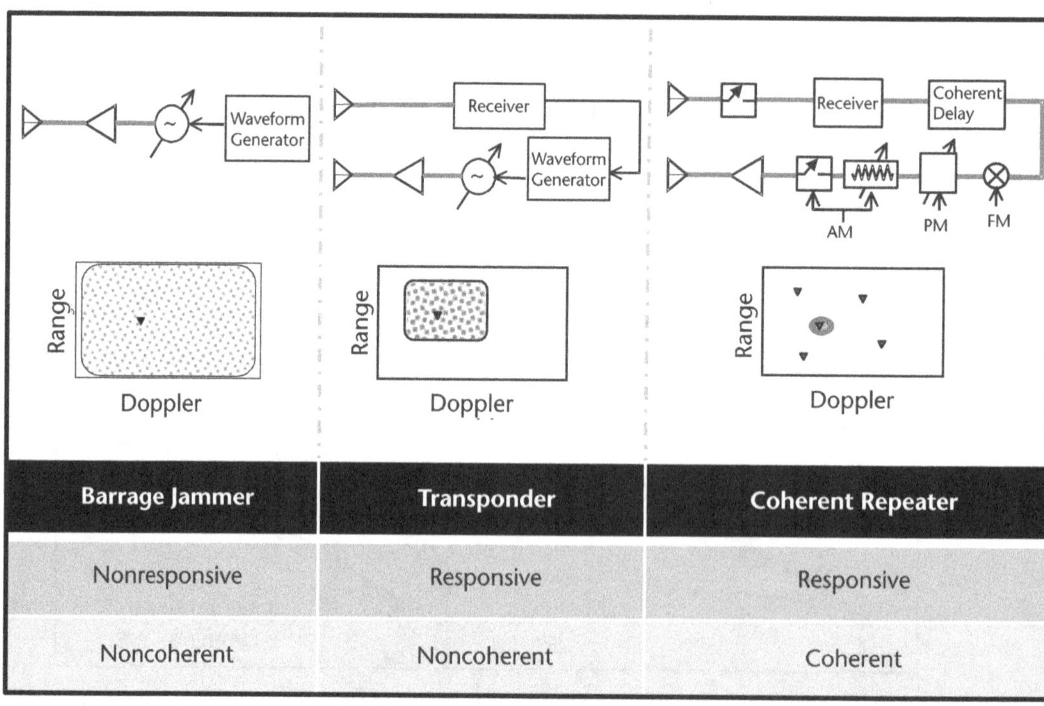

Figure 16.2 Generic jammer types (left to right): barrage jammer, transponder, and coherent repeater.

A transponder system is like the barrage jammer, in the sense that the waveforms it emits are also noncoherent. It does include a receiver channel, however, which means that it can now detect the radar signals and potentially measure its frequency and timing. This jammer class can use its available power more efficiently by concentrating it in range and Doppler around the target return, as indicated by the darker shade of its response in the RDM illustrated in the center of Figure 16.2.

The coherent repeater is the most capable of the three classes. It is not only responsive like the transponder, but it can emit a signal that is a modified version of the radar waveform.

The RRE (as applied to a jammer) is used to calculate the jammer power that the radar sees. Like SNR, the jammer-to-noise ratio (JNR) gives a measure of the jammer strength. Ultimately, however, the jammer effectiveness is best characterized by the jammer-to-signal ratio (JSR). The inverse ratio (SJR) is considered a degraded version of SNR, where the jammer is assumed to be the dominant source of interference. Another generic term is signal-to-interference ratio (SIR), where the interference could be any combination of noise, clutter, and jammer signals.

16.3 Coherent Repeater EA

The coherent repeater illustrated on the right side of Figure 16.2 and reproduced at the top of Figure 16.3 may be implemented in either analog or digital hardware. The time delay associated with range-deception techniques may be accomplished by an analog delay line such as an RF memory loop (RFML) or by downconverting and digitizing the radar pulse, which may then be stored in memory. This is the approach typically followed by digital RF memory (DRFM) devices. Additional modulation (phase or frequency) on the radar pulse may also be implemented in the analog or the digital domain. Functionally, the repeater captures the radar pulse, applies the delay and/or modulation, scales it, and retransmits it. The scaling depends on the repeater operational mode, which is typically either constant-gain or constant-power [3].

In constant-gain mode, the jammer attempts to apply the same gain throughout the receive-transmit loop regardless of the incident radar power density. From the RRE, this power density (power per unit area) is

$$PD_{\text{inc}} = \frac{P_t G_t}{4\pi R^2} \tag{16.1}$$

This is (2.2) neglecting the transmit and atmospheric losses. The total power captured by the jammer receiver is the product of (16.1) and the effective area of the jammer receive antenna, or

$$P_{rj} = PD_{\text{inc}} A_{rj} = \left(\frac{P_t G_t}{4\pi R^2}\right)\left(\frac{G_{rj}\lambda^2}{4\pi}\right) = \frac{P_t G_t G_{rj}\lambda^2}{(4\pi R)^2} \tag{16.2}$$

where G_{rj} is the jammer receive antenna gain. If we define the internal jammer gain and the transmit antenna gain by G_{ij} and G_{tj}, respectively, then the jammer effective radiated power (ERP_j) is

$$ERP_j = P_{tj}G_{tj} = \frac{P_t G_t G_{rj} G_{ij} G_{tj} \lambda^2}{(4\pi R)^2} = \frac{P_t G_t G_{\text{loop}} \lambda^2}{(4\pi R)^2} \tag{16.3}$$

where the loop gain G_{loop} is given by the product ($G_{rj}\, G_{ij}\, G_{tj}$). The received power back at the radar is (16.3) times another one-way spherical propagation loss times the effective area of the radar receive antenna:

$$P_j = \frac{P_t G_t G_{\text{loop}} \lambda^2}{(4\pi R)^2} \left(\frac{1}{4\pi R^2}\right)\left(\frac{G_r \lambda^2}{4\pi}\right) = \frac{P_t G_t G_r \lambda^2}{(4\pi)^3 R^4} \left(\frac{G_{\text{loop}} \lambda^2}{4\pi}\right) \tag{16.4}$$

The last equality of (16.4) is identical to a received target power equation if we interpret the term in parenthesis as an "equivalent jammer RCS," σ_j, which now we formally define as

$$\sigma_j = \frac{G_{\text{loop}} \lambda^2}{4\pi} \tag{16.5}$$

This is how we will scale the incoming radar pulses in our first-order functional jammer models operating in constant-gain mode.

In constant-power mode, however, the transmit power P_{tj} is set to some fixed value $P_{\max}$ independent of the received power. The input-output characteristics for both operating modes are illustrated at the bottom of Figure 16.3. Notice that from the previous definitions, the constant gain applied (the slope of the curve) is G_{ij} and not G_{loop}.

16.3.1 Example Coherent EA Techniques

The techniques described in this section all assume SSJ operation, as it greatly simplifies the timing of the jammer transmissions relative to the radar pulse reflections off the target. The simplest technique is arguably for the jammer to behave like a bent pipe and simply retransmit everything that it receives, with some time delay. We refer to this technique as a straight-through repeater (STR). The time delay relative to the target reflection translates to a range offset relative to the true target range and is therefore a range-deception technique. The time delay is understood to be fixed and is never zero, as there is always a finite delay for the radar pulse to work its way through the jammer electronics.

The next step in technique complexity is for the time delay to be time-varying. This capability leads to a family of range-deception techniques collectively known as range-gate stealer (RGS) jamming. The objective of these techniques is to attack the radar tracker, by creating track errors and/or break the radar track

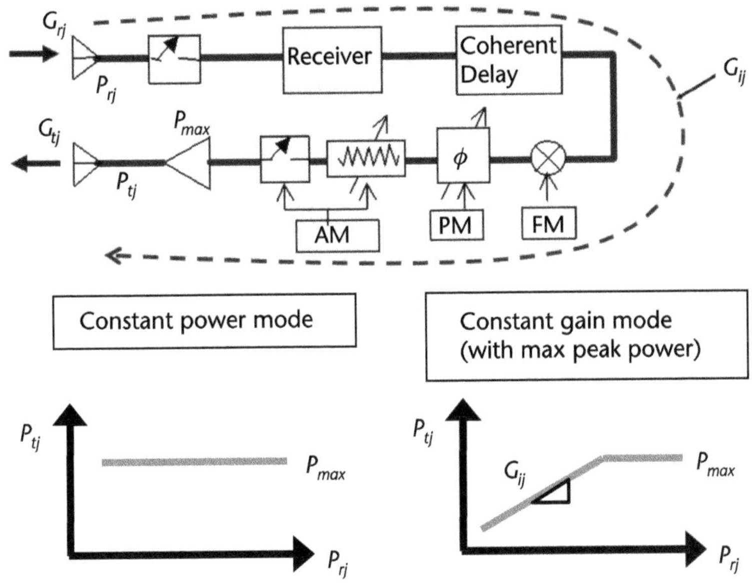

Figure 16.3 Coherent repeater operational modes: constant power (left) and constant gain (right).

on the target. A typical RGS motion follows some or all of the following phases in a repeating cycle:

- *Dwell:* Jammer signal sits at the target cell to capture the radar tracking gate;
- *Walk:* Jammer slowly moves away from the target, with the objective of deceiving the radar into following it;
- *Hold:* Jammer stops at a maximum offset from the target, notionally placing the target outside the radar view;
- *Off:* Jammer may turn off, forcing the radar to restart target search or reacquisition.

These phases are illustrated in Figure 16.4. Within the generic RGS technique, related terms include range-gate pull-off (RGPO) when the jammer walks down-range (ranges longer than the target range), and range-gate pull-in (RGPI) when it walks in-range.

Gate-stealer jamming may occur in either the range or Doppler (velocity) dimension, so Figure 16.1 may represent either an RGS or a velocity gate stealer (VGS). A VGS is implemented by adding a frequency offset to the radar pulse, rather than a time delay. Given an incident radar waveform

$$s_{\text{in}}(t) = A(t)e^{j2\pi f(t)t} \tag{16.6}$$

the VGS will add a frequency offset $\Delta f = 2(\Delta V)/\lambda$, resulting in an output waveform

$$s_{\text{out}}(t) = B(t)e^{j2\pi(f(t)+\Delta f)t} \tag{16.7}$$

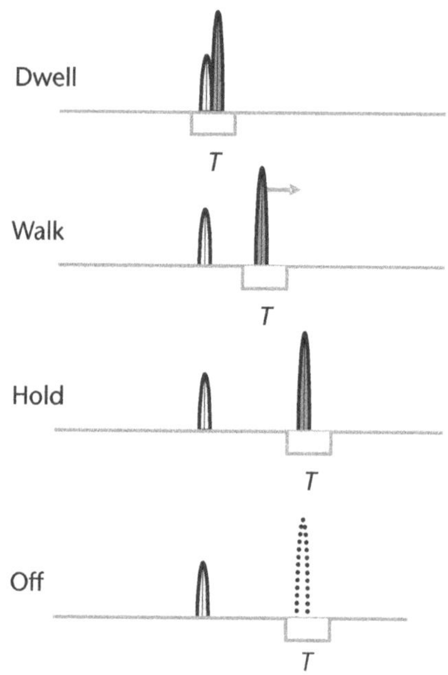

Figure 16.4 Typical phases of a gate-stealer EA technique. The box represents the captured radar tracking gate.

Notice that the jammer must be able to measure or estimate the radar center frequency (i.e., wavelength) in order to calculate the frequency offset for a desired velocity offset ΔV.

A jammer that pulls simultaneously in range and velocity is called a coordinated gate stealer (CGS).

Any of the above single false-target (SFT) techniques may be replicated multiple times, resulting in a multiple false-target (MFT) technique. For specific range or Doppler deception, the MFT technique may also be termed multiple false range targets (MFRT) and multiple false Doppler targets (MFDT), respectively.

16.3.2 Example Coherent EA Models

This section illustrates the functional modeling of coherent repeaters in constant-gain mode, implementing some of the techniques described in Section 16.3.1. These first-order models may be upgraded to improve their fidelity as needed. Figure 16.5 shows a Simulink implementation of a simple STR jammer.

STR_Model takes the radar pulses and sends them back with some minimum fixed delay and an amplitude scaling based on the equivalent RCS as defined by (16.5). A straightforward update for implementing an RGS technique is to vary the delay over time. One block suitable for this purpose is the Repeating Sequence Stair, which repeats a user-defined time sequence. This block facilitates the implementation of an arbitrary dwell-walk-hold-repeat cycle, as illustrated in Figure 16.6 for the model *RGPO_Model*.

A second scope in the jammer subsystem shows the range-walk cycle created with the repeating sequence; the time duration of each step in the staircase was

(a)

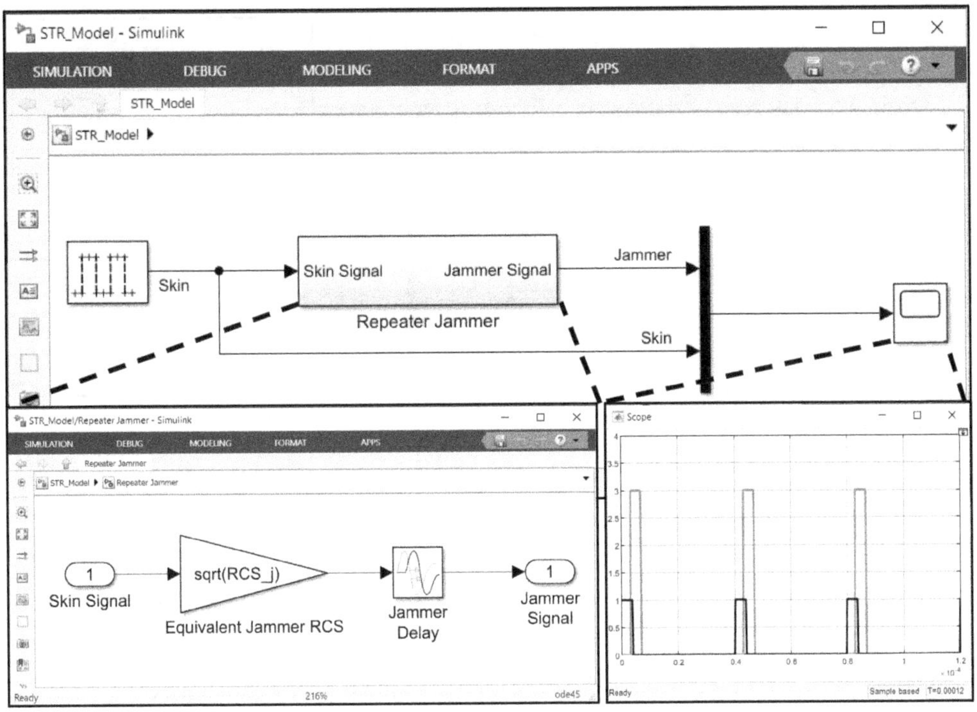

(b) (c)

Figure 16.5 Simulink implementation of STR technique with fixed delay: (a) top-level diagram, (b) jammer subsystem, and (c) resulting target and jammer pulses.

set to one PRI for demonstration purposes only—realistic RGPO jammers would not walk that fast. The jammer initially dwells at the target location for two PRIs, then starts walking toward longer time delays (longer ranges) out to a maximum delay of PRI/2. On the next PRI it snaps back to the target location (i.e., there is no "off" stage in this particular cycle). An "off" stage could be easily implemented with a Pulse Generator block acting like an OOK switch.

A similar setup can implement an RGPI, where the jammer walks in-range. The model for this technique is shown in Figure 16.7.

The trick of implementing the RGPI technique is for the jammer to capture the first pulse, delay it for an entire PRI, and then retransmit it on top of the next pulse. The jammer then reduces the delay of subsequent pulses so that it appears to be walking in front of the target. Again, the step duration is set to one PRI and the maximum delay to PRI/2. Given that the jammer will initially delay its response by one PRI implies that it must measure said PRI. This is an example of a technique that requires ES functionality for accurate timing.

The last range-deception technique is simply a superposition of multiple STR jammers at different range delays. The MFRT model is shown in Figure 16.8. *MFRT_Model* creates four false targets around the real target.

We now consider Doppler or velocity deception, a different type of deception technique. The first example technique creates a single false target with a fixed

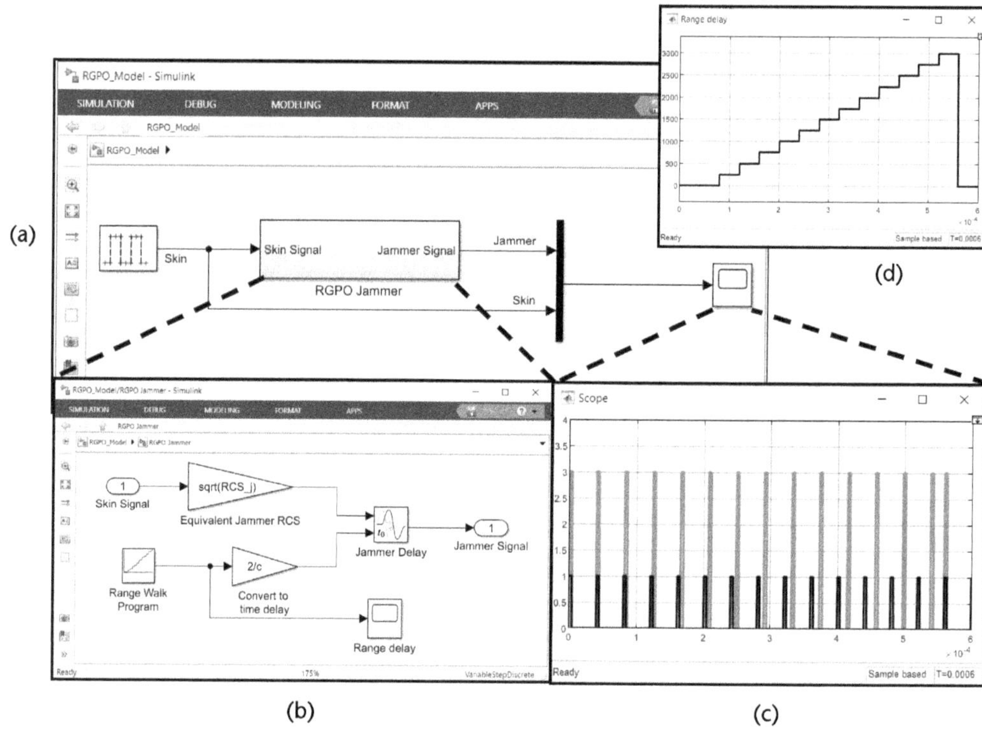

Figure 16.6 Simulink implementation of RGPO technique: (a) top-level diagram, (b) jammer subsystem, (c) resulting target and jammer pulses, and (d) range-walk program.

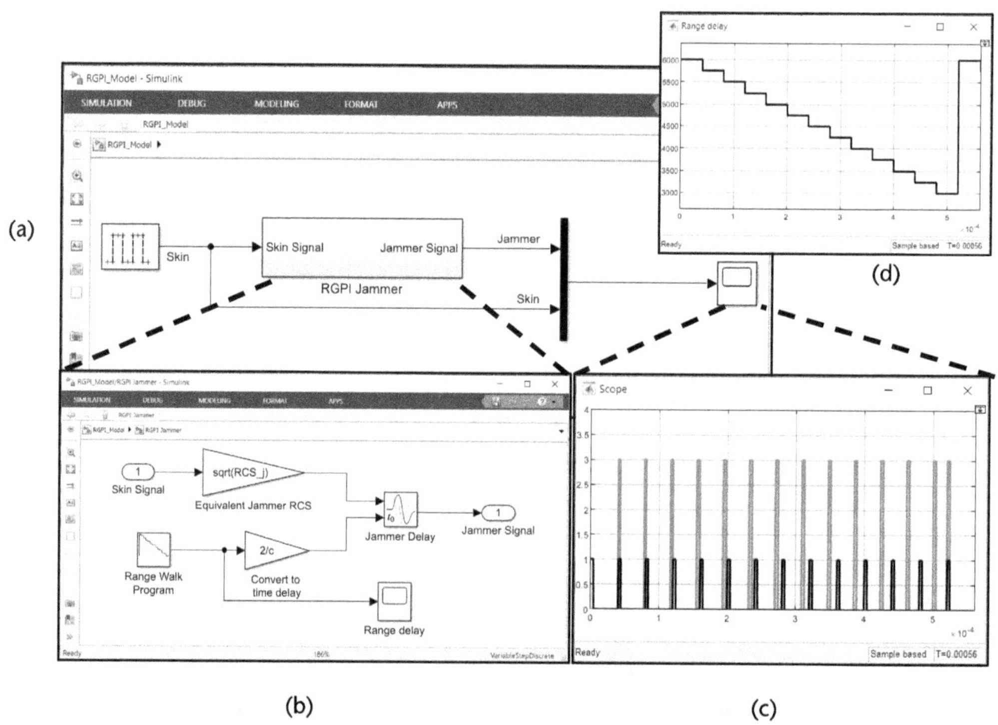

Figure 16.7 Simulink implementation of RGPI technique: (a) top-level diagram, (b) jammer subsystem, (c) resulting target and jammer pulses, and (d) range walk program.

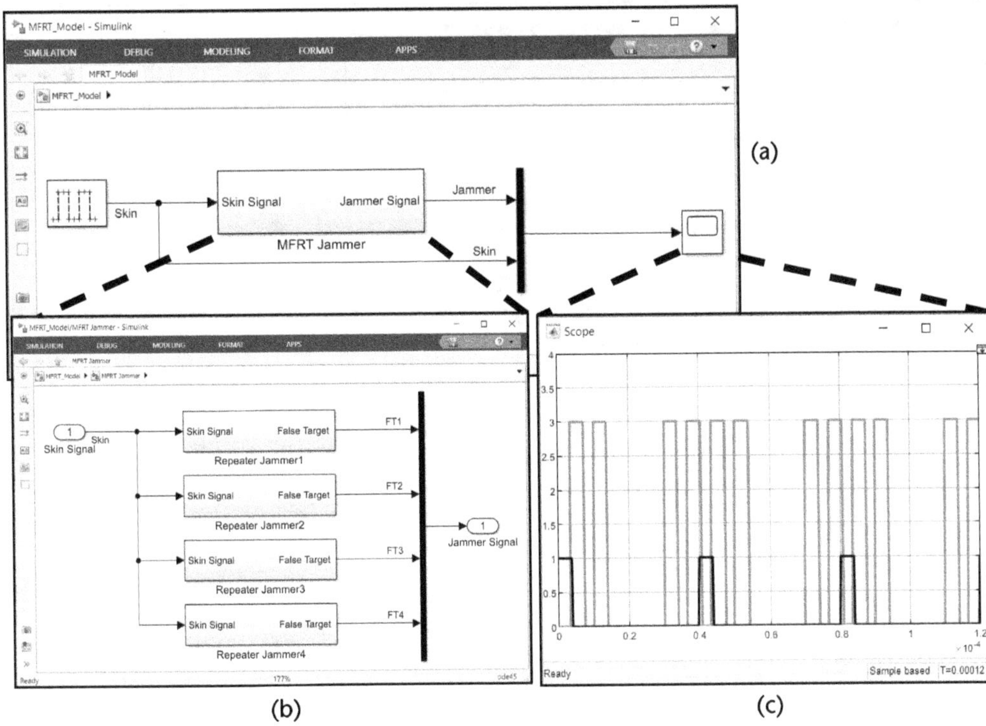

Figure 16.8 Simulink implementation of MFRT technique: (a) top-level diagram, (b) jammer subsystem, and (c) resulting target and jammer pulses.

Doppler offset. The Simulink model is called *VGS_Model*; even though it is not, strictly speaking, a gate-stealer technique, it can easily be turned into one by varying the offset over time. Figure 16.9 shows the details of the jammer model and two example results.

The jammer signal is blocked from being added to the target signal by the zero-gain block (circled in the block diagram in Figure 16.9). Changing the gain to one effectively turns the jammer on. The model adds Doppler shifts as arguments of complex exponentials that multiply both the target and jammer signals $s(t)$ and $j(t)$:

$$\begin{aligned} s(t) &\rightarrow s(t)e^{j2\pi f_{\mathrm{Dop}}t} \\ j(t) &\rightarrow j(t)e^{j2\pi(f_{\mathrm{Dop}}+\Delta f)t} \end{aligned} \tag{16.8}$$

The target Doppler f_{Dop} is set to PRF/4 so when the jammer is off, it shows up in filter 8 out of 32 (displayed as index 9 in the figure since MATLAB indexes are one-based). The jammer signal has an additional offset Δf equal to two Doppler bins so it shows up in filter 10 (index 11 in the figure). The FFT includes a Chebyshev window function, which broadens the jammer mainlobe such that it obscures the target. (What happens if we use uniform weights instead? Try it!)

The VGS model serves as a building block for the MFDT technique. The *MFDT_Model* block diagram and example results are shown in Figure 16.10, featuring four false targets at different Doppler offsets, two below and two above the target.

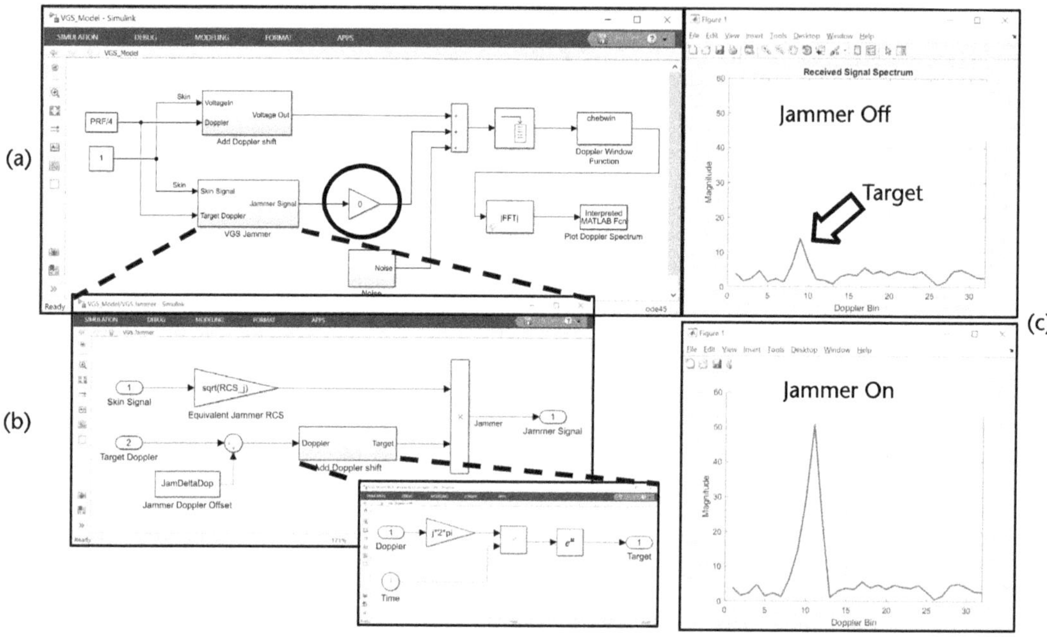

Figure 16.9 Simulink implementation of VGS technique: (a) top-level diagram, (b) jammer subsystem, and (c) resulting target and jammer spectra.

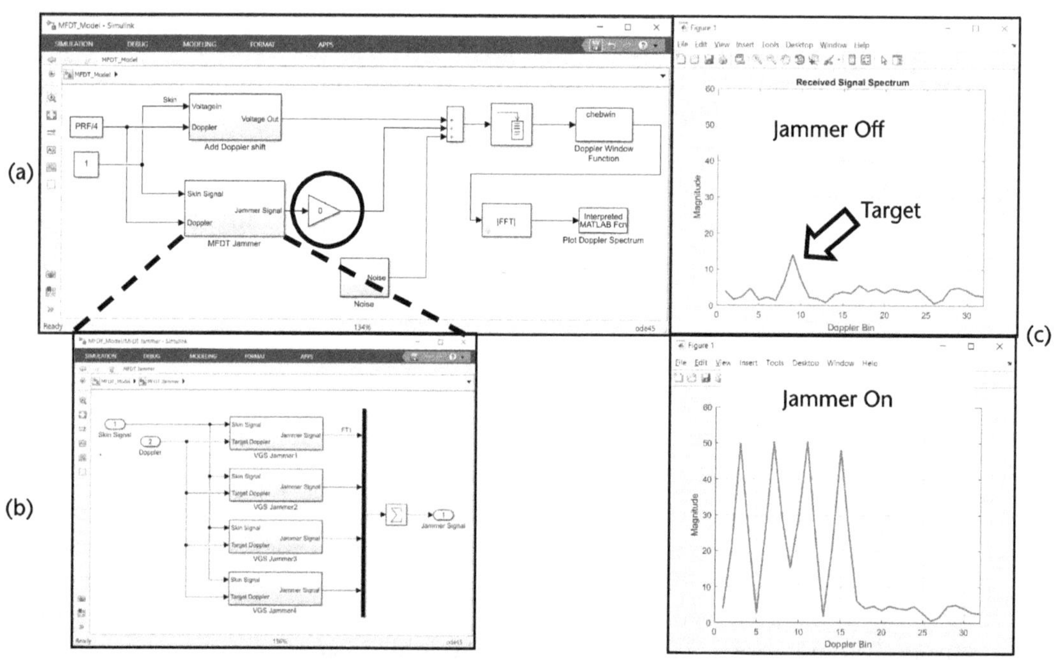

Figure 16.10 Simulink implementation of MFDT technique: (a) top-level diagram, (b) jammer subsystem, and (c) resulting target and jammer spectra.

We will now insert a few of these stand-alone models in the Engagement model and discuss their effects on the radar.

16.4 Engagement Simulation with Coherent Repeater EA

Figure 16.11 shows how *Engagement_v7* is modified by the addition of a jammer subsystem, in this case a range-deception SSJ. The updated model, now renamed *Engagement_v7_ssj*, takes a copy of the signal out of the target subsystem, sends it to the jammer subsystem, and reinserts the jammer output into the return path using a Mux.

For an SSJ geometry, the truth data coming out of the Target Generator applies to both target and jammer signals. Also, there is a Jam preprocessor subsystem prior to the jammer subsystem. This subsystem removes the target RCS from the target signal, since it will be scaled by the equivalent jammer RCS inside the jammer subsystem.

This version of the RGS jammer is able to implement either an STR technique (fixed-range delay) or an RGPO (time-varying) via a user-defined flag that is set in the model Initialization Callback. The Callback runs the m-file *engagement_v7_ssj_init.m*. The flag `JammerType` in line 93 of the m-file is set to zero for STR, or to one for the RGPO. Results for both are shown in Figure 16.12. Under either technique, the skin target is covered by the range sidelobes of the jammer.

Since there is no deception in velocity or angle, those errors are close to zero. The range errors are circled in Figure 16.12. The STR is at a fixed range offset of 60m, which equals two range bin widths. The RGPO walk cycle only goes out to the far edge of the range window, then snaps back to the target location. Almost four RGPO cycles are completed during the simulation run, which are mirrored

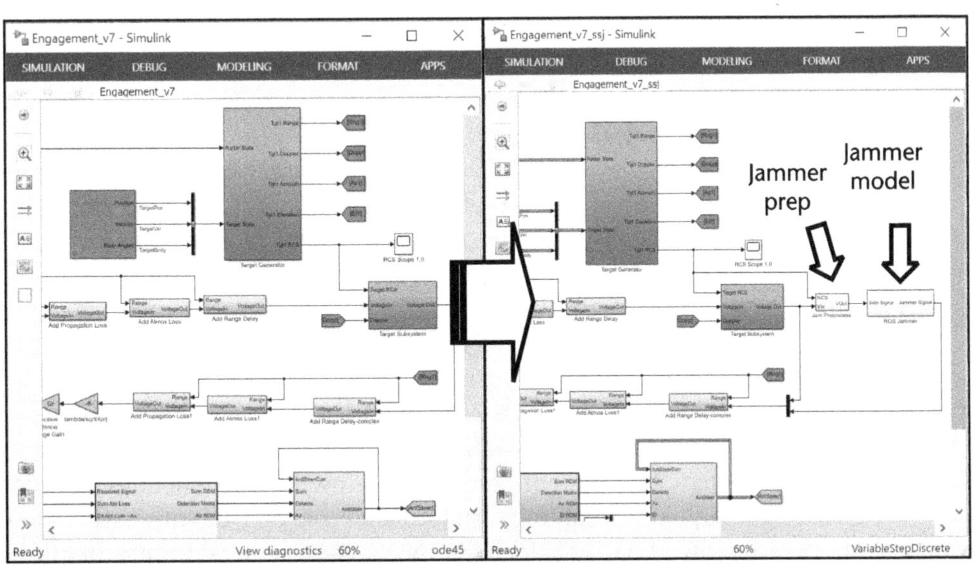

Figure 16.11 Engagement model with EA upgrade: original EA-free model (left) and upgraded version with jammer subsystem in an SSJ configuration (right).

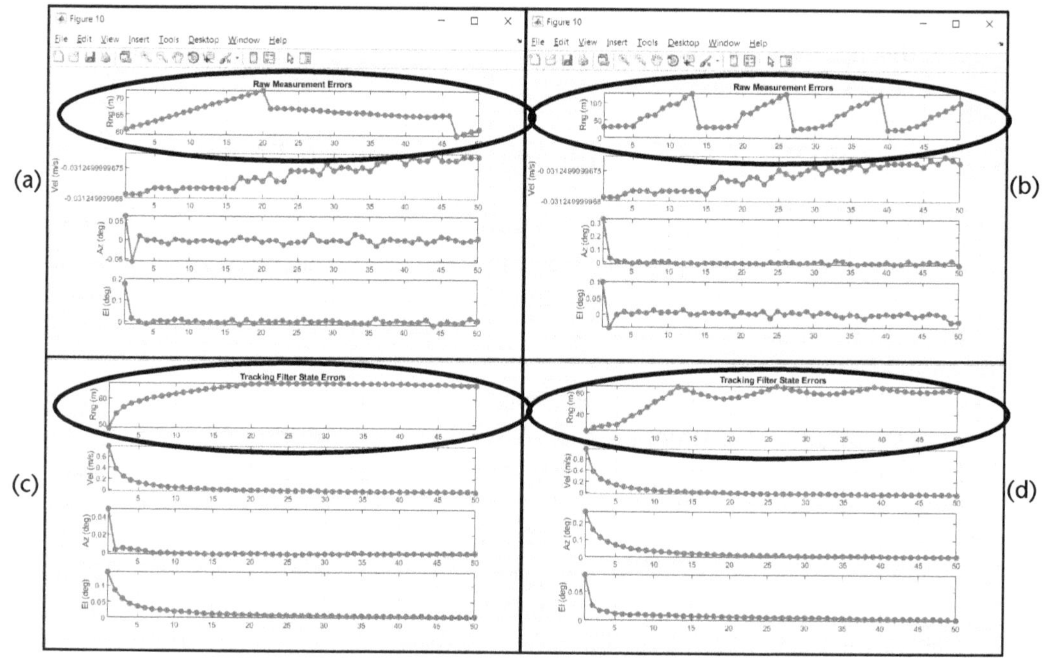

Figure 16.12 Engagement simulation results with range-deception EA: (a, c) STR and (b, d) RGPO. The plots in (a, b) are raw measurement errors, and (c, d) show track state errors.

by the measurement errors. The state errors are basically smoother versions of the measurement errors.

Another version of the engagement model incorporates a velocity deception technique. *Engagement_v7_ssj_mfdt* features the MFDT model first shown in Figure 16.10. Results are shown in Figure 16.13.

Since the tracker only supports a single target, the target at the lowest Doppler (located at Doppler bin 5) was arbitrarily chosen to be the one reported to the tracker. This target has a velocity offset of −58.6 m/s relative to the true target (which is in Doppler bin 10). The measurement and track errors shown on the right of Figure 16.13 capture this error very accurately.

For the last example, we shift from an SSJ geometry into OBJ configurations. To simulate this, a second set of Target Dynamics and Target Generator subsystems must be added to the engagement model. *Engagement_v7_obj* allows us to represent any of the OBJ geometries depicted in Figure 16.1. Two different engagement geometries and their results are shown in Figure 16.14.

The first scenario in Figure 16.14 is that of a towed decoy; this is a jammer pod connected to the protected platform by a tow line. Since the decoy hangs below and behind the target, it causes an elevation angle measurement offset as well as a range offset. For the geometry depicted on the left of Figure 16.14, a two-degree offset is expected. This is confirmed by the raw measurement errors plotted on Figure 16.14's right (the negative sign indicates two degrees down from the radar). Although the towed decoy is technically a self-protect technique, it still qualifies as OBJ because of the separate physical location from the target.

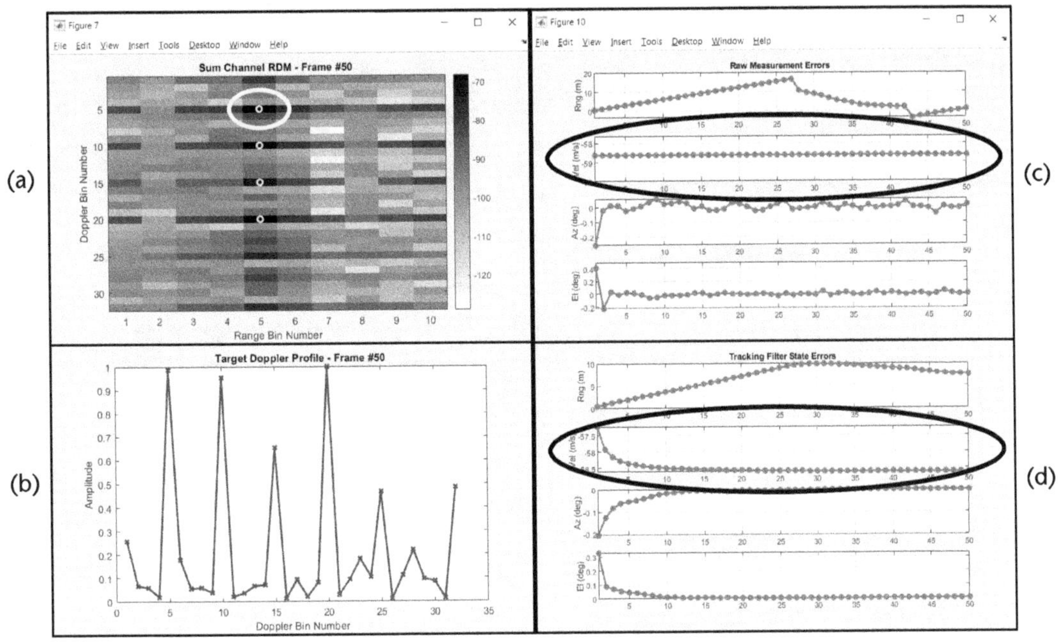

Figure 16.13 Engagement simulation results with Doppler-deception EA (MFDT): (a) RDM, (b) Doppler profile, (c) raw measurement errors, and (d) tracker state errors for false target in Doppler bin 5 (highlighted in RDM).

The bottom scenario in Figure 16.14 depicts an EJ geometry, where the jamming platform flies in formation with the target, such that a three-degree azimuth error is induced. Again, the raw measurement plots show the simulation correctly reflects this (the positive sign indicates three degrees to the right of the radar). As was the case with SSJ techniques, these techniques also prevented target detection due to the jammer range sidelobes.

The examples discussed in this section demonstrated the ease of illustrating the effects of coherent EA in target detection and tracking using our engagement simulation. Section 16.5 addresses noncoherent (i.e., noise) EA.

16.5 Noise EA

As discussed in Section 16.2, the main objective of noise EA is to deny or delay target detection by masking the target return. Noise EA elevates the noise floor in the radar receiver, thereby degrading SNR. This affects both detection and tracking performance. Noise EA is relatively simple to produce (compared to coherent EA), and since it is able to protect multiple aircraft/vehicles simultaneously, it can be extremely efficient and effective. With noise EA, the name of the game is power—and therefore JSR.

The effective radiated power density (ERPD) of a noise jammer is

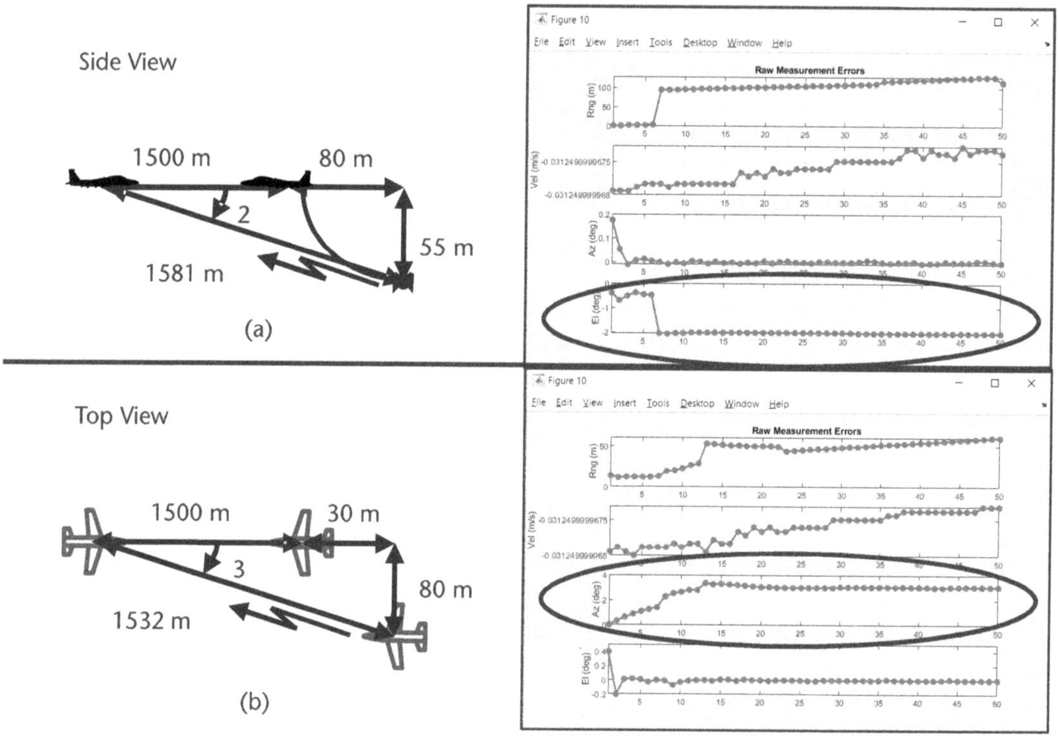

Figure 16.14 Engagement simulation results with angle-deception (and range deception) EA: (a) towed decoy jamming creating elevation errors, and (b) escort jamming creating azimuth errors.

$$ERPD = \frac{P_{tj} G_{tj}}{B_j} \tag{16.9}$$

where B_j is the jammer bandwidth, and the jammer transmit power P_{tj} is assumed to be fixed (i.e., not a function of the incident radar waveform power).

The power per unit area back at the radar receiver is

$$PSD_{\mathrm{R}} = ERPD\left(\frac{1}{4\pi R_{\mathrm{J}}^2}\right)\left(\frac{G_{\mathrm{RJ}}\lambda^2}{4\pi}\right) \tag{16.10}$$

where R_J is the range to the jammer, and G_{RJ} is the gain of the radar antenna in the direction of the jammer. (Uppercase subscripts are used here to distinguish from the receive jammer antenna gain G_{rj}, which uses lowercase subscripts.) Last, the total in-band jammer power at the input to the receiver is

$$P_j = PSD_{\mathrm{R}} B_{\mathrm{R}} = \left(\frac{P_{tj} G_{tj}}{4\pi R_{\mathrm{J}}^2}\right)\left(\frac{G_{\mathrm{RJ}}\lambda^2}{4\pi}\right)\left(\frac{B_{\mathrm{R}}}{B_j}\right) \tag{16.11}$$

Comparing (16.11) to (16.4), we note that the range dependency of the coherent jammer power in constant gain mode is $1/R^4$, the same as a target return; however, a noise jammer (or a coherent jammer in constant-power mode) has a $1/R^2$ dependency. This means that as range gets shorter, the target power grows faster than the noise jammer power. Once the target power exceeds the noise power, it can be detected. This event, known as target burn-through, is a key figure of merit in EW analysis. The burn-through range is defined as the range where JSR = 1, or 0 dB. The received target power is given in Chapter 2 by (2.5). A modified version, defined at the input of the receiver, is

$$P_{\mathrm{T}} = \frac{P_{\mathrm{R}} G_{\mathrm{RT}}^2 \lambda^2 \sigma}{(4\pi)^3 R_{\mathrm{T}}^4} \tag{16.12}$$

where G_{RT} is the gain of the radar antenna in the direction of the target (assumed identical for both transmit and receive). The atmospheric and system losses are assumed negligible for simplicity. The JSR at the input of the receiver is thus the ratio of (16.11) to (16.12). At the output of the signal processor, the JSR is reduced by the signal-processing SNR gain, G_{sp}. Adding this term to the denominator and simplifying with (16.9) results in

$$JSR = \frac{ERPDG_{\mathrm{RJ}}(4\pi) R_{\mathrm{T}}^4 B_{\mathrm{R}}}{P_{\mathrm{R}} G_{\mathrm{RT}}^2 \sigma R_{\mathrm{J}}^2 G_{sp}} \tag{16.13}$$

The burn-through range is found by setting JSR = 1 in (16.13) and solving for target range. Thus,

$$R_{\mathrm{BT}} = \sqrt[4]{\frac{P_{\mathrm{R}} G_{\mathrm{RT}}^2 \sigma R_{\mathrm{J}}^2 G_{sp}}{ERPDG_{\mathrm{RJ}}(4\pi) B_{\mathrm{R}}}} \tag{16.14}$$

For the specific case of an SSJ, $G_{RJ} = G_{RT}$, and $R_J = R_T$, resulting in the simpler expression

$$R_{\mathrm{BT}} = \sqrt{\frac{P_{\mathrm{R}} G_{\mathrm{RT}} \sigma G_{sp}}{ERPD(4\pi) B_{\mathrm{R}}}} \tag{16.15}$$

Another key parameter in modeling responsive jammers (whether noise or not) is the range at which the jammer will turn on upon detecting the presence of the radar. This occurs when the jammer detects a minimum power level, commonly known as the receiver sensitivity, S. The radar power received at the jammer, P_{rj}, is (16.2). Once this power exceeds the receiver sensitivity, it is detectable by the jammer, and the jammer may then respond. Setting $P_{rj} = S$ in (16.2) and solving for R results in

$$R_{\text{turn-on}} = \sqrt{\frac{P_t G_t G_{rj} \lambda^2}{S(4\pi)^2}} \tag{16.16}$$

A simple Simulink implementation of this turn-on condition is illustrated in Figure 16.15. The model *JamSensitivity* assumes a turn-on range of 5 km—calculated based on (16.15). This value is compared against the simulation range, and the result of this Boolean operation (a logical '1' or '0') multiplies the noise. Thus, the comparison operation acts as a switch for passing or blocking the jammer signal. The scope plots show the desired results of the jammer responding at ranges less than or equal to 5 km.

Notice that the noise jammer model in this example is simply a Gaussian random number generator. This is a simple, yet reasonable model for a broadband white-noise waveform. The samples are statistically independent from one another, which means they are also uncorrelated. For a zero-mean random sequence x_n, the autocorrelation function (ACF) is

$$R_{xx}(k) = E\left[x_{n+k}x_n\right] = \sigma_x^2\delta(k) \tag{16.17}$$

where $\delta(k)$ is the discrete delta function (or Kronecker delta function, named after German mathematician Leopold Kronecker), and σ_x^2 is the variance of the Gaussian process. The total average power of the noise is its mean-squared value, or second moment $E[x_n^2]$, found from the ACF evaluated at zero-lag ($k = 0$)

$$R_{xx}(0) = E\left[x_n^2\right] = \sigma_x^2 \tag{16.18}$$

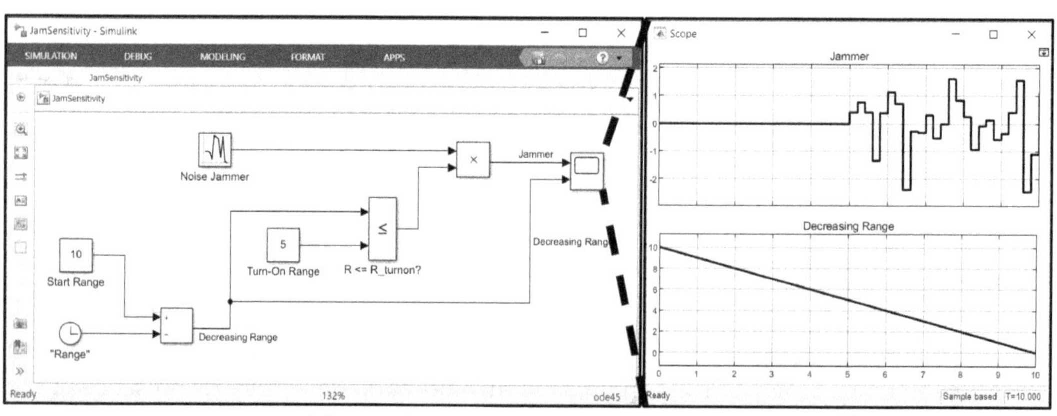

Figure 16.15 Simulink implementation of a responsive noise jammer with a 5-km turn-on range: (a) block, and (b) results.

Thus, in this signal model the power of the noise jammer is set by the noise variance. In the more general case of a non-zero mean noise signal (e.g., in the presence of a DC bias), the total power would be divided between AC power (variance) and DC power (square of mean) components.

The white noise terminology comes from the white light analogy of having components across all frequencies in the visible spectrum. Statistically, white noise has equal power across all frequency components within the sampling bandwidth –Fs/2 to +Fs/2 (or $-\pi$ to $+\pi$ in angular frequency units). This is verified by calculating the PSD of the noise signal. The PSD is the Fourier transform of the ACF defined in (16.17). It is straightforward to show that PSD(f) = σ_x^2 for all frequencies within –Fs/2 to +Fs/2. Thus, sampled white noise has a bandwidth equal to Fs. This bandwidth may be narrowed to some arbitrary fraction of Fs by filtering. If the spectral density is unchanged, the total power will decrease by an amount equal to the bandwidth reduction.

The model *NoiseFiltExamp*, shown in Figure 16.16, demonstrates this concept. The noise signal is low-pass–filtered using a Digital Filter Design block from the DSP System Toolbox. The filter passband is set to 5% of the model sampling frequency; this represents a 13-dB power reduction.

Two Spectrum Analyzer blocks show the noise spectra before and after filtering. The model also calculates the sample variances before and after filtering and takes their ratio. The calculated value of 0.05194 (–12.84 dB) is very close to the expected result.

One consequence of the decrease in noise bandwidth is that the colored noise at the output of the filter (as it is sometimes called) is now partially correlated. In fact, the ACF of the filtered noise is the autocorrelation of the filter impulse response times the variance of the white noise.

16.5.1 Example Noise EA Techniques

The most common technique is arguably barrage noise (BN) jamming (BNJ). BNJ is typical of a nonresponsive jammer that must cover multiple frequencies at once, since it does not know the radar frequency. One downside of this technique is the need for the jammer to spread its power across a wide frequency extent, making it relatively less powerful at any one frequency. Spot-noise jamming focuses its energy on a much narrower frequency extent but can be defeated by frequency-agile radars. An improvement over simple spot-noise jamming is responsive spot noise (RSN) jamming. This technique is typical of transponders that are able to measure the radar frequency and focus their power at that location. The jammer must periodically interrupt its emissions in order for the receiver to listen to the incoming radar frequency to see if it has changed, then hop to that frequency to jam it. We refer to this listening time as a look-through period. RSN jammers can follow frequency-agile radars but with some lag, depending on their look-through scheduling.

Another approach to covering multiple frequencies is the swept-spot jamming technique, where the jammer center frequency shifts over time. The sweeping motion jams over a wide frequency interval, although not all at the same time.

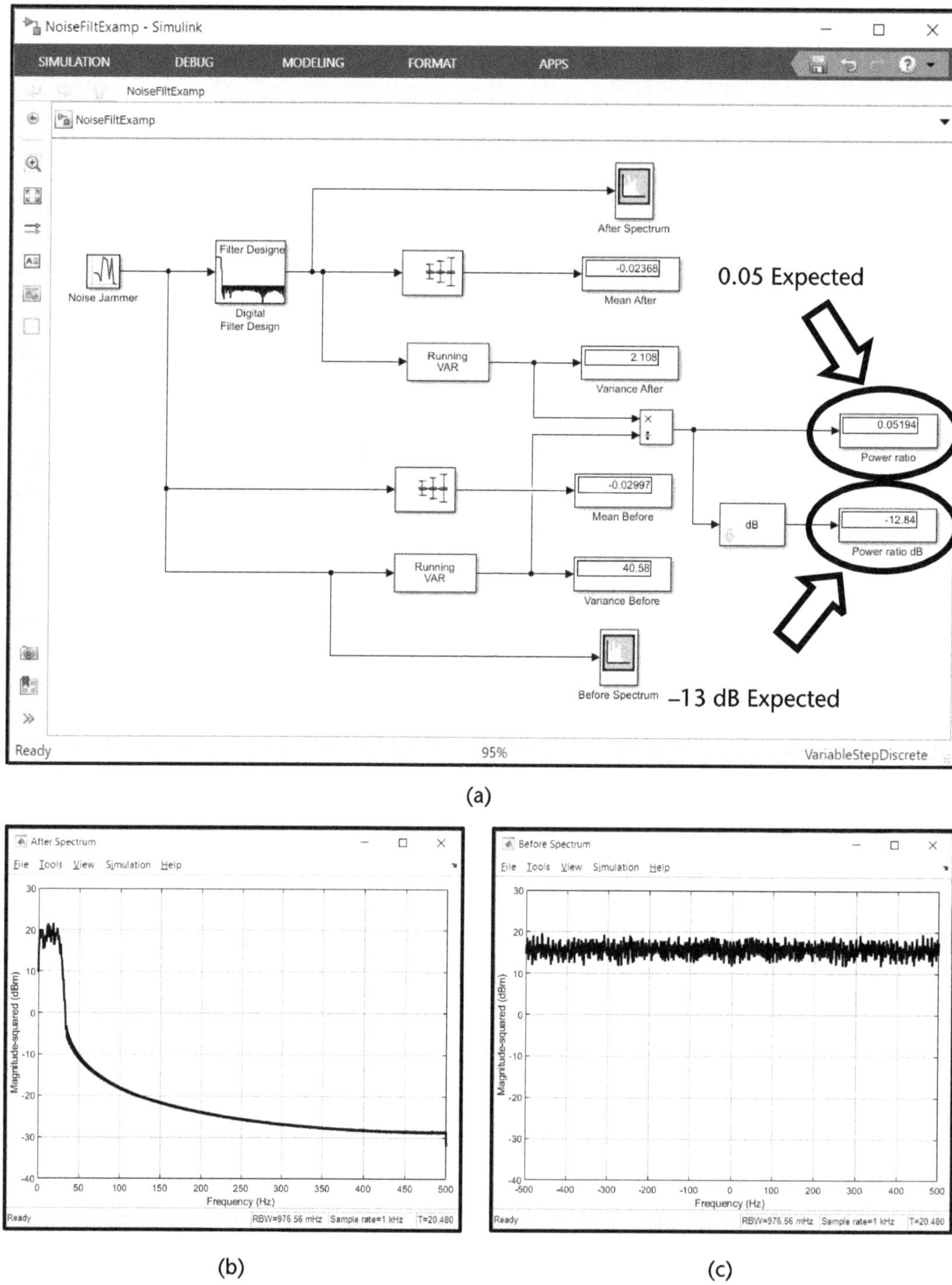

Figure 16.16 Simulink implementation of noise jammer power reduction via filtering: (a) block diagram, (b) noise spectrum after filtering, and (c) noise spectrum before filtering.

Noise jammers may be on all the time (other than during look-through periods), or may turn on and off based on some prescribed program; these are called blinking noise jammers. Cooperative blinking noise (CBN) jamming occurs when multiple jammers synchronize their blinking behavior to confuse the radar.

16.6 Engagement Simulation with Noise EA

We now turn our attention back to our Engagement simulation, this time integrated with noise jammers. We make the following observations related to the simulation assumptions and limitations when modeling noise EA:

- Since the simulation lacks modeling of the analog (RF and IF) sections of the receiver, there are no filtering effects as described in the previous sections. Therefore, the simulation is unable to represent potential real-world issues such as front-end saturation.
- Consistent with a baseband model that starts at the ADC, the jammer noise signal is modeled as white noise over the ADC sampling bandwidth Fs. This assumption corresponds to either BNJ or RSN jamming whose bandwidth exceeds the radar bandwidth.
- Some of the more complex EA techniques that employ frequency agility require additional timing logic for proper characterization. One example is the swept-spot technique discussed in Section 16.5. Proper modeling of this technique could be achieved by adding OOK that matches the timing of the jammer sweeping over the radar location. A blinking jammer may be treated in a similar manner. A Pulse Generator block serving as a gating function for the noise jammer may be a suitable candidate for incorporating this functionality.

Figure 16.17 shows the addition of a BNJ model to the *Engagement_v7_bnj* model. The jammer subsystem simply creates complex Gaussian noise samples, with real and imaginary components that are independent, identically distributed with power equally divided among them.

The results from running this simulation are not too exciting, as the noise jammer prevents the detection of the target in all 50 CPIs. The jammer power was set to yield approximately 20-dB JSR. Setting the target at a shorter range (or conversely, lowering the jammer power) allows us to observe burn-through. As an exercise for the reader, we suggest setting the jammer ERPD (line 94 of the init m-file *engagementv7_bnj_init.m*) to 0.25e-7 and rerunning the simulation. (Try it!)

Yet a more insightful experiment is to examine the angle statistics of a noise jammer. One key difference between a noise jammer and receiver noise (other than the absolute power level) is that the noise waveform originates from a distinct location in space. Thus, it is *spatially correlated*, meaning its angle of arrival can be measured, and it can be tracked. The angle statistics of a noise jammer may be used to detect its presence. Specifically, the angle variance of a noise jammer is much smaller than the angle variance of receiver noise.

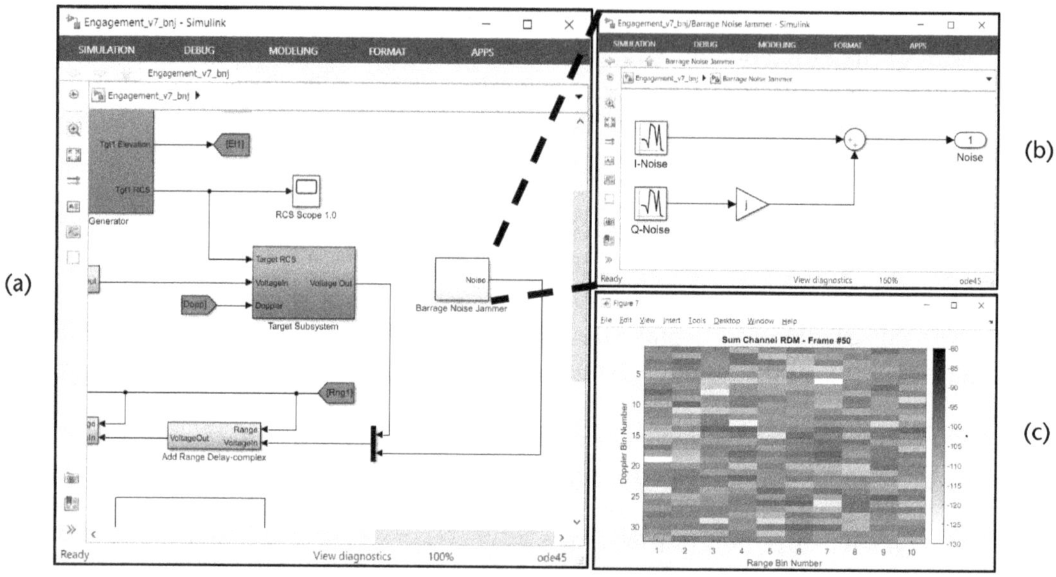

Figure 16.17 Simulink implementation of barrage noise jamming: (a) top-level diagram, (b) BNJ model, and (c) resulting RDM.

This approach requires angle calculations for a significant number of cells within the RDM; fortunately, our monopulse subsystems already calculate angles for the entire 32 × 10 RDM. The model *Engagement_v7_bnj_anglestats* includes the calculation of the mean and variance of both azimuth and elevation angles for all 320 cells, as seen in Figure 16.18.

Figure 16.18(a) shows the azimuth monopulse subsystem after running the model with the jammer turned off, so the angle calculations correspond to receiver noise samples. The variance of 6.14 deg^2 shown on the Display block corresponds to the last radar CPI; the time history shown in the Scope plot on the upper right of Figure 16.18 shows similar values fairly consistently over all 50 CPIs. These statistics do include the target cells, but those are relatively few so no effort was made to exclude them.

Once the jammer is turned on (Figure 16.18(b)), the variance drops to near zero. This dramatic reduction may serve as a reliable indicator of noise jamming; one approach might be to measure the receiver noise angle variance at the beginning of the mission, store it in memory and use it to form a threshold for comparison at subsequent times.

Once the jammer has been detected based on the angle variance, the angle mean can be used to track the jammer. In this case the mean value is consistently near zero, as the jammer is pretty much on the radar antenna boresight throughout the simulation run. To verify the accuracy of the angle mean, the reader may add a lateral offset to the jammer position and rerun the simulation. For example, setting the initial position in the north direction to −50 m results in a +1.9 deg azimuth offset. (Try it!)

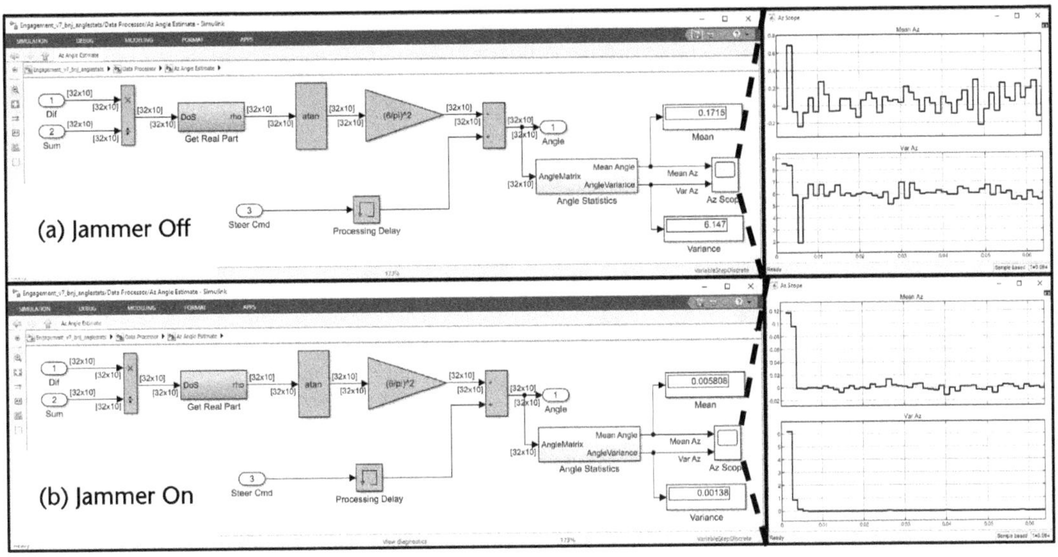

Figure 16.18 Spatially correlated angle results due to barrage noise EA: (a) receiver noise mean and variance, and (b) BNJ mean and variance.

16.7 Concluding Remarks

The preceding examples of both coherent and noise EA are simple, functional models meant to illustrate system effects. Higher fidelity may be added with relative ease, resulting in more realistic behavior. A few examples of potential upgrades include the following:

- *Jammer sensitivity check:* Turn-on based on detecting radar emissions;
- *Jammer antenna patterns:* More realistic directivity in the radar direction;
- *Jammer saturation:* Cap constant gain mode to some maximum transmit power;
- *Technique agility:* Ability to select and/or switch among multiple techniques based on radar waveform parameters and internal jammer logic.

The internal jammer logic can be quite complex; however, consistent with our crawl-walk-run philosophy, a simple rule-based scheme may be implemented initially and then refined over time.

References

[1] *Joint Electromagnetic Spectrum Operations,* Joint Publication 3-85 (JP 3-85), May 22, 2020.

[2] *Electronic Warfare,* Joint Publication 3-13.1 (JP 3-13.1), February 8, 2012.

[3] Partizian, A. A., "Coherent Jammer Formulas," in *Principles of Modern Radar, Volume II: Advanced Techniques,* W. L. Melvin and J. A. Scheer (eds.), Raleigh, NC: Scitech Publishing, Inc., 2013, pp. 549–552.

CHAPTER 17

M&S of Electromagnetic Protection

17.1 Introduction

Electromagnetic protection (EP), the second pillar of EW, is the defense against EA. Radar EP methods run the gamut of purely H/W-based approaches, including waveform and signal-processing methods, down to software-driven logic in the data processor and the tracker, and anything in between. Complex approaches often involve combined H/W-S/W solutions. Development of EP solutions is limited only by the creativity of the radar designer, within existing system constraints.

We may categorize radar EP into two main types: intrinsic and responsive. Intrinsic EP refers to radar design features that are always in effect; they are built into the system because they are consistent with good radar design, even in a non-EA environment. Thus, while they may be inherently resistant to EA, this is not the primary reason why they are employed; they provide an EP benefit for free. An antenna with low sidelobes is a good example of intrinsic EP. Good antenna design calls for maximizing directivity/gain in the antenna mainlobe, while minimizing the sidelobe gains relative to the mainlobe. Low sidelobes reduce the impact of sidelobe clutter, which can be a significant source of interference against target detection. The EP benefit of course is better rejection of sidelobe jamming.

Another example is the use of pulse-compression waveforms, which, as discussed in Chapter 8, improve range resolution without reducing detection performance. The EP benefit is that pulse compression does provide SNR processing gain; this helps against noise EA.

Responsive EP, on the other hand, must be purposely turned on (i.e., it requires logic to enable it when needed). In general, it is undesirable to implement it if not required, as its operation typically sacrifices some radar performance. An example might be a signal processing–based solution that is computationally intensive, to the point that other radar functions are degraded.

Responsive EP requires detecting the presence of EA (to decide when to turn it on), as well as real-time effectiveness assessment (to decide when to turn it off or to change tactics). Responsive EP is reactive by its very nature, so it will tend to lag in a dynamic EA environment. An adaptive canceler is an example of responsive EP. As will be described later in the chapter, once enabled, the canceler will take some amount of time to collect environmental data to calculate adaptive weights. It will then apply those weights to subsequent data in order to mitigate the jammer effects. The adaptive-weight calculation adds to the computational load of the radar processor.

Another approach to categorizing EP is by the radar subsystem that implements it. Figure 17.1 shows a block diagram of a generic radar system, with the main subsystems highlighted: antenna, transmitter/exciter, receiver, signal processor, and data processor.

A few EP techniques are listed in Figure 17.1, associated with one or more subsystems. The list is by no means exhaustive, but it conveys the point that we could discuss EP by focusing on one subsystem at a time. Some waveform techniques involving modulation and agility options do require concurrent implementation in multiple subsystems.

Partizian [1] provides a good discussion of subsystem-based EP with various example approaches. Here we focus on a single subsystem: antenna-based EP. This is arguably the most critical subsystem to overall system resistance to EA, as the antenna represents the first line of defense against external interference sources.

The rest of the chapter is organized as follows. Section 17.2 focuses on basic antenna EP concepts and terminology. Section 17.3 describes the modeling of two specific antenna-based EP techniques, the sidelobe blanker (SLB) and the sidelobe canceler (SLC). We incorporate these methods in the Simulink Engagement model presented in Chapter 15 and observe their effect against jamming. Section 17.4 discusses additional modeling considerations involving wideband arrays and digital beamforming concepts. Section 17.5 concludes the chapter with some final remarks.

17.2 Antenna EP Concepts

The first technique used by the radar to reduce EA is to utilize an antenna with higher gain and narrower beamwidth. As discussed in Section 3.7, both can be

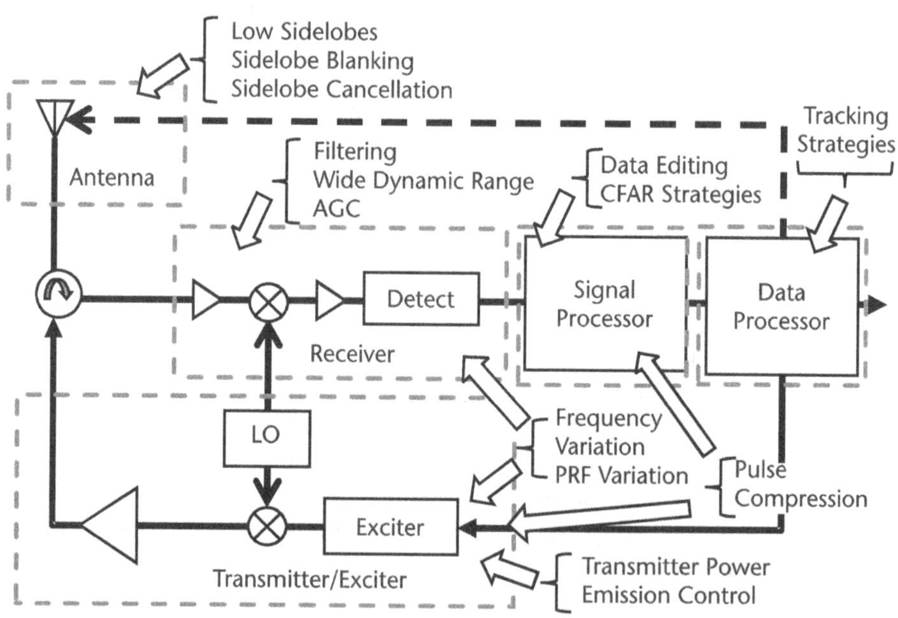

Figure 17.1 Generic radar block diagram with sample EP techniques associated with the major subsystems.

achieved by increasing the antenna aperture area. As shown in Figure 17.2 if the EA is incident on the antenna from off-boresight angles, the higher gain antenna forces the EA to enter through the sidelobe regions, and hence its amplitude is reduced.

To further improve EA mitigation, antenna sidelobe reduction is used. As discussed in Sections 3.7 and 3.12, amplitude tapers can be implemented in the radar antenna and sidelobes can be reduced by 10–30 dB as shown in Figure 17.3. In Figure 17.3(a), the pattern plot is for an ideal antenna, while in Figure 17.3(b) it is for a pattern with simulated real-world errors. Achieving very low sidelobes is costly given the tight tolerances required to minimize the amplitude and phase errors in the antenna. Sidelobe reduction has practical limits and cannot totally eliminate off axis EA.

The gain patterns for radar antennas can be synthesized to improve desired performance. This pattern-shaping can be implemented in the main beam as well as the sidelobe angle regions. Figure 17.4(a) presents an unshaped pattern, while Figure 17.4(b) shows a shaped pattern where the main beam shape similar to a cosecant-squared or shaped fan beam was implemented along with a section of reduced sidelobe levels. Figure 17.4(c) presents modeled performance of a high-gain pattern with sidelobe-notching. Shaping techniques use the amplitude and phase distributions across the antenna, and their performance (e.g., sidelobe reduction level) is dependent on the actual hardware errors in amplitude and phase.

The above sidelobe reduction and sidelobe notching techniques help in the reduction of EA, but they may not provide enough improvement if this was the sole technique for EP. Additional techniques of SLB and SLC have been commonly used to provide further improvement. These both require the addition of one or more auxiliary antennas and signal processing chains, which are often referred to as auxes or aux channels.

Figure 17.5 depicts the gain patterns of a radar antenna and an Aux antenna for the application of a SLB. The key features of the Aux antenna pattern are that (1) the Aux peak gain is below that of the radar main beam, and (2) that the Aux gain is higher than a majority of the radar pattern sidelobes. Figure 17.5 also shows positions of a target aircraft in the main beam and a jammer in the sidelobe region of the radar antenna. Note that while the radar antenna has many sidelobes, only the one in the direction of the jammer is shown. Figure 17.5 includes a notional

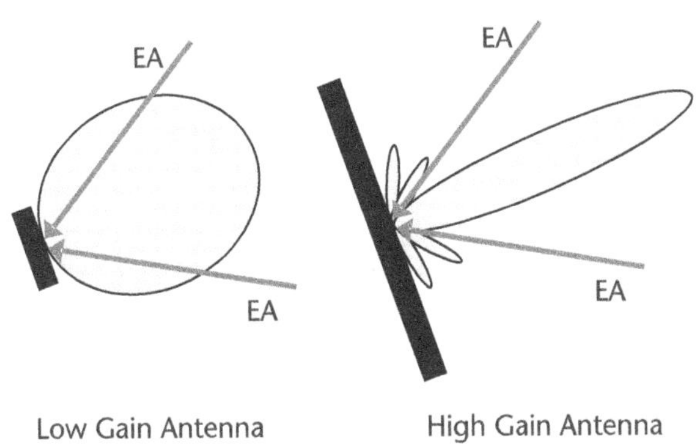

Figure 17.2 Higher-gain antennas reduce the amplitude of EA signals by presenting lower-gain sidelobes and nulls at the off-boresight angles.

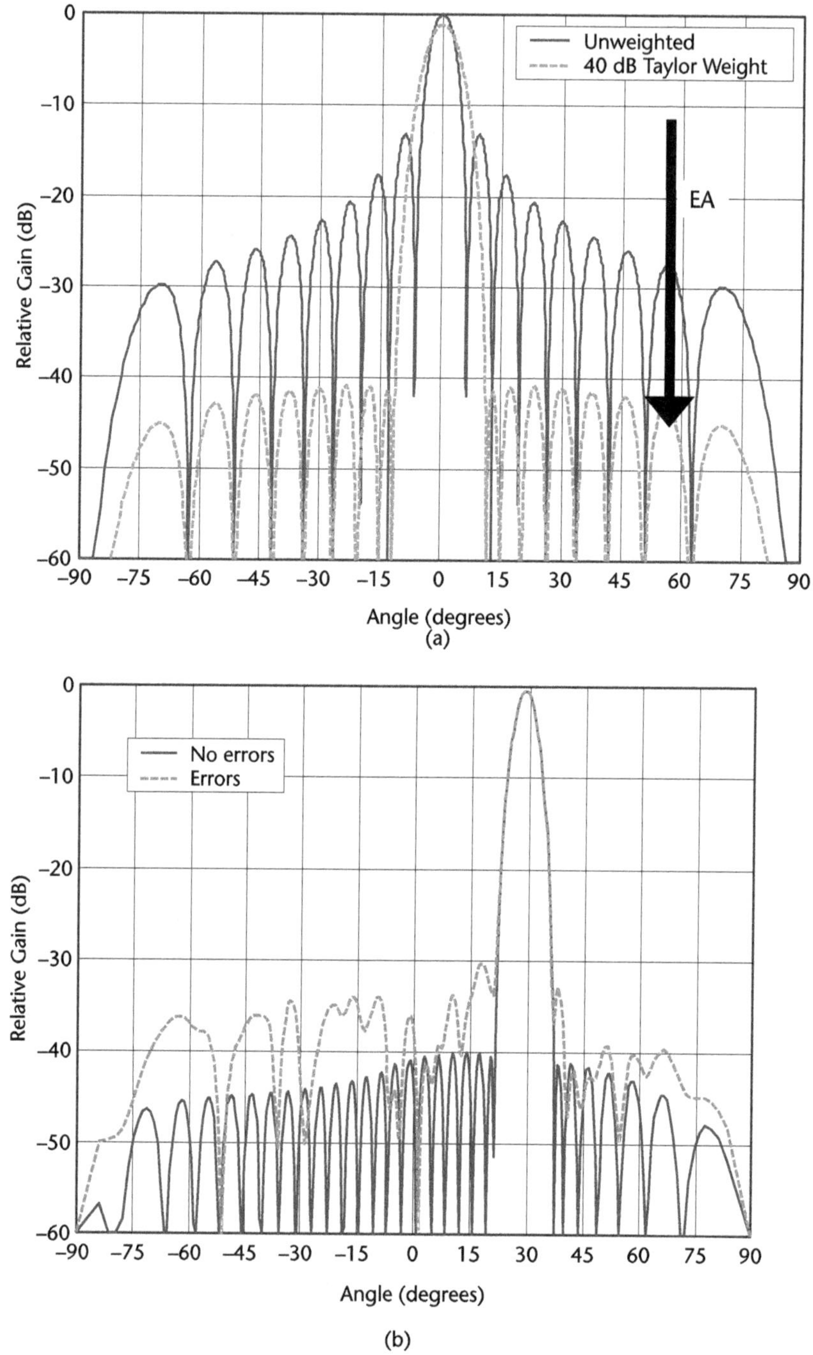

Figure 17.3 The use of sidelobe reduction in radar antennas to reduce the amplitude of EA: (a) pattern for an ideal aperture with and without weighting for sidelobe reduction and (b) scanned, weighted patterns with errors (which increase sidelobe levels) and without.

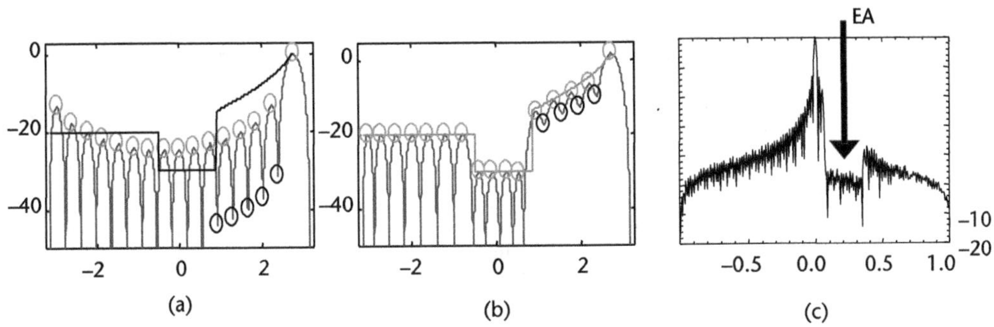

Figure 17.4 Examples of a shaped antenna patterns, which can include main beam as well as sidelobe region notching: (a) unshaped pattern, (b) pattern with sidelobe sector notch and shaped main beam, and (c) example pattern with a priori sidelobe notching.

analog circuit implementation showing how the signals are compared between those coming in both the radar and Aux antennas. The target signal is higher in the main channel than in the Aux, whereas the jammer signal is higher in the Aux than in the main channel. For any case where a signal strength is higher in the Aux channel, the comparator gates (or blanks out) the signal in the main channel. If the jammer is pulsed, this results in any signal in the main channel being blanked for the duration of the pulsed EA. The radar operates as it was at all other times. If the jammer is CW, the blanking operation can significantly degrade the radar system performance.

The question often arises as to what is the appropriate gain pattern of the Aux antenna. Figure 17.6 presents the basic trade space. To cover all the radar sidelobe peaks, the Aux antenna pattern needs to be broad, and the Aux antenna therefore needs to be low gain. The peak sidelobes of a high-gain radar pattern

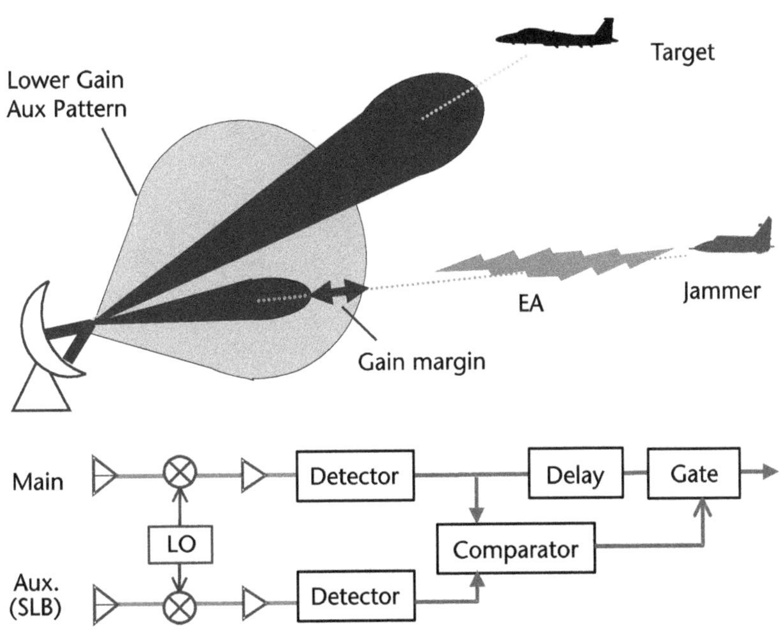

Figure 17.5 SLB concept and block diagram.

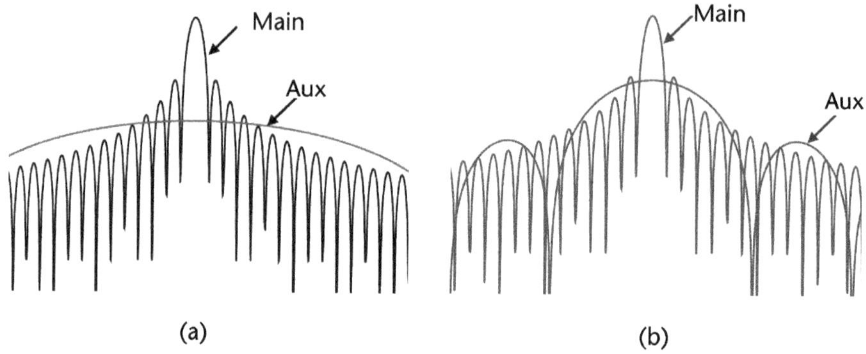

Figure 17.6 The trade space of the Aux antenna pattern and resulting sidelobe coverage: (a) Broad SLB beam suffers against near-in sidelobes, and (b) narrow SLB beam has nulls and suffers in far sidelobes.

can, however, be significantly higher than that of a low gain Aux. This is shown in Figure 17.6(a). This choice of Aux gain would result in poor performance when the jammer is in the higher near-in sidelobes of the radar pattern. Conversely, if the gain of the Aux channel were increased to cover the near-in radar pattern sidelobes, then the Aux channel will have sidelobes and nulls in its pattern. The blanking performance would be poor in the nulls of the higher gain Aux pattern. This is shown in Figure 17.6(b).

Better performance can be obtained by canceling the jammer signal. This can be achieved by implementing a SLC, as shown in Figure 17.7. The architecture of

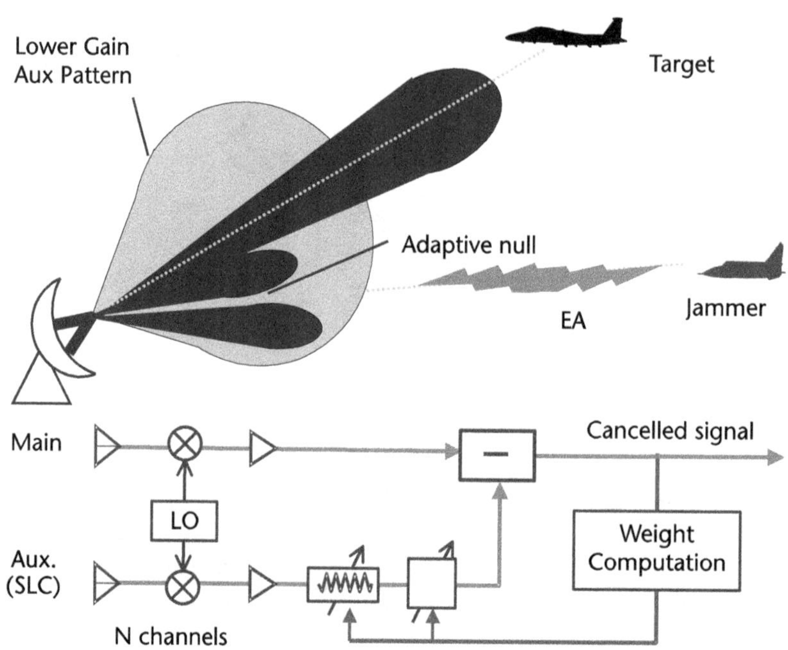

Figure 17.7 SLC concept and block diagram.

the SLC is very similar to that of the SLB. It requires a similar Aux antenna and Aux channel. The signal processing (whether achieved in analog or digital) is significantly different, however. With the SLC, the jammer signal in the Aux channel is weighted to have equal amplitude to match the magnitude of the jammer in the main channel. Its phase is then changed to be 180° as compared to the phase in the main channel. This forms a null in the main channel pattern, and as long as the jammer stays at that same angle, its signal is reduced, ideally to that of the noise level. The pattern null created by the SLC is referred to as an adaptive null.

A significant complication of the use of phase shifters within an array antenna is that it limits the instantaneous bandwidth. This happens because the operator must pick a single frequency for which to calculate the phase settings. Modern waveforms usually require some instantaneous frequency bandwidth. When frequencies radiated are different than that of the phase calculation, the main beam, sidelobes, and nulls are steered to different angles. This is referred to as frequency squint, and it is presented in Figure 17.8. The patterns presented in Figure 17.8 are for a 25-element ULA-scanned to 30° with smaller and larger bandwidths. The amount of beam squint, $\Delta\theta$, is (17.1) where the term Δf represents the total instantaneous frequency bandwidth. Equation (17.2) presents a modified AF that incorporates the beam squint. Equation (17.3) shows a simplified version that allows the user to calculate the angle to which the beam squints. In all of these: λ_{STEER} is the wavelength selected by the operator for the phase shifter calculations; λ is the actual wavelength used for the pattern calculation; θ_{STEER} is the desired main beam steering angle selected by the operator; and θ is the angle where the beam peak actually steers because of the frequency squint. The beam-squint phenomena degrades the ability to effectively cancel jammer signals that use wide instantaneous bandwidths.

$$\Delta\theta = -\frac{\Delta f}{f_{\text{STEER}}}\tan\left(\theta_{\text{STEER}}\right) \tag{17.1}$$

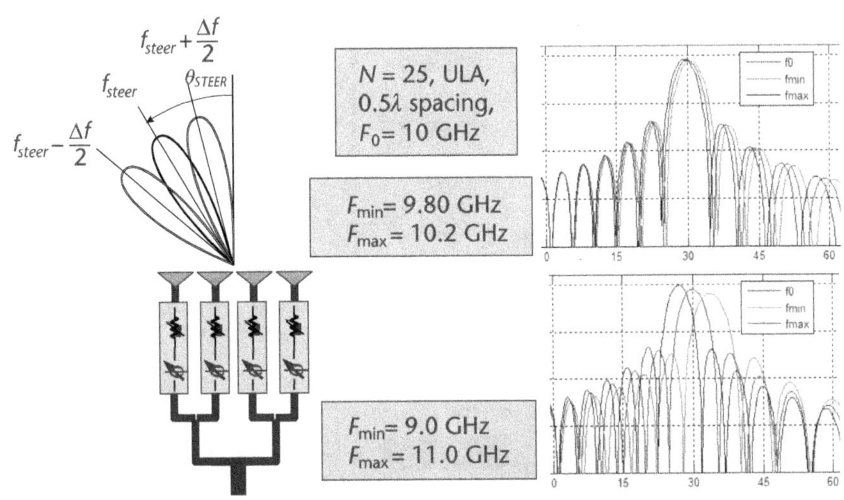

Figure 17.8 Frequency squint with an array antenna caused by the use of phase shifters.

$$AF(\theta) = \sum_{n=1}^{N} e^{j2\pi n\Delta x\left(\frac{\sin\theta}{\lambda} - \frac{\sin\theta_{\text{STEER}}}{\lambda_{\text{STEER}}}\right)} \tag{17.2}$$

$$\frac{\sin\theta}{\lambda} = \frac{\sin\theta_{\text{STEER}}}{\lambda_{\text{STEER}}} \tag{17.3}$$

Modern array radars employ digital beamforming (DBF), as it provides significant benefits for EP. There are multitudinous papers and many good books that describe the benefits and intricacies of digital arrays and array signal processing [2–6]. The first implementations for DBF that have been fielded are actually hybrid analog/digital or subarray-level DBF. In a hybrid array, elements within a subarray use analog T/R modules with amplitude and phase control and are combined within a subarray with an analog beamformer. In the hybrid approach, digitization occurs at the subarray level. The hybrid architecture reduces the number of digital channels and the complexity that arises with large numbers of digital channels. In an element-level (or per-element) digital array, each radiator has an amplifier T/R module, but the signal is then digitized at every element. Amplitude and phase control are implemented digitally in the per-element DBF array. A simplified block diagram of the two digital array architectures is shown in Figure 17.9. Note that in addition to the signal data processor (SDP) additional computation is required to facilitate the digital beamformer. Signals are generated and received in the digital receiver-exciter (DREX). The DBF algorithms are usually hosted in separate processors as the number of high-speed computations is very large.

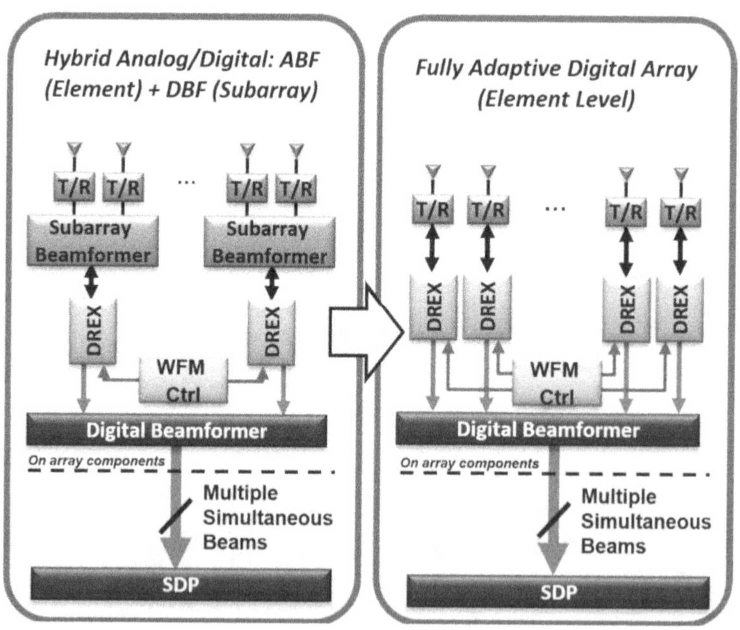

Figure 17.9 Architectures for hybrid subarray level as well as per-element level DBF arrays.

Figure 17.10 presents the architecture for a linear DBF array. Equation (17.4) shows the vector of steering phases for the linear array, $\mathbf{v}(\theta)$. Equation (17.5) shows the vector of individual channel signals $\boldsymbol{x}(t)$, including the desired signal $s(t)$, multiple, jammer signals $a_k(t)$, and random noise, $\mathbf{n}$. The bold notation here indicates that the variable is a vector (for a linear array) or a two-dimensional matrix (for a 2-D array) with a value for every array element channel. The time variable represents the sample time. Each channel is then weighted by a complex weight, w_m^*, and then summed, resulting in $y(t)$. Figure 17.10 and (17.4) to (17.6) are for a linear array with a total number of elements M.

$$v(\theta) = \left[1 \quad e^{j2\pi\left(\frac{d}{\lambda}\right)\sin\theta} \quad \cdots \quad e^{j2\pi(M-1)\left(\frac{d}{\lambda}\right)\sin\theta} \right]^T \tag{17.4}$$

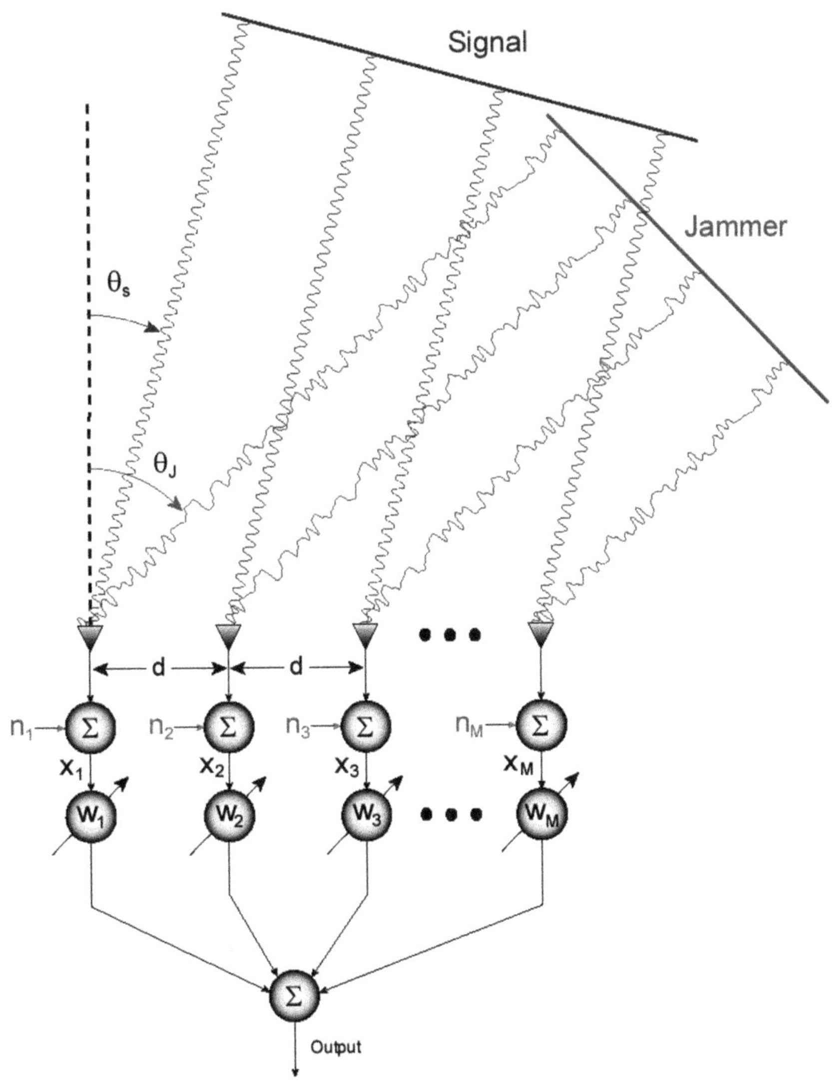

Figure 17.10 Architectures for a linear DBF array.

$$\boldsymbol{x} = \begin{bmatrix} x_1 \\ \vdots \\ x_M \end{bmatrix} = s(t)v(\theta_{\text{STEER}}) + \sum_{k=1}^{K} a_k(t)v(\theta_k) + \boldsymbol{n} \tag{17.5}$$

$$y(t) = \sum_{m=1}^{M} w_m^* x_m(t) = \boldsymbol{w}^H \boldsymbol{x} \tag{17.6}$$

The primary advantage of a DBF array used in a radar is to be able to calculate adaptive weights that simultaneously optimize the main beam in a desired direction while minimizing or canceling the sidelobes in directions of jammers or interference. A secondary advantage of a DBF array is that it can listen in receive-only mode and calculate incident AoAs of external signals (i.e., jammers). Well-known AoA techniques of MUSIC and ESPRIT were presented by Schmidt [7] and Roy and Kailath [8]. The basic DBF concept is shown in Figure 17.11, where the key is in the calculation of the weights, $\boldsymbol{w}$. The weights can be determined by estimating the covariance matrix, $\boldsymbol{R}$, of the received signals, $\boldsymbol{x}$, inverting it and multiplying by the desired main beam-steering vector, v, as shown in (17.7). This results from a derivation of minimization of the mean squared error or a form of the Wiener filter.

The exponential H represents the Hermitian operation, the complex conjugate transpose of the matrix.

$$w = c\boldsymbol{R}^{-1}v(\theta), \text{ where } \boldsymbol{R} = \frac{1}{K}\,\boldsymbol{x}\boldsymbol{x}^H, \text{ and } c \text{ is a scaling constant} \tag{17.7}$$

In the covariance matrix, $\boldsymbol{R}$, noise contributes solely to the terms on the diagonal of the matrix. The off-diagonal matrix elements are due to plane-wave sources and contain information about the AoA. It is the information in the off-diagonal matrix elements that allows the adaptive algorithms to cancel jammers and provide AoA information.

Adaptive calculations of the weights, w, are performed to provide pattern optimization and jammer mitigation, based on different algorithmic approaches and evaluation criteria. The approaches used include matrix inversion, minimizing the

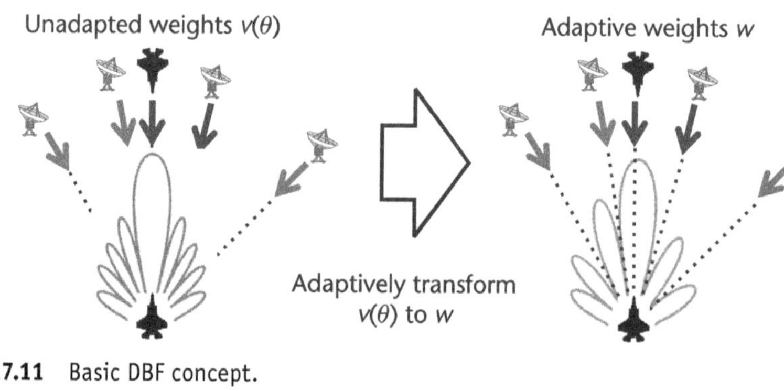

Figure 17.11 Basic DBF concept.

output power while constraining the direction of the main beam in linearly constrained minimum variance (LCMV), the minimum mean-square error (MMSE), and maximizing the signal-to-interference and -noise ratio (SINR). Some of the techniques (e.g., matrix inversion) average a block of K time samples, while others [e.g., least mean square (LMS)] iterate one sample at a time to converge to the optimal solution.

Performance metrics are used to evaluate the improvement of the adapted patterns. The cancelation ratio is the ratio of the residual interferer power after cancelation normalized by the initial interferer power. Examples of unadapted and adapted patterns are presented in Figure 17.12. With the cancelation ratio metric, the potential negative impact to the desired signal is ignored.

The SINR is a better metric to use, as it measures the improvement of the ability to receive the desired signal while mitigating jamming. It is defined in (17.8). Optimizing SINR requires some knowledge of the desired signal or the angle to the desired signal to constrain the optimization.

$$SINR = \frac{Signal\ Power}{Interference\ Power + Noise\ Power} \tag{17.8}$$

The LCMV solution minimizes the output power of the beamformer by varying weights to minimize (17.9). One of the possible outcomes of (17.9) is to set all values of $w = 0$. To ensure that this does not happen, a set of constraints is applied, using (17.10). Using the method of Lagrange multipliers, the constrained optimization problem becomes that of (17.11)

$$\min_w w^H R w \tag{17.9}$$

$$C^H W = g \tag{17.10}$$

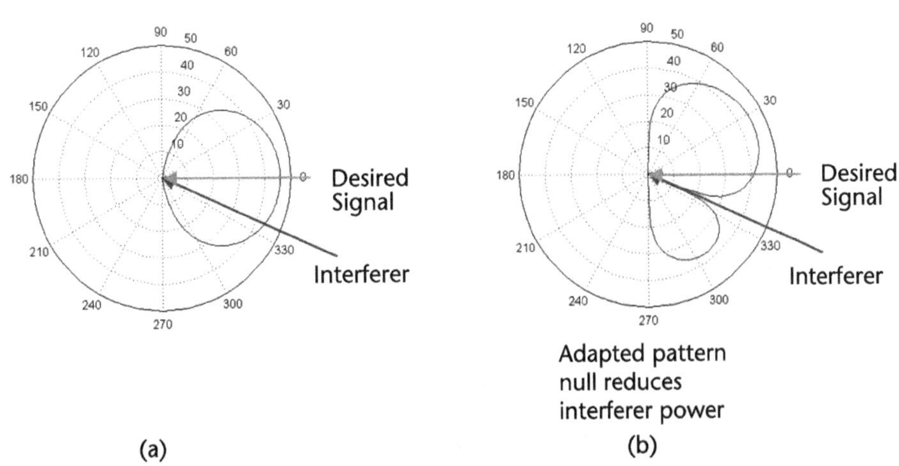

Figure 17.12 Example of (a) unadapted and (b) adapted antenna patterns with a single interfering signal.

$$w_{\mathrm{LCMV}} = R^{-1}C\left(C^{H}R^{-1}C\right)^{-1}g \tag{17.11}$$

The purpose of the constraint is to force the pattern to have a specified maximum in the desired direction of the main beam. If C is equal to the steering vector $v(\theta_{\mathrm{STEER}})$, and g is set to 1, the pattern is assured to have a beam formed in the desired direction. When $g = 1$ and $C = v(\theta_{\mathrm{STEER}})$, the LCMV reduces to be equivalent to the minimum-variance distortionless response (MVDR) shown in (17.12). A good description of the history and multiple numerical approaches to digital beamforming and weight calculations can be found in Van Veen and Buckley [6].

$$w_{\mathrm{MVDR}} = \frac{R^{-1}v\left(\theta_{\mathrm{STEER}}\right)}{v^{H}\left(\theta_{\mathrm{STEER}}\right)R^{-1}v\left(\theta_{\mathrm{STEER}}\right)} \tag{17.12}$$

17.3 Modeling of Antenna EP

We now present models for SLB and SLC, two key antenna-based EP methods presented in Section 17.2. They do share some similarities, such as the need for an auxiliary (or guard) antenna aperture and corresponding receiver channel. They both handle EA threats coming through the antenna sidelobes. It thus begs the question, how to decide when to use one versus the other? The good news is that they are quite complementary.

The SLB is better against EA with a *low duty factor* (i.e., EA that is not present the majority of the time). Otherwise, too much of the main channel signal would be blanked, and the target of interest could itself be wiped out. A good example of low duty factor EA is a false target technique.

By contrast, the SLC is most effective against *high duty factor* EA. The more samples of the unwanted signal are collected, the better the estimation of the adaptive weights, resulting in better cancelation of the EA. A typical example of high duty–factor EA is barrage noise, which may be present during the entire collection interval (100% duty factor).

Next, Section 17.3.1 presents two models for the SLB implemented in our Simulink engagement simulation.

17.3.1 SLB Modeling

Given the main and auxiliary channel data, the SLB is a conceptually simple operation. It consists of comparing the signal amplitudes on both channels, then discarding (blanking) the main channel signal if it is lower than the auxiliary channel signal. Key considerations in a practical implementation include the following:

- Where to implement the signal comparison and blanking in the existing processing chain;
- How to prevent/reduce false alarm blanking due to noise.

Figure 17.13 shows a notional block diagram of a pulse-Doppler processing chain, such as the one implemented in our Simulink model. It points to three candidate locations for the implementation of the SLB signal comparison and gating functions.

Location #1 prior to the MF is not a good choice, since the single-sample SNR will typically be too low for reliable comparison between the two channels. Location #3 after Doppler processing maximizes SNR gain, so it is a better choice. The blanking of RDM cells, however, results in holes that can affect the noise statistics and thereby degrade the CFAR detector performance. Location #2 between the MF and the FFT does include some partial processing gain, and the dropped samples due to blanking will be partially covered by the FFT processor. Thus, we deem this location as the best compromise for the location of the SLB.

This discussion illustrates the trades that are often made as part of the design process. We make no claim on the optimality of the location choice, however. In fact, alternate implementations are discussed at the end of this section.

Figure 17.14 shows the lower portion of the block diagram for a modified version of the Simulink OBJ engagement simulation presented in Chapter 16. *Engagement_v7_obj_slb* features a Guard receiver antenna and channel. (Guard is another common term for the auxiliary channel.)

Figure 17.14(a) shows the Guard antenna gain incorporated as a LUT block and going into the receiver block as an additional channel. Figure 17.14(b) shows the Guard gain pattern compared to that of the Main (i.e., sum) channel. The pattern is idealized in the sense that it covers the entire sidelobe pattern of the main antenna; as discussed in Section 17.3, it is not straightforward to achieve this in a real system. Once in the receiver subsystem, the Guard channel parallels the sum channel processing; indeed, up to the SLB processor, the paths should be closely matched in amplitude and phase to ensure the channel comparison is accurate.

The SLB block is located at the output of the MF blocks for each channel following the buffering of range-gated data. The block diagram for the SLB subsystem is displayed in Figure 17.15.

The input to the SLB subsystem is a 2-D data array (dimensions are range bins and slow-time or pulse number). The subsystem forms an element-by-element magnitude ratio of Main channel to Guard channel samples. The ratios of magnitudes are then compared to a threshold, resulting in a binary (0,1) Blanking Mask array. The Blanking Mask multiplies the Main channel array. If a Main channel sample magnitude is greater than the corresponding Guard sample, the result is a logical

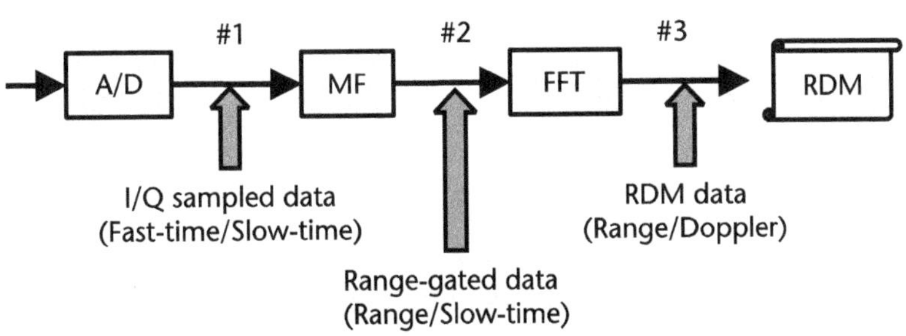

Figure 17.13 Candidate locations for SLB processing in a pulse-Doppler system.

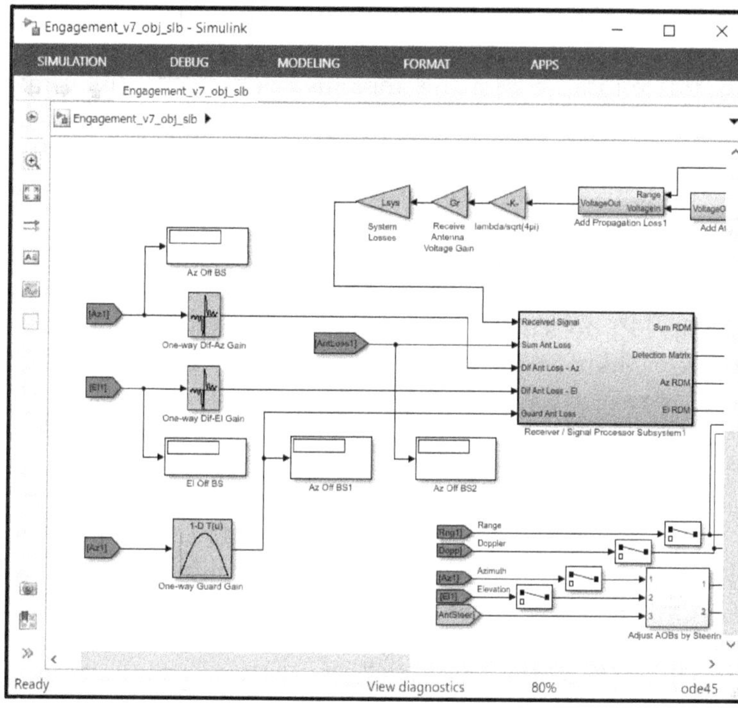

(a)

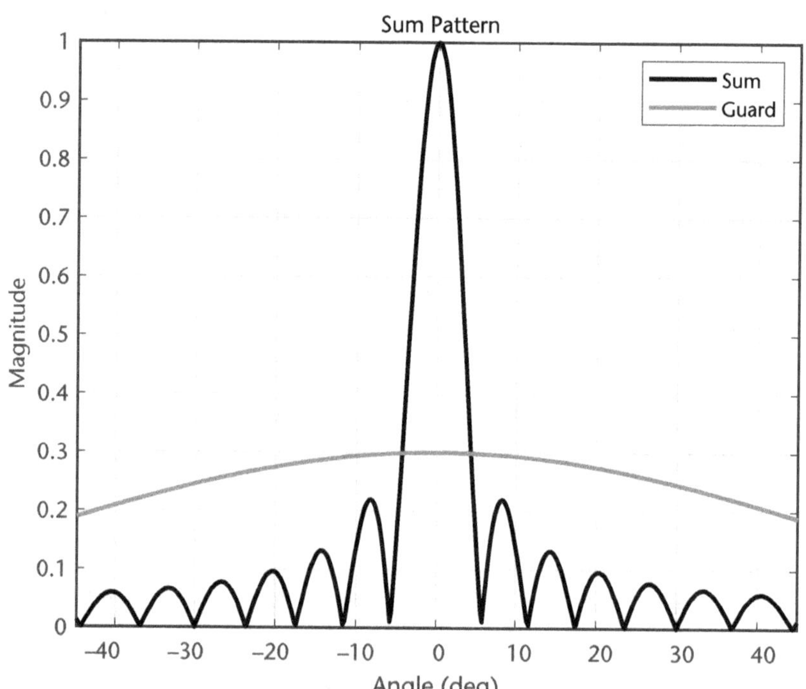

(b)

Figure 17.14 Engagement simulation with SLB modifications: (a) lower portion of block diagram showing Guard antenna LUT and (b) Guard antenna gain compared to Sum (Main) antenna gain.

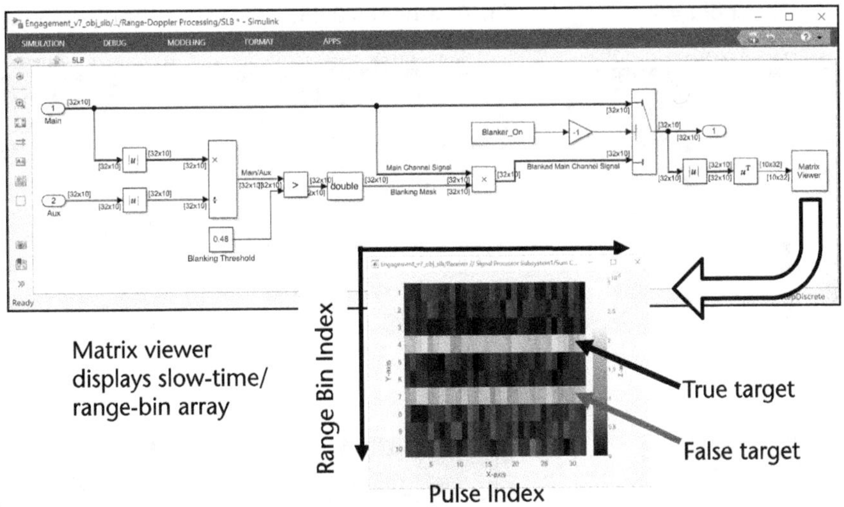

Figure 17.15 SLB implementation in engagement simulation.

'1' and the original complex sample is preserved; otherwise, it is blanked (zeroed out). Ultimately, a Switch block passes either the original data array or the masked one, based on a user-defined flag `Blanker _ On` set in the model Init Callback. A Matrix Viewer block was added to provide visibility on the effectiveness of the Blanker at this stage of the processing, prior to the Doppler processor. The resulting image from this block displayed in Figure 17.5 shows the target pulses lining up at range bin #4. The EA in this scenario is a SFT at a range delay equal to three range bins; its pulses line up at range bin #7 (below the target in the figure). The SLB is disabled in this example.

We now address the selection of the threshold value, which is shown in Figure 17.5 to be set to 0.48. Setting the threshold to 1 will simply blank the Main channel if the Guard channel signal is stronger. This choice, however, leads to excessive blanking of the Main channel data array. Both the real target return and the SFT signal are only present for a short amount of time; for a large fraction of the data array, there is only receiver noise on both channels. Blanking decisions for noise-only samples become basically a coin flip, which leads to too much unnecessary blanking. This will affect the noise estimation in the CFAR processor, impacting performance. This is the false alarm blanking issue alluded to earlier.

One way to mitigate this effect is to lower the threshold value, thereby allowing for some margin in the amplitude of the Guard channel signal exceeding the Main channel before blanking occurs. The value of 0.48 was chosen empirically as a value that resulted in acceptable performance for this scenario. A lower value of 0.4 resulted ineffective, as not enough of the SFT EA was blanked and it was still detected by the CFAR processor. Figure 17.16 shows simulation results under the conditions discussed.

The simulation EA/EP conditions are controlled by the aforementioned `Blanker _ On` flag (set in line 119 of the *engagement_v7_obj_slb_init.m* m-file) and a Manual Switch block at the output of the jammer subsystem, which either blocks the jamming signal or adds it to the return path. The simulation is set to run for

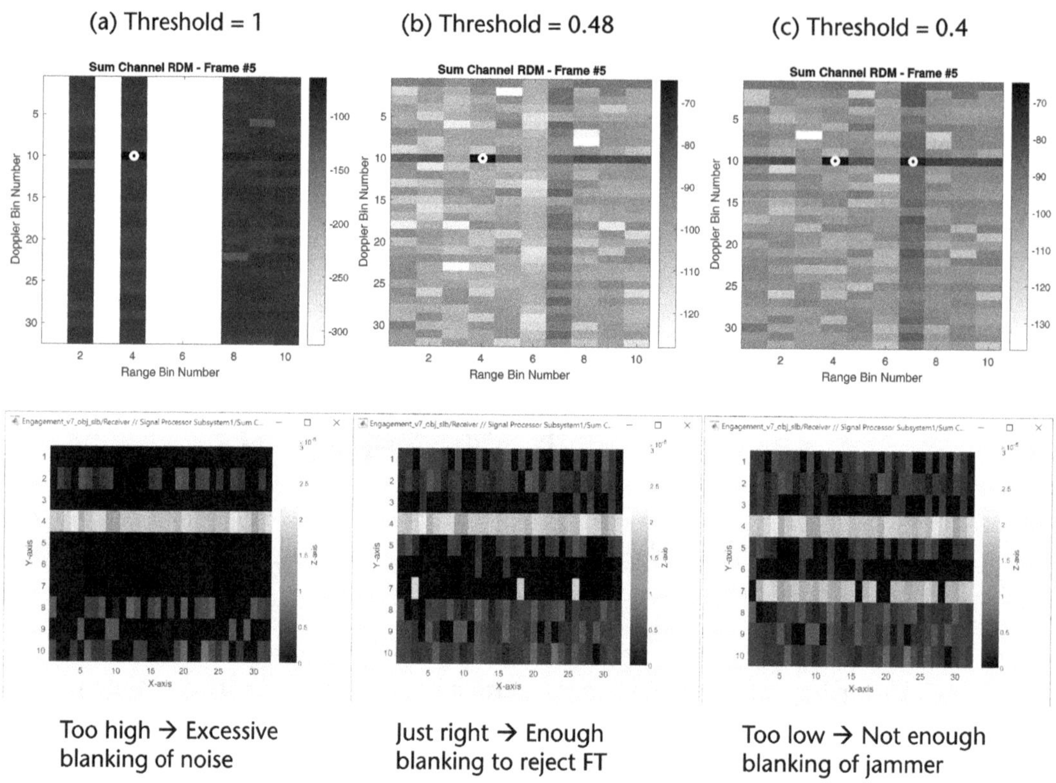

Figure 17.16 SLB performance under different threshold values. The top row shows RDM (output of FFT), and the bottom row shows range bin-pulse number data array (input to FFT).

only five CPIs. Figure 17.17 shows representative single-CPI results under the three conditions of interest (no EA, EA with no EP, and EA with EP).

Figure 17.17(a) displays CPI #5 results with no EA, with the target visible at range bin #4. The jammer is then turned on, showing the SFT in range bin #7 in Figure 17.17(b). The CFAR processor detects both the target and the SFT, as indicated by the highlight circles in the RDM. Figure 17.17(c) shows the results when the SLB is enabled (setting `Blanker_On` = 1). Although the blanking is not perfect, as indicated by the residual energy observed in the range profile, it eliminates enough of the SFT such that the CFAR no longer detects it.

This particular scheme was relatively easy to implement, but it lacks robustness. The main issue is the indiscriminate comparison and blanking of signal samples, whether there are signals of interest present or not. This leads to potentially excessive blanking, even with the current threshold setting. One way to avoid this is to run a predetector on the Main channel data, then perform the Guard comparison test only on the threshold crossings. One risk of this approach is that without the Doppler processing gain, the target SNR may not be high enough for reliable detection with acceptable false-alarm rate. A better choice is to fully process the data into separate sum and Guard RDMs, then perform the Guard test only on the sum cells detected by the CFAR. One key advantage of this approach is that there is no longer a need to actually blank the cells that fail the test, since the signal processing is now completed; those cells may simply be ignored (i.e., not included in

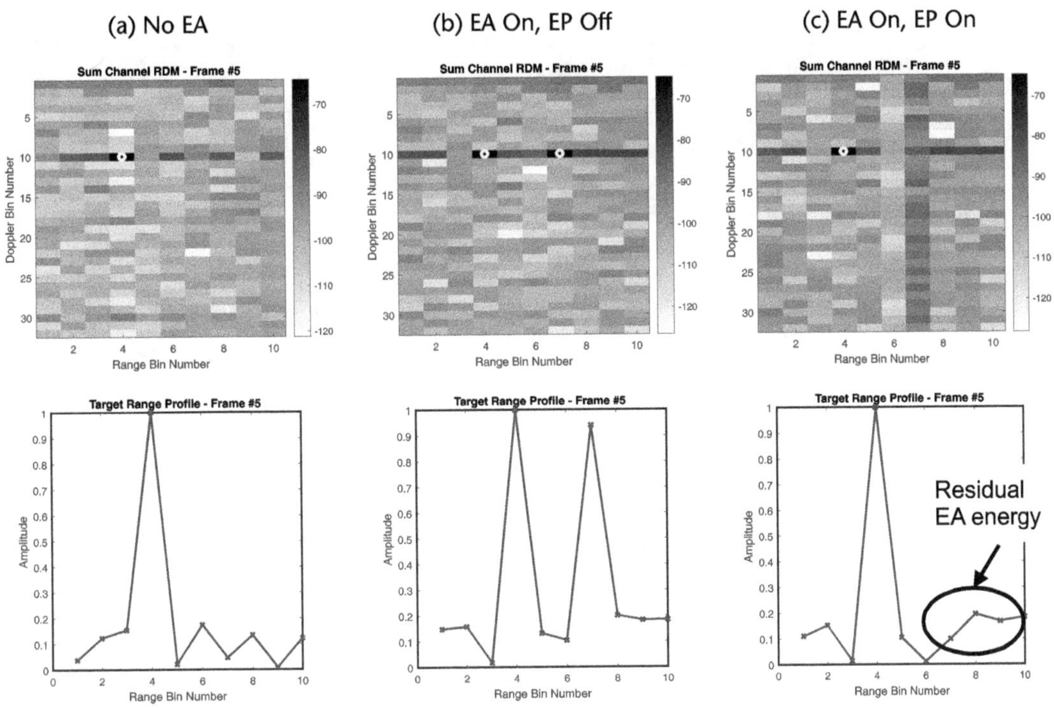

Figure 17.17 Engagement simulation results with SLB implementation. The top row shows main (sum) channel RDMs for CPI #5, and the bottom row shows corresponding range profiles at target Doppler bin.

the detection report sent to the Data Processor). This is the approach described in [2]; there are fielded radar systems that operate their SLB in exactly this manner. A second version of the SLB model (*Engagement_v7_obj_slb_v2*) implements this approach; results are shown in Figure 17.18.

This implementation requires a full receiver channel and signal processor for the Guard data, plus additional logic in the Data Processor subsystem, implemented in MATLAB through an Interpreted MATLAB Function block. Notice that SLB processing after channel comparison consists of tagging the sidelobe SFT and ignoring it from further processing; no actual blanking of data takes place.

17.3.2 SLC Modeling

We now turn our attention to modeling a SLC for suppression of a single jammer. To gain some insight, we first consider a mathematical model of the signals of interest and the desired outcome of the SLC. This will provide us with a notional but intuitive solution to the SLC weight calculation.

Let the signals in the main and auxiliary/guard be $s_m(t)$ and $s_a(t)$, respectively. Ignoring receiver noise, clutter, or any other interference source for now, we may write them as

$$\begin{aligned} s_m(t) &= G_{mt}v_t(t) + G_{mj}v_j(t) \\ s_a(t) &= G_{at}v_t(t) + G_{aj}v_j(t) \end{aligned} \tag{17.13}$$

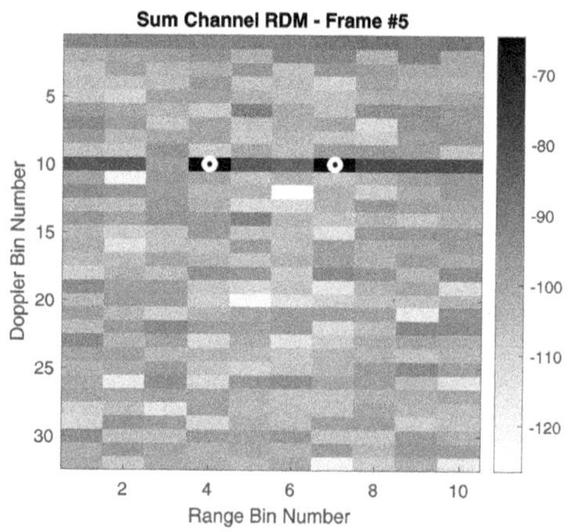

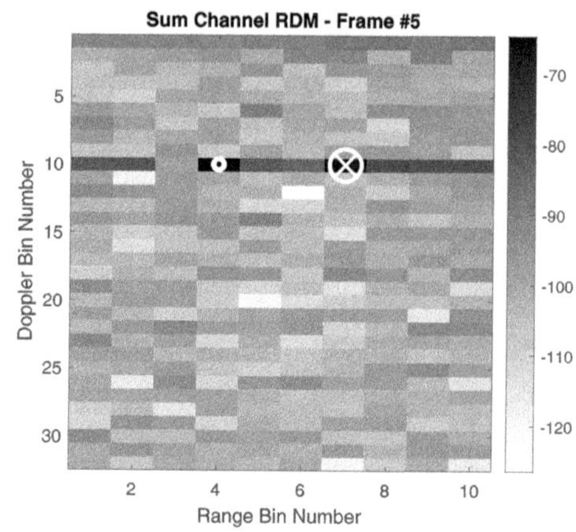

Figure 17.18 Alternate SLB implementation results: (a) RDM with SFT and no SLB and (b) SLB turned on results with SFT flagged as failing Guard test.

where the components are defined in Table 17.1.

From Figure 17.7, the output of the SLC is

$$s_{\text{out}}(t) = s_m(t) - w(t)s_a(t) \tag{17.14}$$

where $w(t)$ is the adaptive weight intended to cancel the jamming signal. Ideally, we would like the SLC output to be

$$s_{\text{out}}(t) = G_{mt}v_t(t) \tag{17.15}$$

We now use (17.13) to rewrite (17.14) as

$$s_{\text{out}}(t) = \left(G_{mt}v_t(t) + G_{mj}v_j(t)\right) - w(t)\left(G_{at}v_t(t) + G_{aj}v_j(t)\right) \tag{17.16}$$

Assuming that the target signal in the auxiliary channel is negligible compared to the others, let

$$G_{at}v_t(t) \approx 0$$

then (17.16) simplifies to

$$s_{\text{out}}(t) = G_{mt}v_t(t) + G_{mj}v_j(t) - w(t)G_{aj}v_j(t) \tag{17.17}$$

Comparing (17.17) to the desired output in (17.15), we must have

Table 17.1 Signal Components in an SLC Mathematical Model

Symbol	*Definition*
$v_t(t)$	Target voltage
$v_j(t)$	Jammer voltage
G_{mt}	Main antenna gain in the direction of the target
G_{mj}	Main antenna gain in the direction of the jammer
G_{at}	Auxiliary antenna gain in the direction of the target
G_{aj}	Auxiliary antenna gain in the direction of the jammer

$$G_{mj}v_j(t) - w(t)G_{aj}v_j(t) = 0, \text{ or } w(t) = \frac{G_{mj}v_j(t)}{G_{aj}v_j(t)} \tag{17.18}$$

With the simplifying assumptions as stated, the adaptive weight is the ratio of the jammer signal in the main channel over that of the auxiliary channel.

A mathematically rigorous solution to the problem is achieved by using a LMS criterion, where the optimal weight is found by solving [9]

$$\left.\frac{\partial E\left\{\left|s_m(t) - w(t)s_a(t)\right|^2\right\}}{\partial w(t)}\right|_{w(t)=w_{\text{opt}}(t)} = 0 \tag{17.19}$$

with the solution

$$w_{\text{opt}}(t) = \frac{E\left\{s_m(t)s_a^*(t)\right\}}{E\left\{\left|s_a(t)\right|^2\right\}} \tag{17.20}$$

where $*$ denotes the complex conjugate. We note that (17.20) somewhat resembles (17.18) (i.e., a signal ratio of main channel over auxiliary channel). A practical implementation of (17.20) would employ time averages instead of statistical expectations. We thus define a realizable weight calculation as

$$\tilde{w}_{\text{opt}}(t) = \frac{\left(\frac{1}{N}\right)\sum_{n=0}^{N-1} s_m(n)s_a^*(n)}{\left(\frac{1}{N}\right)\sum_{n=0}^{N-1}\left|s_a(n)\right|^2} \tag{17.21}$$

As previously mentioned, one measure of the SLC effectiveness is the cancelation ratio (CR), defined as the ratio of jammer powers before and after cancelation. It has been shown [10] that the jammer can only be canceled down to the receiver noise level (i.e., CR ≈ JNR). For an actual cancelation down to 0 dB JNR, the total interference (jammer + noise) may therefore be up to 3-dB higher than the receiver noise power alone, as the two would add noncoherently.

The SLC was implemented in a modified version of the engagement simulation, at the same location where the SLB was implemented (between the MF and the FFT). Figure 17.19 shows the block diagram of the SLC subsystem in the *Engagement_v7_obj_slc* Simulink model.

It is a straightforward implementation of (17.21). The location in the model signal-processing chain was chosen because it is the point where an entire CPI worth of samples is available for the weight calculation—prior to the MF, the model only propagates a single PRI worth of samples at a time, which would not be enough for accurate averaging. The number of samples in the simulation was doubled in both fast time and slow time to provide better estimates of the signals levels; so 20 range gates and 64 Doppler filters are implemented. The jammer model is the same BN model implemented in the BN jamming example from Chapter 16 (Figure 16.17). The jammer is placed 15° off the antenna boresight in azimuth, while the target is in the center of the beam. Simulation results are shown in Figure 17.20.

Figure 17.20(a) shows a representative RDM with no EA. The target SNR is about 25 dB. Once the jammer turns on in Figure 17.20(b), the target is no longer detectable. The JNR is about 18 dB, which implies about a –7 dB JSR—still strong enough to prevent target detection. Once the SLC is enabled in 17.20(c), the jammer is canceled down to about the receiver noise level, and the target is once again detectable with an SINR of about 18 dB. For cancelation down to the noise level, the SINR should have been about 22 dB, which points to some partial target cancelation. This can happen if the target is present during the data collection used in the weight calculation, as is the case in the simulation. A best practice is to calculate the weight with data collected prior to transmission, to avoid including target returns. This also avoids clutter and other returns that may degrade the SLC performance.

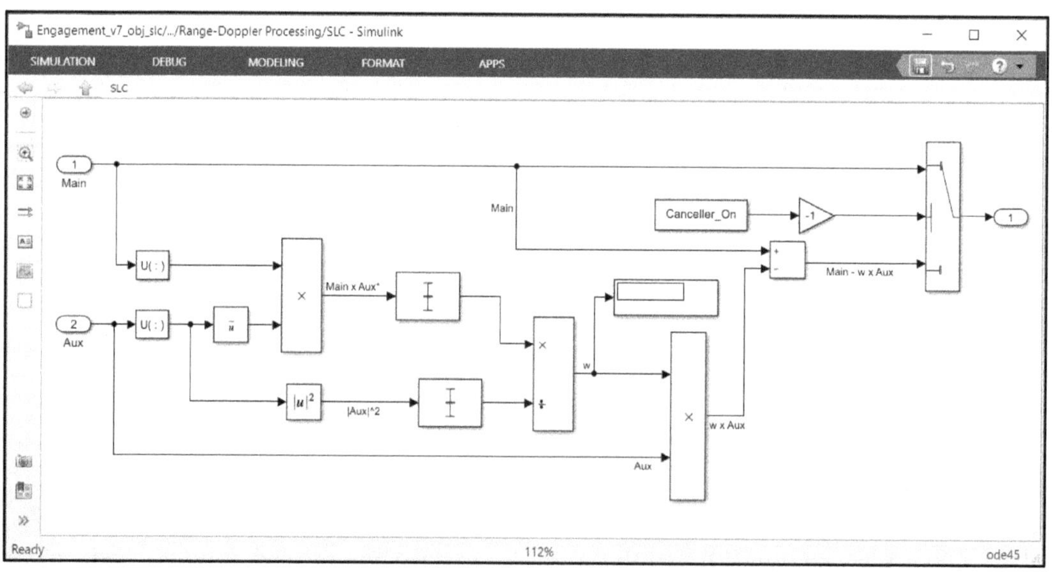

Figure 17.19 Block diagram of SLC subsystem in Simulink engagement simulation.

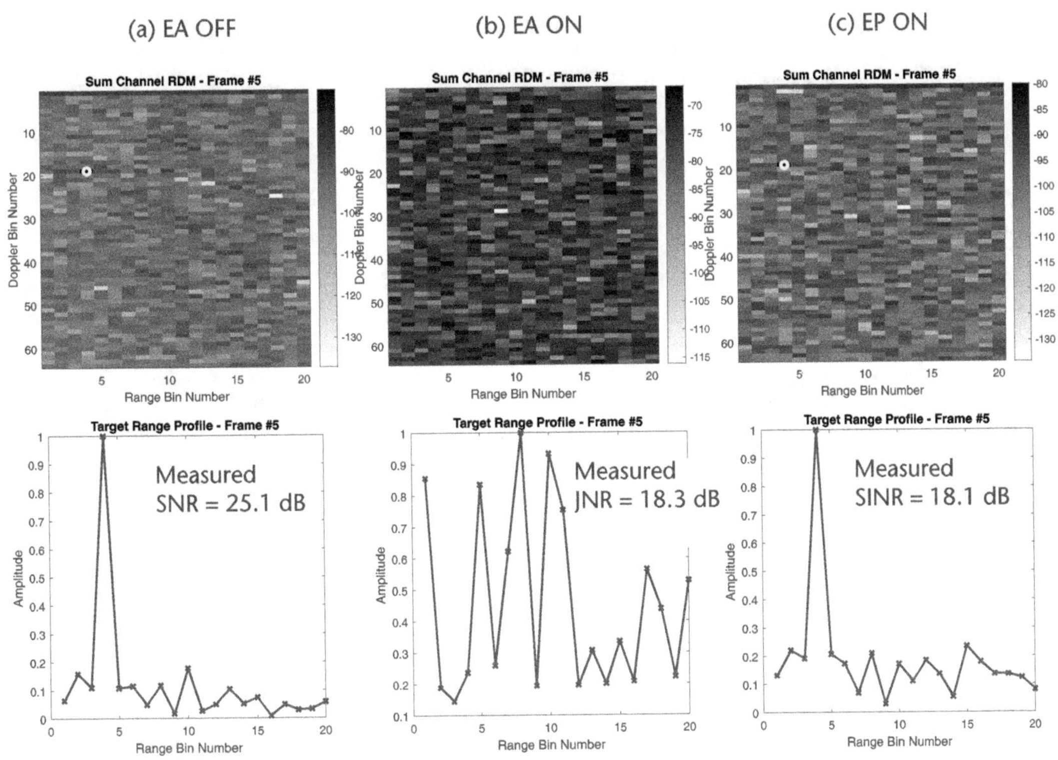

Figure 17.20 Engagement simulation results with SLC implementation. The top row shows main (sum) channel RDMs for CPI #5, and the bottom row shows corresponding range profiles at target Doppler bin.

We also present a MATLAB version of the SLC model. Similar to the Simulink model, the script m-file *radar_SLC_example.m* implements an engagement between a pulse-Doppler radar and a single target along with a sidelobe noise jammer. This MATLAB model only runs for one CPI, however. Furthermore, the radar model collects data over the entire PRI [i.e., it forms enough range gates to cover the entire unambiguous range interval (15 km for the 10-kHz PRF used)]. There are therefore many more data samples available to calculate the cancelation weight. The SLC is implemented before the MF in this model. The main simulation controls consist of two flags `Add_Jam` and `SLC_On` in lines 10 and 13 of the script. Figure 17.21 shows the results.

The target range is 25 km, and in the no-jamming case of Figure 17.21(a) it is detected at an ambiguous range of 10 km, with about an 18-dB SNR. In Figure 17.21(b), the jammer is much stronger than in the Simulink example, with a JNR of about 34 dB. Once the SLC is enabled in Figure 17.21(c), the jammer is canceled down to the noise level, and the target is once again detected, this time with a SINR of about 18 dB.

The SLC can also be viewed as an adaptive beamforming technique, specifically in the manipulation of the sidelobe pattern. The process may be described as a scaling of the Guard pattern, where the scaling is designed to match the gain level on the Main antenna pattern at the jammer location. By subtracting the scaled Guard from the Main, a null is created at that angular location. This is

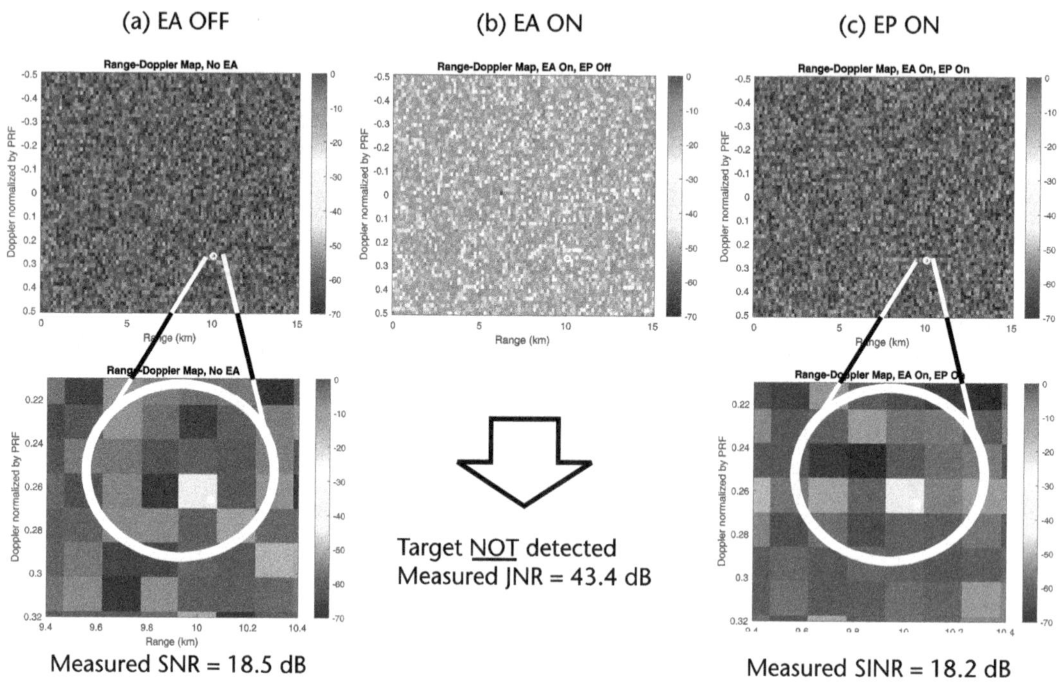

Figure 17.21 MATLAB engagement simulation results with SLC implementation. The top row shows main (sum) channel RDMs, and the bottom row shows close-ups of target location in the RDM.

best visualized by plotting the main channel antenna patterns, before and after adaptation. These are shown in Figure 17.22.

The default simulation parameters form the antenna beam with a uniform illumination function, resulting in a sinc pattern. Before cancelation, the jammer is at the peak of the second sidelobe. Upon cancelation, a sidelobe null is formed at that location, as desired; however, the remaining sidelobes become quite high relative to the mainlobe peak. This is an undesirable side effect; to mitigate it, a window function is applied to lower the precancelation sidelobes. Setting the flag `use _ taylor` to 1 in line 18 of the MATLAB script (Try it!), we obtain the patterns shown in Figure 17.22(b). The sidelobes are much lower to begin with; therefore, even after the increase due to the cancelation, they remain at suitably low levels. Notice that with the lower sidelobes from the Taylor weights, the JNR is about 20-dB lower than with uniform illumination. It is still high enough (about a −2 dB JSR) to prevent target detection, so the SLC is still needed.

The SLC weight calculation presented here was an open-loop configuration where all the data was collected, and then the weight was calculated and applied. Many systems follow a closed-loop approach instead, where the weight is refined over time using gradient search techniques. This is the well-known Howells-Applebaum algorithm [11, 12].

As a last comment on the topic, we point out that the model presented here implements a single SLC (i.e., a single auxiliary channel) to protect against a single jammer. In general, to cancel N jammers, a system needs N auxiliary antennas/

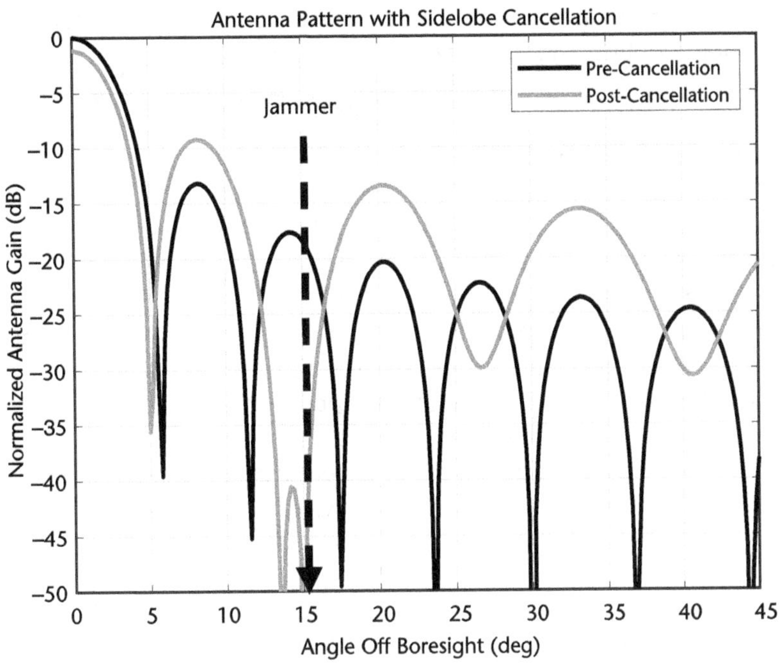

(a)

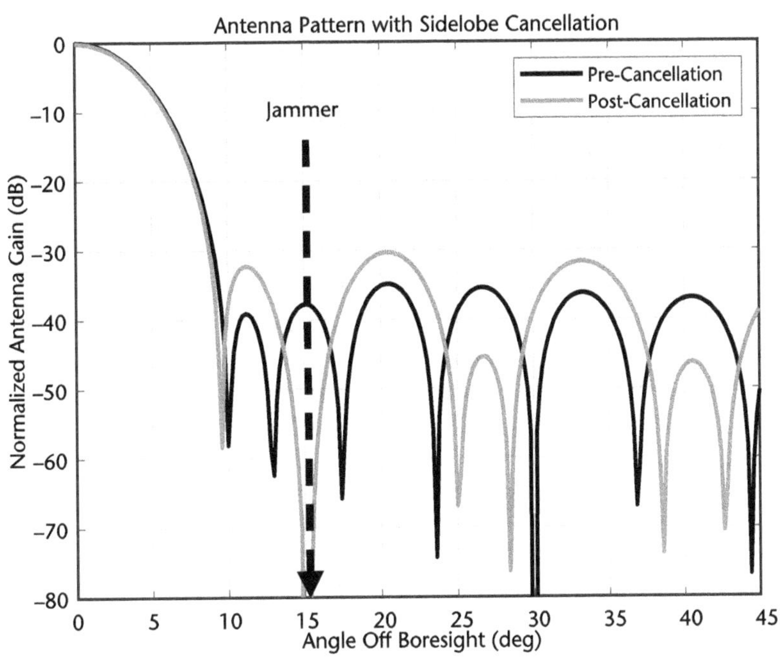

(b)

Figure 17.22 Main antenna patterns before and after cancelation: (a) uniform antenna illumination and (b) Taylor window illumination for lower sidelobes.

channels. If the jammers outnumber the auxiliaries, some partial cancelation may be obtained, but performance will be suboptimal.

17.4 Adaptive Beamforming

Mitigating wideband jammers is complicated by the fact that the radar antenna pattern changes versus frequency. The correct weighting for part of the jammer spectrum may not provide enough nulling in the rest of the spectrum. Use of time delay in the beam steering does provide some improvement, and hence this approach has been employed in both analog and digital architectures. Time delay has historically been unavailable on a per-element level in both analog and digital circuits, and hence it has been applied at the subarray level as shown on the left in Figure 17.9. With this hybrid architecture, another significant problem arises in the form of subarray grating lobes, which appear as potentially high sidelobes. The hybrid architecture uses phase shift at a per-element level, and this phase shift causes beam squint of the subarray pattern versus frequency. How the elements within a subarray are grouped modifies the subarray grating lobe effect. Figure 17.23 presents diagrams of subarrays within line arrays with the subarray grouping arrangements of Regular Periodic, 2:1 Overlapped, 3:1 Overlapped, Fully Overlapped, and Randomized.

Take the first case, where the elements are grouped in regular, periodic subarrays. In Figure 17.24(a), the AF (in the darker trace) has grating lobes caused by the inter-subarray spacing being multiple wavelengths. These AF grating lobes align with the nulls of the subarray pattern (in the lighter trace) and are canceled by the nulls of the subarray pattern—but only at the exact center frequency that is used to calculate the phase shift settings. The resulting complete array pattern is shown in the darker trace in the pattern in Figure 17.24(b). In Figure 17.24(c), the frequency of $f_0 - \Delta f/2$ is used. The AF grating lobes (darker trace) remain at the same angles they were at previously, because the subarrays are steered using time delay. The subarray pattern (lighter trace) squints, hence moving the subarray nulls off of the angles of the subarray grating lobes. The resulting pattern is the one shown in Figure 17.24(d) in the darker trace. The subarray grating lobes are no

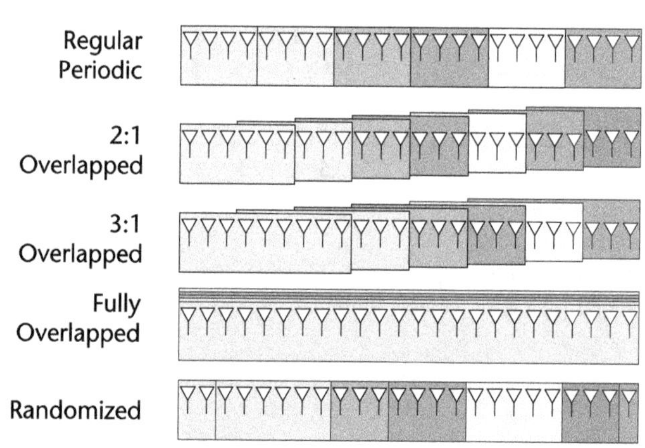

Figure 17.23 Subarray grouping strategies.

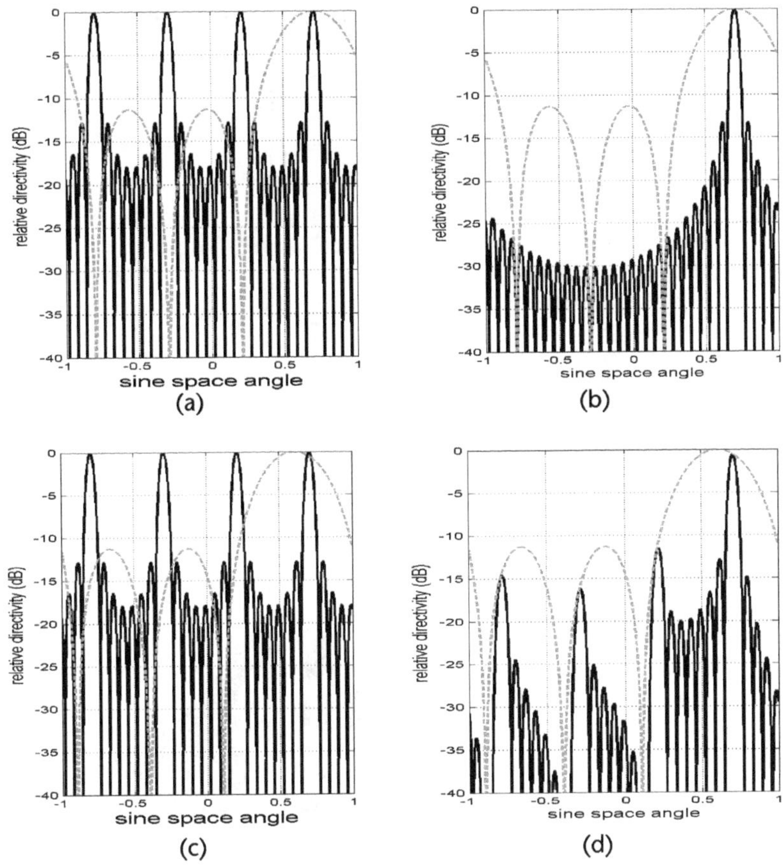

Figure 17.24 Patterns for regular periodic subarrays resulting from use of time delay on a subarray basis, phase shift on an element basis, and a wideband waveform. Note the high regular periodic sidelobes in (d). Please refer to the text for a detailed description of these pattern data plots.

longer canceled, and they exhibit as very high periodic sidelobes that result from use of a frequency other than the one used to calculate the phase shifter settings. This result is contrary to the goal of achieving low sidelobes as an EP technique.

To reduce the high periodic sidelobes, the subarray architectures shown in Figure 17.23 have been used. Figure 17.25 shows the patterns that result from using aperiodic, randomized subarrays. At the center frequency, the sidelobes are all low and well-behaved as shown in the dark trace. At frequencies other than the center frequency, the subarray grating lobes do not cohere. Rather the energy that would have been in those grating lobes is randomly scattered throughout the sidelobe region in the pattern in the lighter trace, again increasing the RMS sidelobe levels as shown in the figure.

Arguably, the best performance for the hybrid architecture can be achieved when 3:1 overlapped subarrays are utilized. In this concept every single element channel is connected simultaneously to three subarrays. Taylor amplitude tapers are implemented on the array factor as well as within the subarray for sidelobe reduction in the array factor and subarray patterns. The low sidelobe levels and tighter subarray pattern main beam work together to mitigate the subarray level

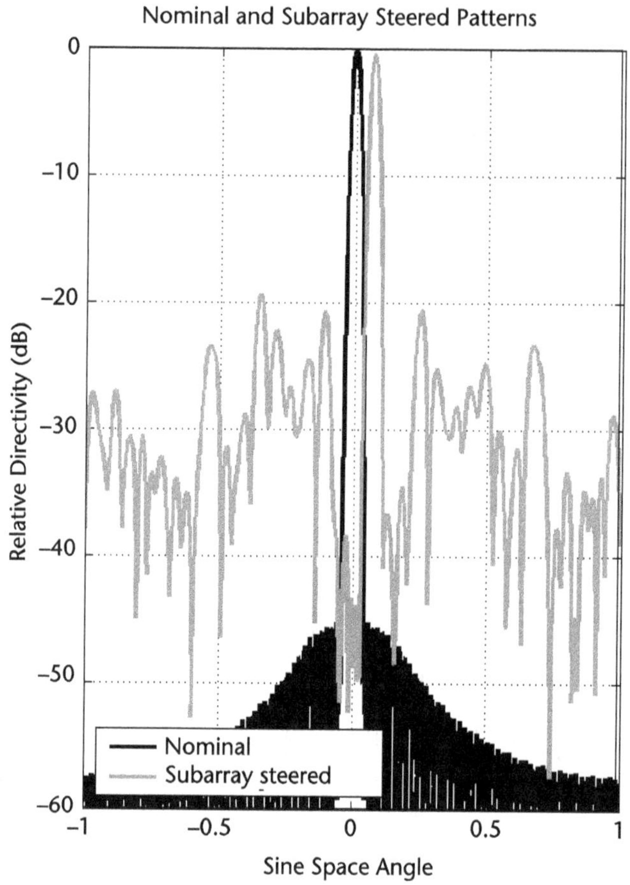

Figure 17.25 Patterns for random subarrays resulting from use of time delay on a subarray basis, phase shift on an element basis, and a wideband waveform. Note the increased RMS sidelobes.

grating lobes and to maintain low sidelobe performance throughout the pattern. In Figure 17.26(a) the lighter trace is the array factor, and the darker trace is the subarray pattern. In Figure 17.26(b), the lighter trace shows the final pattern at a different frequency than that used for calculation of the phase shifters, with the subarray level grating lobes mitigated. This architecture requires a significant increase in hardware and subarray channels, but it does provide the designer with a path to high-performing low sidelobes with increased instantaneous bandwidth support.

17.5 Concluding Remarks

The example EP techniques discussed in this chapter are but a fraction of the myriad possibilities built-in in military radar systems. The detailed implementations are system-specific and more often than not are classified. Thus, material in the open literature tends to be generic and broad. In addition to Partizian [1], another good overview reference is Schleher [10], who explicitly discusses different EP approaches for search/surveillance radars versus tracking radars. Aalfs [2] and Budge [9] provide excellent discussions specifically on the SLC and SLB techniques.

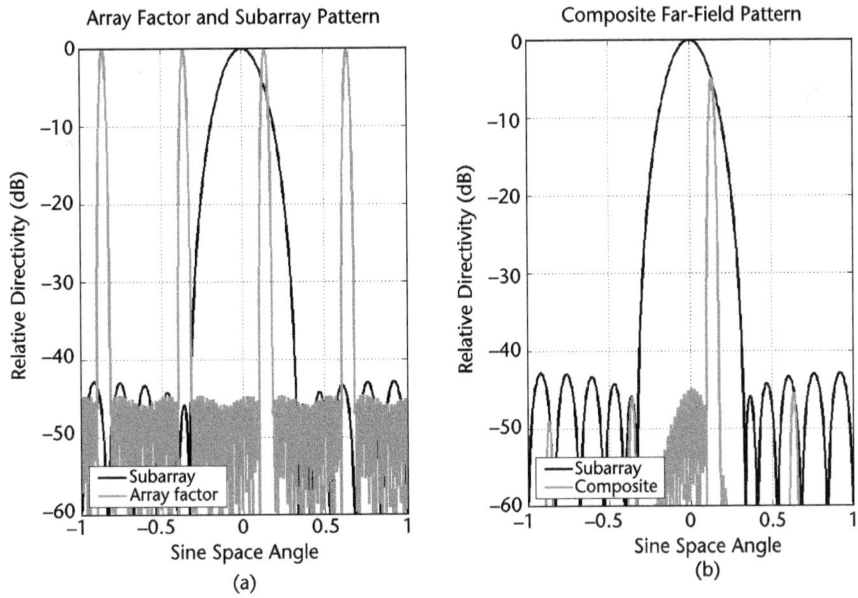

Figure 17.26 Patterns for a 3:1 overlapped subarrays resulting from use of time delay on a subarray basis, phase shift on an element basis, and a wideband waveform: (a) array factor and subarray pattern, and (b) composite far-field pattern. Note the ability to maintain very low sidelobe levels with a wide instantaneous bandwidth.

References

[1] Partizian, A. A., "EP Overview," in *Principles of Modern Radar, Volume II: Advanced Techniques*, W. L. Melvin and J. A. Scheer (eds.), Raleigh, NC: Scitech Publishing, Inc., 2013, pp. 553–581.

[2] Aalfs, D., "Adaptive Digital Beamforming," in *Principles of Modern Radar, Volume II: Advanced Techniques*, W. L. Melvin and J. A. Scheer (eds.), Raleigh, NC: Scitech Publishing, Inc., 2013, pp. 401–452.

[3] Monzingo, R. A., R. L. Haupt, and T. W. Miller, *Introduction to Adaptive Arrays*, Second Edition, Raleigh, NC: Scitech Publishing, Inc., 2011.

[4] Gross, F. B., *Smart Antennas for Wireless Communications*, McGraw Hill, 2005.

[5] Farina, A., *Antenna-Based Signal Processing Techniques for Radar Systems*, Norwood, MA: Artech House, 1992.

[6] Van Veen B. D., and K. M. Buckley, "Beamforming: A Versitile Approach to Spatial Filtering," *IEEE Acoustics, Speech, and Signal Processing Magazine*, Vol. 5, No. 2, 1988.

[7] Schmidt, R., "Multiple Emitter Location and Signal Parameter Estimation," *IEEE Transactions on Antenna Propagation*, Vol. AP-34, No. 2, March 1986, pp. 276–280.

[8] Roy, R., and T. Kailath, "ESPRIT—Estimation of Signal Parameters Via Rotational Invariance Techniques," *IEEE Transactions on ASSP*, Vol. 37, No. 7, July 1989, pp. 984–995.

[9] Budge, M. C., and S. R. German, *Basic Radar Analysis*, Second Edition, Norwood, MA: Artech House, 2020, pp. 747–773.

[10] Schleher, D. C., *Electronic Warfare in the Information Age*, Norwood, MA: Artech House, 1999, pp. 120–126 and 279–286.

[11] Applebaum, S. P., "Adaptive Arrays," *IEEE Transactions on Antennas & Propagation*, Vol. 24, No. 5, September 1976, pp. 585–598.

[12] Gabriel, W. F., "Adaptive Arrays—An Introduction," *Proceedings of the IEEE*, Vol. 64, No. 2, Feb. 1976, pp. 239–272.

CHAPTER 18

M&S of ES

18.1 Introduction

This chapter focuses on electromagnetic support (ES), the third and last pillar of EW. ES entails the search and collection of electromagnetic data from potentially hostile sources for intelligence purposes. It is therefore related to the disciplines of Signals Intelligence (SIGINT) and Electronic Intelligence (ELINT), which is a subcomponent of SIGINT. ELINT is intelligence collected from *noncommunications* EM radiations [1], such as threat radars. There are key differences, however, between ELINT and ES. ELINT is more of a *strategic* operation where the data gathered is used for detailed, off-line analysis of threat systems and their capabilities for future planning purposes. ES, on the other hand is a *tactical* operation where the data collected is intended to inform real-time, near-term decisions, which may involve deployment of kinetic (weapons) or nonkinetic (EA) attacks.

ES systems are typically passive receivers monitoring the EM environment to increase situational awareness about emitters operating in an area of interest. Nowadays, however, it is also common to find ES functionality integrated in both EA and EP systems. On the EA side, information on the emitter waveform enables the jammer to tailor its response for maximum effectiveness. On the radar side, accurate characterization of the EA signal likewise allows the radar to initiate the appropriate EP technique.

Basic functions of ES include:

- Emitter search (in frequency and sometimes angle);
- Waveform measurement and feature extraction;
- Emitter sorting and classification.

Figure 18.1 shows a functional block diagram of an ES receiver. Upon searching and detecting the presence of an emitter signal, the feature extraction block measures relevant waveform parameters that enable the identification of the emitter and a subsequent response to its presence. Measured parameters may include RF frequency, pulse width, pulse repetition interval/frequency (PRI/PRF), amplitude, and angles of arrival. More capable systems may measure additional features such as signal polarization and intrapulse modulation (LFM, Barker code, etc.). The modeling examples presented in this chapter will focus on the detection and feature extraction portions of the block diagram, as highlighted in the figure.

Some of the signal parameters are measured directly, while others are derived; for example, the pulse width is derived from two direct measurements: the pulse

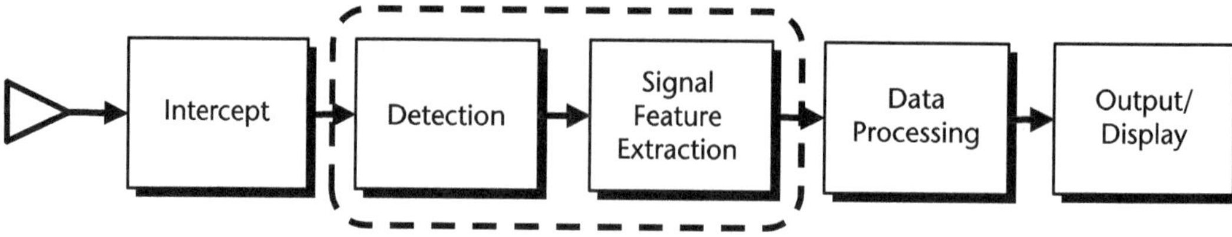

Figure 18.1 Generic block diagram of ES receiver.

time-of-arrival (TOA) and its subsequent time-of-departure (TOD). TOA and TOD are the times when the ES system senses the incoming pulse leading edge and trailing edge, respectively; their difference forms the estimate of the pulse width. Similarly, the PRI is estimated by taking the difference between successive TOAs (or TODs).

The pulse-level measurements are often grouped into a parameter vector commonly known as the Pulse Descriptor Word (PDW) [2]. The PDW vector may be used for identification of the emitter source, by comparing it to a database of PDWs for known waveforms. A model of a simple PDW generator will be presented later in the chapter.

ES systems are expected to operate in congested environments; therefore, another important processing feature is the ability to sort multiple signals arriving at the receiver asynchronously. Algorithms for deinterleaving multiple pulse streams using PDW information are constantly in development [3, 4]. Once the pulses have been sorted into different buckets, their PRIs may then be estimated.

There are different types of receiver architectures used in ES systems. They vary from low cost/low complexity architectures such as the Crystal Video Receiver (CVR), to medium complexity systems such as the Superheterodyne and Instantaneous Frequency Measurement (IFM) receivers, up to high cost but very capable systems such as the Channelized and the Microscan/Compressive receivers. There are also hybrid architectures that combine some of these technologies. An in-depth review of these receiver types is outside our scope, but a good overview is found in [5]. In this chapter we focus on the IFM receiver as a modeling example.

The IFM receiver was first developed in the 1950s as a result of the Korean War [6]. An IFM receiver is capable of measuring the frequency of an incoming signal within a very short time, typically in the order of nanoseconds (thus the designation of "instantaneous"). The IFM operation may be performed in either the analog or digital domain. An analog IFM is typically very early in the RF chain, and the measured frequency is used to tune the filters and mixers further downstream in the receiver to place the downconverted pulses within a common IF passband. In contrast, a digital IFM, or DIFM, may simply be one of many processing stages in the feature extraction section of the receiver, operating on the downconverted, digitized pulses.

In addition to speed, other advantages of the IFM include a relatively simple design, wide bandwidth, and good frequency resolution. Disadvantages include relatively poor sensitivity—due to the wide bandwidth—and an inability to measure frequencies of simultaneous signals (pulse-on-pulse arrivals, for example).

One operation often performed with ES system measurements is geolocation of the detected emitters (i.e., the estimation of their geographic position). A radar may use range, azimuth, and elevation measurements as a form of spherical coordinates, and then transform them into Cartesian coordinates relative to some reference coordinate system. A passive ES receiver, however, must resort to indirect measurements to accomplish this. One subset of methods and techniques under the more general topic of geolocation is known as multilateration (MLAT). MLAT literally means measurements from multiple sides, implying multiple observations. These may come from multiple sensors, or from a single sensor taking observations at multiple locations at different times. These measurements may be taken in different dimensions, the most common being time and frequency.

Arguably, the most common MLAT method is the time difference of arrival (TDOA). In TDOA processing, signals arrive to an array of passive sensors at different TOAs based on the emitter-sensor geometry. The TDOAs are proportional to the difference in range between the emitter and the sensors. This information is used to extract an estimate of the emitter position based on these measurements. Approaches for calculating this estimate have been well investigated, as the topic has wide applicability in areas such as navigation and surveillance [7]. This method is explored later in the chapter, and MATLAB models that address different aspects of it are presented.

The rest of the chapter is organized as follows. Section 18.2 covers IFM modeling, including a mathematical representation as well as a MATLAB implementation. Section 18.3 takes a more comprehensive look at ES functionality, and describes a generic ES model capable of measuring some key PDWs. A Simulink implementation of the ES system is also presented. Section 18.4 follows with an overview of TDOA-based geolocation and related MATLAB models. Lastly, Section 18.5 presents additional modeling considerations and concluding remarks.

18.2 Instantaneous Frequency Measurement Modeling

Figure 18.2 shows a functional block diagram of an instantaneous frequency measurement (IFM) circuit. The signal $x(t)$ is the product of the input signal and a delayed version of it. The output is low-pass filtered, yielding a DC signal (constant amplitude), from which the frequency estimate is extracted. We now present the mathematical formulation of this estimate.

Let the received signal be

$$s(t) = A\cos(2\pi ft + \theta) \tag{18.1}$$

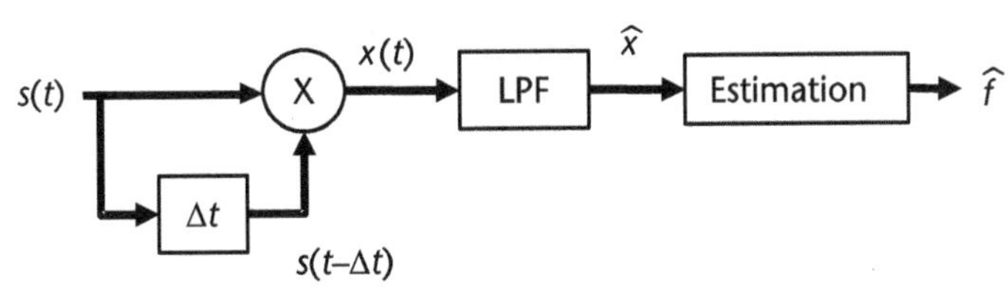

Figure 18.2 Generic block diagram of an IFM processor.

The product $x(t)$ of the received signal and its delayed version is then

$$x(t) = s(t)s(t-\Delta t) = A^2\cos(2\pi ft + \theta)\cos(2\pi f(t-\Delta t) + \theta) \tag{18.2}$$

Recall the trigonometric identity

$$\cos(X)\cos(Y) = \frac{1}{2}\left[\cos(X+Y) + \cos(X-Y)\right] \tag{18.3}$$

Applying (18.3) to (18.2) results in

$$x(t) = \frac{A^2}{2}\left[\cos(2\pi(2ft - f\Delta t) + 2\theta) + \cos(2\pi f\Delta t)\right] \tag{18.4}$$

The LPF eliminates the high-frequency sum term and retains the difference term at baseband,

$$\hat{x} = \frac{A^2}{2}\cos(2\pi f\Delta t) \tag{18.5}$$

which may be solved for f. Thus, the frequency estimate is

$$\hat{f} = \frac{1}{2\pi\Delta t}\cos^{-1}\left(\frac{2\hat{x}}{A^2}\right) \tag{18.6}$$

The MATLAB function m-file *myIFM.m* implements this IFM-based frequency estimator. Given the true frequency and IFM parameters, the function creates the cosine time samples and implements (18.1) to (18.6). The LPF function is effectively implemented by averaging all the time samples corresponding to the baseband signal from (18.5). The function inputs are defined in Table 18.1. The array variable `input` is explicitly passed as an input argument, while the remaining parameters are passed as global variables.

The test script m-file *test_myIFM* sets up inputs and evaluates the estimate outputs for frequencies ranging from 2 GHz to 12 GHz. The results are shown in Figure 18.3.

The maximum measurement error over all frequencies is slightly above 3 MHz, which represents less than 0.2% error; however, we note these are ideal (noiseless) results. We also observe that errors do become smaller as the true frequency increases; this behavior holds even in the presence of noise.

One key parameter in the design of an IFM circuit is the value of the time delay Δt applied to the input signal. The delay must support unambiguous estimation over the frequency range of interest; this implies that the argument of the cosine term in (18.5) must meet the condition

Table 18.1 Inputs to *myIFM* MATLAB Model

MATLAB Variable	*Description*	*Default Value*
`input`	2-D array with pulse envelope samples (row 1) and time values (row 2)	Various
`deltat`	Time delay Δt applied to input signal, in sec	2e-11
`Fs`	Sampling rate in Hz	100e9
`pw`	Pulse width portion that is used for the estimate, in sec	100e-9
`fc`	True RF frequency in Hz	Various

$$2\pi f \Delta t < \pi \quad \text{or} \quad \Delta t < \frac{1}{2f} \tag{18.7}$$

This condition is most constrained at the maximum value of the frequency of interest. For the default value of Δt = 2e-11 sec, $f_{\max}$ is 25 GHz. This is best illustrated in Figure 18.4.

The plot in Figure 18.4(a) shows the monotonicity is only preserved out to $f\Delta t = 0.5$. For a 20 ps time delay (the default value in the MATLAB model), the maximum unambiguous frequency is 25 GHz. Notice that a frequency of 30 GHz will yield the same value as (i.e., will alias into) 20 GHz.

There are many variations to the basic IFM design. Rather than treating the incoming waveform as a real signal (as was the case in our example), many systems form a complex signal by applying the Hilbert Transform. This results in the formation of I/Q signal components; one benefit of this approach is that the inverse cosine operation becomes an inverse tangent of the Q/I ratio. Measuring the pulse amplitude—*A* in (18.6)—is no longer needed, as it is canceled when the sine/cosine ratio is formed.

Another variation is to add an actual LPF to the circuit to reject the high-frequency component, as opposed to the average operation used as a surrogate.

To resolve frequency ambiguities, multiple taps of different length are often used, in a manner not too dissimilar to the range ambiguity resolution using the Chinese Remainder Theorem [6].

18.3 Generic ES Processor Modeling

This section focuses on a first-order Simulink model for a generic ES system. The system works against a single emitter transmitting a pulse waveform of constant parameters, and performs the following steps:

- Signal (pulse-level) detection;
- TOA and TOD estimation—Based on the leading and trailing edges of each pulse;

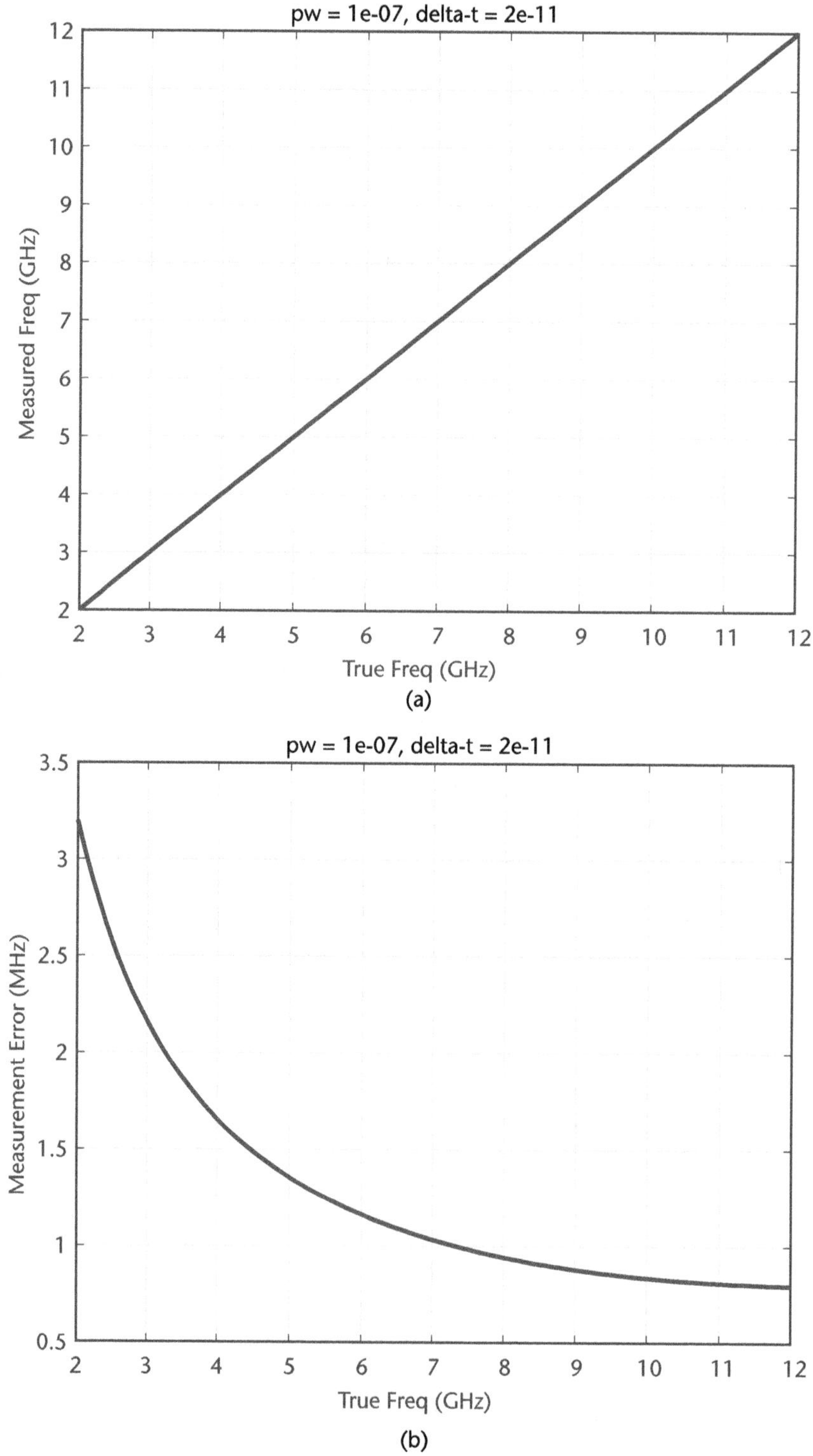

Figure 18.3 Results from IFM model: (a) estimated vs. true frequency; (b) measurement error vs. true frequency.

(a)

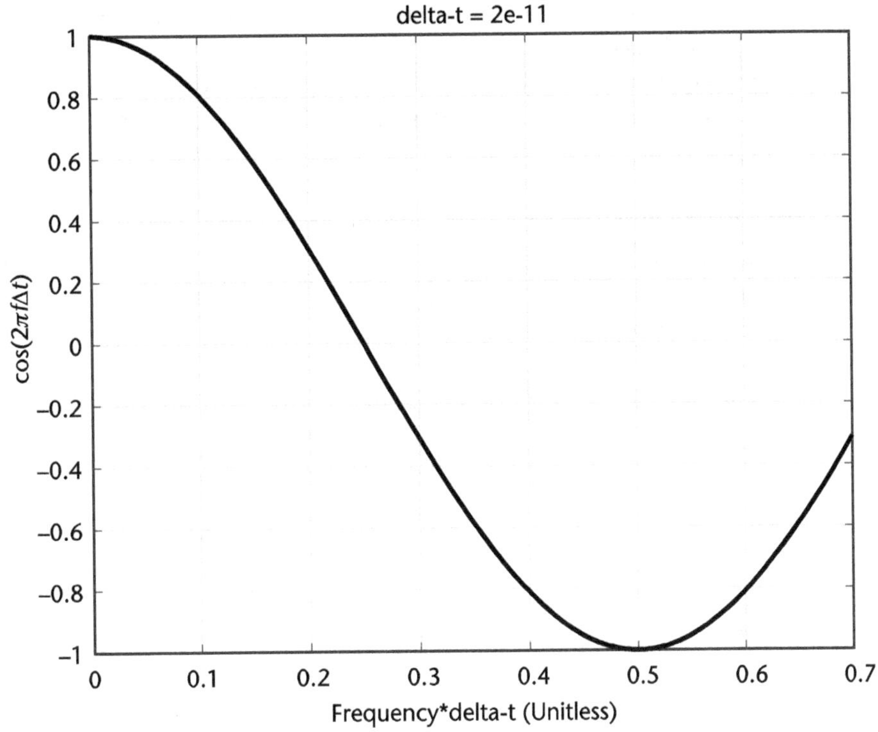

(b)

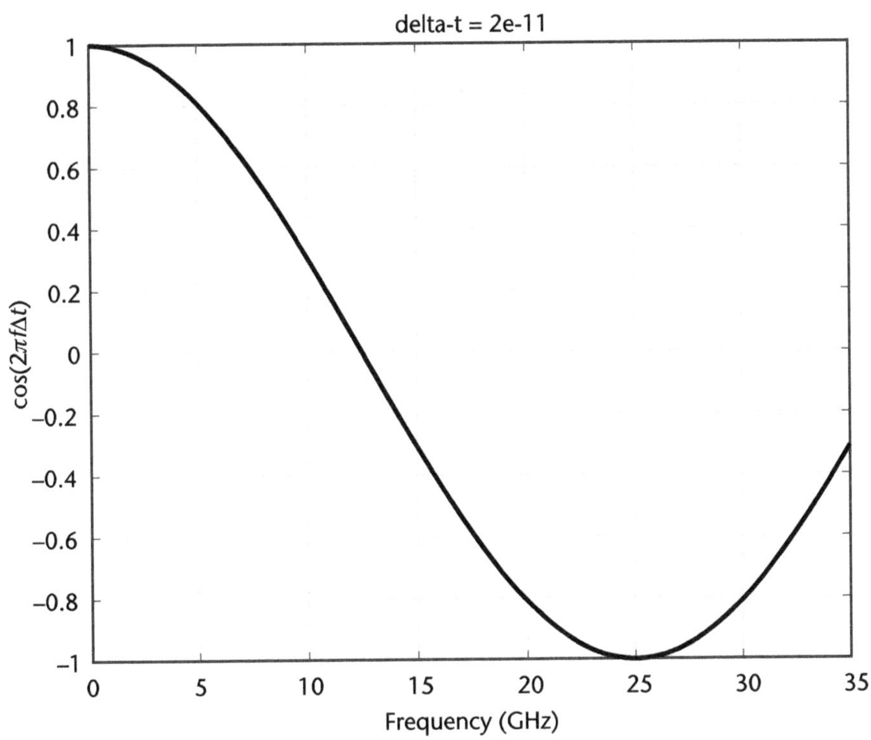

Figure 18.4 Size of time delay determines the maximum unambiguous frequency estimate. (a) Baseband term is monotonic out to $(f\Delta t) = 0.5$. (b) A delay of 20 ps results in a maximum unambiguous frequency of 25 GHz.

- Pulse width estimation—Based on the separation between contiguous TOAs and TODs;
- PRI estimation—Based on the separation between successive TOAs (or TODs);
- Frequency estimation—Based on the IFM model presented in Section 18.2.

18.3.1 Signal Detection

Pulse-level detection requires the incoming signal to be above some minimum, system-specific power level. This minimum detectable signal is known as the receiver sensitivity, and was initially introduced in Chapter 16 when discussing responsive jammers.

The receiver sensitivity S is a key parameter commonly found in the system specifications. It is calculated as [8]:

$$S = \left(kT_0B_nF\right)SNR_{\text{min}} \tag{18.8}$$

where (kT_0B_nF) is the receiver noise power whose components were defined in Chapter 2 (Table 2.1) and SNR_{min} is the SNR required to achieve a desired detection performance, quantified by a specified probability of detection. Figure 18.5 shows the detection scheme implemented in the Simulink model.

The receiver sensitivity is defined at the input of the receiver, after the antenna port. The gain G in the diagram represents the composite signal gains and losses between this point and the signal detector. The signal $x(t)$ is then compared against the detection threshold T (the square root implies a linear detector is assumed). The output of the detector is a binary (0,1) signal designated as a pulse present indicator or PPIN. The PPIN is used for all subsequent measurements.

18.3.2 Pulse Width and PRI Estimation

Figure 18.6 shows a notional realization of the PPIN signal for N detected pulses.

From examining the sequence of ones and zeros, it should be apparent that a leading edge is marked by a $0 \rightarrow 1$ transition, while a trailing edge is indicated by

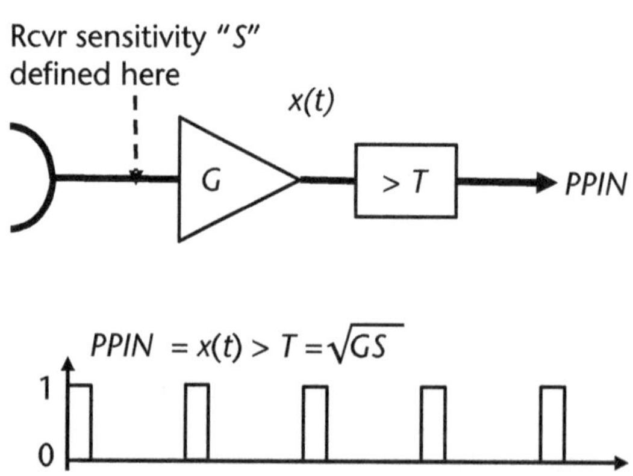

Figure 18.5 Sensitivity-based detection threshold is used to create PPIN signal in ES receiver.

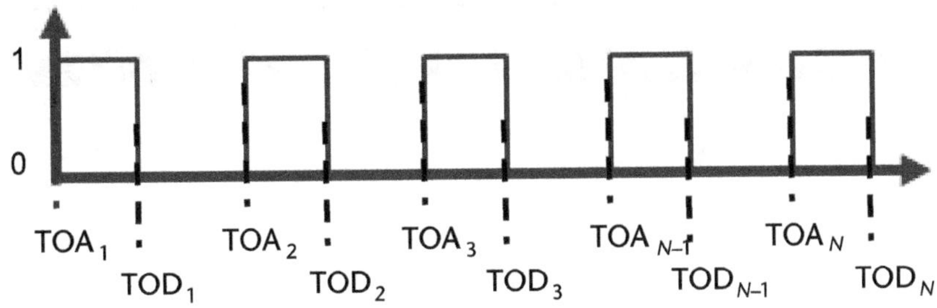

Figure 18.6 Notional PPIN signal with TOAs and TODs used for pulse width and PRI estimation.

a $1 \rightarrow 0$ transition. Thus, the rules for declaring TOAs and TODs are simply the times at which the following conditions hold:

$$\begin{aligned} TOA&: \left(PPIN_{n+1} - PPIN_n\right) = +1 \\ TOD&: \left(PPIN_{n+1} - PPIN_n\right) = -1 \end{aligned} \tag{18.9}$$

These differences are calculated using the MATLAB `diff` function.

Given a sequence of TOAs and TODs, a pulse width estimate is calculated by averaging the differences between them, as

$$\hat{\tau} = \frac{1}{N}\sum_{n=1}^{N}\left(TOD_n - TOA_n\right) \tag{18.10}$$

Similarly, a PRI estimate is calculated by averaging the difference between consecutive TOAs and TODs:

$$\begin{aligned} \widehat{PRI} &= \frac{1}{N}\sum_{n=1}^{N-1} PRI_n \\ PRI_n &= 0.5\left[\left(TOA_{n+1} - TOA_n\right) + \left(TOD_{n+1} - TOD_n\right)\right] \end{aligned} \tag{18.11}$$

18.3.3 Simulink Model

The Simulink model *ESM_Example2* implements the operations described in the preceding pages, and is tested against a pulsed waveform with a 10 μs pulse width, 100 μs PRI, and 10 GHz center frequency. The Simulink block diagram is shown in Figure 18.7. Receiver noise is added to the emitter signal before entering the ES measurement subsystem. Only four pulses are created in this example.

The key functions in the measurement subsystem are highlighted in gray, and are mostly implemented in MATLAB m-files via Interpreted MATLAB Function blocks. In the upper signal path, the threshold comparison creates the PPIN signal.

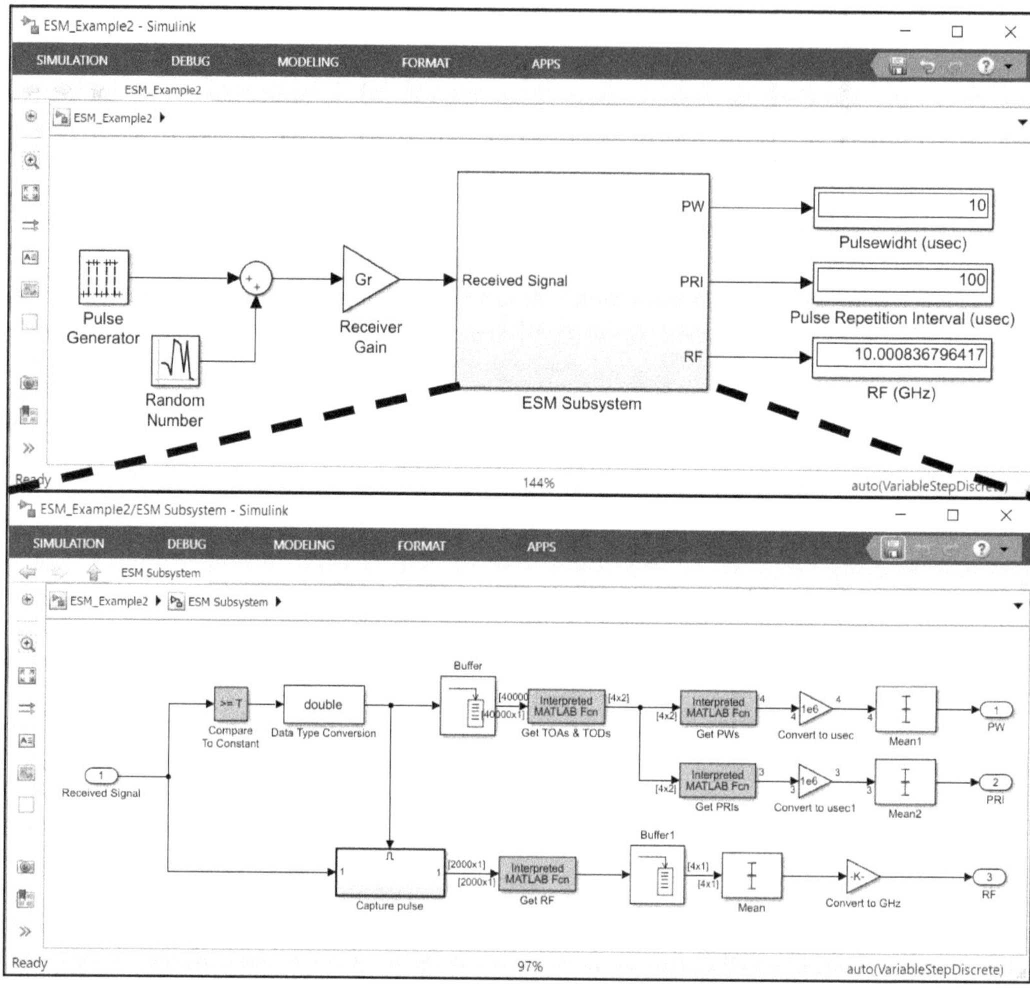

Figure 18.7 Simulink model of generic ES system: Top-level block diagram (top); measurement subsystem (bottom). Signal dimensions are shown.

A Buffer block collects the samples into a vector, and sends it to the first gray block to measure the TOAs and TODs. The output TOA and TOD array is then sent to subsequent blocks to estimate pulse widths and PRIs. The four pulse width estimates and three PRI estimates are averaged, and the results sent to Display blocks.

In the lower signal path, the PPIN signal is used to gate the input samples that are sent to the IFM function (in the block labeled Get RF) to ensure only pulse samples are evaluated. In a manner similar to the upper path, the four output frequency estimates are collected into a vector via a Buffer block, then averaged. The calculated average moves on to the output display. The Display blocks at the top-level diagram show that the three quantities are estimated very accurately.

This model is a good demonstration of feature extraction functionality under benign conditions. Admittedly, it is geared towards analysis of a single waveform, and it would not do very well against a dense signal environment. Multiple, overlapping pulse trains require an additional sorting or deinterleaving step, whose implementation can be quite complex. Another limitation of the model is the fact that the PRI/pulse width estimation calculations are based on a batch processing of data (albeit of only a few pulses). While these may be suitable for non–real time applications, tactical ES implementations may require much faster responses, as well as the ability to handle waveform agility. PRI agility schemes, such as PRI stagger, can be particularly challenging. For PRI estimation in particular, it is common to implement a PRI tracker that maintains a running average by ingesting one pulse at a time. In integrated EA/ES systems, The PRI tracker would also be able to predict the arrival of the next pulse, for precise timing of the EA response.

18.4 TDOA Modeling

Figure 18.8 shows a general layout depicting the nature of the problem to be solved. In the figure, the emitter is at an unknown location transmitting a signal $h(t)$. Multiple distributed passive receivers intercept the signal, which arrives with a distinct time delay t_n at the nth receiver. If there is relative motion, the received signal will also have a distinct Doppler shift f_{Dn}, although we will ignore that for now. The question is whether the emitter location can be determined from these signals. The receivers must communicate among each other, as they need to share their respective measurements. Data must be transferred from each sensor platform to a central processing location, whether it is one of the receiver platforms themselves or a separate processing station altogether.

The passive receivers are unable to measure the absolute time delays t_n, as the time of transmission is unknown. They can only measure the times of arrival, TOA_n. The $TDOA$ between receivers 1 and 2 is

$$TDOA_{21} = TOA_2 - TOA_1 = t_2 - t_1 = \frac{(R_2 - R_1)}{c} \tag{18.12}$$

where R_n is the range from the emitter to the nth receiver.

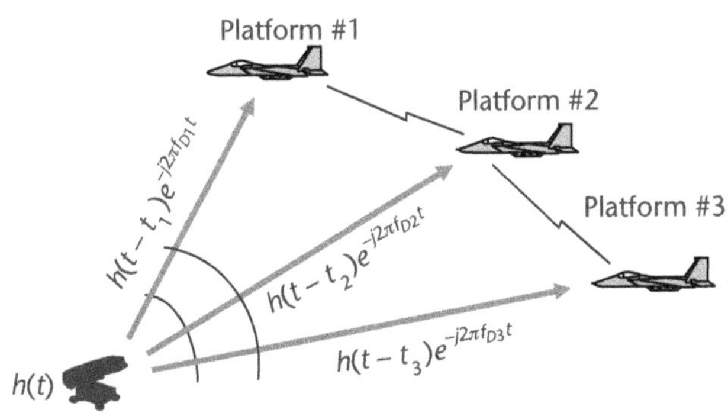

Figure 18.8 The passive geolocation problem, solved via TDOA.

18.4.1 Estimation of Emitter Location

As a graphical illustration of TDOA-based geolocation, consider the 2-D example shown in Figure 18.9. The emitter location is (x, y) in east-north coordinates. From (18.12), the TDOA between the two receivers shown is a constant. Unfortunately, there is an infinite number of solutions to this equation. It can be shown that any point that lies in the hyperbola displayed as a solid black line in 18.9(a) satisfies (18.12). As a side note, the hyperbola shown as a *dotted* line corresponds to $-TDOA_{21}$. We thus require additional information, obtained by adding a third receiver.

The new receiver pair 2-3 forms a new hyperbola, shown in Figure 18.9(b) as a solid gray line. This new 2-3 hyperbola intersects the 1-2 hyperbola, yielding a

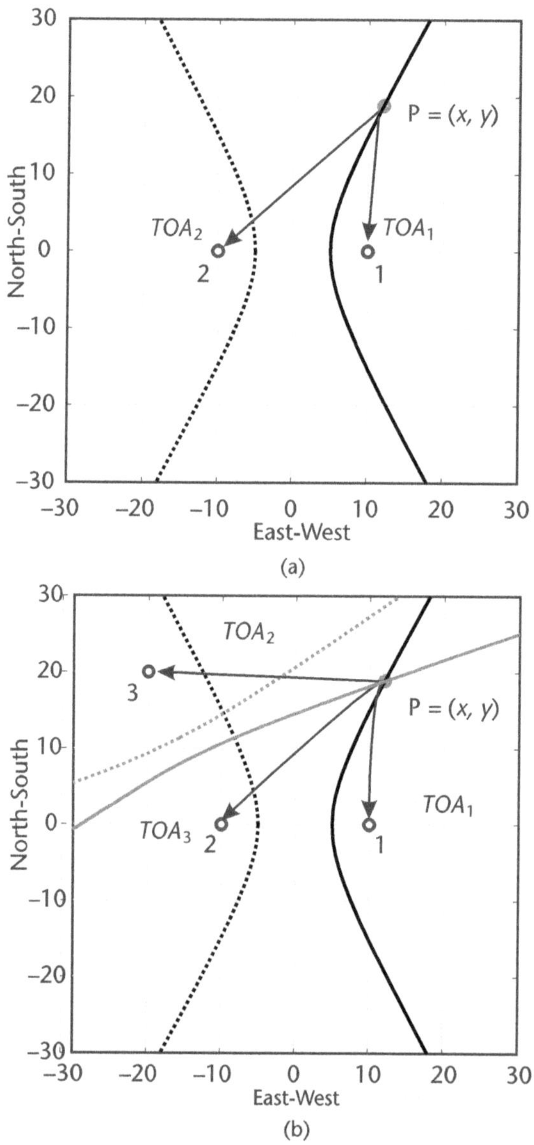

Figure 18.9 Graphical solution to TDOA geolocation: (a) Two receivers yield an infinite number of solutions. (b) Three receivers are required for a unique solution.

unique solution. We note that the third hyperbola 1-3 (not shown) does not provide any additional information. We conclude that at least three receivers are required to solve the 2-D problem. Under some adverse geometries, a fourth receiver might be required to resolve a location ambiguity (as the two hyperbolas could intersect in two places).

We now describe the mathematical approach to the TDOA geolocation solution. The first step is to form N TDOA equations from the $N + 1$ TOA measurements:

$$TDOA_n = \Delta t_n = TOA_n - TOA_0, \quad n = 1,\ldots,N \tag{18.13}$$

Notice that this is a generalization of (18.12), where the zero-index sensor is designated as the reference. We also note from (18.12) that the *TDOAs* are the differences between the true ranges, scaled by the speed of light. We can write the range from the emitter to the nth receiver in Cartesian coordinates as

$$R_n = \sqrt{(x_n - x)^2 + (y_n - y)^2 + (z_n - z)^2}, \quad n = 0,\ldots,N \tag{18.14}$$

where (x_n, y_n, z_n) and (x, y, z) are the nth receiver and emitter coordinates, respectively. Using (18.12), the TDOA equation (18.13) becomes

$$\begin{aligned} cTDOA_n &= \sqrt{(x_n - x)^2 + (y_n - y)^2 + (z_n - z)^2} \\ &\quad - \sqrt{(x_0 - x)^2 + (y_0 - y)^2 + (z_0 - z)^2} \\ n &= 1,\ldots,N \end{aligned} \tag{18.15}$$

Four receivers are required to create three equations in the form of (18.15) to solve for the three unknowns (x, y, z). These are nonlinear equations, which are often solved using recursive methods such as nonlinear least squares (NLLS). A well-known approach is the Gauss-Newton method [9], which is described as follows. Let $\underline{P} = [x, y, z]$ be the position vector to be estimated. We then define a set of functions h_n to be minimized, as

$$\begin{aligned} h_n(\underline{P}) = &\sqrt{(x_n - x)^2 + (y_n - y)^2 + (z_n - z)^2} \\ &-\sqrt{(x_0 - x)^2 + (y_0 - y)^2 + (z_0 - z)^2} \\ &-cTDOA_n \quad n = 1,\ldots,N \end{aligned} \tag{18.16}$$

Each function h_n defined by (18.16) becomes the nth element of the N-dimensional vector $\underline{h}$. Given an initial guess $\underline{P}_0$, the recursive NLLS equation is

$$\underline{P}_n = \underline{P}_{n-1} - \underline{J}^{-1}\underline{h}(\underline{P}_{n-1}) \tag{18.17}$$

where $\underline{J}$ is the $N \times 3$ Jacobian matrix of h. The (i, j)th element of this matrix is defined as

$$\left(\underline{J}\right)_{ij} = \frac{\partial h_i(\underline{P})}{\partial \underline{P}(j)} \tag{18.18}$$

and $\underline{P}(j)$ is the jth element of $\underline{P}$ (i.e., x, y or z). These partial derivatives are given by

$$\begin{aligned}
\frac{\partial h_i(\underline{P})}{\partial \underline{P}(1)} &= \frac{\partial h_i(\underline{P})}{\partial x} = \frac{x - x_i}{\sqrt{(x_i - x)^2 + (y_i - y)^2 + (z_i - z)^2}} - \frac{x - x_0}{\sqrt{(x_0 - x)^2 + (y_0 - y)^2 + (z_0 - z)^2}} \\
\frac{\partial h_i(\underline{P})}{\partial \underline{P}(2)} &= \frac{\partial h_i(\underline{P})}{\partial y} = \frac{y - y_i}{\sqrt{(x_i - x)^2 + (y_i - y)^2 + (z_i - z)^2}} - \frac{y - y_0}{\sqrt{(x_0 - x)^2 + (y_0 - y)^2 + (z_0 - z)^2}} \\
\frac{\partial h_i(\underline{P})}{\partial \underline{P}(3)} &= \frac{\partial h_i(\underline{P})}{\partial z} = \frac{z - z_i}{\sqrt{(x_i - x)^2 + (y_i - y)^2 + (z_i - z)^2}} - \frac{z - z_0}{\sqrt{(x_0 - x)^2 + (y_0 - y)^2 + (z_0 - z)^2}}
\end{aligned} \tag{18.19}$$

Equation (18.17) is iterated until the total change ΔP from the previous iteration falls below a user-defined value.

The MATLAB m-file *test_NLLS* implements a 2-D version of this algorithm to solve the three-receiver scenario depicted in Figure 18.9(b). Under these conditions the Jacobian matrix has dimensions 2×2. The emitter is located at $\underline{P}$ = [12.0 km, 18.9 km]. The initial guess is set to the centroid of all three receivers. The stopping criterion is set to

$$\Delta P = \left\| \underline{P}_n - \underline{P}_{n-1} \right\| < 0.1 \tag{18.20}$$

where the 0.1 is in units of km (i.e., 100m). The results of the recursive estimation are depicted in Figure 18.10. With errorless TDOA measurements, the solution is achieved in only five iterations, as summarized by Table 18.2. Notice that the true estimation error is much smaller than the iteration change, and is already in the order of 1m by the 4th iteration.

18.4.2 TDOA Measurement Approaches

There are two general approaches to the actual TDOA measurement:

- *Leading edge detection:* At each receiver, measure the TOA of each pulse. TDOA is simply the difference of two such measurements. The main advantage of this approach is that it is very simple to compute, and it minimizes the amount of information that must be transferred over the data link to the central processing location. The main disadvantage is that this measurement

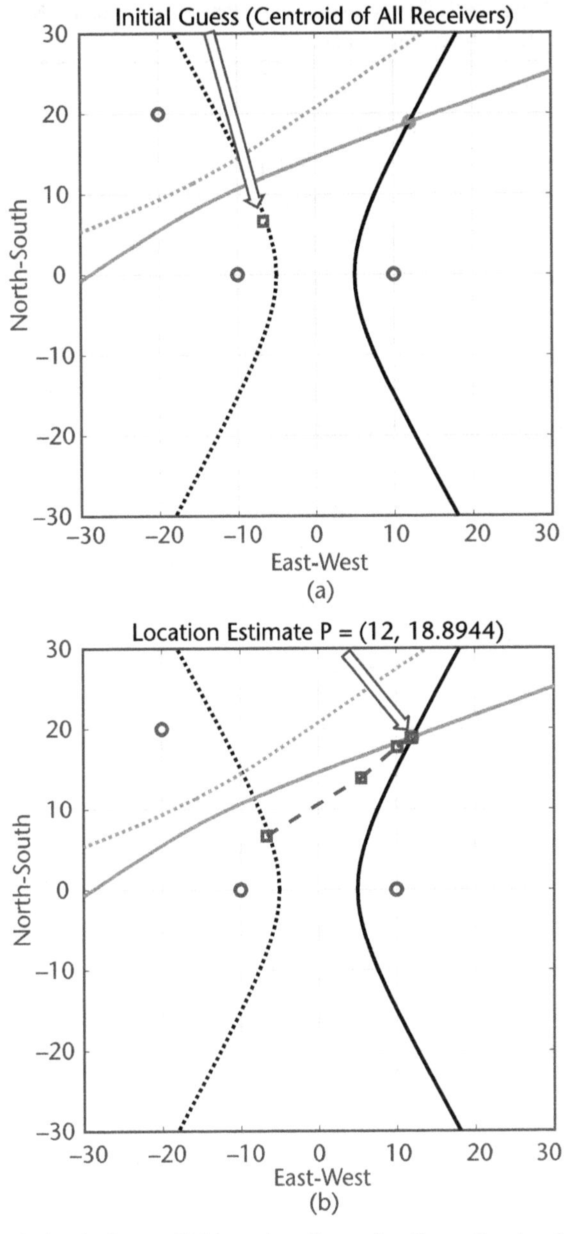

Figure 18.10 Numerical solution to TDOA geolocation using Gauss-Newton NLLS algorithm: (a) Initial guess using centroid of all three receivers. (b) Final estimate after five iterations.

has limited precision, as it is more vulnerable to noise-induced measurement errors.

- *Cross-correlation:* Each receiver collects a portion of I/Q data samples corresponding to the received signal. The TDOA is obtained from the peak of the cross-correlation between any two received signals; this corresponds to the maximum likelihood (ML) estimate. ML estimation can achieve extremely good precision, at the expense of higher data transfer requirements, as raw

Table 18.2 TDOA Geolocation Results using Gauss-Newton Algorithm (All Quantities Are in km)

Iteration	*Estimated Location*	ΔP	*True Error*
0	[–6.6667, 6.6667]	NA	22.3151
1	[5.4923, 13.8745]	14.1349	8.2189
2	[10.1230, 17.7099]	6.0128	2.2195
3	[11.8387, 18.8062]	2.0361	0.1838
4	[11.9988, 18.8939]	0.1825	0.0013
5	[12.0000, 18.8944]	0.0013	6.7338e-08

data records must now be transferred to a common location to perform the cross-correlation operation.

Cross-correlation is the more robust of the two approaches. Generally speaking, accurate TOA measurement requires a very short rise time on the pulse, which implies a wide bandwidth receiver. Wide bandwidth, however, often means lower sensitivity due to increased noise; a stronger pulse is needed for reliable detection and measurement. Another potential source of measurement error is any mismatch between receiver bandwidths, as illustrated in Figure 18.11.

The cross-correlation approach provides a more precise TDOA estimate under most conditions. The cross-correlation operation itself increases the SNR, thereby reducing errors due to noise. It is also less vulnerable to bandwidth mismatches, as shown in Figure 18.12. These are the results obtained from running the MATLAB m-file *test_TDOA_CAF*. The CAF term in the file name refers to the cross (or complex) ambiguity function, which is a generalized form of the cross-correlation between two signals $s_1(t)$ and $s_2(t)$. The CAF is defined as [10]:

$$A(\tau, f) = \int_0^T s_1(t) s_2^*(t-\tau) e^{j2\pi ft} dt \tag{18.21}$$

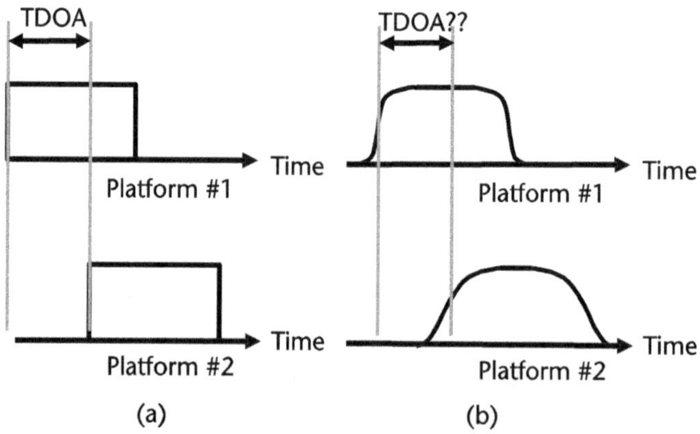

Figure 18.11 TDOA estimation by leading edge detection: (a) Short rise times enable accurate measurements; and (b) receiver bandwidth mismatches causing different rise times to increase the TDOA measurement uncertainty.

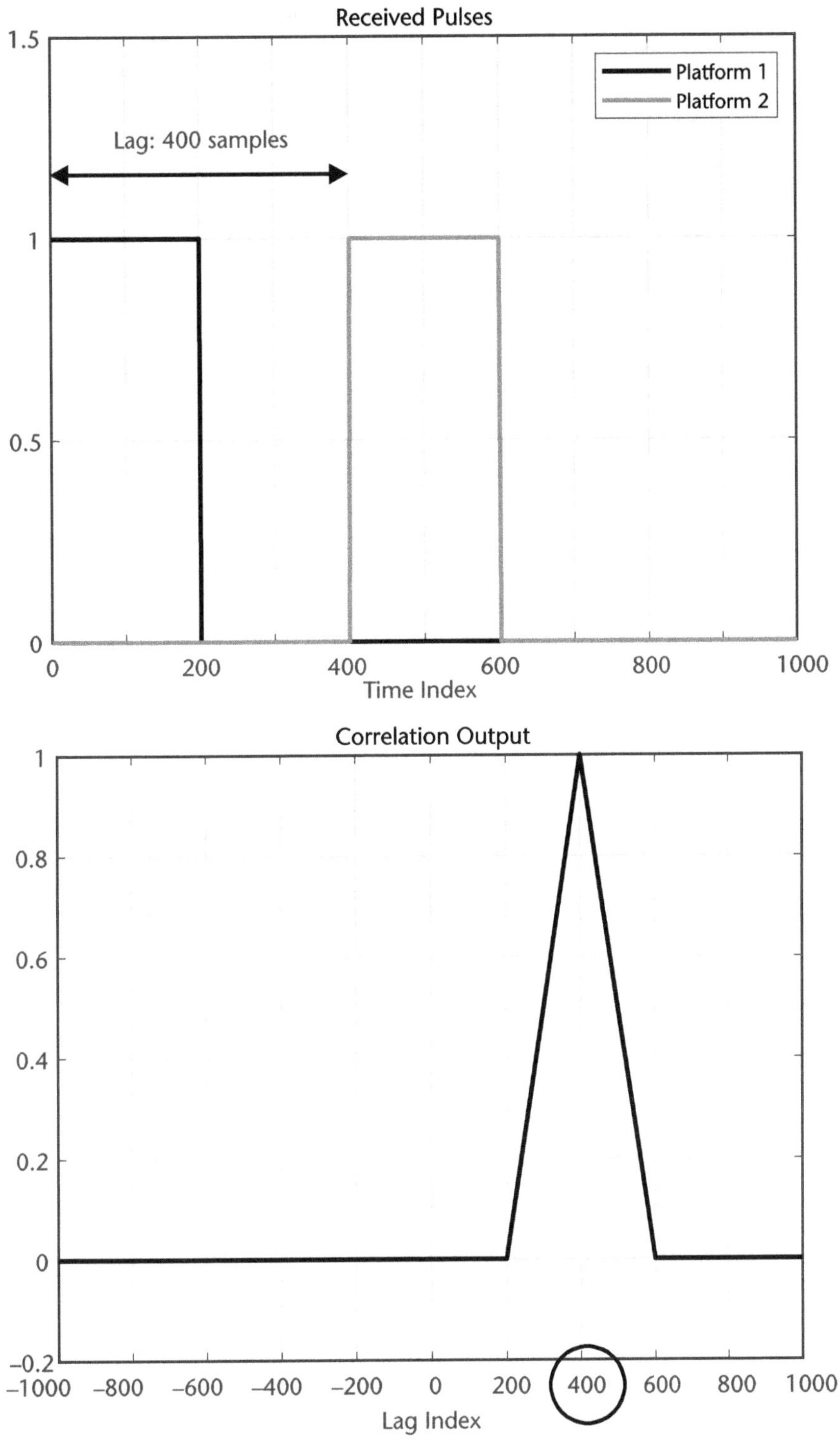

(a)

Figure 18.12 Example results of TDOA estimation by correlation: (a) pulses with short rise times and (b) pulses with long and mismatched rise times due to different receiver bandwidths.

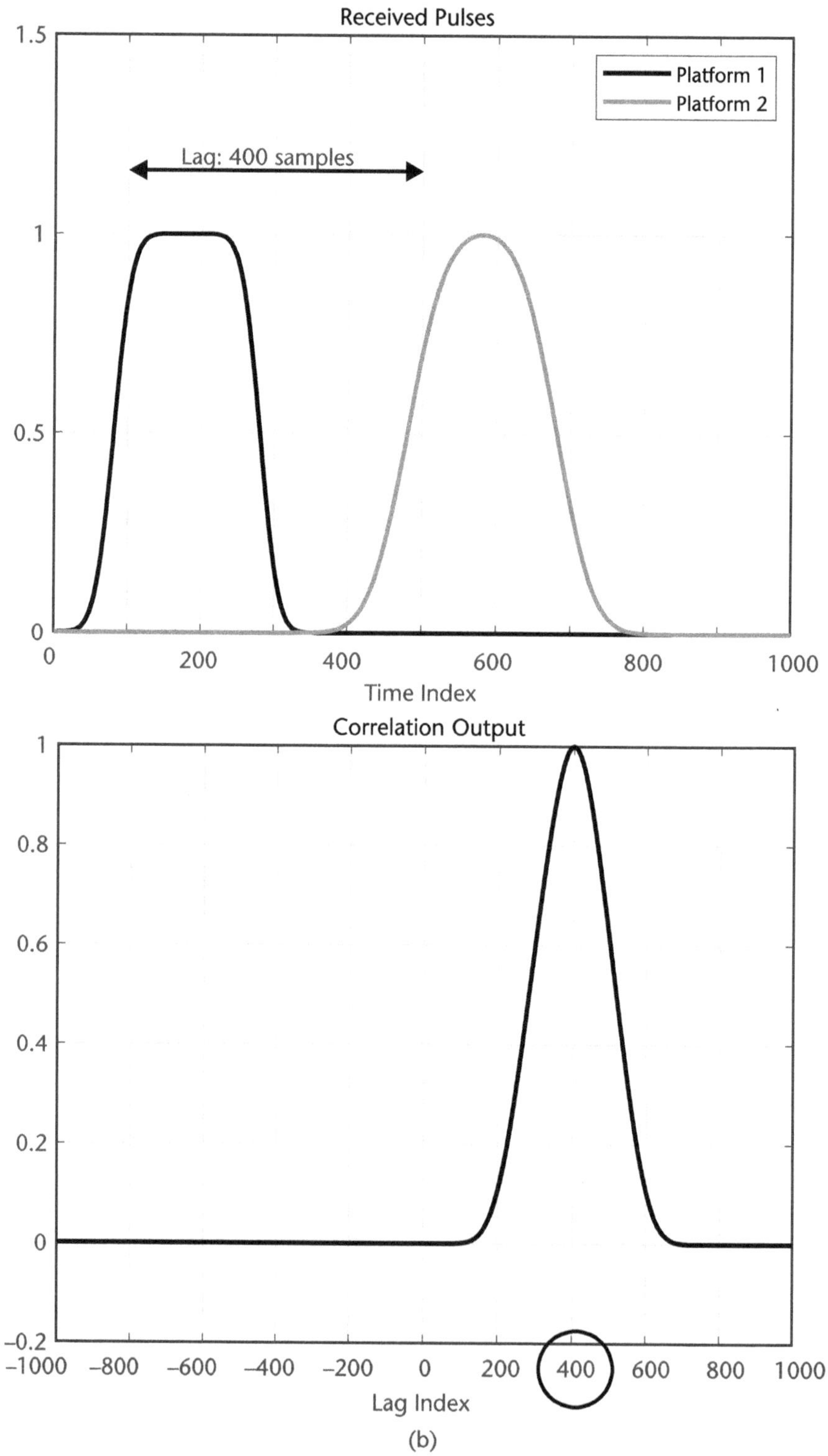

Figure 18.12 *(Continued)*

and may be interpreted as an extension of the cross-correlation function that operates over the time and frequency domains. It not only accounts for the time lag between the two signals, but also for a difference in frequency as well. Thus, it is also a frequency difference of arrival (FDOA) estimator. We note here that the CAF is a form of the radar ambiguity function [11], which is a widely used analytical tool for radar performance assessment, in particular for waveform design.

The time and frequency values that maximize the CAF yield the best (ML) estimate of the signals TDOA and FDOA. The example illustrated in Figure 18.12 does not use any frequency information, however (TDOA estimation only).

The MATLAB model creates two received squared pulses with a time lag of 400 samples. Their cross-correlation accurately peaks at the correct value, as seen at the bottom of Figure 18.12(a). The model then defines two different LPF impulse response functions and convolves the pulses, thereby creating nonsquare pulses of significantly different rise times. The impulse responses are modeled as having Gaussian shape; the response for the *i*th LPF is given by

$$h(t,\alpha_i) = e^{-\frac{1}{2}\left(\frac{t}{\alpha_i}\right)^2} \tag{18.22}$$

It can be shown that the transfer function of the filter is also a Gaussian function, of variance

$$\sigma_i^2 = \frac{1}{(2\pi\alpha_i)^2} \tag{18.23}$$

Thus, different values of α result in filters of varying bandwidths. For this example, the α values were set to 20 and 40, so the bandwidths are mismatched by a factor of two. The bottom of Figure 18.12(b) shows that the CAF output is able to estimate the correct TDOA value despite these differences.

These results are based on noiseless data. The model does support adding receiver noise to the pulse samples. Setting the flag `add_noise` equal to 1 in line 3 adds noise corresponding to 15-dB SNR (specified in line 4). Estimation results are still quite good. (Try it!) A comprehensive measurement error analysis of TDOA is found in [12].

18.5 Concluding Remarks

ES capabilities can be quite complex depending on the system. Their modeling and simulation often entails representing a combination of hardware subsystems and software algorithms. The models presented in this chapter are intended as a selective sampling of representative functionality, rather than a broad simulation capability as was the goal with the radar engagement model presented in Chapter 15. Nevertheless, it is hoped that they provide a foundation upon which a more comprehensive functionality can be built. Some potential enhancements were discussed throughout the chapter, but to sum up, the authors' suggestion would be to focus on real-time operation and processing—as opposed to batch processing—and to incorporate deinterleaving functionality to support the sorting of multiple, simultaneous emitters.

References

[1] *Electronic Warfare,* Joint Publication 3-13.1 (JP 3-13.1), February 8, 2012.

[2] Wiley, R. G., *ELINT: The Interception and Analysis of Radar Signals,* Norwood, MA: Artech House, 2006, pp. 317–334.

[3] Ata'a, A. W., and S. N. Abdullah, "Deinterleaving of Radar Signals and PRF Identification Algorithms," *IET Radar, Sonar and Navigation,* Vol. 1, No. 5, 2007, pp. 340–347.

[4] Fitzgerald, D. F., et al., "Real-Time Deinterleaving of Radar Pulses Trains on Embedded Platforms," *Proceedings of the MSS Tri-Service Radar Symposium,* SENSIAC, July 21–23, 2015.

[5] Wiley, R. G., *ELINT: The Interception and Analysis of Radar Signals,* Norwood, MA: Artech House, 2006, pp. 56–94.

[6] East, P. W., "Fifty Years of Instantaneous Frequency Measurement," *IET Radar, Sonar and Navigation,* Vol. 6, No. 2, 2012, pp. 112–122.

[7] Smith, J., and J. Abel, "Closed-Form Least-Squares Source Location Estimation from Range-Difference Measurements," *IEEE Transactions on Acoustics, Speech, and Signal Processing,* Vol. 35, No. 12, 1987, pp. 1661–1669.

[8] *Electronic Warfare and Radar Systems Engineering Handbook,* NAWCWD TP 8347, Fourth Edition, Naval Air Warfare Center-Weapons Division, Point Mugu, CA, October 2013, pp. 5-2.1–5-2.10.

[9] Björck, A., *Numerical Methods for Least Squares Problems,* Philadelphia, PA: SIAM, 1996, pp. 339–358.

[10] Schatzman, J., "The Cross-Ambiguity Function for Emitter Location and Radar–Practical Issues for Time Discretization," *2015 IEEE Signal Processing and Signal Processing Education Workshop (SP/SPE),* Salt Lake City, UT, August 9–12, 2015, pp. 243–248.

[11] Barton, D. K., *Modern Radar System Analysis,* Norwood, MA: Artech House, 1988, pp. 209–220.

[12] Torrieri, D. J., "Statistical Theory of Passive Location Systems," *IEEE Transactions on Aerospace and Electronic Systems,* Vol. AES-20, No. 2, March 1984, pp. 183–198.

APPENDIX A

Common Sources of Discrepancy and Confusion in Radar M&S

This appendix describes some of the potential pitfalls in radar M&S. Each of these examples has tripped up the authors at one point or another through the course of their careers. The goal here is to highlight the importance of clearly stating and documenting modeling assumptions, to avoid needlessly chasing an apparent 3-dB error.

A.1 Signal Amplitude (Voltage) and Power

Let a square pulse have a peak amplitude of A. This is typically interpreted as a voltage; thus it has units of volts. What is the peak power of this signal? One sensible answer may simply be

$$P = A^2 \tag{A.1}$$

This is a very common modeling assumption, and the one we adopt in this book. Electrical engineers may object to this, however, since from basic principles it is known that

$$P = \frac{A^2}{Z} \tag{A.2}$$

where Z is the circuit impedance in ohms (Ω). In the context of an electrical circuit, this is the power delivered to a load resistance ($Z = R$ in this case) when A is the voltage across the resistor. Obviously, (A.2) will result in a lower power than that predicted by (A.1) for the same voltage A; the two can be made equal—and preserve unit consistency in (A.1)—by assuming $Z = 1\Omega$. The only time Z may need to be anything other than unity is if the model interacts with HWIL equipment (i.e., it either ingests a signal voltage from a real system or generates simulated signal values that are delivered to the real system).

A.2 Peak and RMS Voltages

For an RF pulse, the peak amplitude A is the peak voltage of the underlying carrier sinusoid. While the peak power of the pulse as defined by either (A.1) or

(A.2) is correct for a DC voltage, the conventional definition for the peak power of a sinusoidal (AC) signal uses the RMS voltage instead, V_{RMS}. For a sinusoidal voltage, it is known that

$$V_{RMS} = \frac{A}{\sqrt{2}}, \quad \text{thus} \quad P = \frac{V_{RMS}^2}{Z} = \frac{A^2}{2Z} \tag{A.3}$$

This is the result of averaging the instantaneous power $P(t)$ over the pulse width τ:

$$P = \frac{1}{\tau} \int_{T_0}^{T_0+\tau} P(t)\,dt \tag{A.4}$$

where T_0 is the start time of the pulse. Blake [1] refers to this average power over the pulse width as the *effective* pulse power. It should not be confused with the average power associated with the PRI that was defined in (2.14).

Equation (A.3) results in a calculated peak power that is lower by a factor of 2 (–3 dB) than that of (A.2) for the same values of A and Z. Either definition is perfectly valid to use in a simulation; the key points are to (a) be consistent and (b) document the assumption(s).

A.3 Noise Power Spectral Density and Noise Bandwidth

A third example is found in the definition of noise bandwidth. It is less an issue of differing assumptions, and more of inconsistent definitions in the literature.

We formally define the noise bandwidth B_n from (2.6) as the bandwidth of a perfect rectangular filter (brick wall filter) that passes the same amount of power (i.e., that has the same area under the curve) as an actual, realizable filter. This definition does not specify whether the bandwidth is one-sided or two-sided, however, and in fact, there are conflicting definitions in the literature. Again, the key to using these definitions correctly is consistency: Use a one-sided bandwidth with a one-sided spectral density, and so on.

The noise power spectral density (PSD) N_0 defined in (2.6) is a *one-sided* PSD, therefore we must interpret B_n as a *one-sided* bandwidth. We thus define B_n as

$$B_n = \frac{1}{\left|H(f_c)\right|^2} \int_0^{\infty} \left|H(f)\right|^2 df \tag{A.5}$$

where $H(f)$ is the frequency response of the receiver (typically dominated by the narrowest filter in the receiver chain), and $H(f_c)$ corresponds to the peak value, usually at the center of the passband. This definition matches that of Blake [1] and is consistent with RF circuit theory and applications where measurements are made only over positive frequencies [2]. The one-sided interpretation is also consistent with the original work by Johnson, who studied the thermal nature of receiver noise [3].

If a *two-sided* bandwidth is assumed instead by setting the limits of integration in (A.5) from $-\infty$ to $+\infty$ (we may refer to it as $B' = 2B_n$), then the PSD must also be two-sided. We thus define the two-sided PSD as $N_0/2$, such that $P_n = (N_0/2)(2B_n)$ equals (2.6).

In the context of matched filtering, Nathanson [4], Skolnik [5], and DiFranco and Rubin [6] all define the receiver noise power as

$$P_n = \frac{N_0}{2}\int_{-\infty}^{\infty}|H(f)|^2\,df \tag{A.6}$$

which is two-sided consistent. Unfortunately, there are seemingly inconsistent statements in many of these references. While Chapter 10 of Skolnik presents a form of (A.6) [Equation (10.9)], Chapter 2 [5] presents a two-sided version of (A.5) multiplying the one-sided PSD kT. Similarly, Chapter 8 of Nathanson presents a form of (A.6), but in Chapter 2 [4], the discussion on the bandwidth of (2.6) implies a two-sided bandwidth. As a last example, Barton [7] also explicitly defines a two-sided bandwidth as in (A.6), but uses it to multiply the one-sided PSD N_0 from (2.6).

A.3.1 Optimum Bandwidth for Rectangular Pulse

It is well known [1, 5–8] that the optimum (maximum SNR) receiver bandwidth matched to a rectangular pulse of length τ is approximately

$$B_n \cong \frac{1}{\tau} \tag{A.7}$$

In light of the previous discussion in Section A.3, it is fair to inquire whether (A.7) represents a one-sided or two-sided bandwidth. Let us first consider the ideal (but unrealizable) matched filter. The frequency response of such a filter is proportional to the Fourier transform (FT) of the input signal [4–6]. For a rectangular pulse of length τ, its (normalized) FT is the sinc function

$$H(f) = \frac{\sin(\pi f\tau)}{\pi f\tau} \tag{A.8}$$

which was first defined in (8.11). The zero-crossings in this function occur at integer multiples of $1/\tau$.

Applying the definition for one-sided noise bandwidth from (A.5) to (A.8), the noise bandwidth is calculated as

$$B_n = \frac{1}{2\tau} \tag{A.9}$$

This result can be calculated in MATLAB for a normalized ($\tau = 1$) version of (A.8) using the following commands in which `sinc` is the built-in MATLAB sinc function. (Try it!):

- `sinc2 = @(x) sinc(x).^2;`
- `Bn = integral(sinc2,0,inf)`

These commands result in `Bn` = 0.5. We therefore see that the optimum bandwidth $1/\tau$ is actually a two-sided quantity.

An approximation to the matched filter performance may be achieved with a rectangular band pass filter [5, 6, 9] of HPBW B, also a two-sided bandwidth. A value of B = $1.37/\tau$ results in the smallest SNR loss of 0.85 dB for this type of filter relative to the matched filter. Other values such as B = $1.2/\tau$ have also been suggested [6, 8]. Regardless of the actual value, these are all close to the standard $1/\tau$ value.

We thus seem to have a discrepancy of a one-sided PSD of (kT_s) being multiplied by a two-sided bandwidth of $1/\tau$. One way to reconcile this is to recognize that in most modern radars, the last stage of the receiver features I/Q demodulation as depicted in Figure 1.5, such that there are actually two real channels involved. Each channel has a one-sided noise bandwidth B_n, so the noise power on each is given by (2.6). The total noise power from both channels is therefore

$$P_n = P_{nI} + P_{nQ} = (kT_s)(2B_n) = (kT_s)\left(\frac{1}{\tau}\right) \tag{A.10}$$

Although this may seem much ado about nothing (since we ultimately got the correct result), it should be noted that an incorrect interpretation of these relationships could easily result in a 3-dB error in noise power (and therefore in SNR). For example, if simulating a system intended to handle a 1 μs pulse waveform, one may be inclined to define the filters on each of the I/Q channels as having a one-sided bandwidth of 1 MHz. This would be incorrect, as the preceding discussion shows.

Last, recall that the noise bandwidth and the HPBW are defined differently and therefore in general are not the same. For a class of single-tuned low-pass filters such as Butterworth filters, however, they are related. The magnitude-squared of the frequency response of an nth order Butterworth filter is [10]

$$\left|H_n(f)\right|^2 = \frac{1}{1 + \left(f/f_0\right)^{2n}} \tag{A.11}$$

where f_0 is the cutoff frequency (one-sided HPBW) of the filter. For a first order filter ($n = 1$), the one-sided noise bandwidth is

$$B_n = \frac{\pi}{2} f_0 \tag{A.12}$$

As the filter order increases, the scaling factor approximates unity, as illustrated by Table A.1; for $n \geq 4$, the error is less than 3%. Thus, for sufficiently higher filter order, $B_n \approx f_0$.

A.4 Oversampling and Signal Processing Gains

A key parameter in any computer simulation is the choice of sampling frequency, F_s. It is common practice to tie this value to the highest frequency f_0 of the simulated signals. The sampling should satisfy the Nyquist criterion, which dictates that to accurately capture all the necessary information from a sinusoid waveform, sampling should be at least twice its frequency; or $Fs \geq 2 f_0$. This is true for real signals such as sines and cosines. For complex signals such as the complex exponential $\exp(j2\pi f_0 t)$, each complex sample results in two real values (one each for the real and imaginary parts); therefore the Nyquist sampling criterion for a complex signal is met by defining

$$F_s = f_0 \tag{A.13}$$

The frequency value f_0 in (A.13) may refer to the frequency of a pure complex exponential, or it may refer to the actual signal bandwidth. Setting the sampling frequency according to (A.13) is referred to as critical sampling. Physical radar systems often operate this way, as it minimizes data requirements. In M&S, however, oftentimes it is beneficial to oversample the signals for analysis purposes, as it may provide additional details that are often insightful. An oversampling factor (*osf*) may then be defined as

Table A.1 Relation Between Noise BW and 3-dB BW as a Function of Filter Order

n	B_n/f_0
1	1.5708 ($\pi/2$)
2	1.1107
3	1.0472
4	1.0262
5	1.0166
10	1.0041

$$osf = \frac{F_s}{f_0} \tag{A.14}$$

From the previous discussion, $osf = 1$ for critically sampled data.

A couple of situations, described in Sections A.4.1 and A.4.2, may arise when processing oversampled data; either of these can affect SNR calculations if not handled properly.

A.4.1 Oversampling May Inflate Signal Processing Gain

There are many instances where the radar processor sums signal samples together; examples include matched filtering and Doppler processing. For simplicity, assume that a sampled received signal $x(k) = A + n(k)$ consists of a target with constant amplitude A, plus receiver noise $n(k)$. Assume that the signal duration is N samples when critically sampled. An oversampled signal sequence will thus have a length of $(osf)N$. After summation, the target amplitude is $S = (osf)NA$.

Now consider the summation of noise. The noise samples $n(k)$ are typically assumed to be i.i.d. with zero mean and variance σ^2. Under these assumptions, the samples are uncorrelated, so the variance (power) of their sum is

$$\sigma_S^2 = E\left[\left(\sum_{k=0}^{(osf)N-1} n(k)\right)^2\right] = \sum_{k=0}^{(osf)N-1} \sigma^2 = (osf)N\sigma^2 \tag{A.15}$$

The SNR at the output of the summation operation is therefore

$$SNR = \frac{P_{\text{signal}}}{P_{\text{noise}}} = \frac{S^2}{\sigma_S^2} = \frac{\left((osf)NA\right)^2}{(osf)N\sigma^2} = (osf)N\left(\frac{A^2}{\sigma^2}\right) \tag{A.16}$$

where the term (A^2/σ^2) is the single-sample SNR. Thus, the SNR is overstated by a factor of *osf* relative to the critically sampled case. This is undesirable if the oversampling is solely intended as an analysis artifact and not meant to be part of the processing gain. Not only is the SNR gain off, so are the individual signal and noise gains. Moreover, they are off by different amounts, so simply scaling the output does not fix the individual signal levels.

This situation may be handled in a few ways, described as follows:

1. If the relevant output of the model is the SNR, and the individual values of signal and noise powers are not critical, one may simply adjust the SNR calculated from (A.16) by dividing by *osf*.
2. If not only the SNR but also the individual powers are required, one may keep the target signal and noise sequences apart in the simulation, then perform the summations separately. The output of the signal summation

may be divided by *osf*, and the output of the noise summation may be divided by $\sqrt{osf}$. The two output sequences may then be added and processed thereafter as a single sequence; the power levels (and SNR) will be correct for both.

3. If option (2) is not feasible, or if there are other signals such as clutter or semicoherent EA that may not conform to either scaling, the last resort may be to simply down-sample the received signal sequence to critical sampling, then perform the summation. If modeling an actual radar, one could also down-sample to match the actual radar sampling, which might be slightly higher than critical.

A.4.2 Oversampling Correlates Receiver Noise So That It No Longer Behaves as White Noise

Receiver noise is often assumed to be white, meaning that the time samples are uncorrelated. The output of the summing operation discussed in Section A.4.1 will not just have a variance scaled by *osf*, but will also be spatially correlated due to the filtering function implicit in the summation. The impulse response of the filter defines the autocorrelation function of the output noise sequence. As an example, consider the LPF of (two-sided) bandwidth B_n depicted after the last mixer, shown in Figure 1.5. A perfect (or close to) rectangular frequency response has an impulse response given by a sinc function with zero crossings at multiples of $1/B_n$. This is also the autocorrelation function of the output noise, given a white noise input, as depicted in Figure A.1.

If the sampling frequency is set to $F_s = (osf)B_n$, then the sampling interval is $T_s = 1/F_s = 1/((osf)B_n)$. In Figure A.1, $osf = 10$. The first 10 time lags in Figure A.1 (in gray) are on the mainlobe of the autocorrelation function, meaning the samples within those time lags are highly correlated. If there is no oversampling ($F_s = B_n$, or $osf = 1$) the samples lie on zero crossings of the function (one shown in black in Figure A.1 at $t/T_s = 1$) so the output noise sequence is still uncorrelated.

One downside of this effect is that any further integration processing downstream will result in gains higher than expected of white noise. For a subsequent integration of M correlated samples, the noise power gain will be somewhere between M, the noncoherent gain obtained from uncorrelated noise, and M^2, the gain expected from a fully coherent signal.

Depending on the application, this may not be an issue; however, if an accurate link budget and SNR estimation is desired, it is best to avoid oversampling unless necessary.

A.5 Noise Factor

The radar community tends to employ the term noise figure or noise factor, and the associated variable F, somewhat loosely. (As Table 2.1 noted, the factor and figure modifiers represent linear and decibel units, respectively.) Confusion arises because the same term and variable are used in two very different ways.

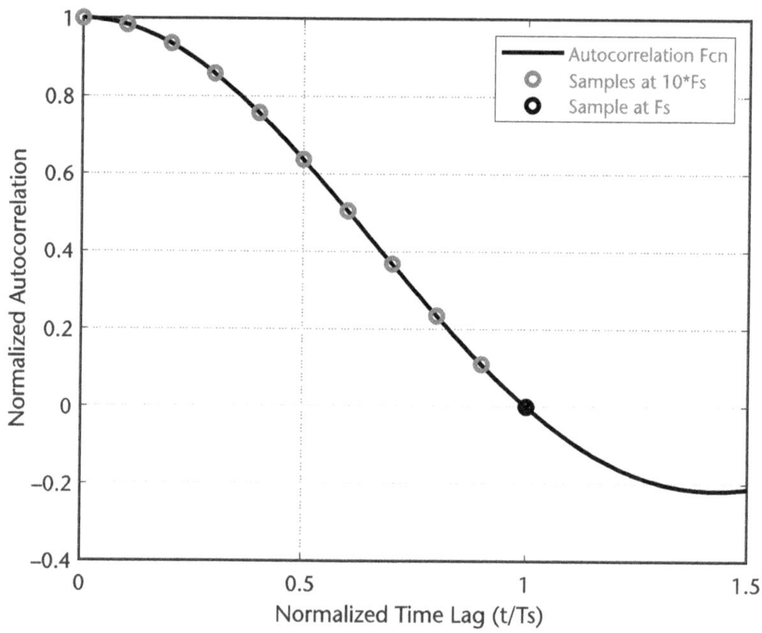

Figure A.1 Autocorrelation function of output noise from a rectangular filter.

The most common application for noise factor is to represent noise power in the RRE (2.10). While sometimes referred to as a *receiver* noise factor, this F is an alternative method for representing system noise temperature T_s [8], as evidenced in (2.7), and so is more properly referred to as the *system* noise factor. Quite simply, the role of F is to scale the reference temperature T_0 (290K) to the system noise temperature T_s. The system noise temperature includes the contributions of all noise sources that limit target SNR in the receiver, including the receiver, but also antenna, radome, and environmental contributions. Modern radars employing active electronically scanned arrays have their LNA stage in each element at the antenna, with one or more subsequent amplification stages further back; all contribute to system noise. The system noise factor, in turn, captures all of these contributions. The rigorous prediction of system noise temperature is a tedious exercise; good guides for this chore are found in the communications literature [11, 12]. Once the system noise temperature is determined, division by the reference temperature yields the system noise factor. M&S practitioners are rewarded from all this hard work by our RF and systems engineer colleagues with a single number; for example, system noise figure $F = 4.5$ dB.

Pegging the system noise figure to a reference temperature of 290K is arbitrary and due perhaps to the fact that many early radar systems, particularly those that were ground-based, operated their receiver close to room temperature. Under such conditions, the system noise temperature is guaranteed to be greater than the reference temperature, and so $F \geq 1$. Indeed, that is usually the case in modern radar systems, but not always. Space surveillance radars may employ cryogenically cooled receivers to support low noise temperatures. For example, the L-band Millstone Hill radar (MHR) has a system noise temperature of 110K while the X-band

Haystack system has a 246K system noise temperature [13]. Normalizing by the reference temperature yields noise factors of 0.38 and 0.85, respectively, or noise figures of −4.2 dB and −0.7 dB.

The second application for noise factor in radar is to characterize microwave devices such as mixers amplifiers. Here, noise factor represents the change in SNR through the component [14]

$$F = \frac{S_{in}/N_{in}}{S_{out}/N_{out}} = \frac{1}{G}\frac{N_{out}}{N_{in}} \tag{A.17}$$

where G is the gain on input signal and noise. Real devices cannot remove noise from the signal path, so SNR always decreases, and $F \geq 1$.

Given a chain of microwave components, with F_n the noise factor of the nth component, the overall noise factor may be found as [15]

$$F = F_1 + \frac{F_2 - 1}{G_1} + \frac{F_3 - 1}{G_1 G_2} + \cdots \tag{A.18}$$

The noise factor of a radar receiver is frequently calculated in this manner. Like system noise temperature, radar system M&S tools rarely operate at this level of detail, but may include receiver noise figure F as an input to equations like (2.8) and (2.9) to yield system noise temperature.

A.6 Assorted Factors of Two

The following parameters, listed in no particular order, are some of the many additional potential sources of modeling errors if quantities are not properly defined:

- One-way versus two-way atmospheric attenuation coefficients (textbooks usually list one-way values);
- One-sided versus two-sided antenna beamwidths (typically two-sided BWs are quoted);
- One-way versus two-way antenna gains (patterns are typically one-way);
- Using $10*\log_{10}(\,)$ vs. $20*\log_{10}(\,)$ when defining power in dB ($20*\log_{10}$ is for amplitude ratios).

References

[1] Blake, L. V., "Prediction of Radar Range," in *Radar Handbook*, M. I. Skolnik (ed.), Boston: McGraw-Hill, 1990, pp. 2.5–2.31.

[2] Smith, J., *Modern Communication Circuits*, New York: McGraw-Hill, 1986, pp. 68–71.

[3] Johnson, J. B., "Thermal Agitation of Electricity in Conductors," *Physical Review*, Vol. 32, July 1928, pp. 97–109.

[4] Nathanson, F. E., *Radar Design Principles*, Second Edition, Mendham, NJ: Scitech, 1999, pp. 54–57, 355–360.

[5] Skolnik, M. I., *Introduction to Radar Systems*, Second Edition, New York: McGraw-Hill, 1980, pp. 18–19, 369–374.

[6] DiFranco, J. V., and W. L. Rubin, *Radar Detection*, Norwood, MA: Artech House, 1980, pp. 146–155.

[7] Barton, D. K., *Modern Radar System Analysis*, Norwood, MA: Artech House, 1988, pp. 13–16.

[8] Scheer, J. S., "The Radar Range Equation," in *Principles of Modern Radar, Vol I: Basic Principles*, M. A. Richards, J. A. Scheer, and W.A. Holm (eds.), Raleigh, NC: Scitech Publishing Inc., 2010, pp. 64–65,

[9] Taylor, J. W., "Receivers," in *Radar Handbook*, M. I. Skolnik (ed.), Boston: McGraw-Hill, 1990, pp. 3.19–3.22.

[10] Proakis, J. G., and D. G. Manolakis, *Introduction to Digital Signal Processing*, New York: Macmillan, 1988, pp. 613–615.

[11] Pratt, T., and C. W. Bostian, *Satellite Communications*, New York: John Wiley & Sons, 1986, pp. 114–119.

[12] Reid, M. S., *Low-Noise Systems in the Deep Space Network*, Hoboken, NJ: John Wiley & Sons, 2008, pp. 16–28.

[13] Lee, R. K., "A Cost Optimized High Level Reference Design for a Deep Space Custody Radar," Mission Engineering Analysis for Radar Systems Report under Air Force Contract FA8721-05-C-002, MIT Lincoln Laboratory, November 1, 2017, p. 26.

[14] Bruder, J. A., "Radar Receivers," in *Principles of Modern Radar, Vol I: Basic Principles*, M. A. Richards, J. A. Scheer, and W.A. Holm (eds.), Raleigh, NC: Scitech Publishing Inc., 2010, p. 404.

[15] Skolnik, M. I., *Introduction to Radar Systems*, Second Edition, New York: McGraw-Hill, 1980, pp. 344–345.

List of Acronyms

2-D	Two-dimensional
3-D	Three-dimensional
6DOF	Six-degrees-of-freedom
A/D	Analog-to-digital converter
AC	Alternating current
ACS	Antenna coordinate system
ADC	Analog-to-digital converter
AESA	Active electronically scanned antenna
AF	Array factor
AF/A9	Air Force Studies, Analyses and Assessments Office
AFRL	Air Force Research Laboratory
AFSIM	Advanced Framework for Simulation, Integration, and Modeling
AGC	Automatic gain control
AIAA	American Institute of Aeronautics and Astronautics
AM	Amplitude modulation
AMTA	Antenna Measurement Techniques Association
ANSI	American National Standards Institute
AoA	Angle-of-arrival
BN	Barrage noise
BNJ	Barrage noise jamming
BPF	Band pass filter
BPM	Binary phase modulation
BPSK	Binary phase shift keying
BSC	Beam steering computer
BW	Bandwidth or beamwidth
CA-CFAR	Cell Averaging CFAR
CAF	Complex ambiguity function
CBN	Cooperative blinking noise
CCW	Counterclockwise

CEM	Computational electromagnetics
CFAR	Constant false alarm rate
CG	Center of gravity
CGS	Coordinated gate stealer
CLT	Central limit theorem
CNR	Clutter-to-noise ratio
COHO	Coherent oscillator
CPI	Coherent processing interval
CR	Cancellation ratio
CRLB	Cramer-Rao lower bound
CS	Coordinate system
CTFT	Continuous-time Fourier transform
CUT	Cell under test
CVR	Crystal video receiver
CW	Continuous wave or clockwise
dB	Decibels
dBc	Decibels relative to carrier power
DBF	Digital beamforming
dBi	Decibels relative to isotropic
dBmb	Decibels relative to peak level of main beam
dBsm	Decibels relative to 1 square meter
DC	Direct current
DD	Direct-direct (re: multipath)
DE	Directed energy
DEM	Digital elevation model
DFT	Discrete Fourier transform
DI	Direct-indirect (re: multipath)
DoD	Department of Defense
DOF	Degrees-of-freedom
DREX	Digital receiver-exciter
DRFM	Digital radio frequency memory
DSB	Double sideband
DSIAC	Defense Systems Information Analysis Center
DTFT	Discrete time Fourier transform
DTKF	Discrete time Kalman filter
EA	Electromagnetic attack
EADSIM	Extended Air Defense Simulation
ECCM	Electronic counter countermeasures

ECEF	Earth-centered Earth-fixed
ECM	Electronic countermeasures
EJ	Escort jamming
EKF	Extended Kalman filter
ELINT	Electronic intelligence
EM	Electromagnetic
EMI	Electromagnetic interference
ENU	East-north-up
EP	Electromagnetic protection
ERP	Effective radiated power
ERPD	Effective radiated power density
ES	Electromagnetic support
ESA	Electronically scanned (or steered) antenna
ESJ	Escort support jamming
ESM	Electronic support measures
ESPRIT	Estimation of signal parameters via rational invariance techniques
EW	Electromagnetic warfare
FCR	Fire-control radar
FDOA	Frequency difference of arrival
FDTD	Finite difference time domain
FEM	Finite element method
FFT	Fast Fourier transform
FIR	Finite impulse response
FIT	Finite integration technique
FLAMR	Forward-Looking Advanced Multimode Radar
FM	Frequency modulation
FSR	Full-scale range
FT	Fourier transform
GHz	Gigahertz
GO	Geometric optics
GOCA-CFAR	Greatest-of-cell-averaging constant false alarm rate
GTD	Geometric theory of diffraction
GTRI	Georgia Tech Research Institute
GUI	Graphical user interface
GUIDE	GUI development environment
HF	High frequency
HPBW	Half-power bandwidth (or beamwidth)

HPRF	High pulse repetition frequency
HRR	High range resolution
HWIL	Hardware-in-the-loop
Hz	Hertz
i.i.d.	Independent, identically-distributed
I/Q	In-phase and quadrature
ICM	Internal clutter motion
ICS	Inertial coordinate system
ID	Indirect-direct (re: multipath)
IDFT	Inverse discrete Fourier transform
IF	Intermediate filter
IFFT	Inverse fast Fourier transform
IFM	Instantaneous frequency measurement
II	Indirect-indirect (re: multipath)
IIR	Infinite impulse response
IPP	Interpulse period
ITU	International Telecommunications Union
JNR	Jammer-to-noise ratio
JSR	Jammer-to-signal ratio
KF	Kalman filter
kHz	Kilohertz
LCC	Land Cover Classification
LCMV	Linearly constrained minimum variance
LEO	Low Earth orbit
LF	Low frequency
LFM	Linear frequency modulation
LFMCW	Linear frequency modulation continuous wave
LH	Left-handed
LHCP	Left-hand circular polarization
LHS	Left-hand side
LLCS	Local-level coordinate system
LMS	Least mean square
LNA	Low noise amplifier
LO	Local oscillator
LOS	Line-of-sight
LPF	Low pass filter
LPRF	Low pulse repetition frequency
LRT	Likelihood ratio test

LSB	Least significant bit, or lower sideband
LSI	Linear, shift-invariant
LTI	Linear, time-invariant
LTP	Local tangent plane
LUT	Look-up table
LVC	Live, virtual, and constructive
M&S	Modeling and simulation
MATLAB	Matrix Laboratory
MC	Monte Carlo
MF	Matched filter
MFDT	Multiple false doppler targets
MFRT	Multiple false range targets
MFT	Multiple false targets
MHR	Millstone Hill radar
MHz	Megahertz
ML	Maximum likelihood
MLAT	Multilateration
MMIC	Monolithic microwave integrated circuit
MMSE	Minimum mean-square error
MoM	Method of moments
MOP	Measure-of-performance
MOPA	Master oscillator-power amplifier
MPRF	Medium pulse repetition frequency
MTI	Moving target indication
MUSIC	Multiple signal classification
MVDR	Minimum variance distortionless response
NCI	Noncoherent integration
NE	North-east
NED	North-east-down
NGTS	Next Generation Threat System
NLFM	Nonlinear frequency modulation
NLLS	Nonlinear least squares
NM	Nautical miles
NN	Nearest-neighbor
NP	Neyman-Pearson
OBJ	Off-board jamming
OOK	On-off keying
OS-CFAR	Order-statistics cell-averaging constant false alarm rate

osf	Oversampling factor
PA	Power aperture
PA^2	Power-aperture-squared
PAG	Power-aperture-gain
PAR	Phased array radar
PD	Probability of detection
PDF	Probability density function
PDI	Post-detection integration
PDW	Pulse descriptor word
PFA	Probability of false alarm
PM	Phase modulation
PO	Physical optics
POV	Point-of-view
PPM	Pulse position modulation
PRF	Pulse repetition frequency
PRI	Pulse repetition interval
PSD	Power spectral density
PW	Pulsewidth
RAM	Random access memory
RCS	Radar cross-section
RDM	Range-Doppler map (or matrix)
RF	Radio frequency
RFML	Radio frequency memory loop
RGPI	Range gate pull-in
RGPO	Range gate pull-off
RGS	Range gate stealer
RH	Right-handed
RHCP	Right-hand circular polarization
RHS	Right-hand side
RMS	Root-mean-square
RNG	Random number generator
ROC	Receiver operating characteristic
RRE	Radar range equation
RSN	Responsive spot noise
RWR	Radar warning receiver
SAR	Synthetic aperture radar
SBR	Shooting-and-bouncing rays

SDP	Signal data processor
SFT	Single false target
SIGINT	Signals intelligence
SIJ	Stand-in jamming
SINR	Signal-to-interference and noise ratio
SIR	Signal-to-interference ratio
SJR	Signal-to-jammer ratio
SL	Sidelobe
SLB	Sidelobe blanker
SLC	Sidelobe canceler
SLL	Sidelobe level
SNR	Signal-to-noise ratio
SOCA-CFAR	Smallest-of-cell-averaging constant false alarm rate
SOJ	Stand-off jamming
SPJ	Self-protection jamming
SQNR	Signal-to-quantization noise ratio
SSB	Single sideband
SSJ	Self-screening jamming
SSM	Single sideband mixer
STALO	Stable local oscillator
STAP	Space-time adaptive processing
STC	Sensitivity time control
STORM	Synthetic Theater Operations Research Model
STR	Straight-through repeater
STT	Single target track
T/R	Transmit/receive
TAC	Tactical air combat
TBP	Time-bandwidth product
TDOA	Time difference of arrival
TEMPER	Tropospheric Electromagnetic Parabolic Equation Routine
TFM	Track File Manager
TOA	Time-of-arrival
TOD	Time-of-departure
TSPI	Time-space-position information
TWS	Track-while-scan
ULA	Uniform linear array
USGS	United States Geological Survey

UTD	Uniform theory of diffraction
VCD	Vertical coverage diagram
VGS	Velocity gate stealer
VLSB	Value (or voltage) of least significant bit

About the Authors

Carlos A. Dávila is a principal research engineer at the Georgia Tech Research Institute (GTRI), and the chief engineer at GTRI's ISR Division of the Sensors and Electromagnetic Applications Laboratory (SEAL). At GTRI, he is involved in the modeling and simulation, analysis, and performance evaluation of numerous aspects of ground-based, airborne, and space-based radar systems, including research and development of signal processing algorithms with primary focus on electronic warfare. Prior to joining GTRI, he worked for over 15 years at Hughes Aircraft Company, Raytheon Company, and Alliant Techsystems.

Dr. Dávila received a BS from the University of Puerto Rico, an MS from UCLA, and a PhD from the University of Arizona, all in electrical engineering. He is a Senior Member of the IEEE, and a member of the Association of Old Crows. Dr. Dávila holds one U.S. patent in radar technology. He has been lecturing in Georgia Tech Professional Education training courses on radar and electronic warfare since 2003.

Glenn D. Hopkins is a Fellow and principal research engineer at the GTRI. He is currently the chief engineer of the Antenna Systems Division of the SEAL. In his 36 years at GTRI he has contributed to and led numerous assessments, designs, prototypes, and system developments of radar, electronic warfare, and communications systems hardware. He is a Senior Member of the IEEE and has been very active with the Antennas and Propagation, Aerospace and Electronic Systems, and Microwave Theory and Techniques Societies. He is also a life member of the Association of Old Crows.

Mr. Hopkins earned his BSEE in 1987 and his MSEE in 1991, both from the Georgia Institute of Technology. Since 1991, he has taught in over 300 continuing education course offerings with the Georgia Tech Professional Education.

He currently leads five GTPE courses. Mr. Hopkins holds three U.S. and one European patent in antenna and array technologies. He has received GTRI and IEEE recognition as a mentor, technical innovator, and educator.

Gregory A. Showman is a principal research engineer at the GTRI, a GTRI Fellow, and a Georgia Institute of Technology regents researcher. He has over three decades of experience in advanced radio frequency sensor research and development, with an emphasis on the design and implementation of innovative signal processing techniques for radar imaging, electronic protection, and multidimensional adaptive filtering.

Dr. Showman's accomplishments include development of novel techniques for ultrawideband synthetic aperture radar (SAR) and high-precision turntable inverse SAR image formation, methods for polarimetric SAR calibration, electronic protection (EP) against jamming, and space-time adaptive processing (STAP) algorithms and architectures for airborne and space-based ground moving target indication (GMTI) systems. While at GTRI, he has supported nearly 200 short course offerings on the topics of airborne pulse-Doppler radar, SAR, EP, and advanced radar signal processing. Additionally, Dr. Showman is a Senior Member of the IEEE and author of two chapters in the *Principles of Modern Radar* textbook series on radar imaging and stripmap SAR.

Index

D

Artech House Radar Library

Joseph R. Guerci, Series Editor

Adaptive Antennas and Phased Arrays for Radar and Communications, Alan J. Fenn

Adaptive Radar Detection: Model-Based, Data-Driven, and Hybrid Approaches, Angelo Coluccia

Adaptive Signal Processing for Radar, Ramon Nitzberg

Advanced Techniques for Digital Receivers, Phillip E. Pace

Advances in Direction-of-Arrival Estimation, Sathish Chandran, editor

Airborne Early Warning System Concepts, Maurice W. Long

Airborne Pulsed Doppler Radar, Second Edition, Guy V. Morris and Linda Harkness, editors

Aspects of Modern Radar, Eli Brookner

Aspects of Radar Signal Processing, Frank F. Kretschmer, Bernard Lewis, and Wesley W. Shelton

Basic Radar Analysis, Mervin C. Budge, Jr. and Shawn R. German

Basic Radar Tracking, Mervin C. Budge, Jr. and Shawn R. German

Bayesian Multiple Target Tracking, Second Edition, Lawrence D. Stone, Roy L. Streit, Thomas L. Corwin, and Kristine L Bell

Beyond the Kalman Filter: Particle Filters for Tracking Applications, Branko Ristic, Sanjeev Arulampalam, and Neil Gordon

Cognitive Radar: The Knowledge-Aided Fully Adaptive Approach, Second Edition, Joseph R. Guerci

Coherent Radar Performance Estimation, James A. Scheer

Computer Simulation of Aerial Target Radar Scattering, Recognition, Detection, and Tracking, Yakov D. Shirman, editor

Control Engineering in Development Projects, Olis Rubin

Countermeasures for Aerial Drones, Garik Markarian and Andrew Staniforth

Deep Learning Applications of Short-Range Radars, Avik Santra, and Souvik Hazra

Deep Learning for Radar and Communications Automatic Target Recognition, Uttam K. Majumder, Erik P. Blasch, and David A. Garren

Design and Analysis of Modern Tracking Systems, Samuel Blackman and Robert Popoli

Detectability of Spread-Spectrum Signals, George Dillard

Detecting and Classifying Low Probability of Intercept Radar, Second Edition, Phillip E. Pace

Engineering UAS Applications: Sensor Fusion, Machine Vision, and Mission Management, Jesús García, José M. Molina, Juan Pedro Llerena, Daniel Amigo, and David Sánchez Pedroche

FMCW Radar Design, M. Jankiraman

Fourier Transforms in Radar and Signal Processing, Second Edition, David Brandwood

Fundamentals of Short-Range FM Radar, Igor V. Komarov and Sergey M. Smolskiy

Handbook of Computer Simulation in Radio Engineering, Communications, and Radar, Sergey A. Leonov and Alexander I. Leonov

Handbook of Radar Measurement, David K. Barton

High-Resolution Radar, Second Edition, Donald R. Wehner

High Resolution Radar Cross-Section Imaging, Dean L. Mensa

Highly Integrated Low-Power Radars, Sergio Saponara, Maria Greco, Egidio Ragonese, Giuseppe Palmisano, and Bruno Neri

Introduction to LabVIEW™ FPGA for RF, Radar, and Electronic Warfare Applications, Terry Stratoudakis

An Introduction to Passive Radar, Second Edition, Hugh D. Griffiths and Christopher J. Baker

Introduction to Radar using Python and MATLAB®, Lee Andrew Harrison

Introduction to RF Equipment and System Design, Pekka Eskelinen

Linear Systems and Signals: A Primer, JC Olivier

Meter-Wave Synthetic Aperture Radar for Concealed Object Detection, Hans Hellsten

Methods and Techniques of Radar Recognition, Victor Nebabin

The Micro-Doppler Effect in Radar, Second Edition, Victor C. Chen

Microwave Radar: Imaging and Advanced Concepts, Roger J. Sullivan

Millimeter-Wave Radar Clutter, Nicholas C. Currie, Robert D. Hayes, and Robert Trebits

Millimeter-Wave Radar Targets and Clutter, Gennadiy P. Kulemin

MIMO Radar: Theory and Application, Jamie Bergin and Joseph R. Guerci

Modern Radar Systems, Second Edition, Hamish Meikle

Modern Radar System Analysis, David K. Barton

Modern Radar System Analysis Software and User's Manual, Version 3.0, David K. Barton

Monopulse Principles and Techniques, Second Edition, Samuel M. Sherman and David K. Barton

Motion and Gesture Sensing with Radar, Jian Wang and Jaime Lien

MTI and Pulsed Doppler Radar with MATLAB®, Second Edition, D. Curtis Schleher

Multifunction Array Radar, Dale Billetter

Multiple-Target Tracking with Radar Applications, Samuel S. Blackman

Multitarget-Multisensor Tracking, Applications and Advances, Volume II, Yaakov Bar-Shalom and Robert F. Popoli, editors

Multitarget-Multisensor Tracking: Applications and Advances Volume III, Yaakov Bar-Shalom and William Dale Blair, editors

Non-Line-of-Sight Radar, Brian C. Watson and Joseph R. Guerci

Photonic Aspects of Modern Radar, Edward N. Toughlian and Henry Zmuda

Practical Aspects of Active Phased Array Antenna Development, Ashok K. Agrawal

Practical Phased Array Antenna Systems, Eli Brookner

Practical Simulation of Radar Antennas and Radomes, Herbert L. Hirsch

Precision FMCW Short-Range Radar for Industrial Applications, Boris A. Atayants, Viacheslav M. Davydochkin, Victor V. Ezerskiy, Valery S. Parshin, and Sergey M. Smolskiy

Principles of High-Resolution Radar, August W. Rihaczek

Principles of Radar and Sonar Signal Processing, François Le Chevalier

Radar Cross-Section Analysis & Control, Asoke K. Bhattacharyya and D. L. Sengupta

Radar Cross Section, Second Edition, Eugene F. Knott, et al.

Radar Detection and Tracking Systems, Shaheen A. Hovanessian

Radar and EW Modeling in MATLAB® and Simulink®, Carlos A. Dávila, Glenn D. Hopkins, and Gergory A. Showman

Radar Engineer's Sourcebook, William C. Morchin

Radar Equations for Modern Radar, David K. Barton

Radar Evaluation Handbook, David K. Barton, et al.

Radar for Fully Autonomous Driving, Matt Markel, editor

Radar Meteorology, Henri Sauvageot

Radar Range Performance Analysis, Lamont V. Blake

Radar Reflectivity of Land and Sea, Third Edition, Maurice W. Long

Radar Resolution and Complex-Image Analysis, August W. Rihaczek and Stephen J. Hershkowitz

Radar RF Circuit Design, Second Edition, Nickolas Kingsley and Joseph R. Guerci

Radar Signal Processing and Adaptive Systems, Ramon Nitzberg

Radar Signals: An Introduction to Theory and Application, Charles E. Cook

Radar System Analysis, Design, and Simulation, Eyung W. Kang

Radar System Analysis and Modeling, David K. Barton

Radar System Design and Analysis, Shaheen A. Hovanessian

Radar System Performance Modeling, Second Edition, G. Richard Curry

Radar Technology, Eli Brookner

Radar Technology Encyclopedia, David K. Barton and Sergey A. Leonov, editors

Radar Vulnerability to Jamming, Stephen Lothes

Radars, Volume 1: Monopolse Radar, David K. Barton

Radars, Volume 2: The Radar Equation, David K. Barton

Radars, Volume 3: Pulse Compression, David K. Barton

Radars, Volume 4: Radar Resolution & Multipath Effects, David K. Barton

Radars, Volume 5: Radar Clutter, David K. Barton

Radars, Volume 6: Frequency Agility and Diversity, David K. Barton

Radars, Volume 7: CW and Doppler Radar, David K. Barton

Radio Wave Propagation Fundamentals, Second Edition, Artem Saakian

Range-Doppler Radar Imaging and Motion Compensation, Jae Sok Son, et al.

Robotic Navigation and Mapping with Radar, Martin Adams, John Mullane, Ebi Jose, and Ba-Ngu Vo

Russian-English and English-Russian Dictionary of Radar and Electronics, William F. Barton and Sergey A. Leonov

Secondary Surveillance Radar, Michael C. Stevens

Signal Design for Modern Radar Systems, Mohammad Alaee-Kerahroodi, Mojtaba Soltanalian, Prabhu Babu, and Bhavani Shankar

Signal Detection and Estimation, Second Edition, Mourad Barkat

Signal Processing in Noise Waveform Radar, Krzysztof Kulpa

Signal Processing for Passive Bistatic Radar, Mateusz Malanowski

Solid-State Radar Transmitters, Edward D. Ostroff

Space-Based Radar Handbook, Leopold J. Cantafio, Richard K. Moore, and Earl E. Swartzlander Jr.

Space-Time Adaptive Processing for Radar, Second Edition, Joseph R. Guerci

Spaceborne Weather Radar, Toshiaki Kozu and Robert Meneghini

Special Design Topics in Digital Wideband Receivers, James Tsui

Statistical Signal Characterization, Herbert L. Hirsch

Surface-Based Air Defense System Analysis, Robert H. MacFadzean

Synthetic Array and Imaging Radars, Shaheen A. Hovanessian

Systems Engineering of Phased Arrays, Rick Sturdivant, Clifton Quan, and Enson Chang

Techniques of Radar Reflectivity Measurement, Nicholas C. Currie

Theory and Practice of Radar Target Identification, August W. Rihaczek and Stephen J. Hershkowitz

Time-Frequency Signal Analysis with Applications, Ljubiša Stanković, Miloš Daković, and Thayananthan Thayaparan

Time-Frequency Transforms for Radar Imaging and Signal Analysis, Victor C. Chen and Hao Ling

Transmit Receive Modules for Radar and Communication Systems, Rick Sturdivant and Mike Harris

For further information on these and other Artech House titles, including previously considered out-of-print books now available through our In-Print-Forever® (IPF®) program, contact:

Artech House
685 Canton Street
Norwood, MA 02062
Phone: 781-769-9750
Fax: 781-769-6334
e-mail: artech@artechhouse.com

Artech House
16 Sussex Street
London SW1V HRW UK
Phone: +44 (0)20 7596-8750
Fax: +44 (0)20 7630-0166
e-mail: artech-uk@artechhouse.com

Find us on the World Wide Web at: www.artechhouse.com

www.ingramcontent.com/pod-product-compliance
Lightning Source LLC
LaVergne TN
LVHW080644210926
848485LV00003B/326
9781630819064